CW00394525

Complete
Chemistry
for Cambridge IGCSE®

For the updated syllabus

Third edition

RoseMarie Gallagher

Paul Ingram

Oxford excellence for Cambridge IGCSE®

OXFORD

OXFORD
UNIVERSITY PRESS

Great Clarendon Street, Oxford OX2 6DP

Oxford University Press is a department of the University of Oxford.
It furthers the University's objective of excellence in research, scholarship, and education by publishing worldwide in

Oxford New York

Auckland Cape Town Dar es Salaam Hong Kong Karachi
Kuala Lumpur Madrid Melbourne Mexico City Nairobi
New Delhi Shanghai Taipei Toronto

With offices in

Argentina Austria Brazil Chile Czech Republic France Greece
Guatemala Hungary Italy Japan Poland Portugal Singapore
South Korea Switzerland Thailand Turkey Ukraine Vietnam

© RoseMarie Gallagher and Paul Ingram 2014

The moral rights of the authors have been asserted

Database right Oxford University Press (maker)

First published in 2006
Third edition published 2014

All rights reserved. No part of this publication may be reproduced, stored in a retrieval system, or transmitted, in any form or by any means, without the prior permission in writing of Oxford University Press, or as expressly permitted by law, or under terms agreed with the appropriate reprographics rights organization. Enquiries concerning reproduction outside the scope of the above should be sent to the Rights Department, Oxford University Press, at the address above.

You must not circulate this book in any other binding or cover and you must impose this same condition on any acquirer

British Library Cataloguing in Publication Data

Data available

ISBN 978-0-19-839914-8

10 9 8 7 6 5 4 3

MIX
Paper from responsible sources
FSC www.fsc.org FSC® C007785

Printed in Great Britain by Bell & Bain Ltd, Glasgow

Paper used in the production of this book is a natural, recyclable product made from wood grown in sustainable forests. The manufacturing process conforms to the environmental regulations of the country of origin.

Acknowledgments

®IGCSE is the registered trademark of Cambridge International Examinations.

The publisher would like to thank Cambridge International Examinations for their kind permission to reproduce past paper questions.

p2t: Radius Images/Alamy; p2r: Suhendri/Bigstock; p2l: Dan Gair/Photolibrary/Getty Images; p3l: OUP; p3r: OUP; p4: Philippe Plailly/Science Photo Library; p5t: Pegasus/Visuals Unlimited Inc./Science Photo Library; p5b: IrinaK/Shutterstock; p6t: Yuri Samsonov/ Shutterstock; p6m: Alexandr Vlassyuk/Shutterstock; p6b: KC Slagle/Shutterstock; p8t: Jose AS Reyes/Shutterstock ; p8m: Ubik/MARKA/Alamy; p8b: Cebas/Shutterstock; p9t: Africa Studio/ Shutterstock ; p9b: Bobby Yip/Reuters; p12t: Shutterstock NS; p12b: Shutterstock NS ; p13: Rich Legg/Istockphoto; p14l: Mariusz S. Jurgielewicz/Shutterstock ; p14r: Kid Stock/Blend Images/Corbis; p15: OUP;p16: Imagestock/Istockphoto; p17t: Wloven/Istockphoto; p17m: OUP; p17b: OUP; p18: Robert Francis/Robert Harding Picture Library Ltd/Alamy; p20: Yuri Arcurs/Dreamstime; p22: LorenRodgers/Shutterstock ; p23t: Mtr/Shutterstock; p23b: PhilAugustavo/Istockphoto; p26l: OUP; p26m: Air009/Shutterstock; p26r: OUP; p26bl: The Gallery Collection/Corbis; p26br: Andri Tambunan/Corbis; p27: Hulton Archive/Stringer/Getty Images ; p28: Iofoto/Shutterstock; p30: U.S. Dept. of Energy/Science Photo Library; p31l: Ashley Cooper/Alamy; p31t: Hank Morgan/Science Photo Library; p31r: Roberto Caucino/ Shutterstock ; p32: Baron/Stringer/Hulton Archive/Getty Images; p33: Andrew Lambert Photography/Science Photo Library; p34t: Georgios Kollidas/Shutterstock; p34m: Stephen Kiers/Shutterstock; p34b: Science Photo Library; p35t: Sheila Terry/Science Photo Library; p35b: Philippe Plailly/Science Photo Library; p36t: Science Photo Library; p36b: Mark Yuill/ Shutterstock ; p37r: A. Barrington Brown/Science Photo Library; p37l: CERN; p38: Imagedb. com/Shutterstock; p39l: Gannet77/Istockphoto; p39r: Dutourdumonde Photography/ Shutterstock; p39b: Kushch Dmitry/Shutterstock; p42l: Shutterstock NS;p42m: Shutterstock NS; p42r: Shutterstock NS; p43: Tlorna/Shutterstock; p44tl: Martyn F. Chillmaid/Science Photo Library; p44tr: Joerg Beuge/Shutterstock; p44m: Photocritical/Shutterstock; p44b: Xtrekx/Shutterstock; p46: OUP; p48: Kuleczka/Shutterstock; p49: OUP; p50: Shutterstock NS ; p51t: OUP; p51m: OUP; p51b: OUP; p52t: OUP; p52mt: OUP; p52mb: OUP; p52b: OUP; p53t: OUP; p53m: OUP; p53b: OUP;p54l: Shutterstock NS; p54r: Daniel Taeger/Shutterstock ; p54bl: Stefan Glebowski/Shutterstock ; p54br: Dima Galanternik/Istockphoto; p55t: A. Barrington Brown/Science Photo Library; p55b: Stefan Redel/Shutterstock ; p56t: Dmitry Kalinovsky/ Shutterstock ; p56b: Shutterstock NS; p57t: OUP; p57b: Shawn Hempel/Shutterstock ; p58t: Andrew Lambert Photography/Science Photo Library; p58l: Shutterstock NS; p58r: Zbynek Burival/Shutterstock ; p59t: TonyV3112/Shutterstock; p59b: NigelSpiers/Shutterstock; p62: Robert Llewellyn/Imagestate Media Partners Limited - Impact Photos/Alamy; p63: Paul Rapson/Science Photo Library; p64: NASA; p65t: OUP; p65b: OUP; p68t: OUP; p68b: Jean-Loup Charmet/Science Photo Library; p69: Iain McGillivray/Shutterstock; p72l: OUP; p72m: OUP; p72r: OUP; p74: OUP; p75t: OUP; p75b: OUP; p76: OUP; p77: Vulcan.wr.usgs.gov; p80: OUP; p81: Ron Kloberdanz/Shutterstock; p82: Geoff Tompkinson/Science Photo Library; p85t: AJ Photo/Hop Americain/Science Photo Library;p85b: Dlewis33/Istockphoto; p84t: Thor Jorgen Udvang/Shutterstock; p84b: Chepko/Istockphoto; p85: Tina Lorien/Vetta/Getty Images; p88t: OUP; p88l: Jordache/Shutterstock; p88r: Claudio Arnese/Istockphoto; p89t: Ulga/Shutterstock; p89b: More Pixels/Istockphoto; p91: Charles D. Winters/Science Photo Library; p92t: OUP; p92b: OUP; p93t: OUP; p93b: OUP; p95m: Sheftsoff/Shutterstock; p94: Andrew Lambert Photography/Science Photo Library; p95b: Photographee.eu/Shutterstock; p95bl: Jim Varney/ Science Photo Library; p95t: Andrew Lambert Photography/Science Photo Library; p98l: OUP; p98r: Muellek Josef/Shutterstock; p99l: OUP; p99m: OUP; p99r: OUP; p104t: Aptyp_koK/ Shutterstock; 104b: OUP; p105l: Andrzej5003/Istockphoto; 105r: Mona Makela/Shutterstock; p107l: Ryan Pyle/Corbis; p107r: OUP; p110l: OUP; p110m: OUP; p110r: OUP; p111l: OUP; p111m: Science Photos/Alamy; mp111r: OUP; p112: Charles D. Winters/Science Photo Library; p114l: Jiri Jura/Shutterstock; p114m: Oleksandr Pakhay/Dreamstime; p114r: Chris G. Parkhurst/Shutterstock; p115t: Carolina K. Smith MD/Shutterstock; p115b: Kristina Postnikova/Shutterstock; p117: 2005 American Honda Motor Co., Inc.; p118t: Schulte Productions/Istockphoto; p118b: Hywit Dimyadi/Shutterstock; p119t: StepStock/Shutterstock; p119mt: Design56/Shutterstock; p119mb: Agita Leimane/Shutterstock; p119b: Konstantin Chagin/Shutterstock; p120l: OUP; p120r: OUP; p121: Peter Albrektsen/Shutterstock; p126tl: OUP; p126tm: Jim Lopes/Shutterstock; p126tr: Charles Schug/Istockphoto; p126bl: Best Images/Shutterstock; p126bm: Valeria73/Shutterstock; p126br: Seraficus/Shutterstock; p127l: OUP; p127m: OUP; p127r: OUP; p128: OUP; p130: OUP; p131t: Varela/Istockphoto; p131b: Christine Glade/Istockphoto; p132l: OUP; p132r: OUP; p133t: Tyler Stableford/Stone/Getty Images; p133b: STR News/Reuters; p136: PRILL/Shutterstock; p137t: James L. Amos/Corbis; p137b: Gabi Moisa/Shutterstock; p138t: Knorre/Shutterstock; p138m: Grasko/Shutterstock; p138b: OUP; p138l: Volker Steger/Science Photo Library; p139t: Shannon Matteson/ Shutterstock; 139m: Jinga80/Fotolia; p139b: NOAA Photo Library; p140t: Power and Syred/ Science Photo Library; p140b: Nigel Catlin/Visuals Unlimited/Corbis;p141: OUP; p144tl: OUP; p144tm: OUP; p144tr: OUP; p144m: OUP; p144b: OUP; p146t: Dino Osmic/Shutterstock; p146b: Robcruse/Istockphoto; p147: Helene Rogers/Art Directors & TRIP/Alamy; p148t: OUP; p148m: OUP; p148b: OUP; p149t: Bilderbox/INSADCO Photography/Alamy; p149b: Ivaschenko Roman/Shutterstock; p150: Andrew Lambert Photography/Science Photo Library; p151: RMG; p153t: OUP; p153b: Jodi Jacobson/Istockphoto; p154: canismaior/Shutterstock; p155: Lightpoet/Shutterstock;p156: OUP; p157t: Brenda Carson/Shutterstock; p157l: Niall McDiarmid/Alamy; p157m: Marianne Rosenstiehl/Sygma/Corbis; p159l: Andrew Lambert Photography/Science Photo Library;p159r: Andrew Lambert Photography/Science Photo Library; p159b: Nredmond/Istockphoto; p163t: Photodisc/OUP ; p163b: Kazki/Alamy; p164: OUP; p165: OUP; p166: Dlewis33/Istockphoto; p167: Andrew Lambert Photography/Science Photo Library; p169l: Amy Nichole Harris/Shutterstock; p169r: OUP; p169b: Oneclearvision/ Istockphoto;p170l: OUP; p170m: OUP; p170r: OUP; p170b: Collection of 'Bizarre' series ceramics (ceramic), Cliff, Clarice (1899-1972)/Private Collection/Photo © Bonhams, London, UK/The Bridgeman Art Library; p171: Kozmoat98/E/Getty Images; p173t: Maria Skaldina/ Shutterstock; p173b: John Keith/Shutterstock; p174t: March Cattle/Shutterstock; p174b: Skyhawk x/Shutterstock; p175l: Time & Life Pictures/Getty Images ; p175r: Webelements.com; p175m: Dmitry Yashkin/Shutterstock; p175b: G. Victoria/Shutterstock; p178: Fer Gregory/ Shutterstock; p179tl: Track5/Istockphoto; p179tr: Pgiam/Istockphoto; p179l: JeremyRichards/ Shutterstock; p179m: Charles d. Winters/Science Photo Library; p179r: OUP; p181: OUP; p184t: Martin Anderson/Shutterstock; p184b: Sergey Peterman/Shutterstock; p185: Dean Conger/Corbis; p186: The Metropolitan Council; p187t: Ian Cartwright/LGPL/Alamy p187b: George Peters /Istockphoto; p190t: STS-114 Crew/NASA; p190m: Thorsten Rust/Shutterstock; 190m: Zoran Karapancev/Shutterstock; p191l: OUP; p191m: OUP; p191r: OUP; p191b: Lee Prince/Shutterstock; p192: Paul Fleet/Shutterstock; p193t: Maximilian Stock Ltd/Science Photo Library; p193b: Tamara Kulikova/Dreamstime; p194t: Crown Copyright/Health & Safety Laboratory/Science Photo Library; p194b: Sergey Zavalnyuk/Dreamstime; p195: David Reilly/ Shutterstock; p196tl: Andreas Reh/Istockphoto; p196tm: Luoman/Istockphoto; p196tr: Ngataringa/Istockphoto; p196bl: OUP; p196bm: OUP; mp196br: lightphoto/Istockphoto p197t: Maximilian Stock Ltd/Science Photo Library; p197b: Nolimitpictures/Istockphoto; p198l: Christian Darkin/Science Photo Library; p198m: Tracy Hebden/Alamy; p198r: Vasily Smirnov/ Shutterstock; p199: Ruslan Grumble/Shutterstock; p200l: OUP; p200m: Anton Foltin/ Shutterstock; p200r: Svetlana Lukienko/Shutterstock; p200bl: Miguel Malo/Istockphoto; p200br: Masterfile; p201l: Rihardzz/Shutterstock; p201r: Oleg - F/Shutterstock; p203l: Erhan Dayi/Shutterstock; p203r: Cobalt/Istockphoto; p203m: Eoghan McNally/Shutterstock; p203b: GeorgiosArt/Istockphoto; p206r: Elisei Shafer/Shutterstock; p206l: NASA; p207t: Eraxion/ Istockphoto; p207b: Viridis/Istockphoto; p208: Pasquale Sorrentino/Science Photo Library; p209l: Mediscan/Medical-on-Line/Alamy; p209r: Jordache/Shutterstock; p209b: Jonathan Feinstein/Shutterstock; p210t: OUP; p210b: Ian Bracegirdle/Shutterstock; p211: Barnaby Chambers/Shutterstock; p212l: VT750/Shutterstock; p212r: Piotr Tomicki/Shutterstock; p212b: Kolbjorn/Istockphoto; p213l: Tourdottk/Shutterstock; p213r: Robert Buchanan Taylor/ Shutterstock; p213b: RicAguiar/Istockphoto; p214t: OUP; p214b: Kris Jacobs/Shutterstock; p216l: NASA; p216r: NASA; p216b: NASA; p217: NASA; p220: Mohammed Yousuf/THE HINDU; p221: OUP; p222: Shutterstock NS p224l: P Wei/Istockphoto; p224m: Aggphotographer/ Bigstock; p224r: Adam Hart Davis/Science Photo Library; p224m: Maxine Adcock/Science Photo Library; p224b: OUP; p225t: Yara International ASA; p225b: HeikeKampe/Istockphoto; p226t: OUP; p226b: OUP; p227t: OUP; p227l: Kuttelvaserova Stuchelova/Shutterstock; p227r: Alessandro Minguzzi/Thinkstock; p228t: Manamana/Shutterstock; p228b: Barry Barnes/ Shutterstock; p229t: Martyn F. Chillmaid/Science Photo Library; p229b: WizData/Bigstock; p230b: Jurie/Shutterstock; p231t: Elwynn/Shutterstock p231b: Maxisport/Shutterstock; p232t: Audiohead/Deamstime; p232b: Don Hammond/Design Pics/Corbis; p233t: Scott Lomenzo/ Shutterstock; p233l: ranplett/Istockphoto; p233r: Tratong/Shutterstock; p234: Javarman/ Shutterstock; p235t: Marc Van Vuren/Shutterstock; p235b: Plinney/Istockphoto; p236: C Jones/Shutterstock; p237t: Emmanuel R Lacoste/Shutterstock; p237b: Ruslan Gilmanshin/ Dreamstime; p240l: Paul Rapson/Science Photo Library; p240m: Oleksandr Kalinichenko/ Shutterstock p240r: Hywit Dimyadi/Shutterstock; p241l: Lya_Cattel/Istockphoto; p241m: OUP; p241r: OUP; p241b: Eyeidea/Depositphoto; p243: Kozmoat98/Istockphoto; p244t: Zybr78/ Shutterstock; p244b: Christian Lagereek/Dreamstime; p245t: Buzbuzzer/Istockphoto; p245b: MirAgreb/Istockphoto; p246: Rtimages/Shutterstock; p248: Science Photos/Alamy; p249: Mark Evans/Vetta/Getty Images; p250: Nfsphoto/Shutterstock; p251: Aastock/Shutterstock; p252t: Originalpunkt/Shutterstock; p252b: Tish1/Shutterstock; p253: Celso Pupo/Shutterstock; p254: Gordana Sermek/Shutterstock; p255: OUP; p258l: OUP; p258m: OUP; p258r: Gelpi/Canstock; p258b: Shutterstock NS; p259l: Highviews/Istockphoto; p259m: Michaeljung/Shutterstock; p259r: Gabi Moisa/Shutterstock; p259b: Chris Hellyar/Shutterstock; p260t: Miguel Malo/ Istockphoto; p260b: Andriscam/Shutterstock; p261: Angelo Gilardelli/Shutterstock; p262t: Charles D. Winters/Science Photo Library; p262b: Marbo/Dreamstime; p263t: OUP; p263b: Elnur Amikishiyev/123rf; p264t: Oksix/Shutterstock; p264b: Radius Images/Alamy; p265t: Tinabelle/Istockphoto; p265l: Alexey U/Shutterstock; p265r: Bluestocking/Istockphoto; p266t: Izusek/IStockphoto; p266l: Andreas Weiss/Shutterstock; p266r: Andy Davies/Science Photo Library; p267t: Sheila Terry/Science Photo Library; p267b: Peter Yates/Science Photo Library; p269t: Nattika/Shutterstock; p269m: Juris Sturainis/Shutterstock; p269b: Michal Krakowiak/ Vetta/Getty Images; p269l: Shutterstock; p270: NorGal/Shutterstock; p271t: Anita Patterson Peppers/Shutterstock; p271b: Jiri Hera/Shutterstock; p274: Radu Razvan/Shutterstock; p275tl: Ktsdesign/Shutterstock; p275tm: Science Photo Library; p275tr: OUP; p275bl: OUP; p275bm: Steven Collins/Shutterstock; p275br: Arieliona/Shutterstock; p276: Martyn F. Chillmaid/Science Photo Library; p277: OUP; p280l: Science Photo Library; p280r: Andrew Lambert Photography/Science Photo Library; p280b: Laboko/Shutterstock; p282r: OUP; p282l: Science Photo Library; p283t: Yeko Photo Studio/Shutterstock; p283b: Andrew Lambert Photography/Science Photo Library;

Cambridge International Examinations bears no responsibility for the example answers to questions taken from its past question papers which are contained in this publication.

Introduction

If you are taking IGCSE chemistry, using the Cambridge International Examinations syllabus 0620, then this book is for you. It covers the syllabus fully.

Finding your way around the book

The contents list on the next page shows how the book is organised. Take a look. Note the extra material at the back of the book too: for example the questions from past exam papers, and the glossary.

Finding your way around the chapters

Each chapter is divided into two-page units. Some colour coding is used within the units, to help you use them properly:

Core syllabus content
If you are following the Core syllabus content you can ignore any material with a red line beside it.

Extended syllabus content
For this, you need *all* the material on the white pages, including the material marked with a red line.

Extra material
Pages of this colour contain extra material for some topics. We hope that you will find it interesting – but it is not needed for the exam.

Chapter checkups
There is a revision checklist at the end of each chapter, and also a set of exam-level questions about the chapter, on a coloured background.

Making the most of the book and support website

We want you to understand chemistry, and do well in your exams. This book, and the website, can help you. So make the most of them!

Work through the units   The two-page units will help you build up your knowledge and understanding of the chemistry on your syllabus.

Use the glossary   If you come across a chemical term that you do not understand, try the glossary. You can also use the glossary to test yourself.

Answer the questions   It is a great way to get to grips with a topic. This book has lots of questions: at the end of each unit and each chapter, and questions from past exam papers at the end of the book.

Answers to all questions, except those from exam papers, are at the back of the book. Your teacher can provide answers for the exam questions.

Use the website   The website has an interactive test for each chapter, advice on revision, exam-style questions, and more.

And finally, enjoy!   Chemistry is an important and exciting subject. We hope this book will help you to enjoy it, and succeed in your course.

RoseMarie Gallagher
Paul Ingram

Contents

1 States of matter

1.1	Everything is made of particles	2
1.2	Solids, liquids, and gases	4
1.3	The particles in solids, liquids, and gases	6
1.4	A closer look at gases	8
	Checkup on Chapter 1	10

2 Separating substances

2.1	Mixtures, solutions, and solvents	12
2.2	Pure substances and impurities	14
2.3	Separation methods (part I)	16
2.4	Separation methods (part II)	18
2.5	More about paper chromatography	20
	The chromatography detectives	22
	Checkup on Chapter 2	24

3 Atoms and elements

3.1	Atoms and elements	26
3.2	More about atoms	28
3.3	Isotopes and radioactivity	30
3.4	How electrons are arranged	32
	How our model of the atom developed	34
	The atom: the inside story	36
3.5	The metals and non-metals	38
	Checkup on Chapter 3	40

4 Atoms combining

4.1	Compounds, mixtures, and chemical change	42
4.2	Why do atoms form bonds?	44
4.3	The ionic bond	46
4.4	More about ions	48
4.5	The covalent bond	50
4.6	Covalent compounds	52
4.7	Comparing ionic and covalent compounds	54
4.8	Giant covalent structures	56
4.9	The bonding in metals	58
	Checkup on Chapter 4	60

5 Reacting masses, and chemical equations

5.1	The names and formulae of compounds	62
5.2	Equations for chemical reactions	64
5.3	The masses of atoms, molecules, and ions	66
5.4	Some calculations about masses and %	68
	Checkup on Chapter 5	70

6 Using moles

6.1	The mole	72
6.2	Calculations from equations, using the mole	74
6.3	Reactions involving gases	76
6.4	The concentration of a solution	78
6.5	Finding the empirical formula	80
6.6	From empirical to final formula	82
6.7	Finding % yield and % purity	84
	Checkup on Chapter 6	86

7 Redox reactions

7.1	Oxidation and reduction	88
7.2	Redox and electron transfer	90
7.3	Redox and changes in oxidation state	92
7.4	Oxidising and reducing agents	94
	Checkup on Chapter 7	96

8 Electricity and chemical change

8.1	Conductors and insulators	98
8.2	The principles of electrolysis	100
8.3	The reactions at the electrodes	102
8.4	The electrolysis of brine	104
8.5	Two more uses of electrolysis	106
	Checkup on Chapter 8	108

9 Energy changes, and reversible reactions

9.1	Energy changes in reactions	110
9.2	Explaining energy changes	112
9.3	Energy from fuels	114
9.4	Giving out energy as electricity	116
	The batteries in your life	118
9.5	Reversible reactions	120
9.6	Shifting the equilibrium	122
	Checkup on Chapter 9	124

10 The speed of a reaction

10.1	Rates of reaction	126
10.2	Measuring the rate of a reaction	128
10.3	Changing the rate of a reaction (part I)	130
10.4	Changing the rate of a reaction (part II)	132
10.5	Explaining rates	134
10.6	Catalysts	136
	More about enzymes	138
10.7	Photochemical reactions	140
	Checkup on Chapter 10	142

11 Acids, bases, and salts

11.1	Acids and alkalis	144
11.2	A closer look at acids and alkalis	146
11.3	The reactions of acids and bases	148
11.4	A closer look at neutralisation	150
11.5	Oxides	152
11.6	Making salts	154
11.7	Making insoluble salts by precipitation	156
11.8	Finding concentrations by titration	158
	Checkup on Chapter 11	160

12 The Periodic Table

12.1	An overview of the Periodic Table	162
12.2	Group I: the alkali metals	164
12.3	Group VII: the halogens	166
12.4	Group VIII: the noble gases	168
12.5	The transition elements	170
12.6	Across the Periodic Table	172
	How the Periodic Table developed	174
	Checkup on Chapter 12	176

13 The behaviour of metals

13.1	Metals: a review	178
13.2	Comparing metals for reactivity	180
13.3	Metals in competition	182
13.4	The reactivity series	184
13.5	Making use of the reactivity series	186
	Checkup on Chapter 13	188

14 Making use of metals

14.1	Metals in the Earth's crust	190
14.2	Extracting metals from their ores	192
14.3	Extracting iron	194
14.4	Extracting aluminium	196
14.5	Making use of metals and alloys	198
14.6	Steels and steel-making	200
	Metals, civilisation, and you	202
	Checkup on Chapter 14	204

15 Air and water

15.1	What is air?	206
15.2	Making use of air	208
15.3	Pollution alert!	210
15.4	The rusting problem	212
15.5	Water supply	214
	Living in space	216
	Checkup on Chapter 15	218

16 Some non-metals and their compounds

16.1	Hydrogen, nitrogen, and ammonia	220
16.2	Making ammonia in industry	222
16.3	Fertilisers	224
16.4	Sulfur and sulfur dioxide	226
16.5	Sulfuric acid	228
16.6	Carbon and the carbon cycle	230
16.7	Some carbon compounds	232
16.8	Greenhouse gases, and global warming	234
16.9	Limestone	236
	Checkup on Chapter 16	238

17 Organic chemistry

17.1	Petroleum: a fossil fuel	240
17.2	Refining petroleum	242
17.3	Cracking hydrocarbons	244
17.4	Families of organic compounds	246
17.5	The alkanes	248
17.6	The alkenes	250
17.7	The alcohols	252
17.8	The carboxylic acids	254
	Checkup on Chapter 17	256

18 Polymers

18.1	Introducing polymers	258
18.2	Addition polymerisation	260
18.3	Condensation polymerisation	262
18.4	Making use of synthetic polymers	264
18.5	Plastics: here to stay?	266
18.6	Natural polymers in food (part I)	268
18.7	Natural polymers in food (part II)	270
	Checkup on Chapter 18	272

19 In the lab

19.1	Chemistry: a practical subject	274
19.2	Example of an experiment	276
19.3	Working with gases in the lab	278
19.4	Testing for ions in the lab: cations	280
19.5	Testing for ions in the lab: anions	282
	Checkup on Chapter 19	284

Answers for Chapters 1 – 19	286

Cambridge IGCSE Chemistry past exam questions

Exam questions for Paper 3	302
Exam questions for Paper 4	308
Exam questions from Paper 6	314

Reference

Glossary	320
The Periodic Table and atomic masses	324
Index	326

Support website: www.oxfordsecondary.com/9780198399148

1.1 Everything is made of particles

Particles everywhere

Rock, air, and water look very different. But they have one thing in common: they are made of tiny pieces. Let's call these pieces **particles**.

In fact everything around you is made of particles – and so are you!

In rock and other solids, the particles are not free to move. But in liquids and gases, they are always moving – and colliding with each other.

Proof from pollen

For centuries, people had guessed that water and air were made of tiny particles. But since they could not see them, they could not prove it.

Then in 1827, a botanist called Robert Brown noticed something strange. He was studying pollen from a flower, in water, under a microscope. He saw particles from the pollen jiggling around. They were not alive – so why were they moving?

78 years later, Albert Einstein came up with the answer. The particles moved because they were being struck by tiny invisible moving water particles. In fact their movement proved that water is made of particles.

Brownian motion

Today the **random motion** of particles that you can see with the naked eye, or under a microscope, is called **Brownian motion**. The particles follow a zig-zag path, because they are being struck by tiny invisible particles. Look at the drawing on the right.

More evidence

Everyone now accepts that things are made of particles. Most are far too small to be seen under a microscope. But there is plenty of evidence for them. Let's look at more examples.

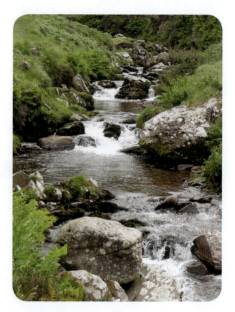

▲ All made of particles!

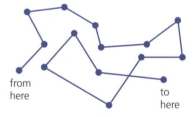

from here — to here

▲ The random motion of a particle. This is what Robert Brown observed.

Outside the lab

1　In sunlit rooms, you sometimes see dust dancing in the air. It dances because the dust particles are being bombarded by the tiny moving particles in air. This is an example of Brownian motion.

2　Cooking smells spread. The 'smells' are due to tiny particles, which spread because they are bombarded by the particles in air. This is an example of diffusion. (See next page.) Some of them end up in your nose!

In the lab

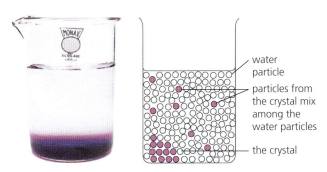

water particle
particles from the crystal mix among the water particles
the crystal

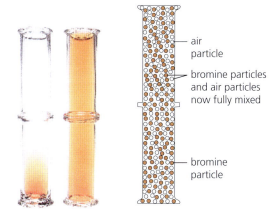

air particle
bromine particles and air particles now fully mixed
bromine particle

3 Place a crystal of purple potassium manganate(VII) in a beaker of water. The colour spreads through the water, because the particles leave the solid crystal – it **dissolves** – and spread among the water particles.

4 Place an open gas jar of air upside down on an open gas jar containing a few drops of red-brown bromine. The colour spreads upwards because particles of bromine vapour mix among the gas particles in the air.

Diffusion

Look again at the two examples above. The particles mix and spread by colliding with other moving particles, and bouncing off in all directions. This mixing process is called **diffusion**.

The end result is that the particles spread from where they are more concentrated to where they are less concentrated, until they are evenly mixed. Look at the drawings on the right.

So what are these particles?

In examples 2, 3, and 4, we cannot see the moving particles – even with the most powerful microscope. They are far too small. So what are they?

- The smallest particles, that we cannot not break down further in chemical reactions, are called **atoms**.

- In some substances, the particles are just single atoms. For example air contains single argon atoms.

- In many substances, the particles consist of two or more atoms joined together. These particles are called **molecules**. Water and bromine exist as molecules. Air is mostly nitrogen and oxygen molecules.

- In other substances the particles are atoms or groups of atoms that carry a charge. These particles are called **ions**. In example 3 above, the particles in potassium manganate(VII) are ions.

You'll find out more about all these particles in Chapters 2 and 3.

But what about the particles Robert Brown observed? They contained thousands of atoms. That's why they were big enough to be seen with a microscope. Dust particles in air contain thousands of atoms too.

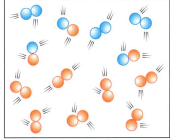

The moving particles collide, then bounce apart in all directions.

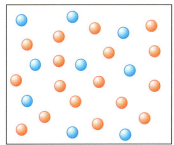

In the end they are all mixed up. (But they keep on moving.)

Q

1 The particles that Robert Brown observed showed *random motion*. What does that mean, and why did it occur?

2 Why was *Brownian motion* given that name?

3 Example 3 above proves that potassium manganate(VII) is made of particles. Explain why.

4 Bromine vapour is heavier than air. Even so, it spreads upwards in example 4 above. Why?

5 a What is *diffusion*?
  b Use the idea of diffusion to explain how the smell of perfume travels across a room.

1.2 Solids, liquids, and gases

What's the difference?

It is easy to tell the difference between a solid, a liquid, and a gas:

A solid has a fixed shape and a fixed volume. It does not flow. Think of all the solid things around you: their shapes and volumes do not change.

A liquid flows easily. It has a fixed volume, but its shape changes. It takes the shape of the container you pour it into.

A gas does not have a fixed volume or shape. It spreads out to fill its container. It is much lighter than the same volume of solid or liquid.

Water: solid, liquid and gas

Water can be a solid (ice), a liquid (water), and a gas (water vapour or steam). Its state can be changed by heating or cooling:

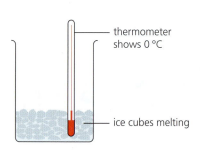

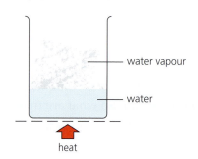

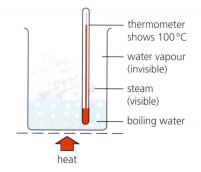

1 Ice slowly changes to **water**, when it is put in a warm place. This change is called **melting**. The thermometer shows 0 °C until all the ice has melted. So 0 °C is called its **melting point**.

2 When the water is heated its temperature rises, and some of it changes to **water vapour**. This change is called **evaporation**. The hotter the water gets, the more quickly it evaporates.

3 Soon bubbles appear in the water. It is **boiling**. The water vapour shows up as steam. The thermometer stays at 100 °C while the water boils off. 100 °C is the **boiling point** of water.

And when steam is cooled, the opposite changes take place:

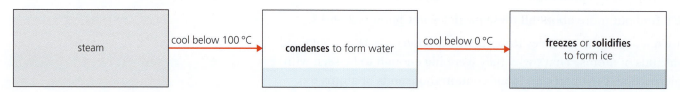

You can see that:

- condensing is the opposite of evaporating
- freezing is the opposite of melting
- the freezing point of water is the same as the melting point of ice, 0 °C.

Other things can change state too

It's not just water! Nearly all substances can exist as solid, liquid, and gas. Even iron and diamond can melt and boil! Some melting and boiling points are given below. Look how different they are.

Substance	Melting point/°C	Boiling point/°C
oxygen	−219	−183
ethanol	−15	78
sodium	98	890
sulfur	119	445
iron	1540	2900
diamond	3550	4832

▼ Molten iron being poured out at an iron works. Hot – over 1540°C!

Showing changes of state on a graph

Look at the graph on the right below. It shows how the temperature changes as a block of ice is steadily heated. First the ice melts to water. Then the water gets warmer and warmer, and eventually turns to steam.

This type of graph is called a **heating curve**.

Look at the step where the ice is melting. Once melting starts, the temperature stays at 0°C until *all* the ice has melted.

Then when the water starts to boil, the temperature stays at 100°C until *all* the water has turned to steam.

So the melting and boiling points are clear and sharp.

You can draw a heating curve for any substance.

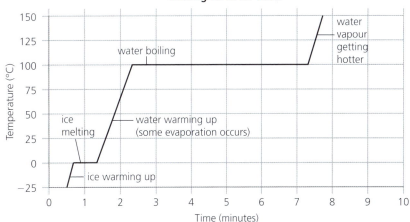

Heating curve for water

Sublimation: a special change

Some substances go straight from solid to gas, when they warm up. This change is called **sublimation**.

For example, if you leave solid (frozen) carbon dioxide sitting at room temperature, it will **sublime**. And if you warm some iodine crystals gently on a clock glass, they will give off a purple vapour.

Naphthalene, which is used in old-fashioned moth balls, also sublimes. The vapour kills moths.

▲ Carbon dioxide subliming. Solid carbon dioxide is also called *dry ice*.

Q

1 Write down two properties of a solid, two of a liquid, and two of a gas.
2 Which word means the opposite of:
   a boiling?   b melting?
3 Which has a lower freezing point, oxygen or ethanol?
4 Which has a higher boiling point, oxygen or ethanol?

5 Look at the heating curve above.
   a About how long did it take for the ice to melt, once melting started?
   b How long did boiling take to complete, once it started?
   c Try to think of a reason for the difference in **a** and **b**.
6 What does *sublimation* mean? Give an example.

1.3 The particles in solids, liquids, and gases

How the particles are arranged

Water can change from solid to liquid to gas. Its *particles* do not change.
They are the same in each state. But their *arrangement* changes.
The same is true for all substances.

State	How the particles are arranged	Diagram of particles
Solid 	The particles in a solid are arranged in a fixed pattern or **lattice**. Strong forces hold them together. So they cannot leave their positions. The only movements they make are tiny vibrations to and fro.	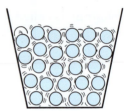
Liquid 	The particles in a liquid can move about and slide past each other. They are still close together, but not in a lattice. The forces that hold them together are weaker than in a solid.	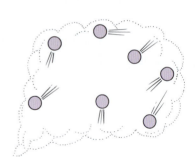
Gas	The particles in a gas are far apart, and they move about very quickly. There are almost no forces holding them together. They collide with each other and bounce off in all directions.	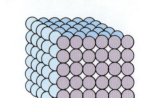

Changing state

So why do substances change state when you heat them? It is because
the particles take in heat energy – and this changes how they move.

Melting When a solid is heated, its particles get more energy and vibrate
more. This makes the solid **expand**. At the melting point, the particles vibrate
so much that they break away from their positions. The solid turns liquid.

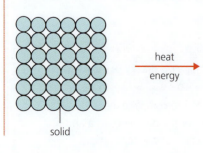
solid

heat
energy

the vibrations get larger

heat energy at
melting point

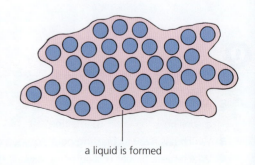
a liquid is formed

Boiling When a liquid is heated, its particles get more energy and move faster. They bump into each other more often, and bounce further apart. This makes the liquid expand. At the boiling point, the particles get enough energy to overcome the forces between them. They break away to form a gas:

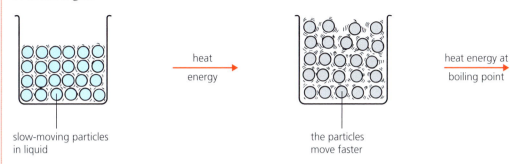
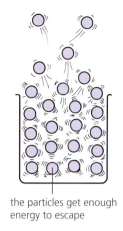

slow-moving particles in liquid

heat energy

the particles move faster

heat energy at boiling point

the particles get enough energy to escape

Evaporating Some particles in a liquid have more energy than others. Even well below the boiling point, some have enough energy to escape and form a gas. This is called **evaporation**. It is why puddles of rain dry up in the sun.

How much heat is needed?

The amount of heat needed to melt or boil a substance is different for every substance. That's because the particles in each substance are different, with different forces between them.

The stronger the forces, the more heat energy is needed to overcome them. So the higher the melting and boiling points will be.

Reversing the changes

You can reverse those changes again by cooling. As a gas cools, its particles lose energy and move more slowly. When they collide, they do not have enough energy to bounce away. So they stay close, and form a liquid. On further cooling, the liquid turns to a solid. Look:

The kinetic particle theory

Look at the key ideas you have met:

- A substance can be a solid, a liquid, or a gas, and change from one state to another.
- It has different characteristics in each state. (For example, solids do not flow.)
- The differences are due to the way its particles are arranged, and move, in each state.

Together, these ideas make up the **kinetic particle theory**.
(*Kinetic* means *about motion*.)

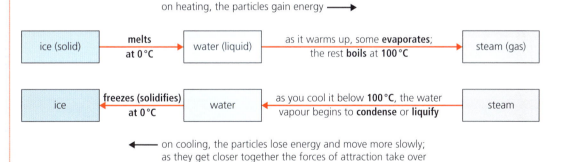

on heating, the particles gain energy ⟶

| ice (solid) | **melts** **at 0 °C** | water (liquid) | as it warms up, some **evaporates**; the rest **boils** at **100 °C** | steam (gas) |

| ice | **freezes (solidifies)** **at 0 °C** | water | as you cool it below **100 °C**, the water vapour begins to **condense** or **liquify** | steam |

⟵ on cooling, the particles lose energy and move more slowly; as they get closer together the forces of attraction take over

Q

1 Draw diagrams to show the arrangement of particles in a solid, a liquid, and a gas. Add notes to each diagram.
2 Using the idea of particles, explain why:
   a you cannot pour solids **b** you can pour liquids
3 Draw a diagram to show what happens to the particles, when a liquid cools to a solid.

4 See if you can explain what happens to the particles, when:
   a a solid sublimes **b** wet clothes dry in the sunshine
5 Oxygen is the gas we breathe in. It can be separated from the air. It freezes at −219 °C and boils at −183 °C.
   a In which state is oxygen, at: **i** 0 °C? **ii** −200 °C?
   b How would you turn oxygen gas into solid oxygen?

1.4 A closer look at gases

What is gas pressure?

When you blow up a balloon, you fill it with air particles. They collide with each other. They also hit the sides of the balloon, and exert **pressure** on it. This pressure keeps the balloon inflated.

In the same way, *all* gases exert a pressure. The pressure depends on the **temperature** of the gas and the **volume** it takes up, as you'll see below.

When you heat a gas

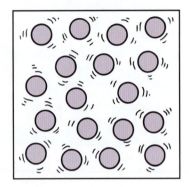 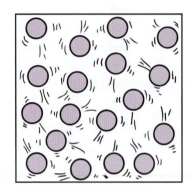

The particles in this gas are moving fast. They hit the walls of the container and exert pressure on them. If you now heat the gas . . .

. . . the particles take in heat energy and move even faster. They hit the walls more often, and with more force. So the gas pressure increases.

The same happens with all gases:
When you heat a gas in a closed container, its pressure increases.
That is why the pressure gets very high inside a pressure cooker.

When you squeeze a gas into a smaller space

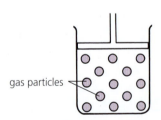 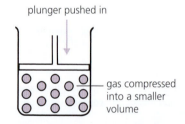

plunger pushed in

gas particles

gas compressed into a smaller volume

There is a lot of space between the particles in a gas. You can **compress** the gas, or force its particles closer, by pushing in the plunger …

… like this. Now the particles are in a smaller space – so they hit the walls more often. So the gas pressure increases.

The same thing is true for all gases:
When a gas is compressed into a smaller space, its pressure increases.

All gases can be compressed. If enough force is applied, the particles can be pushed so close that the gas turns into a liquid. But liquids and solids cannot be compressed, because their particles are already very close together.

▲ The harder you blow, the greater the pressure inside the balloon.

▲ In a pressure cooker, water vapour (gas) is heated to well over 100 °C. So it is at high pressure. You must let a pressure cooker cool before you open it!

▲ When you blow up a bicycle tyre, you compress air into the inner tube.

The rate of diffusion of gases

On page 3 you saw that gases **diffuse** because the particles collide with other particles, and bounce off in all directions. But gases do not all diffuse at the same rate, every time. It depends on these two factors:

1 The mass of the particles

The particles in hydrogen chloride gas are twice as heavy as those in ammonia gas. So which gas do you think will diffuse faster? Let's see:

- Cotton wool soaked in ammonia solution is put into one end of a long tube (at A below). It gives off ammonia gas.
- *At the same time*, cotton wool soaked in hydrochloric acid is put into the other end of the tube (at B). It gives off hydrogen chloride gas.
- The gases diffuse along the tube. White smoke forms where they meet:

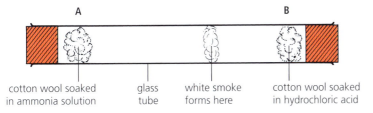

| cotton wool soaked in ammonia solution | glass tube | white smoke forms here | cotton wool soaked in hydrochloric acid |

The white smoke forms closer to B. So the ammonia particles have travelled further than the hydrogen chloride particles – which means they have travelled *faster*.
The lower the mass of its particles, the faster a gas will diffuse.

That makes sense when you think about it. When particles collide and bounce away, the lighter particles will bounce further.

The particles in the two gases above are molecules. The mass of a molecule is called its **relative molecular mass**. So we can also say:
The lower its relative molecular mass, the faster a gas will diffuse.

2 The temperature

When a gas is heated, its particles take in heat energy, and move faster. They collide with more energy, and bounce further away. So the gas diffuses faster. **The higher the temperature, the faster a gas will diffuse.**

▲ The scent of flowers travels faster in a warm room. Can you explain why?

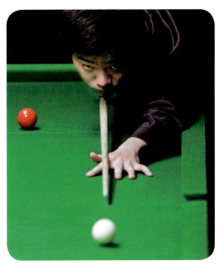

▲ The faster a particle is moving when it hits another, the faster and further it will bounce away. Just like snooker balls!

Q

1 What causes the *pressure* in a gas?
2 Why does a balloon burst if you keep on blowing?
3 A gas is in a sealed container. How do you think the pressure will change if the container is cooled? Explain your answer.
4 A gas flows from one container into a larger one. What do you think will happen to its pressure? Draw a diagram to explain.

5 a Why does the scent of perfume spread?
  b Why does the scent of perfume wear off faster in warm weather than in cold?
6 Of all gases, hydrogen diffuses fastest at any given temperature. What can you tell from this?
7 Look at the glass tube above. Suppose it was warmed a little in an oven, before the experiment. Do you think that would change the result? If so, how?

Checkup on Chapter 1

Revision checklist

Core syllabus content

Make sure you can …

☐ state that everything is made of particles, and say what these particles are

☐ explain what *Brownian motion* is

☐ explain why these show that particles exist:
  – the movement of particles from pollen, in water
  – the movement of dust in air

☐ explain what *diffusion* is, and how it happens

☐ name the three states of matter, and give their physical properties (hard, fixed shape, and so on)

☐ describe, and sketch, the particle arrangement in each state of matter

☐ explain, and use, these terms:

melt	*boil*	*evaporate*
condense	*melting point*	*boiling point*
freezing point	*sublimation*	

☐ sketch, and label, a heating curve

☐ explain why a gas exerts a pressure

☐ explain why the pressure increases when you:
  – heat the gas
  – push it into a smaller space

Extended syllabus content

Make sure you can also …

☐ give examples of Brownian motion

☐ say that:
  – the random movement of small light particles suspended in water is due to their bombardment by water molecules
  – the random movement of dust particles in air is due to their bombardment by gas molecules

☐ describe how a substance changes state when you heat it, and explain this using the idea of particles and their movement

☐ describe an experiment to show that a gas will diffuse faster than another gas that has heavier particles

☐ say how, and why, the temperature affects the rate at which a gas diffuses

Questions

Core syllabus content

1 Give the correct name for each change of state:
  a solid $\longrightarrow$ liquid **b** liquid $\longrightarrow$ solid
  c gas $\longrightarrow$ liquid **d** solid $\longrightarrow$ gas

2 A large crystal of potassium manganate(VII) was placed in the bottom of a beaker of cold water, and left for several hours.

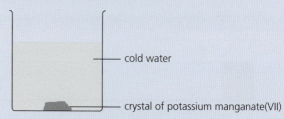

  a Describe what would be seen:
    i after five minutes **ii** after several hours
  b Explain your answers using the idea of particles.
  c Name the two processes that took place during the experiment.

3 Use the idea of particles to explain why:
  a solids have a definite shape
  b liquids fill the bottom of a container
  c you can't store gases in open containers
  d you can't squeeze a sealed plastic syringe that is completely full of water
  e a balloon expands as you blow into it.

4 Below is a heating curve for a pure substance. It shows how the temperature rises over time, when the substance is heated until it melts, then boils.

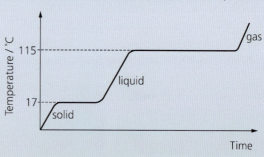

  a What is the melting point of the substance?
  b What happens to the temperature while the substance changes state?
  c The graph shows that the substance takes longer to boil than to melt. Suggest a reason for this.
  d How can you tell that the substance is not water?
  e Sketch a rough heating curve for pure water.

5 A **cooling curve** is the opposite of a heating curve. It shows how the temperature of a substance changes with time, as it is cooled from a gas to a solid. Here is the cooling curve for one substance:

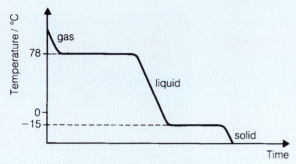

a What is the state of the substance at room temperature (20 °C)?

b Use the list of melting and boiling points on page 5 to identify the substance.

c Sketch a cooling curve for pure water.

6 Using the idea of particles, explain each of these.

a When two solids are placed on top of each other, they do not mix.

b Heating a gas in a closed container will increase its pressure.

c Poisonous gases from a factory chimney can affect a large area.

d In a darkened cinema, dust particles appear to dance in the beam of light.

7 a Pick out an example of diffusion.

   i a helium-filled balloon floating in the air

   ii an ice lollipop melting

   iii a creased shirt losing its creases when you iron it with a hot iron

   iv a balloon bursting when you stick a pin in it

   v a blue crystal forming a blue solution, when it is left sitting in a glass of water

   vi brushing paint onto a wall from a paint tin

   vii the yolk hardening, when you boil an egg

b For your choice in **a**, draw a diagram showing the particles before and after diffusion.

c Explain why this diffusion occurs.

Extended syllabus content

8 In order to demonstrate Brownian motion, some fine powdered charcoal is added to water. The suspension is viewed under a microscope.

a What will be observed, under the microscope?

b Explain this observation.

c Why would you *not* use copper(II) sulfate crystals for this demonstration?

9 You can measure the rate of diffusion of a gas using this apparatus. The gas enters through the thin tube:

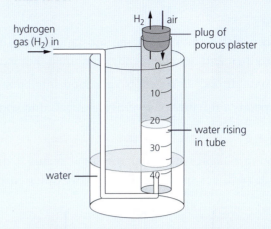

The measuring tube is sealed at the top with a plug of porous plaster. Air and other gases can diffuse in and out through the tiny holes in the plug.

The water rises in the measuring tube if the chosen gas diffuses out through the plug faster than air diffuses in. Air is mainly nitrogen and oxygen.

a When you use hydrogen gas, the water rises in the measuring tube. Why?

b What does this tell you about the rate of diffusion of hydrogen, compared with the gases in air?

c Explain your answer to **b** Use the term *mass*!

d The molecules in carbon dioxide are heavier than those in nitrogen and oxygen. So what do you think will happen to the water in the measuring tube, when you use carbon dioxide? Explain your answer.

10

Gas	Formula	Relative atomic or molecular mass
methane	CH_4	16
helium	He	4
oxygen	O_2	32
nitrogen	N_2	28
chlorine	Cl_2	71

Look at the table above.

a Which two gases will mix fastest? Explain.

b Which gas will take least time to escape from a gas syringe?

c Would you expect chlorine to diffuse more slowly than the gases in air? Explain.

d An unknown gas diffuses faster than nitrogen, but more slowly than methane. What you can say about its relative molecular mass?

2.1 Mixtures, solutions, and solvents

Mixtures

A **mixture** contains more than one substance. The substances are just mixed together, and not chemically combined. For example:

- air is a mixture of nitrogen, oxygen, and small amounts of other gases
- shampoo is a mixture of several chemicals and water.

Solutions

When you mix sugar with water, the sugar seems to disappear. That is because its particles spread all through the water particles, like this:

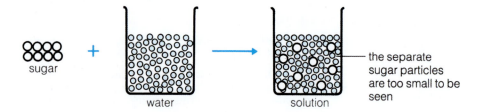

the separate
sugar particles
are too small to be
seen

The sugar has **dissolved** in the water, giving a mixture called a **solution**. Sugar is the **solute**, and water is the **solvent**:

solute + solvent = solution

You can't get the sugar out again by filtering.

Not everything dissolves so easily

Now think about chalk. If you mix chalk powder with water, most of the powder eventually sinks to the bottom. You can get it out again by filtering.

Why is it so different for sugar and chalk? Because their particles are very different! How easily a substance dissolves depends on the particles in it. Look at the examples in this table:

Compound	Mass (g) dissolving in 100 g of water at 25 °C
silver nitrate	241.3
calcium nitrate	102.1
sugar (glucose)	91.0
potassium nitrate	37.9
potassium sulfate	12.0
calcium hydroxide	0.113
calcium carbonate (chalk)	0.0013
silver chloride	0.0002

decreasing solubility

So silver nitrate is much more soluble than sugar – but potassium nitrate is a lot less soluble than sugar. It all depends on the particles.

Look at calcium hydroxide. It is only very slightly or **sparingly soluble** compared with the compounds above it. Its solution is called **limewater**.

Now look at the last two substances in the table. They are usually called **insoluble** since so very little dissolves.

▲ A mixture of sugar and water. This mixture is a solution.

▲ A mixture of chalk powder and water. This is not a solution. The tiny chalk particles do not separate and spread through the water particles. They stay in clusters big enough to see. In time, most sink to the bottom.

What's soluble, what's not?
- The solubility of every substance is different.
- But there are some overall patterns. For example *all* sodium compounds are soluble.
- Find out more on page 156.

Helping a solute dissolve

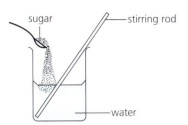

sugar — stirring rod

water

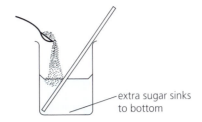

extra sugar sinks to bottom

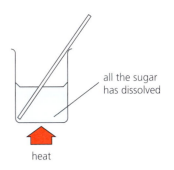

all the sugar has dissolved

heat

Sugar dissolves quite slowly in water at room temperature. If you stir the liquid, that helps. But if you keep on adding sugar …

… eventually no more of it will dissolve, no matter how hard you stir. The extra sinks to the bottom. The solution is now **saturated**.

But look what happens if you heat the solution. The extra sugar dissolves. Add more sugar and it will dissolve too, as the temperature rises.

So sugar is **more soluble** in hot water than in cold water.

A soluble solid usually gets more soluble as the temperature rises.

A solution is called *saturated* when it can dissolve no more solute, at that temperature.

Water is not the only solvent

Water is the world's most common solvent. A solution in water is called an **aqueous solution** (from *aqua*, the Latin word for water).

But many other solvents are used in industry and about the house, to dissolve substances that are insoluble in water. For example:

Solvent	It dissolves
white spirit	gloss paint
propanone (acetone)	grease, nail polish
ethanol	glues, printing inks, the scented substances that are used in perfumes and aftershaves

All three of these solvents evaporate easily at room temperature – they are **volatile**. This means that glues and paints dry easily. Aftershave feels cool because ethanol cools the skin when it evaporates.

▲ Nail polish is insoluble in water. It can be removed later by dissolving it in propanone.

About volatile liquids

- A **volatile** liquid is one that evaporates easily.
- This is a sign that the forces between its particles are weak.
- So volatile liquids have low boiling points too. (Propanone boils at 56.5 °C.)

Q

1 Explain each term in your own words:
   a soluble   b insoluble   c aqueous solution
2 Look at the table on page 12.
   a Which substance in it is the most soluble?
   b About how many times more soluble is this substance than potassium sulfate, at 25 °C?
   c The substance in a gives a colourless solution. What will you see if you add 300 g of it to 100 g of water at 25 °C?
   d What will you see if you heat up the mixture in c?

3 Now turn to the table at the top of page 156.
   a Name two metals that have *no* insoluble salts.
   b Name one other group of salts that are *always* soluble.
4 See if you can give three examples of:
   a solids you dissolve in water, at home
   b insoluble solids you use at home.
5 Name two solvents other than water that are used in the home. What are they used for?
6 Many gases dissolve in water. Give some examples.

2.2 Pure substances and impurities

What is a pure substance?

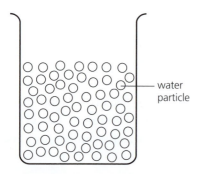

water particle

This is water. It has only water particles in it, and nothing else. So it is 100% **pure**.

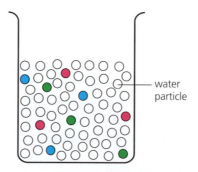

water particle

This water has particles of other substances mixed with it. So it is not pure.

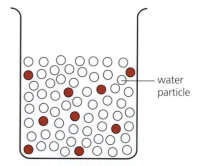

water particle

This water has particles of a harmful substance in it. So it is not pure – and could make you ill.

A pure substance has no other substance mixed with it.

In real life, very few substances are 100% pure. For example tap water contains small amounts of many different particles (such as calcium ions and chloride ions). The particles in it are not usually harmful – and some are even good for you.

Distilled water is much purer than tap water, but still not 100% pure. For example it may contain particles of gases, dissolved from the air.

Does purity matter?

Often it does not matter if a substance is not pure. We wash in tap water, without thinking too much about what is in it. But sometimes purity is very important. If you are making a new medical drug, or a flavouring for food, you must make sure it contains nothing that could harm people.

An unwanted substance, mixed with the substance you want, is called an **impurity**.

▲ Baby foods and milk powder are tested in the factory, to make sure they contain no harmful impurities.

▲ Getting ready for a jab. Vaccines and medicines must be safe, and free of harmful impurities. So they are tested heavily.

How can you tell if a substance is pure?

Chemists use some complex methods to check purity. But there is one simple method *you* can use in the lab: **you can check melting and boiling points**.

- A pure substance has a definite, sharp, melting point and boiling point. These are different for each substance. You can look them up in tables.

- When a substance contains an impurity:
  - its melting point falls and its boiling point rises
  - it melts and boils over a range of temperatures, not sharply.

- The more impurity there is:
  - the bigger the change in melting and boiling points
  - the wider the temperature range over which melting and boiling occur.

For example:

Substance	sulfur	water
Melts at (°C)	119	0
Boils at (°C)	445	100

These are the melting and boiling points for two pure substances: sulfur and water.

This sulfur sample melts sharply at 119 °C and boils at 445 °C. So it must be pure.

This water freezes around −0.5 °C and boils around 101 °C. So it is not pure.

ID check!
- Every substance has a unique pair of melting and boiling points.
- So you can also use melting and boiling points to **identify** a substance.
- First, measure them. Then look up data tables to find out what the substance is.

Separation: the first step in obtaining a pure substance

When you carry out a reaction, you usually end up with a *mixture* of substances. Then you have to separate the one you want.

The table below shows some separation methods. These can give quite pure substances. For example when you filter off a solid, and rinse it well with distilled water, you remove a lot of impurity. But it is just not possible to remove every tiny particle of impurity, in the school lab.

Method of separation	Used to separate...
filtration	a solid from a liquid
crystallisation	a solute from its solution
evaporation	a solute from its solution
simple distillation	a solvent from a solution
fractional distillation	liquids from each other
paper chromatography	different substances from a solution

There is more about these methods in the next three units.

▲ At the end of this reaction, the beaker may contain several products, plus reactants that have not reacted. Separating them can be a challenge!

Q

1 What does *a pure substance* mean?

2 You mix instant coffee with water, to make a cup of coffee. Is the coffee an *impurity*? Explain.

3 Explain why melting and boiling points can be used as a way to check purity.

4 Could there be impurities in a gas? Explain.

2.3 Separation methods (part I)

Separating a solid from a liquid

Which method should you use? It depends on whether the solid is dissolved, and how its solubility changes with temperature.

1 By filtering

For example, chalk is insoluble in water. So it is easy to separate by filtering. The chalk is trapped in the filter paper, while the water passes through. The trapped solid is called the **residue**. The water is the **filtrate**.

▲ Filtering in the kitchen …

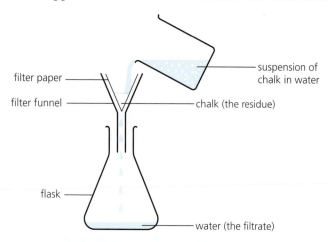

filter paper
filter funnel
flask
suspension of chalk in water
chalk (the residue)
water (the filtrate)

Saturated solutions

- Remember, most solutes get *more* soluble as the temperature rises – so *less* soluble as it falls!
- A saturated solution can hold no more solute, at that temperature.

2 By crystallisation

You can obtain many solids from their solutions by letting crystals form. The process is called **crystallisation**. It works because soluble solids tend to be *less soluble at lower temperatures*. For example:

heat

heat

1 This is a solution of copper(II) sulfate in water. You want to obtain solid copper(II) sulfate from it.

2 So you heat the solution to evaporate some of the water. It becomes more concentrated.

3 Eventually the solution becomes **saturated**. If you cool it now, crystals will start to form.

glass rod
microscope slide

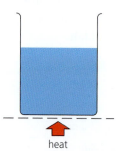

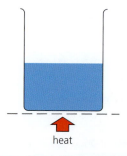

blue crystals of copper(II) sulfate

dilute copper(II) sulfate solution

4 Check that it is ready by placing a drop on a microscope slide. Crystals will form quickly on the cool glass.

5 Leave the solution to cool. Crystals start to form in it, as the temperature falls.

6 Remove the crystals by filtering. Then rinse them with distilled water and dry them with filter paper.

◄ Making a living from crystallisation. Seawater is led into shallow ponds. The water evaporates in the sun. He collects the sea salt, and sells it.

3 By evaporating all the solvent

For some substances, the solubility changes very little as the temperature falls. So crystallisation does not work for these. Salt is an example.

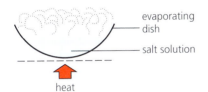

evaporating dish

salt solution

heat

the water evaporates leaving the salt behind

To obtain salt from an aqueous solution, you need to keep heating the solution, to evaporate the water.

When there is only a little water left, the salt will start to appear. Heat carefully until it is dry.

▲ Evaporating the water from a solution of salt in water.

Separating a mixture of two solids

To separate two solids, you could choose a solvent that will dissolve just one of them.

For example, water dissolves salt but not sand. So you could separate a mixture of salt and sand like this:

1 Add water to the mixture, and stir. The salt dissolves.
2 Filter the mixture. The sand is trapped in the filter paper, but the salt solution passes through.
3 Rinse the sand with water, and dry it in an oven.
4 Evaporate the water from the salt solution, to give dry salt.

Water could *not* be used to separate salt and sugar, because it dissolves both. But you could use ethanol, which dissolves sugar but not salt. Ethanol is flammable, so should be evaporated over a water bath, as shown here.

▲ Evaporating the ethanol from a solution of sugar in ethanol, over a water bath.

1 What does this term mean? Give an example.
  a filtrate   **b** residue
2 You have a solution of sugar in water. You want to obtain the sugar from it.
  a Explain why filtering will not work.
  b Which method will you use instead?

3 Describe how you would crystallise potassium nitrate from its aqueous solution.
4 How would you separate salt and sugar? Mention any special safety precaution you would take.
5 Now see if you can think of a way to get clean sand from a mixture of sand and little bits of iron wire.

2.4 Separation methods (part II)

Simple distillation

This is a way to obtain the *solvent* from a solution.
The apparatus is shown on the right. It could be used to
obtain water from salt water, for example. Like this:

1 Heat the solution in the flask. As it boils, water
   vapour rises into the condenser, leaving salt behind.
2 The condenser is cold, so the vapour condenses to
   water in it.
3 The water drips into the beaker. It is called **distilled
   water**. It is almost pure.

You could get drinking water from seawater, in this way.
Many countries in the Middle East obtain drinking water
by distilling seawater in giant distillation plants.

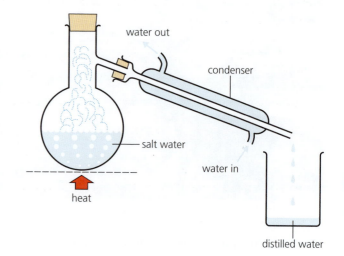

Fractional distillation

This is used to separate *a mixture of liquids* from each other.
It makes use of their different boiling points. You could use it
to separate a mixture of ethanol and water, for example.
The apparatus is shown on the right.

These are the steps:

1 Heat the mixture in the flask. At about 78 °C, the ethanol
   begins to boil. Some water evaporates too. So a mixture of
   ethanol and water vapours rises up the column.
2 The vapours condense on the glass beads in the column,
   making them hot.
3 When the beads reach about 78 °C, ethanol vapour no longer
   condenses on them. Only the water vapour does. So water
   drips back into the flask. The ethanol vapour goes into the
   condenser.
4 There it condenses. Pure liquid ethanol drips into the beaker.
5 Eventually, the thermometer reading rises above 78 °C –
   a sign that all the ethanol has gone. So you can stop heating.

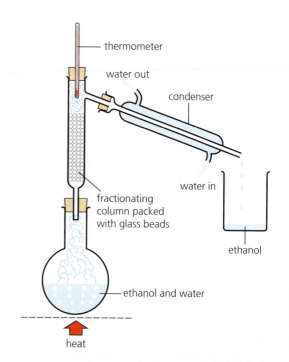

Fractional distillation in industry

Fractional distillation is very important in industry. It is used:

● in the petroleum industry, to **refine** crude oil into petrol and
  other groups of compounds. The oil is heated and the vapours
  rise to different heights, up a tall steel fractionating column.
  See page 243.

● in producing **ethanol**. The ethanol is made by fermentation,
  using sugar cane or other plant material. It is separated from
  the fermented mixture by fractional distillation. Ethanol is
  used as a solvent, and as car fuel. See page 252.

● to separate the gases in **air**. The air is cooled until it is liquid,
  then warmed up. The gases boil off one by one. See page 208.

▲ A petroleum refinery. It produces
petrol and many other useful substances,
with the help of fractional distillation.

Paper chromatography

This method can be used to separate a mixture of substances.
For example, you could use it to find out how many different dyes there are in black ink:

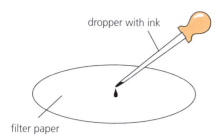

dropper with ink

filter paper

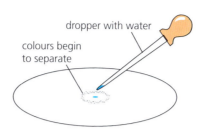

dropper with water

colours begin to separate

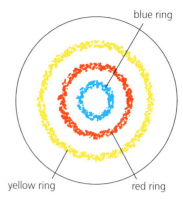

blue ring

yellow ring   red ring

1 Place a drop of black ink in the centre of some filter paper. Let it dry. Then add three or four more drops on the same spot, in the same way.

2 Now drip water onto the ink spot, one drop at a time. The ink slowly spreads out and separates into rings of different colours.

3 Suppose there are three rings: yellow, red, and blue. This shows that the ink contains three dyes, coloured yellow, red, and blue.

The dyes in the ink have different solubilities in water. So they travel across the paper at different rates. (The most soluble one travels fastest.) That is why they separate into rings. The filter paper with the coloured rings is called a **chromatogram**. (*Chroma* means *colour*.)

Paper chromotography can also be used to **identify** substances. For example, mixture **X** is thought to contain substances **A**, **B**, **C**, and **D**, which are all soluble in propanone. You could check the mixture like this:

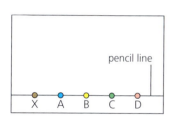

pencil line

X A B C D

glass tank with lid

clip

propanone

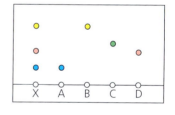

X A B C D

1 Prepare concentrated solutions of **X**, **A**, **B**, **C**, and **D**, in propanone. Place a spot of each along a line, on chromatography paper. Label them.

2 Stand the paper in a little propanone, in a covered glass tank. The solvent rises up the paper. When it's near the top, remove the paper.

3 **X** has separated into three spots. Two are at the same height as **A** and **B**, so **X** must contain substances **A** and **B**. Does it also contain **C** and **D**?

Note that you must use a pencil to draw the line on the chromatography paper. If you use a biro or felt-tipped pen, the ink will run.

Q

1 How would you obtain pure water from seawater? Draw the apparatus, and explain how the method works.
2 Why are *condensers* called that? What is the cold water for?
3 You would not use *exactly* the same apparatus you described in **1**, to separate ethanol and water. Why not?

4 Explain how fractional distillation works.
5 In the last chromatogram above, how can you tell that **X** does *not* contain substance **C**?
6 Look at the first chromatogram above. Can you think of a way to separate the coloured substances from the paper?

2.5 More about paper chromatography

How paper chromatography works

Paper chromatography depends on how the substances in a mixture interact with the chromatography paper and the solvent.

chromatography paper

1 These coloured dots represent a mixture of two substances. The mixture is dissolved in a suitable solvent.

2 The two substances travel over the paper at different speeds, because of their different solubilities in the solvent, and attraction to the paper.

3 Eventually they get completely separated from each other. Now you can identify the substances – and even collect them if you wish.

The more soluble a substance is in the solvent, the further it will travel up the chromatography paper.

Making use of paper chromatography

You can use paper chromatography to:

- identify a substance
- separate mixtures of substances
- purify a substance, by separating it from its impurities.

Example: Identify substances in a colourless mixture

On page 19, paper chromatography was used to identify *coloured* substances. Now for a bigger challenge!

Test-tubes **A–E** on the right below contain five *colourless* solutions of amino acids, dissolved in water. The solution in **A** contains several amino acids. The other solutions contain just one each.

Your task is to identify *all* the amino acids in **A–E**.

1 Place a spot of each solution along a line drawn in pencil on slotted chromatography paper, as shown below. (The purpose of the slots is to keep the samples separate.) Label each spot in pencil at the *top* of the paper.

▲ Amino acids coming up! When you digest food, the proteins in it are broken down to amino acids. Your body needs 20 different amino acids to stay healthy.

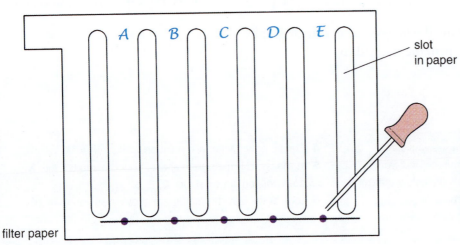

A B C D E

slot in paper

filter paper

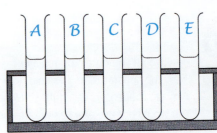

A B C D E

▲ The five mystery solutions.

2 Place a suitable solvent in the bottom of a beaker. (For amino acids, a mixture of water, ethanoic acid, and butanol is suitable.)

3 Roll the chromatography paper into a cylinder and place it in the beaker. Cover the beaker.

4 The solvent rises up the paper. When it has almost reached the top, remove the paper.

5 Mark a line in pencil on it, to show where the solvent reached. (You can't tell where the amino acids are, because they are colourless.)

6 Put the paper in an oven to dry out.

7 Next spray it with a **locating agent** to make the amino acids show up. **Ninhydrin** is a good choice. (Use it in a fume cupboard!) After spraying, heat the paper in the oven for 10 minutes. The spots turn purple. So now you have a proper chromatogram.

8 Mark a pencil dot at the centre of each spot. Measure from the base line to each dot, and to the line showing the final solvent level.

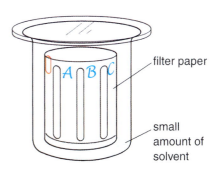

filter paper

small amount of solvent

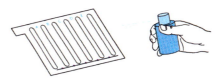

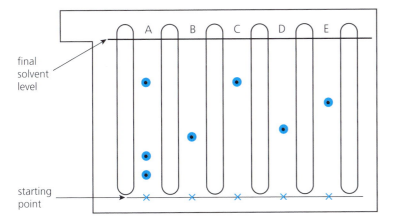

9 Now work out the R_f value for each amino acid. Like this:

$$R_f \text{ value} = \frac{\text{distance moved by amino acid}}{\text{distance moved by solvent}}$$

10 Finally, look up R_f tables to **identify** the amino acids. Part of an R_f table for the solvent you used is shown on the right. The method works because: **the R_f value of a compound is always the same for a given solvent, under the same conditions**.

R_f values for amino acids
(for water/butanol/ethanoic acid as solvent)

amino acid	R_f value
cysteine	0.08
lysine	0.14
glycine	0.26
serine	0.27
alanine	0.38
proline	0.43
valine	0.60
leucine	0.73

Q

1 Explain in your own words how paper chromatography works.

2 a What do you think a *locating agent* is?
   b Why would you need one, in an experiment to separate amino acids by chromatography?

3 What makes R_f values so useful?

4 For the chromatogram above:
   a Were any of the amino acids in **B–E** also present in **A**? How can you tell at a glance?
   b Using a ruler, work out the R_f values for the amino acids in **A–E**.
   c Now use the R_f table above to name them.

The chromatography detectives

▲ After a crime, the forensic detectives move in, looking for fingerprints and other samples they can use in evidence.

The key ideas in chromatography.

Much of chromatography is detective work. You have already met paper chromatography. There are many other kinds too. But the key ideas are always the same.

- You need two phases:
  - a non-moving or **stationary** phase, such as filter paper
  - a moving or **mobile** phase. This consists of the mixture you want to separate, dissolved in a solvent.
- The substances in the mixture separate because each has different levels of attraction to the solvent and the stationary phase. Look at the diagram on the right.
- You can then identify each separated substance. Depending on the technique you use, you can also collect them.

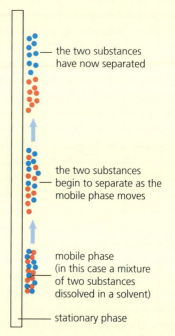

the two substances have now separated

the two substances begin to separate as the mobile phase moves

mobile phase (in this case a mixture of two substances dissolved in a solvent)

stationary phase

▲ How chromatography works.

Ringing the changes

Although those key ideas are always the same, the techniques used for chromatography can be quite different. For example:

The stationary phase could be ...	The mobile phase could be ...	To analyse the substances, you could ...
• paper, as in paper chromatography	• a mixture of substances dissolved in a liquid, as in paper chromatography	• study the coloured spots on the chromatogram, as in paper chromatography
• a thin coat of an adsorbent substance on a glass plate, or inside a tube	• a mixture of gases, carried in an **inert** (unreactive) gas; this is called **gas chromatography**	• pass them through a machine that will help you analyse them
• plastic beads packed into a tube		

Chromatography and crime detection

Chromatography is widely used in crime detection. For example it is used to analyse samples of fibre from crime scenes, check people's blood for traces of illegal drugs, and examine clothing for traces of explosives.

This shows how a blood sample could be analysed, for traces of illegal drugs, or a poison, using gas chromatography:

2 A sample of blood is injected into the carrier gas.

3 The mixture goes into a hot oven, where the blood sample forms a vapour.

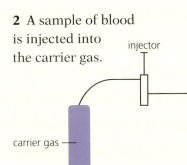
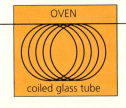

injector

carrier gas

OVEN

coiled glass tube

mass spectrometer

recorder

5 The separated substances pass into a mass spectrometer, where they are analysed.

6 The data is fed into a recorder. The police study it. They might make an arrest …

1 The carrier gas is fed in. It could be helium or nitrogen, for example.

4 The vapour moves over the stationary phase: an adsorbent substance lining a coiled glass tube.

Other uses

Chromatography can be used on a small scale in the lab, or on a very large scale in industry. For example it is used on a small scale to:

- identify substances (such as amino acids, on page 271)
- check the purity of substances
- help in crime detection (as above)
- identify pollutants in air, or in samples of river water.

It is used on a large scale to:

- separate pure substances (for example for making medical drugs or food flavourings) from tanks of reaction mixtures, in factories
- separate individual compounds from the groups of compounds (fractions) obtained in refining petroleum.

So chromatography is a really powerful and versatile tool.

▲ Injecting a sample into the carrier gas, at the start of gas chromatography.

WARNING POLLUTED WATER Unsafe for Drinking or Recreational Use

◀ Collecting water samples, to analyse for pollutants. The factories that produce them could then be identified – and fined.

Checkup on Chapter 2

Revision checklist

Core syllabus content

Make sure you can …

☐ define and use these terms:

mixture   *solute*   *solvent*

solution   *aqueous solution*

☐ give at least three examples of solvents

☐ state that most solids become more soluble as the temperature of the solvent rises

☐ explain what these terms mean:

pure substance   *impurity*

☐ give examples of where purity is very important

☐ say how melting and boiling points change, when an impurity is present

☐ decide whether a substance is pure, from melting and boiling point data

☐ describe these methods for separating mixtures, and sketch and label the apparatus:

filtration

crystallisation

evaporation to dryness

simple distillation

fractional distillation

paper chromatography

☐ explain why each of those separation methods works

☐ say which method you would choose for a given mixture, and why

☐ identify the coloured substances present in a mixture, using chromatography

Extended syllabus content

Make sure you can also …

☐ explain what a *locating agent* is

☐ describe how to carry out chromatography, to identify colourless substances

☐ define R_f *value*

☐ identify the substances in a mixture, given a chromatogram and a table of R_f values

Questions

Core syllabus content

1 This question is about ways to separate and purify substances. Match each term on the left with the correct description on the right.

A evaporation	**i**	a solid appears as the solution cools
B condensing	**ii**	used to separate a mixture of two liquids
C filtering	**iii**	the solvent is removed as a gas
D crystallising	**iv**	this method allows you to recycle a solvent
E distillation	**v**	a gas changes to a liquid, on cooling
F fractional distillation	**vi**	separates an insoluble substance from a liquid

2 This apparatus can be used to obtain pure water from salt water.

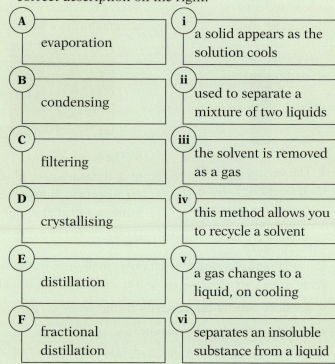

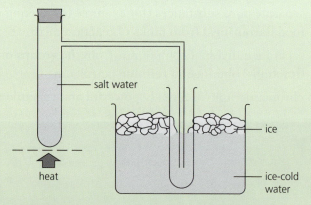

a What is the purpose of the ice-cold water?

b The glass arm must reach far down into the second test-tube. Why?

c Where in the apparatus does this take place?

   i evaporation

   ii condensation

d What is this separation method called?

e What will remain in the first test-tube, at the end of the experiment?

3 Seawater can be purified using this apparatus:

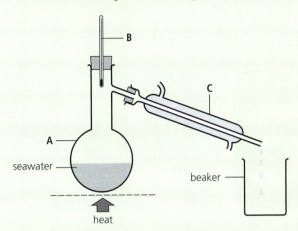

a i What is the maximum temperature recorded on the thermometer, during the distillation?
ii How does this compare to the boiling point of the seawater?
b In which piece of apparatus does evaporation take place? Give its name.
c i Which is the condenser, A, B, or C?
ii Where does the supply of cold water enter?
d Distillation is used rather than filtration, to purify seawater for drinking. Why?

4 Gypsum is insoluble in water. You are asked to purify a sample of gypsum that is contaminated with a soluble salt.
a Which of these pieces of apparatus will you use?
Bunsen burner   filter funnel   tripod
distillation flask   conical flask   pipette
thermometer   condenser   gauze
stirring rod   filter paper   beaker
b Write step-by-step instructions for the procedure.

5 Argon, oxygen, and nitrogen are obtained from air by fractional distillation. Liquid air, at −250 °C, is warmed up, and the gases are collected one by one.
a Is liquid air a mixture, or a pure substance?
b Explain why fractional distillation is used, rather than simple distillation.
c During the distillation, nitrogen gas is obtained first, then argon and oxygen. What can you say about the boiling points of these three gases?

6 A mixture of salt and sugar has to be separated, using the solvent ethanol.
a Draw a diagram to show how you will separate the salt.
b How could you obtain sugar crystals from the sugar solution, *without* losing the ethanol?
c Draw a diagram of the apparatus for **b**.

7 In a chromatography experiment, eight coloured substances were spotted onto a piece of filter paper. Three were the basic colours red, blue, and yellow. The others were unknown substances, labelled A–E. This shows the resulting chromatogram:

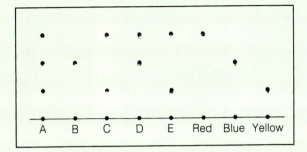

a Which one of substances A–E contains only one basic colour?
b Which contains all three basic colours?
c The solvent was propanone. Which of the three basic colours is the most soluble in propanone?

Extended syllabus content

8 The diagram below shows a chromatogram for a mixture of amino acids.

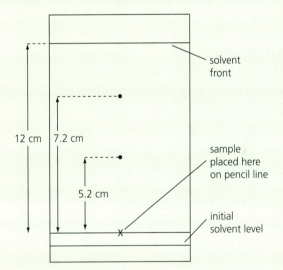

The solvent was a mixture of water, butanol, and ethanoic acid.
a Using the table of R_f values on page 21, identify the two amino acids.
b Which of them is less soluble in the solvent?
c How will the R_f values change if the solvent travels only 6 cm?

9 You have three colourless solutions. Each contains an amino acid you must identify.
Explain how to do this using chromatography.
Use the terms R_f and *locating agent* in your answer, and show that you understand what they mean.

3.1 Atoms and elements

Atoms

Sodium is made of tiny particles called sodium atoms.

Diamond is made of carbon atoms – different from sodium atoms.

Mercury is made of mercury atoms – different again!

Atoms are the smallest particles of matter, that we cannot break down further by chemical means.

Single atoms are far too small to see. Perhaps a million sodium atoms could fit in a line across this full stop. So you can see sodium only if there are enough sodium atoms together in one place!

In fact atoms are mostly empty space. Each consists of a **nucleus** and a cloud of particles called **electrons** that whizz around it. This drawing shows how a sodium atom might look, magnified many millions of times.

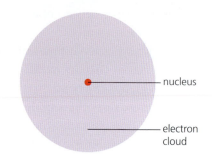

nucleus

electron cloud

The elements

Sodium is made of sodium atoms only, so it is an **element**.
An element contains only one kind of atom.

Around 90 elements have been found in the Earth and atmosphere. Scientists have made nearly 30 others in the lab. Many of the 'artificial' elements are very unstable, and last just a few seconds before breaking down into other elements. (That is why they are not found in nature.)

Symbols for the elements

To make life easy, each element has a symbol. For example the symbol for carbon is C. The symbol for potassium is K, from its Latin name *kalium*. Some elements are named after the people who discovered them.

▲ This painting shows Hennig Brand, who discovered the element phosphorus, in 1669. It glows in the dark!

◄ Collecting the element sulfur from a volcano crater in Indonesia. It is used as an ingredient in many cosmetics.

The Periodic Table

I	II											III	IV	V	VI	VII	VIII
						1 **H** hydrogen 1											2 **He** helium 4
3 **Li** lithium 7	4 **Be** beryllium 9											5 **B** boron 11	6 **C** carbon 12	7 **N** nitrogen 14	8 **O** oxygen 16	9 **F** fluorine 19	10 **Ne** neon 20
11 **Na** sodium 23	12 **Mg** magnesium 24											13 **Al** aluminium 27	14 **Si** silicon 28	15 **P** phosphorus 31	16 **S** sulfur 32	17 **Cl** chlorine 35.5	18 **Ar** argon 40
19 **K** potassium 39	20 **Ca** calcium 40	21 **Sc** scandium 45	22 **Ti** titanium 48	23 **V** vanadium 51	24 **Cr** chromium 52	25 **Mn** manganese 55	26 **Fe** iron 56	27 **Co** cobalt 59	28 **Ni** nickel 59	29 **Cu** copper 64	30 **Zn** zinc 65	31 **Ga** gallium 70	32 **Ge** germanium 73	33 **As** arsenic 75	34 **Se** selenium 79	35 **Br** bromine 80	36 **Kr** krypton 84
37 **Rb** rubidium 85	38 **Sr** strontium 88	39 **Y** yttrium 89	40 **Zr** zirconium 91	41 **Nb** niobium 93	42 **Mo** molybdenum 96	43 **Tc** technetium –	44 **Ru** ruthenium 101	45 **Rh** rhodium 103	46 **Pd** palladium 106	47 **Ag** silver 108	48 **Cd** cadmium 112	49 **In** indium 115	50 **Sn** tin 117	51 **Sb** antimony 122	52 **Te** tellurium 128	53 **I** iodine 127	54 **Xe** xenon 131
55 **Cs** caesium 133	56 **Ba** barium 137	57–71 lanthanoids	72 **Hf** hafnium 178	73 **Ta** tantalum 181	74 **W** tungsten 184	75 **Re** rhenium 186	76 **Os** osmium 190	77 **Ir** iridium 192	78 **Pt** platinum 195	79 **Au** gold 197	80 **Hg** mercury 201	81 **Tl** thallium 204	82 **Pb** lead 207	83 **Bi** bismuth 209	84 **Po** polonium –	85 **At** astatine –	86 **Rn** radon –
87 **Fr** francium –	88 **Ra** radium –	89–103 actinoids	104 **Rf** rutherfordium –	105 **Db** dubnium –	106 **Sg** seaborgium –	107 **Bh** bohrium –	108 **Hs** hassium –	109 **Mt** meitnerium –	110 **Ds** darmstadtium –	111 **Rg** roentgenium –	112 **Cr** copernicium –		114 **Fl** flerovium –		116 **Lv** livermorium –		

	57 **La** lanthanum 139	58 **Ce** cerium 140	59 **Pr** praseodymium 141	60 **Nd** neodymium 144	61 **Pm** promethium –	62 **Sm** samarium 150	63 **Eu** europium 152	64 **Gd** gadolinium 157	65 **Tb** terbium 159	66 **Dy** dysprosium 163	67 **Ho** holmium 165	68 **Er** erbium 167	69 **Tm** thulium 169	70 **Yb** ytterbium 173	71 **Lu** lutetium 175
lanthanoids															
actinoids	89 **Ac** actinium –	90 **Th** thorium 232	91 **Pa** protactinium 231	92 **U** uranium 238	93 **Np** neptunium –	94 **Pu** plutonium –	95 **Am** americium –	96 **Cm** curium –	97 **Bk** berkelium –	98 **Cf** californium –	99 **Es** einsteinium –	100 **Fm** fermium –	101 **Md** mendelevium –	102 **No** nobelium –	103 **Lr** lawrencium –

The table above is called the **Periodic Table**.

- It gives the names and symbols for the elements.

- The column and row an element is in gives us lots of clues about it.
  For example, look at the columns numbered I, II, III …
  The elements in these form families or **groups**, with similar properties.
  So if you know how one element in Group I behaves, for example, you
  can make a good guess about the others.

- The rows are called **periods**.

- Look at the zig-zag line. It separates metals from non-metals, with the
  non-metals on the right of the line, except for hydrogen. So there is a
  change from metal to non-metal, as you go across a period.

Now look at the small numbers beside each symbol. These tell us a lot
about the atoms of the element, as you will soon see.

▲ The element chlorine is a poisonous
gas. It was used as a weapon in World
War I. This soldier was prepared.

Q

1 What is: **a** an atom? **b** an element?

2 If you could look inside an atom, what would you see?

3 The symbols for some elements come from their
  Latin names. See if you can identify the element whose
  Latin name is:
  a natrium **b** ferrum **c** plumbum **d** argentum

4 Which element has this symbol? **a** Ca **b** Mg **c** N

5 See if you can pick out an element named after the
  famous scientist Albert Einstein.

6 From the Periodic Table, name
  a three metals **b** three non-metals
  that you expect to behave in a similar way.

3.2 More about atoms

Protons, neutrons, and electrons

Atoms consist of a **nucleus** and a cloud of **electrons** that move around the nucleus. The nucleus is itself a cluster of two kinds of particles, **protons** and **neutrons**.

All the particles in an atom are very light. So their mass is measured in **atomic mass units**, rather than grams. Protons and electrons also have an **electric charge**:

Particle in atom		Mass	Charge
proton	●	1 unit	positive charge (1+)
neutron	●	1 unit	none
electron	●	almost nothing	negative charge (1−)

Since electrons are so light, their mass is usually taken as zero.

How the particles are arranged

The sodium atom is a good one to start with. It has **11** protons, **11** electrons, and **12** neutrons. They are arranged like this:

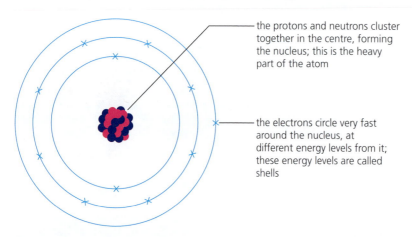

the protons and neutrons cluster together in the centre, forming the nucleus; this is the heavy part of the atom

the electrons circle very fast around the nucleus, at different energy levels from it; these energy levels are called shells

▲ The nucleus is very tiny compared with the rest of the atom. If the atom were the size of a football stadium, the nucleus would be the size of a pea!

> **Note**
> Since they make up the atom, protons, neutrons, and electrons are often called **sub-atomic particles**.

Proton number

A sodium atom has 11 protons. This can be used to identify it, since *only* a sodium atom has 11 protons. Every other atom has a different number. **You can identify an atom by the number of protons it has**.

The number of protons in the nucleus of an atom is called its proton number (or atomic number).
The proton number of sodium is 11.

How many electrons?

The sodium atom also has 11 electrons. So it has an equal number of protons and electrons. The same is true for every sort of atom:
Every atom has an equal number of protons and electrons.
So atoms have no overall charge.

Look at the box on the right. It shows that the positive and negative charges cancel each other, for the sodium atom.

> **The charge on a sodium atom:**
> ●●●● 11 protons
> ●●●● Each has a charge of 1+
> ●●● Total charge 11+
> ×××× 11 electrons
> ×××× Each has a charge of 1−
> ××× Total charge 11−
> Adding the charges: 11+
> 11−
> 0
> The answer is zero.
> The atom has no overall charge.

Nucleon number

Protons and neutrons form the nucleus, so are called **nucleons.**
The total number of protons and neutrons in the nucleus of an atom is called its nucleon number (or mass number).

The nucleon number for the sodium atom is 23. $(11 + 12 = 23)$

So sodium can be described in a short way like this: $^{11}_{23}Na$.

The upper number is the proton number. The other number is the nucleon number. So you can tell straight away that sodium atoms have 12 neutrons. $(23 - 11 = 12)$

Try it yourself!
You can describe any element in a short way like this:

proton number
nucleon number
symbol

For example: $^{8}_{16}O$

The atoms of the first 20 elements

In the Periodic Table, the elements are arranged in order of increasing proton number. Here are the first 20 elements, shown as a list:

Element	Symbol	Proton number	Electrons	Neutrons	Nucleon number (protons + neutrons)
hydrogen	H	1	1	0	1
helium	He	2	2	2	4
lithium	Li	3	3	4	7
beryllium	Be	4	4	5	9
boron	B	5	5	6	11
carbon	C	6	6	6	12
nitrogen	N	7	7	7	14
oxygen	O	8	8	8	16
fluorine	F	9	9	10	19
neon	Ne	10	10	10	20
sodium	Na	11	11	12	23
magnesium	Mg	12	12	12	24
aluminium	Al	13	13	14	27
silicon	Si	14	14	14	28
phosphorus	P	15	15	16	31
sulfur	S	16	16	16	32
chlorine	Cl	17	17	18	35
argon	Ar	18	18	22	40
potassium	K	19	19	20	39
calcium	Ca	20	20	20	40

So the numbers of protons and electrons increase by 1 at a time – and are always equal. What do you notice about the number of neutrons?

Q

1 Name the particles that make up the atom.
2 Which particle has:
   a a positive charge? **b** no charge? **c** almost no mass?
3 An atom has 9 protons. Which element is it?
4 Why do atoms have no overall charge?
5 What does this term mean?
   a proton number   **b** nucleon number
6 Name each of these atoms, and say how many protons, electrons, and neutrons it has:
   $^{6}_{12}C$ $^{8}_{16}O$ $^{12}_{24}Mg$ $^{13}_{27}Al$ $^{29}_{64}Cu$

3.3 Isotopes and radioactivity

How to identify an atom: a reminder

Only sodium atoms have 11 protons.
You can identify an atom by the number of protons in it.

Isotopes

All carbon atoms have *6 protons*. So their proton number is 6. But not all carbon atoms are identical. Some have more *neutrons* than others.

 6 protons
6 electrons
6 neutrons

Most carbon atoms are like this, with **6** neutrons. That makes **12** *nucleons* (protons + neutrons) in total, so it is called **carbon-12**.

 6 protons
6 electrons
7 neutrons

But about one in every hundred carbon atoms is like this, with **7** neutrons. It has **13** nucleons in total, so is called **carbon-13**.

 6 protons
6 electrons
8 neutrons

And a very tiny number of carbon atoms are like this, with **8** neutrons. It has **14** nucleons in total, so is called **carbon-14**.

The three atoms above are called **isotopes** of carbon.
Isotopes are atoms of the same element which have the same proton number, but a different nucleon number.

Most elements have isotopes. For example calcium has six, magnesium has three, iron has four, and chlorine has two.

Some isotopes are radioactive

A carbon-14 atom behaves in a strange way. It is **radioactive**. That means its nucleus is unstable. Sooner or later the atom breaks down naturally or **decays**, giving out **radiation** in the form of rays and particles, plus a large amount of energy.

Like carbon, a number of other elements have radioactive isotopes – or **radioisotopes** – that occur naturally, and eventually decay.

But the other two isotopes of carbon (like most natural isotopes) are **non-radioactive**.

> **Radiation may contain …**
> - **alpha particles** – made up of 2 protons and 2 neutrons
> - **beta particles** – electrons moving at high speed
> - **neutrons**
> - **gamma rays** – high energy rays

> **Decay is a random process**
> We can't tell whether a given atom of carbon-14 will decay in the next few seconds, or in a thousand years. But we *do* know how long it takes for *half* the radioisotopes in a sample to decay. This is called the **half-life.**
>
> The half-life for carbon-14 is 5730 years. So if you have a hundred atoms of carbon-14, 50 of them will have decayed 5730 years from now.
>
> Half-lives vary a lot. For example:
> - for radon-220   55.5 seconds
> - for cobalt-60   5.26 years
> - for potassium-40   1300 million years

▲ Radioisotopes are dangerous. This scientist is using a glove box, for safety.

Radiation can harm you

If the radiation from radioisotopes gets into your body, it will kill body cells. A large dose causes **radiation sickness**. Victims vomit a lot, and feel really tired. Their hair falls out, their gums bleed, and they die within weeks. Even small doses of radiation, over a long period, will cause cancer.

Making use of radioisotopes

Radioisotopes are dangerous – but they are also useful. For example:

To check for leaks   Engineers can check oil and gas pipes for leaks by adding radioisotopes to the oil or gas. If a **Geiger counter** detects radiation outside the pipe, it means there is a leak. Radioisotopes used in this way are called **tracers**.

To treat cancer   Radioisotopes can cause cancer. But they are also used in **radiotherapy** to *cure* cancer – because the gamma rays in radiation kill cancer cells more readily than healthy cells. Cobalt-60 is usually used for this. The beam of gamma rays is aimed carefully at the site of the cancer in the body.

To kill germs and bacteria   Gamma rays kill germs too. So they are used to sterilise syringes and other disposable medical equipment. They also kill the bacteria that cause food to decay. So in many countries, foods like vegetables, fruit, spices, and meat, are treated with a low dose of radiation. Cobalt-60 and caesium-137 are used for this.

▲ Checking for radiation using a Geiger counter. The meter gives a reading, and you may also hear beeps.

▲ Radioisotopes are used as fuel in nuclear power stations, because they give out so much energy when they break down. See page 115 for more.

▲ Another use for radiation: carbon dating. Our bodies contain some carbon-14, taken in food. When we die, we take no more in. But the carbon-14 atoms continue to decay. So scientists can tell the age of ancient remains by measuring the radioactivity from them. This mummy was found to be around 5300 years old.

Q

1 a What are *isotopes*?
  b Name the three isotopes of carbon, and write symbols for them.
2 Carbon-14 is *radioactive*. What does that mean?
3 What is a *radioisotope*? Give two examples.
4 a Radiation can kill us. Why?
  b So why are radioisotopes used to treat cancer?

5 Radioisotopes can be used to check pipes for leaks.
  a Explain how this works.
  b How could you tell that a pipe had no leak?
6 Spices are shipped all over the world, and are often stored for long periods.
  a They are usually treated with radiation. Why?
  b Name two radioisotopes used for this.

3.4 How electrons are arranged

Electron shells

Electrons are arranged in shells around the nucleus.
The first shell, closest to the nucleus, is the lowest energy level.
The further a shell is from the nucleus, the higher the energy level.

Each shell can hold only a certain number of electrons. These are the rules:

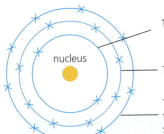

The first shell can hold only 2 electrons. It fills first.

The second shell can hold 8 electrons. It fills next.

The third shell can hold 18 electrons. But it fills up to 8. The next 2 go into the fourth shell (not shown). Then the rest of the third shell fills.

The distribution of electrons in the atom above is written in a short way as **288**. (Or sometimes as **2,8,8** or **2.8.8**.)

▲ The Danish scientist Niels Bohr (1885 – 1962) was the first person to put forward the idea of electron shells.

The electron shells for the first 20 elements

Below are the electron shells for the first 20 elements of the Periodic Table.

The number of electrons increases by 1 each time. (It is the same as the proton number.) The shells fill according to the rules above.

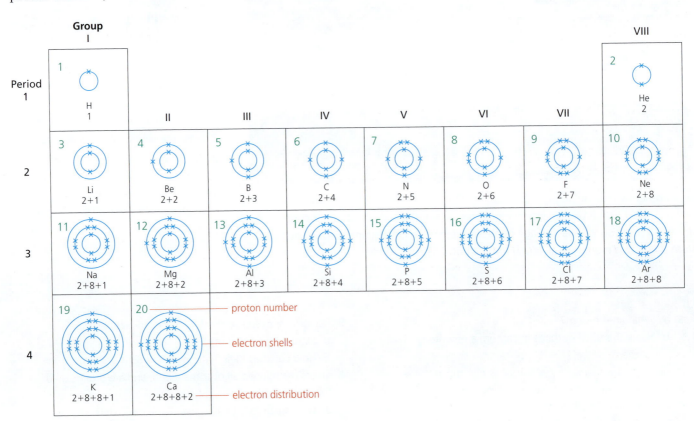

Patterns in the Periodic Table

Note these patterns for the table of the first 20 elements, on page 32:

- The **period number** tells you how many shells there are.
- All the elements in a group have the same number of electrons in their outer shells. So Group I elements have 1, Group II have 2, and so on. These outer-shell electrons are also called the **valency electrons**.
- The **group number** is the same as the number of outer-shell electrons.
- The valency electrons dictate how an element reacts. So the elements in Group I all have similar reactions, for example.

Group VIII, a special group

The elements in Group VIII have a very stable arrangement of electrons. Their atoms all have **8 outer-shell electrons**, *except for* helium, which has **2**. (It has only one shell.) Group VIII is sometimes called Group 0.

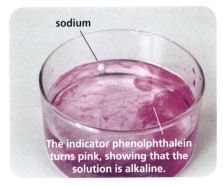

The indicator phenolphthalein turns pink, showing that the solution is alkaline.

▲ Sodium reacts with water to give an alkaline solution. The other Group I metals react in a similar way – because their atoms all have one outer electron.

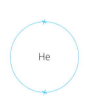

helium atom
full outer shell of 2 electrons
stable

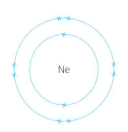

neon atom
full outer shell of 8 electrons
stable

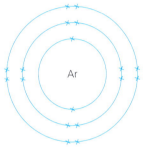

argon atom
outer shell of 8 electrons
stable

This stable arrangement of electrons has a very important result: it makes the Group VIII elements **unreactive**.

The elements after calcium

After the 20th element, calcium, the electron shells fill in a more complex order. But you should be able to answer questions about electron distribution for later elements, if you remember the points above.

Example The element rubidium, Rb, is the 37th element in the Periodic Table. It is in Group I, Period 5. Its proton number is 37. What is its electron distribution?

Group I tells you there is 1 electron in the outer shell.
Period 5 tells you there are five shells.
The proton number is 37, so there are also 37 electrons.
The third shell holds 18 electrons, when full.

So the electron distribution for rubidium is: **2+8+18+8+1**.

Reactions of isotopes

- All isotopes of an element have the same number of valency electrons.
- So their chemical reactions are the same.

Note ...

electron arrangement
electron distribution
electronic structure
electronic configuration
These terms all mean the same: how the electrons are arranged in shells.

Q

1 One element has atoms with 13 electrons.
   a Draw a diagram to show the electron distribution.
   b Write the electron distribution in this form: 2+ ...
   c Name the element.

2 The electron distribution for boron is 2+3. What is it for:
   a lithium?   **b** magnesium?   **c** hydrogen?

3 An element has 5 valency electrons. Which group is it in?

4 How many electron shells do atoms of Period 3 have?

5 The element krypton, Kr, is in Group VIII, Period 4. Its proton number is 36.
   a Write down the electronic configuration for krypton.
   b What can you say about the reactivity of krypton?

How our model of the atom developed

The two big ideas

All chemistry depends on these two big ideas:

- everything is made of particles, and …
- atoms are the simplest particles of an element, that cannot be broken down in a chemical reaction.

But how did chemists find out about atoms? It's a long story.

It began with the Ancient Greeks

In Ancient Greece (around 750 BC–150 BC), the philosophers thought hard about the world around them. Is water continuous matter, or lots of separate bits? Is air just empty space? If you crush a stone to dust, then crush the dust, will you end up with bits that will not break up further?

The philosopher Democritus came up with an answer: everything is made of tiny particles that cannot be divided. He called them **atoms**. He said they came in four colours: white, black, red, and green. And in different shapes and sizes: large round atoms that taste sweet, and small sharp ones that taste sour. White atoms are smooth, and black ones jagged.

He said everything is made up of these atoms, mixed in different amounts.

Other philosophers thought this was nonsense. Aristotle (384–270 BC) believed that everything was made of four elements – earth, air, fire, and water – mixed in different amounts. A stone has a lot of earth but not much water. No matter how much you crush it, each tiny bit will still have the properties of stone.

On to the alchemists

The Greek philosophers did a lot of heavy thinking – but no experiments. The **alchemists** were different. They experimented day and night, mixing this with that. Their main quests were to find the **elixir of life** (to keep us young), and turn common metals into **gold**.

From about 600 AD, the practice of alchemy spread to many countries, including Persia (Iran), India, China, Greece, France, and Britain.

▲ The Greek philosopher Democritus (around 460–370 BC), shown here on a Greek bank note. A lot of thinking – but no experiments!

▲ The alchemists developed many secret recipes.

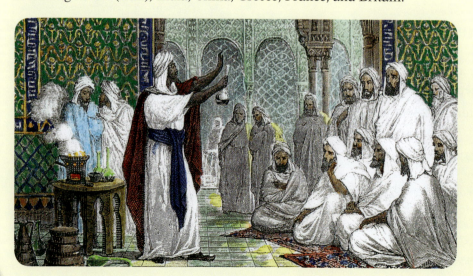

◄ The Persian alchemist Geber (around 721–815 AD) is often called 'the father of chemistry'.

The alchemists did not succeed in making gold. But they made many substances look like gold, by using secret recipes to coat them with other substances. They also developed many of the techniques we use in the lab today, such as distillation and crystallisation.

Make way for us chemists

Some alchemists got a reputation as cheats, who swindled 'grants' from rich men with the promise of gold. In the end, by around 1600 AD, the alchemists gave way to a new breed of **chemists**.

By now the idea of atoms was almost forgotten. But in 1661 the scientist Robert Boyle showed that a gas can be compressed into a smaller space. He deduced that gas is made of particles with empty space between them.

Over 130 years later, in 1799, the French chemist Joseph Louis Proust showed that copper(II) carbonate always contained the same proportions by mass of copper, carbon, and oxygen, no matter how it was made: 5.3 parts of copper to 1 of carbon to 4 of oxygen. This suggested that copper, carbon, and oxygen were made of particles, and these always combined in the same ratios.

▲ Robert Boyle (1627–91). He was born in Ireland but did most of his work in England. He put forward Boyle's Law for gases. And yes, it is a wig.

Dalton's dilemma

The English chemist John Dalton puzzled over these discoveries. In 1803 he concluded that if elements *really were made of indivisible particles* then everything made sense. He called the particles **atoms**, as a tribute to the Greek philosophers. He suggested that atoms of one element could combine with atoms of another element only in a fixed ratio.

Now the idea of atoms caught on really fast, because it fitted with the results from experiments.

Jiggling pollen grains

There was still one problem. Nobody could *prove* that matter was made of separate particles, since they were too small to see.

Then in 1827, the botanist Robert Brown noticed something strange, when he was looking at pollen in water, under a microscope. He saw small particles jiggling around. They were not alive – so why were they moving? 78 years later, Albert Einstein explained. The pollen particles were being struck by water particles. And their motion proved that water was made of particles.

And then …

In 1955 Erwin Müller, an American, developed a machine called a **field-ion microscope**. It could 'see' the tip of a needle, magnified 5 million times! The atoms in the needle showed up as dots.

Today, microscopes are much more powerful. The **scanning tunneling microscope** can trace the surfaces of materials, and gives images magnified by up to 100 million times. We 'see' atoms as blobs.

Meanwhile, for many decades, scientists wondered what was *inside* atoms. And that is another story.

▼ Getting ready to use the scanning tunneling miscroscope.

The atom: the inside story

Bring on the physicists

By 200 years ago, chemists had accepted that everything was indeed made of tiny indivisible particles: atoms. But now we know they are not quite indivisible! In the last 120 years or so, we have learned a great deal about the particles inside atoms, thanks to physicists.

First, the electrons

In 1897, the English physicist J.J. Thomson was investigating **cathode rays**. These mystery rays glowed inside an empty glass tube, when it was plugged into an electric circuit.

He deduced that these rays were streams of charged particles, *much smaller than atoms*. In fact they were bits from atoms. He called them *corpuscles*, but soon the name got changed to **electrons**.

It was a shock to find that atoms were not the smallest particle after all! Thomson imagined that electrons were stuck on the atoms like raisins on a bun. The rest of the atom (the bun) had a positive charge.

Dalton's atom

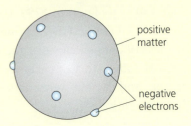

positive matter

negative electrons

Thomson's atom

More strange rays

A year earlier, a French physicist called Becquerel had been working with crystals of a uranium salt. He found that they glowed in the dark. By accident, he left some in a drawer, wrapped in thick paper, on top of a photographic plate. To his surprise, he found an image of the crystals on the plate. They had given out rays of some kind, that could pass through paper! He had discovered **radioactivity**.

Later, the English physicist Ernest Rutherford found that radiation could be separated into **alpha particles, beta particles,** and **gamma rays.** Alpha particles were found to be 7000 times heavier than electrons, with a positive charge. You could speed them up and shoot them like tiny bullets! (We know now they consist of two protons and two neutrons.)

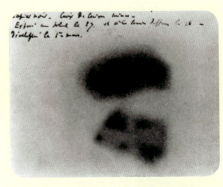

▲ Becquerel's plate, showing the image of the crystals.

◀ The Polish scientist Marie Curie (1867–1934) heard about Becquerel's discovery. She began to look for other radioactive substances – and discovered the elements polonium and radium. Marie Curie spent years searching for ways to use radiation in medicine. Sadly, she herself died from cancer caused by exposure to radium.

The nucleus and protons

In 1911, in England, Ernest Rutherford was experimenting with alpha particles. He shot a stream of them at some gold foil. Most went right through it. But some bounced back!

Rutherford deduced that an atom is mostly empty space, which the alpha particles can pass through. But there is something small and dense at the centre of the atom – and if an alpha particle hits this it will bounce back. He had discovered the **nucleus**. He assumed it was made up of particles of positive charge, and called them **protons**.

Those electron shells

If the nucleus is positive, why don't the negative electrons rush straight into it? In 1913 Niels Bohr came up with the theory of 'electron shells'. It fitted all the experiments.

At last, the neutrons

In 1930, two German physicists, Bothe and Becker, shot alpha particles at beryllium – and knocked a stream of new particles from it. In 1932 the English physicist James Chadwick found that these particles had the same mass as protons, but no charge. He named them **neutrons**.

So finally, 129 years after Dalton proposed the atom, the chemist's model of it was complete.

The whole truth

But now for the whole truth. The model of the atom that we use works well for chemists. It explains how the elements behave. But it is only a **model** – a simplified picture.

In fact atoms are far more complex than our model suggests. Physicists have discovered around 50 different **elementary particles** within atoms. They include the **up** and **anti-up**, the **charm** and **anti-charm**, and the **strange** and **anti-strange**. There may be even more to discover.

So those tiny atoms, far far too small to see, are each a throbbing universe of particles. And you are made up of atoms. Think about that!

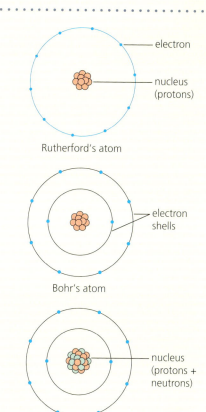

Rutherford's atom

Bohr's atom

Chadwick's atom

electron
nucleus (protons)
electron shells
nucleus (protons + neutrons)

▲ Sir James Chadwick, who gave the neutron its name.

◀ The Large Hadron Collider. Scientists hope this machine will tell them more about the particles inside atoms. It lies in a huge circular tunnel, 27 km across, on the border between France and Switzerland. Protons are accelerated through the pipes to enormous speeds, and allowed to collide.

3.5 The metals and non-metals

Two groups of elements

Look again at the Periodic Table on page 27. The zig-zag line separates the elements into two groups: **metals** and **non-metals**. The non-metals lie to the right of the line, except for hydrogen.

As you can see, there are many more metals than non-metals. In fact over 80% of the known elements are metals.

What is the difference between them?

The metals and non-metals have very different general properties. Look at this table:

General properties of metals	General properties of non-metals
• good conductors of electricity and heat	• do not conduct electricity or heat
• high melting and boiling points – which means they are solid at room temperature	• lower melting and boiling points – many are gases at room temperature
• hard, strong, do not shatter if you hammer them	• solid non-metals break up easily – they are **brittle**
• can be hammered into different shapes (they are **malleable**) and drawn out to make wires (they are **ductile**)	• solid non-metals are not malleable or ductile – they are brittle
• look shiny when they are polished	• look dull, in the solid state
• make a ringing noise when struck – they are **sonorous**	• solid non-metals break up when you strike them
• have high density – they feel 'heavy'	• solid non-metals have low density
• form positive ions when they react. For example sodium forms sodium ions (Na^+). You will learn about ions in Chapter 4.	• often form negative ions when they react. For example oxygen forms oxide ions (O^{2-}).
• react with oxygen to form oxides that are **bases**. (In other words, the oxides can neutralise acids.)	• react with oxygen to form oxides that are **acidic**. (Their aqueous solutions will turn litmus red.)

The properties in the last two rows above are called **chemical properties**, since they are about chemical change. The others are **physical properties.** You will find out more about many of those properties later.

Exceptions to those properties

The properties above are *general* properties of metals and non-metals. But there are exceptions. For example:

- not all metals are hard solids. You can cut sodium and potassium with a knife, and mercury is a liquid at room temperature.

- hydrogen is a non-metal, but forms positive ions (H^+) like metals do.

- carbon is a non-metal, but one form of it (graphite) is a good conductor; another form (diamond) is very hard, with a very high melting point.

▲ Gold: malleable, ductile, attractive, unreactive, scarce – and expensive.

▲ Think of two reasons why metals are used to make drums . . .

▲ . . . and three reasons why they are used for saucepans.

Making use of the metals

Because metals are generally hard and strong, and good conductors, we make great use of them. For example:

- Iron is the most-used metal in the world. It is used in buildings, bridges, cars, tin cans (coated with tin), needles, and nails.
- Copper is used for electrical wiring in homes.
- Aluminium is strong but light. So it is used in planes and space rockets.

Non-metals are everywhere

There are far fewer non-metals than metals. But they are all around us – and inside us.

- Air is almost 80% nitrogen, and about 20% oxygen.
- Water is a compound of hydrogen and oxygen.
- Our bodies are mostly water, plus hundreds of carbon compounds. Many of these contain atoms of other non-metals too, such as nitrogen, phosphorus, and iodine. (Plus metals such as calcium and iron.)
- Sand is mainly the compound silicon dioxide, formed from silicon and oxygen.

▲ Sea, sand, sky, palms – made almost all of non-metals.

Q

1 Without looking at the Periodic Table, see if you can quickly list 30 elements, and give their symbols. Then underline the metals.

2 Explain what these terms mean. (The glossary may help.)

| conductor | ductile | malleable |
| brittle | sonorous | density |

3 Aluminium is used for outdoor electricity cables. See if you can suggest three reasons why.

4 Write down what you think are the three *main* general properties that distinguish metals from non-metals.

5 Give one example of a physical property, and one of a chemical property, for non-metals.

Checkup on Chapter 3

Revision checklist

Core syllabus content

Make sure you can …

☐ define these terms:

  atom   *element*   *compound*

☐ say where in the atom the nucleus is, and which particles it contains

☐ define *proton number* and *nucleon number*

☐ state the number of protons, neutrons and electrons in an atom, from a short description like this: $^{11}_{23}Na$

☐ explain what an *isotope* is

☐ explain what a *radioisotope* is

☐ give one medical and one industrial use, for radioisotopes

☐ sketch the structure of an atom, showing the nucleus and electron shells

☐ state the order in which electrons fill the electron shells

☐ name the first 20 elements of the Periodic Table, in order of proton number, and give their symbols

☐ sketch the electron distribution for any of the first 20 elements of the Periodic Table, when you are given the proton number

☐ show electron distribution in this form:

  2 + 8 + …

☐ define the term *valency electron*

☐ state the connection between the number of valency electrons and the group number in the Periodic Table

☐ state the connection between the number of electron shells and the period number in the Periodic Table

☐ work out the electron distribution for an element, given its period and group numbers

☐ say how many outer-shell electrons there are in the atoms of Group VIII elements

☐ explain why the Group VIII elements are unreactive

☐ point out where the metals and non-metals are, in the Periodic Table

☐ give at least five key differences between metals and non-metals

☐ name and give the symbols for the common metals and non-metals (including metals from the transition block of the Periodic Table)

Questions

Core syllabus content

1

Particle	Electrons	Protons	Neutrons
A	12	12	12
B	12	12	14
C	10	12	12
D	10	8	8
E	9	9	10

The table above describes some particles.

a Which three particles are neutral atoms?

b Which particle is a negative ion? What is the charge on this ion?

c Which particle is a positive ion? What is the charge on this ion?

d Which two particles are isotopes?

e Use the table on page 29 to identify A to E.

2 The following statements are about the particles that make up the atom. For each statement write:

  p if it describes the **proton**

  e if it describes the **electron**

  n if it describes the **neutron**

A the positively-charged particle

B found with the proton, in the nucleus

C the particle that can occur in different numbers, in atoms of the same element

D held in shells around the nucleus

E the negatively-charged particle

F the particle with negligible mass

G the number of these particles is found by subtracting the proton number from the nucleon number

H the particle with no charge

I the particle with the same mass as a neutron

J the particle that dictates the position of the element in the Periodic Table

3 The atoms of an element can be represented by a set of three letters, as shown on the right. $^{y}_{z}X$

a What does this letter stand for?

  i X   **ii** y **iii** z

b How many neutrons are there in these atoms?

  i $^{47}_{107}Ag$ **ii** $^{29}_{63}Cu$ **iii** $^{1}_{1}H$ **iv** $^{10}_{20}Ne$ **v** $^{92}_{238}U$

c Bromine atoms have 45 neutrons and 35 protons. Describe them using the method in **b**.

4 For each of the six elements aluminium (Al), boron (B), nitrogen (N), oxygen (O), phosphorus (P), and sulfur (S), write down:

 a i which period of the Periodic Table it belongs to

 ii its group number in the Periodic Table

 iii its proton number

 iv the number of electrons in its atoms

 v its electronic configuration

 vi the number of outer electrons in its atoms

 b The outer electrons are also called the _____ electrons. What is the missing word? (7 letters!)

 c Which of the above elements would you expect to have similar properties? Why?

5 Boron has two types of atom, shown below.

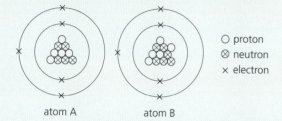

○ proton
⊗ neutron
× electron

atom A   atom B

 a What is different about these two atoms?

 b What name is given to atoms like these?

 c Describe each atom in shorthand form, as in 3.

 d What is the nucleon number of atom A?

 e Is atom B heavier, or lighter, than atom A?

 f i Give the electronic configuration for A and B.

 ii Comment on your answer for **i**.

6 The two metals sodium (proton number 11) and magnesium (proton number 12) are found next to each other in the Periodic Table.

 a Say whether this is the same, or different, for their atoms:

 i the number of electron shells

 ii the number of outer (valency) electrons

 The relative atomic mass of sodium is 23.0.

 The relative atomic mass of magnesium is 24.3.

 b Which of the two elements may exist naturally as a single isotope? Explain your answer.

7 Strontium, proton number 38, is in the fifth period of the Periodic Table. It belongs to Group II. Copy and complete the following.

An atom of strontium has:

 a .... electrons

 b .... shells of electrons

 c .... electrons in its outer shell

8 This diagram represents the electronic arrangement in an atom of an element.

 a i Give the electron distribution for the atom.

 ii What is special about this arrangement?

 b Which group of the Periodic Table does the element belong to?

 c Name another element with the same number of outer-shell electrons in its atoms.

9 Gallium exists naturally as a mixture of two non-radioactive isotopes, gallium-69 and gallium-71. The proton number of gallium is 31.

 a i How many neutrons are there in gallium-69?

 ii How many neutrons are there in gallium-71?

 Gallium also has a radioactive isotope, gallium-67. As gallium-67 decays, it gives out rays called gamma rays.

 b How does the radioactive isotope differ from the non-radioactive isotope?

 c Give two uses, one medical and one non-medical, for radioisotopes.

10 Read this passage about metals.

Elements are divided into metals and non-metals. All metals are <u>electrical conductors</u>. Many of them have a high <u>density</u> and they are usually <u>ductile</u> and <u>malleable</u>. All these properties influence the way the metals are used. Some metals are <u>sonorous</u> and this leads to special uses for them.

 a Explain the underlined terms.

 b Copper is ductile. How is this property useful in everyday life?

 c Aluminium is hammered and bent to make large structures for use in ships and planes. What property allows it to be shaped like this?

 d Name one metal that has a *low* density.

 e Some metals are cast into bells. What property must the chosen metals have?

 f Give the missing word: *Metals are good conductors of .... and electricity*.

 g Choose another physical property of metals, and give two examples of how it is useful.

 h Phosphorus is a solid non-metal at room temperature. What other physical properties would you expect it to have?

 i Explain how the chemical properties of metals and non-metals can be used to tell them apart.

4.1 Compounds, mixtures, and chemical change

Elements: a reminder

An **element** contains only one kind of atom. For example the element sodium contains only sodium atoms.

Compounds

A compound is made of atoms of different elements, bonded together.

The compound is described by a **formula**, made from the symbols of the atoms in it. (The plural of formula is **formulae**.)

There are millions of compounds. This table shows three common ones.

Name of compound	Elements in it	How the atoms are joined	Formula of compound
water	hydrogen and oxygen		H_2O
carbon dioxide	carbon and oxygen		CO_2
ethanol	carbon, hydrogen, and oxygen		C_2H_5OH

Water has two hydrogen atoms joined or bonded to an oxygen atom. So its formula is H_2O. Note where the 2 is written. Now check the formulae for carbon dioxide and ethanol. Are they correct?

Compounds and mixtures: the difference

A mixture contains different substances that are *not* bonded together.
So you can usually separate the substances quite easily, using methods like those you met in Chapter 2. For example:

This is a **mixture** of iron powder and sulfur. You could separate them by dissolving the sulfur in methylbenzene (a solvent), and filtering the iron off.

But if you heat the end of a metal rod in a Bunsen burner, and push it into the mixture, the mixture starts to glow brighly. A **chemical change** is taking place.

The result is a black **compound** called iron(II) sulfide. It is made of iron and sulfur atoms bonded together. Its formula is FeS. It will not dissolve in methylbenzene.

The signs of a chemical change

When you heat a mixture of iron and sulfur, a chemical change takes place. The iron and sulfur atoms bond together to form a compound.

You can tell when a chemical change has taken place, by these three signs:

1 **One or more new chemical substances are formed.**
You can describe the change by a **word equation** like this:

iron + sulfur $\longrightarrow$ iron(II) sulfide

The + means *reacts with*, and the $\longrightarrow$ means *to form*.

The new substances usually look different from the starting substances. For example sulfur is yellow, but iron(II) sulfide is black.

2 **Energy is taken in or given out, during the reaction.**
Energy was needed to start off the reaction between iron and sulfur, in the form of heat from the hot metal rod. But the reaction gave out heat once it began – the mixture glowed brightly.

3 **The change is usually difficult to reverse.**
You would need to carry out several reactions to get the iron and sulfur back from iron sulfide. (But it can be done!)

A chemical change is usually called a chemical reaction.

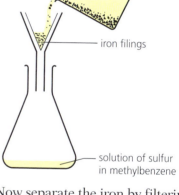

▲ Burning gas, to fry eggs. Are chemical changes taking place?

It is different from physical change

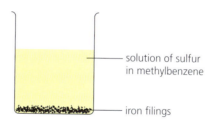

solution of sulfur in methylbenzene

iron filings

iron filings

solution of sulfur in methylbenzene

When you mix iron powder with sulfur, that is a **physical change**. No new substance has formed. If you then dissolve the sulfur …

… in methylbenzene, that is also a physical change. The solvent could be removed again by distilling it. (Danger! It is highly flammable.)

Now separate the iron by filtering. That is a physical change. You can reverse it by putting the iron back into the filtrate again.

No new chemical substances are formed in these changes.
If no new chemical substance is formed, a change is a physical change.

Unlike chemical changes, a physical change is usually easy to reverse.

Q

1 Explain the difference between a *mixture* of iron and sulfur and the *compound* iron sulfide.

2 When you light a piece of magnesium ribbon, it burns with a dazzling white light. A white ash forms. What signs are there that a chemical change has taken place?

3 Is it a chemical change or a physical change? Give reasons.
   a a glass bottle breaking
   b butter and sugar being made into toffee
   c cotton being woven to make sheets
   d coal burning in air

4.2 Why do atoms form bonds?

The reaction between sodium and chlorine

Sodium and chlorine are both **elements**. When sodium is heated and placed in a jar of chlorine, it burns with a bright flame.

The result is a white solid that has to be scraped from the sides of the jar. It looks completely different from the sodium and chlorine.

So a chemical reaction has taken place. The white solid is **sodium chloride**. Atoms of sodium and chlorine have bonded (joined together) to form a compound. The word equation for the reaction is:

sodium + chlorine ⟶ sodium chloride

Why do atoms form bonds?

Like sodium and chlorine, the atoms of most elements form bonds.

Why? We get a clue by looking at the elements of Group VIII, **the noble gases**. Their atoms *do not* form bonds.

This is because the atoms have a very stable arrangement of electrons in the outer shell. This makes the noble gases **unreactive**.

helium atom:
full outer shell of 2
electrons – *stable*

neon atom:
full outer shell of 8
electrons – *stable*

argon atom:
outer shell of 8
electrons – *stable*

2

2 + 8

2 + 8 + 8

And that gives us the answer to our question:

Atoms bond with each other in order to gain a stable arrangement of outer-shell electrons, like the atoms of Group VIII.

In other words, they bond in order to gain 8 electrons in their outer shell (or 2, if they have only one shell).

▲ Neon: the unreactive gas used in light tubes for advertising.

▲ Welding is often carried out in an atmosphere of argon, which will not react with hot metals (unlike oxygen).

How sodium atoms gain a stable outer shell

A sodium atom has just 1 electron in its outer shell. To obtain a stable outer shell of 8 electrons, it loses this electron to another atom. It becomes a **sodium ion**:

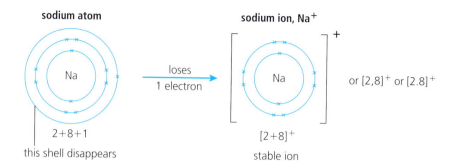

The sodium ion has 11 protons but only 10 electrons, so it has a charge of 1+, as you can see from the panel on the right.

The symbol for sodium is Na, so the symbol for the sodium ion is **Na⁺**.

The + means *1 positive charge*. Na⁺ is a **positive ion**.

> **The charge on a sodium ion**
> | | |
> |---|---|
> | charge on 11 protons | 11+ |
> | charge on 10 electrons | 10− |
> | total charge | 1+ |

How chlorine atoms gain a stable outer shell

A chlorine atom has 7 electrons in its outer shell. It can reach 8 electrons by accepting 1 electron from another atom. It becomes a chloride ion:

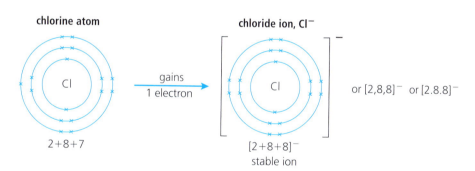

The chloride ion has a charge of 1−, so it is a **negative ion**. Its symbol is **Cl⁻**.

Ions

An atom becomes an ion when it loses or gains electrons.
An ion is a charged particle. It is charged because it has an unequal number of protons and electrons.

> **The charge on a chloride ion**
> | | |
> |---|---|
> | charge on 17 protons | 17+ |
> | charge on 18 electrons | 18− |
> | total charge | 1− |

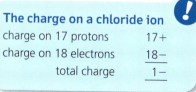

Q

1 Why are the atoms of the Group VIII elements unreactive?
2 Explain why all other atoms are reactive.
3 Draw a diagram to show how this atom gains a stable outer shell of 8 electrons:
  a a sodium atom   **b** a chlorine atom

4 Explain why
  a a sodium ion has a charge of 1+
  b a chloride ion has a charge of 1−.
5 Explain what an *ion* is, in your own words.
6 Atoms of Group VIII elements do *not* form ions. Why not?

4.3 The ionic bond

How sodium and chlorine atoms bond together

As you saw on page 45, a sodium atom must lose one electron, and a chlorine atom must gain one, to obtain stable outer shells of 8 electrons.

So when a sodium atom and a chlorine atom react together, the sodium atom loses its electron *to the chlorine atom*, and two ions are formed.

Here, sodium electrons are shown as • and chlorine electrons as × :

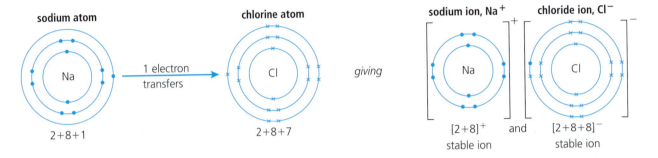

The two ions have opposite charges, so they attract each other. The force of attraction between them is strong. It is called an **ionic bond**.

The ionic bond is the bond that forms between ions of opposite charge.

How solid sodium chloride is formed

When sodium reacts with chlorine, billions of sodium and chloride ions form. But they do not stay in pairs. They form a regular pattern or **lattice** of alternating positive and negative ions, as shown below. The ions are held together by strong ionic bonds.

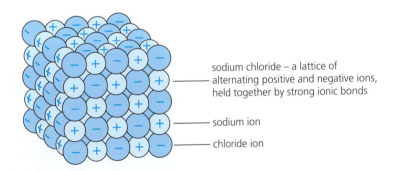

sodium chloride – a lattice of alternating positive and negative ions, held together by strong ionic bonds

sodium ion

chloride ion

The lattice grows to form a giant 3-D structure. It is called 'giant' because it contains a very large number of ions. This giant structure is the compound **sodium chloride**, or **common salt**.

Since it is made of ions, sodium chloride is called an **ionic compound**. It contains one Na^+ ion for each Cl^- ion, so its formula is **NaCl**.

The charges in the structure add up to zero:

the charge on each sodium ion is	$1+$
the charge on each chloride ion is	$1-$
total charge	0

So the compound has no overall charge.

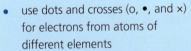

Bonding diagrams

To show the bonding clearly:

- use dots and crosses (o, •, and ×) for electrons from atoms of different elements
- write the symbol for the element in the centre of each atom.

▲ These polystyrene balls were given opposite charges. So they are attracted to each other, and cling together. The same happens with ions of opposite charge.

Other ionic compounds

Sodium is a metal. Chlorine is a non-metal. They react together to form an ionic compound. Other metals and non-metals follow the same pattern.

A metal reacts with a non-metal to form an ionic compound.
The metal atoms lose electrons. The non-metal atoms gain them.
The ions form a lattice. The compound has no overall charge.

Below are two more examples.

Magnesium oxide

A magnesium atom has 2 outer electrons and an oxygen atom has 6. When magnesium burns in oxygen, each magnesium atom loses its 2 outer electrons to an oxygen atom. Magnesium and oxide ions are formed:

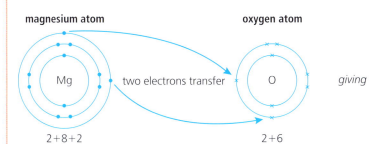

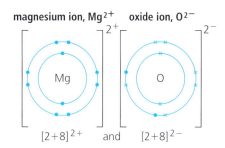

The ions attract each other because of their opposite charges. Like the sodium and chloride ions, they group to form a lattice.

The resulting compound is called **magnesium oxide**. It has one magnesium ion for each oxide ion, so its formula is **MgO**. It has no overall charge.

> **The charge on magnesium oxide**
>
> | | |
> |---|---|
> | charge on a magnesium ion | 2+ |
> | charge on an oxide ion | 2− |
> | total charge | 0 |

Magnesium chloride

When magnesium burns in chlorine, each magnesium atom reacts with *two* chlorine atoms, to form **magnesium chloride**. Each ion has 8 outer electrons:

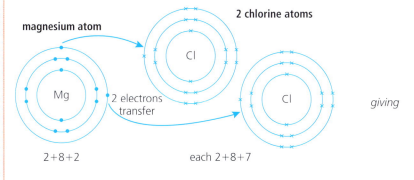

The ions form a lattice with two chloride ions for each magnesium ion. So the formula of the compound is **MgCl$_2$**. It has no overall charge.

---

1 Draw a diagram to show what happens to the electrons, when a sodium atom reacts with a chlorine atom.

2 What is an *ionic bond*?

3 Describe in your own words the structure of solid sodium chloride, and explain why its formula is NaCl.

4 Explain why:
  a a magnesium ion has a charge of 2+
  b the ions in magnesium oxide stay together
  c magnesium chloride has no overall charge
  d the formula of magnesium chloride is MgCl$_2$.

4.4 More about ions

Ions of the first 20 elements

Not every element forms ions during reactions. In fact, out of the first 20 elements in the Periodic Table, only 12 easily form ions. These ions are given below, with their names.

Group I	II	H$^+$ hydrogen	III	IV	V	VI	VII	VIII none
Li$^+$ lithium	Be^{2+} beryllium					O^{2-} oxide	F$^-$ fluoride	none
Na$^+$ sodium	Mg^{2+} magnesium		Al^{3+} aluminium			S^{2-} sulfide	Cl$^-$ chloride	none
K$^+$ potassium	Ca^{2+} calcium	transition elements						

Note that:

- Hydrogen and the metals lose electrons and form **positive ions**. The ions have the same names as the atoms.
- Non-metals form **negative ions**, with names ending in **-ide**.
- The elements in Groups IV and V do not usually form ions, because their atoms would have to gain or lose several electrons, and that takes too much energy.
- Group VIII elements do not form ions: their atoms already have stable outer shells, so do not need to gain or lose electrons.

The names and formulae of ionic compounds

The names   To name an ionic compound, you just put the names of the ions together, with the positive one first:

Ions in compound	Name of compound
K$^+$ and F$^-$	potassium fluoride
Ca^{2+} and S^{2-}	calcium sulfide

The formulae   The formulae of ionic compounds can be worked out using these four steps. Look at the examples that follow.

1   Write down the name of the ionic compound.
2   Write down the symbols for its ions.
3   The compound must have no overall charge, so balance the ions until the positive and negative charges add up to zero.
4   Write down the formula without the charges.

▲ Bath time. Bath salts contain ionic compounds such as magnesium sulfate (Epsom salts) and sodium hydrogen carbonate (baking soda). Plus scent!

Example 1
1   Lithium fluoride.
2   The ions are Li$^+$ and F$^-$.
3   One Li$^+$ is needed for every F$^-$, to make the total charge zero.
4   The formula is LiF.

Example 2
1   Sodium sulfide.
2   The ions are Na$^+$ and S^{2-}.
3   Two Na$^+$ ions are needed for every S^{2-} ion, to make the total charge zero: Na$^+$ Na$^+$ S^{2-}.
4   The formula is Na$_2$S. (What does the $_2$ show?)

Some metals form more than one type of ion

Look back at the Periodic Table on page 27. Look for the block of **transition elements**. These include many common metals, such as iron and copper.

Some transition elements form only one type of ion:

● silver forms only Ag^+ ions

● zinc forms only Zn^{2+} ions.

But most transition elements can form more than one type of ion. For example, copper and iron can each form two:

Ion	Name	Example of compound
Cu^+	copper(I) ion	copper(I) oxide, Cu_2O
Cu^{2+}	copper(II) ion	copper(II) oxide, CuO
Fe^{2+}	iron(II) ion	iron(II) chloride, $FeCl_2$
Fe^{3+}	iron(III) ion	iron(III) chloride, $FeCl_3$

The (II) in the name tells you that the ion has a charge of 2+. What do the (I) and (III) show?

copper(II) oxide

copper(I) oxide

▲ The two oxides of copper.

Compound ions

All the ions you met so far have been formed from single atoms. But ions can also be formed from a **group** of bonded atoms. These are called **compound ions**.

The most common ones are shown on the right. Remember, each is just one ion, even though it contains more than one atom.

The formulae for their compounds can be worked out as before. Some examples are shown below.

NH_4^+, the ammonium ion

OH^-, the hydroxide ion

NO_3^-, the nitrate ion

SO_4^{2-}, the sulfate ion

CO_3^{2-}, the carbonate ion

HCO_3^-, the hydrogen carbonate ion

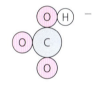

Example 3

1 Sodium carbonate.
2 The ions are Na^+ and CO_3^{2-}.
3 Two Na^+ are needed to balance the charge on one CO_3^{2-}.
4 The formula is Na_2CO_3.

Example 4

1 Calcium nitrate.
2 The ions are Ca^{2+} and NO_3^-.
3 Two NO_3^- are needed to balance the charge on one Ca^{2+}.
4 The formula is $Ca(NO_3)_2$. Note that brackets are put round the NO_3, before the $_2$ is put in.

1 Explain why a calcium ion has a charge of 2+.
2 Why is the charge on an aluminium ion 3+?
3 Write down the symbols for the ions in:
　a potassium chloride　**b** calcium sulfide
　c lithium sulfide　**d** magnesium fluoride
4 Now work out the formula for each compound in **3**.

5 Work out the formula for each compound:
　a copper(II) chloride　**b** iron(III) oxide
6 Write a name for each compound:
　CuCl, FeS, Mg(NO_3)_2, NH_4NO_3, Ca(HCO_3)_2
7 Work out the formula for:　**a** sodium sulfate
　b potassium hydroxide　**c** silver nitrate

4.5 The covalent bond

Why atoms bond: a reminder

As you saw in Unit 4.3, atoms bond in order to gain a stable outer shell of electrons, like the noble gas atoms. So when sodium and chlorine react together, each sodium atom gives up an electron to a chlorine atom.

But that is not the only way. Atoms can also gain stable outer shells by *sharing* electrons with each other.

Sharing electrons

When two non-metal atoms react together, *both need to gain electrons* to achieve stable outer shells. They manage this by sharing electrons.

We will look at **non-metal elements** in this unit, and at **non-metal compounds** in the next unit. Atoms can share only their outer (valence) electrons, so the diagrams will show only these.

Hydrogen

A hydrogen atom has only one shell, with one electron. The shell can hold two electrons. When two hydrogen atoms get close enough, their shells overlap and then they can share electrons. Like this:

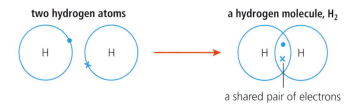

two hydrogen atoms   →   a hydrogen molecule, H₂

a shared pair of electrons

So each has gained a full shell of two electrons, like helium atoms.

The bond between the atoms

Each hydrogen atom has a positive nucleus. Both nuclei attract the shared electrons – and this strong force of attraction holds the two atoms together.

This force of attraction is called a **covalent bond**.
A single covalent bond is formed when atoms share two electrons.

Molecules

The two bonded hydrogen atoms above form a **molecule**.
A molecule is a group of atoms held together by covalent bonds.

Since it is made up of molecules, hydrogen is a molecular element.
Its formula is **H$_2$**. The $_2$ tells you there are 2 hydrogen atoms in each molecule.

Many other non-metals are also molecular. For example:

iodine, I$_2$ oxygen, O$_2$ nitrogen, N$_2$
chlorine, Cl$_2$ sulfur, S$_8$ phosphorus, P$_4$

Elements made up of molecules containing two atoms are called **diatomic**. So iodine and oxygen are diatomic. Can you give two other examples?

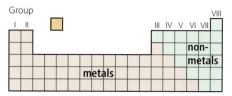

▲ Atoms of non-metals do not *give up* electrons to gain a full shell, because they would have to lose so many. It would take too much energy to overcome the pull of the positive nucleus.

▲ A model of the hydrogen molecule. The molecule can also be shown as H–H. The line represents a single bond.

Chlorine

A chlorine atom needs a share in one more electron, to obtain a stable outer shell of eight electrons. So two chlorine atoms bond covalently like this:

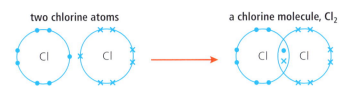

two chlorine atoms   a chlorine molecule, Cl_2

▲ A model of the chlorine molecule.

Since only one pair of electrons is shared, the bond between the atoms is called a **single covalent bond**, or just a **single bond**. You can show it in a short way by a single line, like this: $Cl-Cl$.

Oxygen

An oxygen atom has six outer electrons, so needs a share in *two* more. So two oxygen atoms share two electrons each, giving molecules with the formula O_2. Each atom now has a stable outer shell of eight electrons:

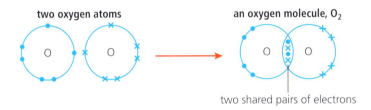

two oxygen atoms   an oxygen molecule, O_2

two shared pairs of electrons

▲ A model of the oxygen molecule.

Since the oxygen atoms share two pairs of electrons, the bond between them is called a **double bond**. You can show it like this: $O=O$.

Nitrogen

A nitrogen atom has five outer electrons, so needs a share in *three* more. So two nitrogen atoms share three electrons each, giving molecules with the formula N_2. Each atom now has a stable outer shell of eight electrons:

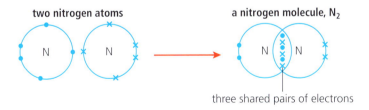

two nitrogen atoms   a nitrogen molecule, N_2

three shared pairs of electrons

▲ A model of the nitrogen molecule.

Since the nitrogen atoms share three pairs of electrons, the bond between them is called a **triple bond**. You can show it like this: $N\equiv N$.

Q

1 a Name the bond between atoms that share electrons.
  b What holds the bonded atoms together?
2 What is a *molecule*?
3 Give five examples of molecular elements.

4 Draw a diagram to show the bonding in:
  a hydrogen   b chlorine
5 Now explain why the bond in a nitrogen molecule is called a *triple* bond.

4.6 Covalent compounds

Covalent compounds

In the last unit you saw that many non-metal elements exist as molecules. A huge number of *compounds* also exist as molecules.

In a molecular compound, atoms of *different* elements share electrons. The compounds are called **covalent compounds**. Here are three examples.

> **!** **Most are molecular …**
> Most non-metal elements and their compounds exist as molecules.

Covalent compound	Description	Model of the molecule
hydrogen chloride, HCl 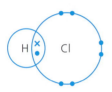 a molecule of hydrogen chloride	The chlorine atom shares one electron with the hydrogen atom. Both now have a stable arrangement of electrons in their outer shells: 2 for hydrogen (like the helium atom) and 8 for chlorine (like the other noble gas atoms).	
water, H₂O 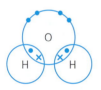 a molecule of water	The oxygen atom shares electrons with the two hydrogen atoms. All now have a stable arrangement of electrons in their outer shells: 2 for hydrogen and 8 for oxygen.	
methane, CH₄ 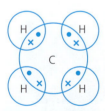 a molecule of methane	The carbon atom shares electrons with four hydrogen atoms. All now have a stable arrangement of electrons in their outer shells: 2 for hydrogen and 8 for carbon.	
ammonia, NH₃ 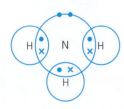 a molecule of ammonia	Each nitrogen atom shares electrons with three hydrogen atoms. So all three atoms now have a stable arrangement of electrons in their outer shells: 2 for hydrogen and 8 for nitrogen.	

The shapes of molecules

The pairs of electrons around an atom repel each other, and move as far apart as they can. This dictates the shapes of molecules.

Look at methane, for example. A carbon atom shares electrons with four hydrogen atoms. The four electron pairs move as far apart as possible, giving the molecule a tetrahedral shape.

In a water molecule, the four electron pairs around oxygen also take up a tetrahedral arrangement. So the water molecule appears to be bent.

More examples of covalent compounds

This table shows three more examples of covalent compounds.

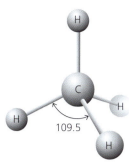

▲ A different way to show the methane molecule. Each stick represents a bond. All the angles are the same.

Covalent compound	Description	Model of the molecule
methanol, CH_3OH 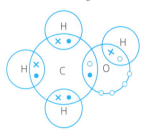 a molecule of methanol	The carbon atom shares electrons with three hydrogen atoms and one oxygen atom. All the atoms gain full outer shells. Look at the shape of the molecule: a little like methane, but changed by the presence of the oxygen atom.	
carbon dioxide, CO_2 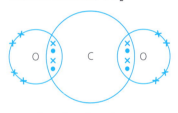 a molecule of carbon dioxide	The carbon atom shares all four of its electrons: two with each oxygen atom. So all three atoms gain stable shells. The two sets of bonding electrons repel each other, giving a **linear** molecule. All the bonds are double bonds, so we can show the molecule like this: $O = C = O$.	
ethene, C_2H_4 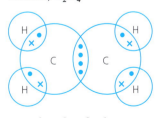 a molecule of ethene	Look how each carbon atom shares its four electrons this time. It shares two with two hydrogen atoms. and two with another carbon atom, giving a carbon-carbon double bond. So the molecule is usually drawn as here:	

Q

1 **a** What is a *covalent compound*?
   b Give five examples, with their formulae.
2 Draw a diagram to show the bonding in a molecule of:
   a methane   **b** water

3 How do the atoms gain stable outer shells, in ammonia?
4 Draw a diagram to show the bonding in carbon dioxide.
5 In ethene, the bonds between the carbon atoms are *double bonds*. Explain what this means, using the term *shares*.

4.7 Comparing ionic and covalent compounds

Remember

Metals and non-metals react together to form **ionic compounds**.
Non-metals react together to form **covalent compounds**.
The covalent compounds you have met so far exist as **molecules**.

Comparing the structures of the solids

In Chapter 1, you met the idea that solids are a **regular lattice** of particles.
In ionic compounds, these particles are **ions**. In the covalent compounds
you have met so far, they are **molecules**. Let's compare their lattices.

A solid ionic compound   Sodium chloride is a typical ionic compound:

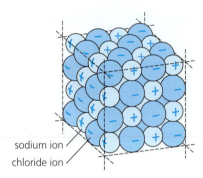

sodium ion
chloride ion

In sodium chloride, the ions are
held in a regular lattice like this.
They are held by strong ionic bonds.

The lattice grows in all directions,
giving a crystal of sodium chloride.
This one is magnified 35 times.

The crystals look white and shiny.
We add them to food, as **salt**, to
bring out its taste.

A solid molecular covalent compound   Water is a molecular covalent
compound. When you cool it below 0 °C it becomes a solid: ice.

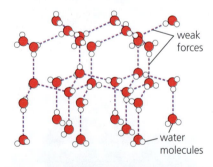

weak
forces

water
molecules

In ice, the water molecules are held
in a regular lattice like this. But the
forces between them are weak.

The lattice grows in all directions,
giving a crystal of ice. These grew
in an ice-tray in a freezer.

We use ice to keep drinks cool,
and food fresh. (The reactions that
cause decay are slower in the cold.)

So both types of compounds have a regular lattice structure in the solid
state, and form crystals. But they differ in two key ways:

- In ionic solids the particles (ions) are charged, and the forces between
  them are strong.

- In molecular covalent solids the particles (molecules) are not charged,
  and the forces between them are weak.

These differences lead to very different properties, as you will see next.

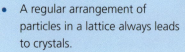

About crystals
- A regular arrangement of
  particles in a lattice always leads
  to crystals.
- The particles can be atoms, ions,
  or molecules.

The properties of ionic compounds

1 Ionic compounds have high melting and boiling points.
For example:

Compound	Melting point/°C	Boiling point/°C
sodium chloride, NaCl	801	1413
magnesium oxide, MgO	2852	3600

This is because the ionic bonds are very strong. It takes a lot of heat energy to break up the lattice. So ionic compounds are solid at room temperature.

Note that magnesium oxide has a far higher melting and boiling point than sodium chloride does. This is because its ions have double the charge (Mg^{2+} and O^{2-} compared with Na^+ and Cl^-), so its ionic bonds are stronger.

2 Ionic compounds are usually soluble in water.
The water molecules are able to separate the ions from each other. The ions then move apart, surrounded by water molecules.

3 Ionic compounds conduct electricity, when melted or dissolved in water.
A solid ionic compound will not conduct electricity. But when it melts, or dissolves in water, the ions become free to move. Since they are charged, they can then conduct electricity.

▲ Magnesium oxide is used to line furnaces in steel works, because of its high melting point, 2852 °C. (By contrast, iron melts at 1538 °C.)

The properties of covalent compounds

1 Molecular covalent compounds have low melting and boiling points.
For example:

Compound	Melting point/°C	Boiling point/°C
carbon monoxide, CO	−199	−191
hexane, C_6H_{14}	−95	69

This is because the attraction between the molecules is low. So it does not take much energy to break up the lattice and separate them from each other. That explains why many molecular compounds are liquids or gases at room temperature – and why many of the liquids are **volatile** (evaporate easily).

2 Covalent compounds tend to be insoluble in water.
But they do dissolve in some solvents, for example tetrachloromethane.

3 Covalent compounds do not conduct electricity.
There are no charged particles, so they cannot conduct.

▲ The covalent compound carbon monoxide is formed when petrol burns in the limited supply of air in a car engine. And it is poisonous.

Q

1 The particles in solids usually form a *regular lattice*. Explain what that means, in your own words.
2 Which type of particles make up the lattice, in:
   a ionic compounds?   **b** molecular compounds?
3 Solid sodium chloride will not conduct electricity, but a solution of sodium chloride will conduct. Explain this.

4 A compound melts at 20 °C.
   a What kind of structure do you think it has? Why do you think so?
   b Will it conduct electricity at 25 °C? Give a reason.
5 Describe the arrangement of the molecules in ice. How will the arrangement change as the ice warms up?

4.8 Giant covalent structures

Not all covalent solids are molecular

In all the solids in this table, the atoms are held together by covalent bonds. But compare their melting points. What do you notice?

Substance	Melting point/°C
ice	0
phosphorus	44
sulfur	115
silicon (IV) oxide (silicon dioxide or silica)	1710
carbon (as diamond)	3550

The first three substances are molecular solids. Their molecules are held in a lattice by weak forces – so the solids melt easily, at low temperatures.

But diamond and silicon (IV) oxide are different. Their melting points show that *they* are not molecular solids with weak lattices. In fact they exist as **giant covalent structures**, or **macromolecules**.

Diamond – a giant covalent structure

Diamond is made of carbon atoms, held in a strong lattice:

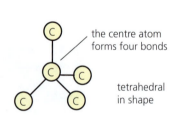

the centre atom forms four bonds

tetrahedral in shape

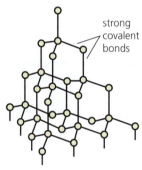

strong covalent bonds

A carbon atom forms covalent bonds to *four* others, as shown above. Each outer atom then bonds to three more, and so on.

Eventually billions of carbon atoms are bonded together, in a giant covalent structure. This shows just a very tiny part of it.

Diamond: so hard that it is used to edge wheels for cutting stone.

The result is a single crystal of diamond. This one has been cut, shaped, and polished, to make it sparkle.

Diamond has these properties:

1 It is very hard, because each atom is held in place by four strong covalent bonds. In fact it is the hardest substance on Earth.
2 For the same reason it has a very high melting point, 3550 °C.
3 It can't conduct electricity because there are no ions or free electrons to carry the charge.

Silicon (IV) oxide is similar to diamond

Silicon (IV) oxide, SiO_2, occurs naturally as **quartz**, the main mineral in **sand**. Like diamond, it forms a giant covalent structure, as shown on the right.

Each silicon atom bonds covalently to four oxygen atoms. And each oxygen atom bonds covalently to two silicon atoms. The result is a very hard substance with a melting point of 1710 °C.

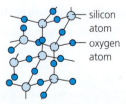

silicon atom

oxygen atom

Silicon (IV) oxide (silicon dioxide or silica) is made up of oxygen atoms ● and silicon atoms ○. Billions of them bond together like this, to give a giant structure.

Graphite – a very different giant structure

Like diamond, graphite is made only of carbon atoms. So diamond and graphite are **allotropes** of carbon – two forms of the same element.

Diamond is the hardest solid on Earth. But graphite is one of the softest! This difference is a result of their very different structures:

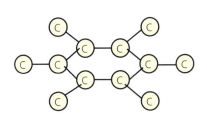

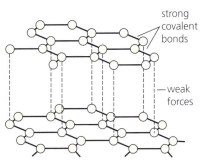

strong covalent bonds

weak forces

In graphite, each carbon atom forms covalent bonds to *three* others. This gives rings of *six* atoms.

The rings form flat sheets that lie on top of each other, held together by weak forces.

Under a microscope, you can see the layered structure of graphite quite clearly.

Graphite has these properties:

1 Unlike diamond, it is soft and slippery. That is because the sheets can slide over each other easily.
2 Unlike diamond, it is a good conductor of electricity. That is because each carbon atom has four outer electrons, but forms only three bonds. So the fourth electron is free to move through the graphite, carrying charge.

Making use of these giant structures

Different properties lead to different uses, as this table shows.

Substance	Properties	Uses
diamond	hardest known substance	in tools for drilling and cutting
	does not conduct electricity	
	sparkles when cut	for jewellery
graphite	soft and slippery	as a lubricant for engines and locks
	soft and dark in colour	for pencil 'lead' (mixed with clay)
	conducts electricity	for electrodes, and connecting brushes in generators
silicon (IV) oxide	hard, can scratch things	in sandpaper
	hard, lets light through	for making glass and lenses
	high melting point	in bricks for lining furnaces

▲ Pencil 'lead' is a mixture of graphite and clay.

Q

1 The covalent compound ethanol melts at –114 °C. Is it a molecular compound, or a giant structure? Explain.
2 Diamond and graphite are *allotropes* of carbon. What does that mean?
3 Why is diamond so hard?
4 Why do diamond and graphite have such very different properties? Draw diagrams to help you explain.

5 a What is *silicon (IV) oxide*?
   b Diamond and silicon (IV) oxide share some properties. See how many you can list.
   c Now explain why they have similar properties.
6 a Silicon (IV) oxide has a high melting point. Why?
   b Silicon (IV) oxide's melting point is not as high as diamond's. See if you can suggest a reason.

4.9 The bonding in metals

Clues from melting points

Compare these melting points:

Structure	Examples	Melting point /°C
molecular	carbon dioxide water	−56 0
giant ionic	sodium chloride magnesium oxide	801 2852
giant covalent	diamond silicon (IV) oxide	3550 1610
metal	iron copper	1535 1083

The table shows clearly that:

- **molecular substances have low melting points.** That is because the forces between molecules in the lattice are weak.

- **giant structures such as sodium chloride and diamond have much higher melting points.** That is because the bonds between ions or atoms within giant structures are very strong.

Now look at the metals. They too have high melting points – much higher than for carbon dioxide or water. This gives us a clue that they too might be giant structures. And so they are, as you'll see below.

The structure of metals

In metals, the atoms are packed tightly together in a regular lattice. The tight packing allows outer electrons to separate from their atoms. The result is a lattice of ions in a 'sea' of electrons that are free to move.

Look at copper:

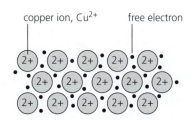

copper ion, Cu²⁺ free electron

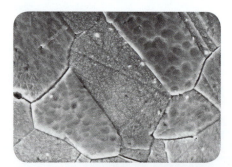

▲ Equipment for measuring melting points in the school lab. It can heat substances up to 300°C – so no good for sodium chloride!

The copper ions are held together by their attraction to the free electrons between them. The strong forces of attraction are called **metallic bonds**.

The regular arrangement of ions results in **crystals** of copper. This shows the crystals in a piece of copper, magnified 1000 times. (They are all at different angles.)

The copper crystals are called **grains**. A lump of copper like this one consists of millions of grains joined together. You need a microscope to see them.

The metallic bond is the attraction between metal ions and free electrons.

It is the same with all metals. The ions sit in a lattice, held together by their strong attraction to the free electrons. And because the ions are in a regular pattern, metals are crystalline.

> **Delocalised electrons**
> The electrons that move freely in the metal lattice are not tied to any one ion. So they are called **delocalised**.

Explaining some key properties of metals

In Unit 3.5 you read about the properties of metals. We can now explain some of those properties. Look at these examples.

1 **Metals usually have high melting points.**

That is because it takes a lot of heat energy to break up the lattice, with its strong metallic bonds. Copper melts at 1083 °C, and nickel at 1455 °C. (But there are exceptions. Sodium melts at only 98 °C, for example. And mercury melts at –39 °C, so it is a liquid at room temperature.)

2 **Metals are malleable and ductile.**

Malleable means they can be bent and pressed into shape. *Ductile* means they can be drawn out into wires. This is because the layers can slide over each other. This diagram represents any metal lattice:

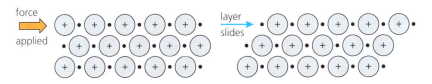

▲ Metals: malleable, ductile, and sometimes very glamorous – like this silver bracelet.

The layers can slide without breaking the metallic bond, because the electrons are free to move too.

3 **Metals are good conductors of heat.**

That is because the free electrons take in heat energy, which makes them move faster. They quickly transfer the heat through the metal structure:

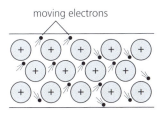

4 **Metals are good conductors of electricity.**

That is because the free electrons can move through the lattice carrying charge, when a voltage is applied across the metal.

Silver is the best conductor of all the metals. Copper is next – but it is used much more than silver because it is cheaper.

▲ What uses of metals can you see in this scene?

1 Describe in your own words the structure of a metal.
2 What is a *metallic bond*?
3 What does *malleable* mean?
4 Explain why metals can be drawn out into wires without breaking.
5 a Explain why metals can conduct electricity.
   b Would you expect molten metals to conduct? Give a reason.

6 Because metals are malleable, we use some of them to make saucepans. Give two other examples of uses of metals that depend on:
   a their malleability   b their ductility
   c their ability to conduct electricity
7 Mercury forms ions with a charge of 2+. It goes solid (freezes) at −39 °C. Try drawing a diagram to show the structure of solid mercury.

Checkup on Chapter 4

Revision checklist

Core syllabus content

Make sure you can …

☐ explain the difference between:
- an *element* and a *compound*
- a *compound* and a *mixture*

☐ say what the signs of a chemical change are

☐ explain why:
- atoms of Group VIII elements do not form bonds
- atoms of other elements do form bonds

☐ explain the difference between an *ionic bond* and a *covalent bond*

☐ draw a diagram to show how an ionic bond forms between atoms of sodium and chlorine

☐ explain what a *molecule* is

☐ say that non-metal atoms form covalent bonds with each other (except for the noble gas atoms)

☐ draw diagrams to show the covalent bonding in:

  hydrogen   *chlorine*   *water*
  methane   *hydrogen chloride*   *ammonia*

☐ describe how these properties differ, for ionic and molecular compounds: melting and boiling points, solubility, and ability to conduct electricity

☐ describe the giant covalent structures of graphite and diamond, and sketch them

☐ explain how their structures lead to different uses for diamond and graphite, with examples

Extended syllabus content

Make sure you can also …

☐ show how ionic bonds form between atoms of other metals and non-metals

☐ describe the lattice structure of ionic compounds

☐ work out the formulae of ionic compounds, from the charges on the ions

☐ draw diagrams to show the covalent bonding in nitrogen, oxygen, methanol, carbon dioxide, and ethene

☐ describe metallic bonding, and draw a sketch for it

☐ explain how the structure and bonding in metals enables them to be malleable, ductile, and good conductors of heat and electricity

☐ describe the structure of silicon dioxide

☐ explain why silicon dioxide and diamond have similar properties

☐ give examples of uses for silicon dioxide

Questions

Core syllabus content

1 This question is about the ionic bond formed between the metal lithium (proton number 3) and the non-metal fluorine (proton number 9).

a How many electrons does a lithium atom have? Draw a diagram to show its electron structure.

b How does a metal atom obtain a stable outer shell of electrons?

c Draw the structure of a lithium ion, and write the symbol for it, showing its charge.

d How many electrons does a fluorine atom have? Draw a diagram to show its electron structure.

e How does a non-metal atom become an ion?

f Draw the structure of a fluoride ion, and write a symbol for it, showing its charge.

g Draw a diagram to show what happens when a lithium atom reacts with a fluorine atom.

h Write a word equation for the reaction between lithium and fluorine.

2 This diagram represents a molecule of a certain gas.

a Name the gas, and give its formula.

b What do the symbols • and × represent?

c Which type of bonding holds the atoms together?

d Name another compound with this type of bonding.

3 Hydrogen bromide is a compound of the two elements hydrogen and bromine. It melts at $-87\,°C$ and boils at $-67\,°C$. It has the same type of bonding as hydrogen chloride.

a Is hydrogen bromide a solid, a liquid, or a gas at room temperature ($20\,°C$)?

b Is hydrogen bromide molecular, or does it have a giant structure? What is your evidence?

c **i** Which type of bond is formed between the hydrogen and bromine atoms, in hydrogen bromide?

  ii Draw a diagram of the bonding between the atoms, showing only the outer electrons.

d Write a formula for hydrogen bromide.

e **i** Name two other compounds with bonding similar to that in hydrogen bromide.

  ii Write formulae for these two compounds.

4 These are some properties of substances A to G.

Substance	Melting point /°C	Electrical conductivity		Solubility in water
		solid	liquid	
A	−112	poor	poor	insoluble
B	680	poor	good	soluble
C	−70	poor	poor	insoluble
D	1495	good	good	insoluble
E	610	poor	good	soluble
F	1610	poor	poor	insoluble
G	660	good	good	insoluble

a Which of the seven substances are metals? Give reasons for your choice.

b Which of the substances are ionic compounds? Give reasons for your choice.

c Two of the substances have very low melting points, compared with the rest. Explain why these could *not* be ionic compounds.

d Two of the substances are molecular. Which two are they?

e **i** Which substance is a giant covalent structure?

  ii What other name is used to describe this type of structure? (Hint: starts with *m*.)

f Name the type of bonding found in:

  i B   **ii** C   **iii** E   **iv** F

Extended syllabus content

5 Aluminium and nitrogen react to form an ionic compound called aluminium nitride. These show the electron arrangement for the two elements:

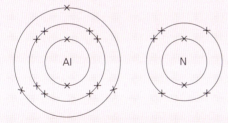

a Answer these questions for an aluminium atom.

  i Does it gain or lose electrons, to form an ion?

  ii How many electrons are transferred?

  iii Is the ion formed positive, or negative?

  iv What charge does the ion have?

b Now repeat **a**, but for a nitrogen atom.

c **i** Give the electron distribution for the ions formed by the two atoms. (2 + …)

  ii What do you notice about these distributions? Explain it.

d Name another non-metal that will form an ionic compound with aluminium, in the same way as nitrogen does.

6 Silicon lies directly below carbon in Group IV of the Periodic Table. Here is some data for silicon, carbon (in the form of diamond), and their oxides.

Substance	Symbol or formula	Melting point/°C	Boiling point/°C
carbon	C	3730	4530
silicon	Si	1410	2400
carbon dioxide	CO_2	(sublimes at −78 °C)	
silicon dioxide	SiO_2	1610	2230

a In which state are the two *elements* at room temperature (20 °C)?

b Which type of structure does carbon (diamond) have: giant covalent, or molecular?

c Which type of structure would you expect to find in silicon? Give reasons.

d In which state are the two oxides, at room temperature?

e Which type of structure has carbon dioxide?

f Does silicon dioxide have the same structure as carbon dioxide? What is your evidence?

7 The compound zinc sulfide has a structure like this:

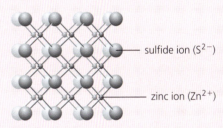
sulfide ion (S^{2-})
zinc ion (Zn^{2+})

a Which does the diagram represent: a giant structure, or a molecular structure?

b Which type of bonding does zinc sulfide have?

c Look carefully at the structure. How many:

  i sulfur ions are joined to each zinc ion?

  ii zinc ions are joined to each sulfur ion?

d **i** From **c**, deduce the formula of zinc sulfide.

  ii Is this formula consistent with the charges on the two ions? Explain your answer.

e Name another metal and non-metal that will form a compound with a similar formula.

8 The properties of metals can be explained by the structure and bonding within the metal lattice.

a Describe the bonding in metals.

b Use the bonding to explain why metals:

  i are good conductors of electricity

  ii are malleable and flexible.

5.1 The names and formulae of compounds

The names of compounds

Many compounds contain just two elements. If you know which elements they are, you can usually name the compound. Just follow these rules:

- When the compound contains a metal and a non-metal:
  - the name of the metal is given first
  - and then the name of the non-metal, but ending with *-ide*.
  Examples: sodium chloride, magnesium oxide, iron sulfide.

- When the compound is made of two non-metals:
  - if one is hydrogen, that is named first
  - otherwise the one with the lower group number comes first
  - and then the name of the other non-metal, ending with *-ide*.
  Examples: hydrogen chloride, carbon dioxide.

But some compounds have 'everyday' names that give you no clue about the elements in them. Water, methane, and ammonia are examples. You just have to remember their formulae!

▲ That very common compound, water. Your body is full of it. Which elements does it contain?

Finding formulae from the structure of compounds

Every compound has a formula as well as a name. The formula is made up of the symbols for the elements, and often has numbers too.

The formula of a compound is related to its structure. For example:

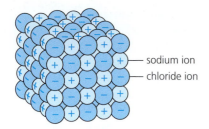

sodium ion
chloride ion

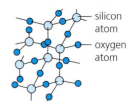

silicon atom
oxygen atom

Sodium chloride forms a giant structure with one sodium ion for every chloride ion. So its formula is NaCl.

Water is made up of molecules in which two hydrogen atoms are bonded to an oxygen atom. So its formula is H_2O.

Silicon (IV) oxide (silicon dioxide or silica) forms a giant structure in which there are two oxygen atoms for every silicon atom. So its formula is SiO_2.

Note the difference:

- In giant structures like sodium chloride and silicon dioxide, the formula tells you the *ratio* of the ions or atoms in the compound.

- In a molecular compound, the formula tells you *exactly* how many atoms are bonded together in each molecule.

Valency

But you don't need to draw the structure of a compound to work out its formula. You can work it out quickly if you know the valency of the elements:

The valency of an element is the number of electrons its atoms lose, gain or share, to form a compound.

Look at this table. (You can check the groups in the Periodic Table on page 27.)

Elements	In forming a compound, the atoms …	So the valency of the element is …	Examples of compounds formed (those in blue are covalent, with shared electrons)
Group I	lose 1 electron	1	sodium chloride, NaCl
Group II	lose 2 electrons	2	magnesium chloride, $MgCl_2$
Group III	lose 3 electrons	3	aluminium chloride, $AlCl_3$
Group IV	share 4 electrons	4	methane, CH_4
Group V	gain or share 3 electrons	3	ammonia, NH_3
Group VI	gain or share 2 electrons	2	magnesium oxide, MgO; water, H_2O
Group VII	gain or share 1 electron	1	sodium chloride, NaCl; hydrogen chloride, HCl
Group 0	(do not form compounds)	–	none
hydrogen	lose or share 1 electron	1	hydrogen bromide, HBr
transition elements	can lose different numbers of electrons	variable	iron(II) chloride, $FeCl_2$; iron(III) chloride, $FeCl_3$ copper(I) chloride, CuCl; copper(II) chloride, $CuCl_2$

Writing formulae using valencies

This is how to write the formula of a compound, using valencies:

1 Write down the valencies of the two elements.
2 Write down their symbols, in the same order as the elements in the name.
3 Add numbers after the symbols if you need to, to balance the valencies.

Example 1 What is the formula of hydrogen sulfide?

1 Valencies: hydrogen, 1; sulfur (Group VI), 2
2 HS (valencies not balanced)
3 The formula is H_2S (2×1 and 2, so the valencies are now balanced)

Example 2 What is the formula of aluminium oxide?

1 Valencies: aluminium (Group III), 3; oxygen (Group VI), 2
2 AlO (valencies not balanced)
3 The formula is Al_2O_3 (2×3 and 3×2, so the valencies are now balanced)

Writing formulae by balancing charges

In an ionic compound, the total charge is zero. So you can also work out the formula of an ionic compound by balancing the charges on its ions. To find out how to do this, turn to Unit 4.4.

▲ Hydrogen sulfide is a very poisonous colourless gas. It smells of rotten eggs.

Q

The Periodic Table on page 27 will help you with these.
1 Write the chemical name for water (ending in *-ide*).
2 Name the compounds containing only these elements:
   a sodium and fluorine   b fluorine and hydrogen
   c sulfur and hydrogen   d bromine and beryllium
3 Why does silicon (IV) oxide have the formula SiO_2?

4 Decide whether this formula is correct. If it is not correct, write it correctly.
   a HBr_2 b ClNa   c Cl_3Ca d Ba_2O
5 Write the correct formula for barium iodide.
6 See if you can give a name and formula for a compound that forms when phosphorus reacts with chlorine.

5.2 Equations for chemical reactions

Equations for two sample reactions

1 **The reaction between carbon and oxygen**   When carbon is heated in oxygen, they react together to form carbon dioxide. Carbon and oxygen are the **reactants**. Carbon dioxide is the **product**. You could show the reaction using a diagram, like this:

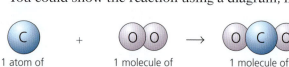

or by a word equation, like this:

carbon + oxygen $\longrightarrow$ carbon dioxide

or by a **symbol equation**, which gives symbols and formulae:

$C + O_2 \longrightarrow CO_2$

2 **The reaction between hydrogen and oxygen**   Hydrogen and oxygen react together to give water. The diagram is:

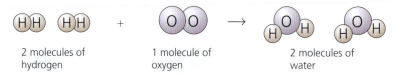

and you can use it to write the symbol equation:

$2H_2 + O_2 \longrightarrow 2H_2O$

Symbol equations must be balanced

Now look at the number of atoms on each side of this equation:

$2H_2 + O_2 \qquad\qquad \longrightarrow \qquad 2H_2O$

On the left:		On the right:
4 hydrogen atoms	$\longrightarrow$	4 hydrogen atoms
2 oxygen atoms		2 oxygen atoms

The number of each type of atoms is the same on both sides of the arrow. This is because atoms do not *disappear* during a reaction – they are just *rearranged*, as shown in the diagram of the molecules, in **2** above.

When the number of each type of atom is the same on both sides, the symbol equation is **balanced**. If it is not balanced, it is not correct.

Adding state symbols

Reactants and products may be solids, liquids, gases, or in solution. You can show their states by adding **state symbols** to the equations:

(*s*) for solid   (*l*) for liquid
(*g*) for gas   (*aq*) for aqueous solution (solution in water)

For the two reactions above, the equations with state symbols are:

$C\ (s) + O_2\ (g) \longrightarrow CO_2\ (g)$
$2H_2\ (g) + O_2\ (g) \longrightarrow 2H_2O\ (l)$

Note
Reactants are sometimes called **reagents**.

▲ The reaction between hydrogen and oxygen gives out so much energy that it is used to power rockets. The reactants are carried as liquids in the fuel tanks.

How to write the equation for a reaction

These are the steps to follow, when writing an equation:

1 Write the equation in words.
2 Now write it using symbols. Make sure all the formulae are correct.
3 Check that the equation is balanced, for each type of atom in turn.
   Make sure you do not change any formulae.
4 Add the state symbols.

Example 1 Calcium burns in chlorine to form calcium chloride, a solid.
Write an equation for the reaction, using the steps above.

1 calcium + chlorine $\longrightarrow$ calcium chloride
2 $\quad$ Ca $\quad$ + $\quad$ Cl_2 $\quad$ $\longrightarrow$ $\quad$ $CaCl_2$
3 Ca: $\quad$ 1 atom on the left and 1 atom on the right.
   Cl: $\quad$ 2 atoms on the left and 2 atoms on the right.
   The equation is balanced.
4 $\quad$ Ca (s) $\quad$ + $\quad$ Cl_2 (g) $\longrightarrow$ $\quad$ $CaCl_2$ (s)

Example 2 Hydrogen chloride is formed by burning hydrogen in
chlorine. Write an equation for the reaction.

1 hydrogen + chlorine $\longrightarrow$ hydrogen chloride
2 $\quad$ H_2 $\quad$ + $\quad$ Cl_2 $\quad$ $\longrightarrow$ $\quad$ HCl
3 H: $\quad$ 2 atoms on the left and 1 atom on the right.
   Cl: $\quad$ 2 atoms on the left and 1 atom on the right.
   The equation is *not* balanced. It needs another molecule of hydrogen
   chloride on the right. So a 2 is put *in front of* the HCl.
   $H_2 + Cl_2 \longrightarrow 2HCl$
   The equation is now balanced. Do you agree?
4 H_2 (g) + Cl_2 (g) $\longrightarrow$ 2HCl (g)

Example 3 Magnesium burns in oxygen to form magnesium oxide,
a white solid. Write an equation for the reaction.

1 magnesium + oxygen $\longrightarrow$ magnesium oxide
2 $\quad$ Mg $\quad$ + $\quad$ O_2 $\quad$ $\longrightarrow$ $\quad$ MgO
3 Mg: 1 atom on the left and 1 atom on the right.
   O: $\quad$ 2 atoms on the left and 1 atom on the right.
   The equation is *not* balanced. Try this:
   $Mg + O_2 \longrightarrow 2MgO$ (The 2 goes *in front of* the MgO.)
   Another magnesium atom is now needed on the left:
   $2Mg + O_2 \longrightarrow 2MgO$
   The equation is balanced.
4 $2Mg$ (s) + O_2 (g) $\longrightarrow$ $2MgO$ (s)

▲ Calcium chloride absorbs water, so it is used to dry gases. The glass cylinder above is packed with calcium chloride, and the gas is piped up through it.

▲ Magnesium burning in oxygen.

Q

1 What do + and $\longrightarrow$ mean, in an equation?
2 Balance the following equations:
   a Na (s) + Cl_2 (g) $\longrightarrow$ NaCl (s)
   b H_2 (g) + I_2 (g) $\longrightarrow$ HI (g)
   c Na (s) + H_2O (l) $\longrightarrow$ NaOH (aq) + H_2 (g)
d NH_3 (g) $\longrightarrow$ N_2 (g) + H_2 (g)
e C (s) + CO_2 (g) $\longrightarrow$ CO (g)
f Al (s) + O_2 (g) $\longrightarrow$ Al_2O_3 (s)
3 Aluminium burns in chlorine to form aluminium chloride,
$AlCl_3$, a solid. Write a balanced equation for the reaction.

5.3 The masses of atoms, molecules, and ions

A standard carbon atom

A single atom weighs almost nothing. You can't use scales to weigh it. But scientists need to compare the masses of atoms. So here's what they did.

First, they chose an atom of carbon-12 to be the standard atom. They fixed its mass as exactly 12 atomic mass units. (It has 6 protons and 6 neutrons, as shown on the right. They ignored the electrons.)

Then they compared all the other atoms with this standard atom, in a machine called a mass spectrometer, and found values for their masses. Like this:

$^{12}_{6}C$ 6 protons, 6 electrons, 6 neutrons

▲ An atom of carbon-12. It is the main isotope of carbon. (See page 30.)

This is the standard atom, $^{12}_{6}C$ or carbon-12. Its mass is taken as exactly 12.

This magnesium atom is twice as heavy as the carbon-12 atom. So its mass is 24.

This hydrogen atom has only one-twelfth the mass of the carbon-12 atom. So its mass is 1.

But what about isotopes?

As you saw on page 30, most elements have isotopes which occur naturally: atoms with different numbers of neutrons. That means they have different masses too.

We looked at just one isotope of magnesium, above, and one of hydrogen. But magnesium has three naturally occurring isotopes, with masses of 24, 25, and 26. Hydrogen also has three, with masses 1, 2, and 3.

So scientists had to take isotopes into account, and find *average* masses. The result is the relative atomic mass, or A_r, as shown below.

Relative atomic mass, A_r

The relative atomic mass, A_r for an element is the average mass of its naturally-occurring isotopes, relative to the mass of a carbon-12 atom.

The small **r** stands for *relative to the mass of a carbon-12 atom*.

Taking isotopes into account, the A_r of magnesium is 24.305. This is usually rounded off to 24, to make life easier. The A_r of hydrogen is 1.008, usually rounded off to 1.

In fact A_r values are usually rounded off. See the table on the next page. (But note the value for chlorine!)

Finding the A_r for chlorine

75% of naturally-occurring chlorine atoms have mass 35, and 25% have mass 37.
So A_r for chlorine = 75% × 35 + 25% × 37

$$= \frac{75}{100} \times 35 + \frac{25}{100} \times 37 \text{ (changing \% to fractions)}$$

$$= 26.25 + 9.25$$

$$= \textbf{35.5}$$

Because this is halfway between two whole numbers, it is not rounded off.

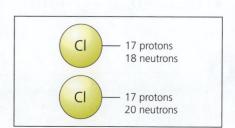

▲ The two isotopes of chlorine. They are called chlorine-35 and chlorine-37. Which is which?

A_r values for some common elements

Element	Symbol	A_r	Element	Symbol	A_r
hydrogen	H	1	chlorine	Cl	35.5
carbon	C	12	potassium	K	39
nitrogen	N	14	calcium	Ca	40
oxygen	O	16	iron	Fe	56
sodium	Na	23	copper	Cu	64
magnesium	Mg	24	zinc	Zn	65
sulfur	S	32	iodine	I	127

> **Finding the mass of an ion**
> mass of sodium atom = 23, so
> mass of sodium ion = 23
> since a sodium ion is just a sodium
> atom minus an electron (which has
> negligible mass).
>
> **An ion has the same mass as the atom from which it is made.**

Finding the masses of molecules and ions

Using A_r values, it is easy to work out the mass of any molecule or group of ions. Read the blue panel on the right above, then look at these examples:

mass of ion = mass of its atom

Hydrogen gas is made of molecules. Each molecule contains 2 hydrogen atoms, so its mass is 2. ($2 \times 1 = 2$)

The formula for water is H_2O. Each water molecule contains 2 hydrogen atoms and 1 oxygen atom, so its mass is 18. ($2 \times 1 + 16 = 18$)

Sodium chloride (NaCl) forms a giant structure with 1 sodium ion for every chloride ion. So the mass of a 'unit' of sodium chloride is 58.5. ($23 + 35.5 = 58.5$)

If the substance is made of molecules, its mass found in this way is called the **relative molecular mass**, or M_r. So the M_r for hydrogen is 2, and for water is 18.

But if the substance is made of *ions*, its mass is called the **relative formula mass**, which is also M_r for short. So the M_r for NaCl is 58.5.

This table gives two more examples of how to calculate M_r values.

Substance	Formula	Atoms in formula	A_r of atoms	M_r
ammonia	NH_3	1N 3H	N = 14 H = 1	$1 \times 14 = 14$ $3 \times 1 = 3$ Total = **17**
magnesium nitrate	$Mg(NO_3)_2$	1Mg 2N 6O	Mg = 24 N = 14 O = 16	$1 \times 24 = 24$ $2 \times 14 = 28$ $6 \times 16 = 96$ Total = **148**

1 a What does *relative atomic mass* mean?
 b Why does it have the word *relative*?
2 What is the A_r of the iodide ion, I^-?
3 The relative molecular mass and formula mass are both called M_r for short. What is the difference between them?

4 Work out the M_r for each of these, and say whether it is the relative molecular mass or the relative formula mass:
 a oxygen, O_2 **b** iodine, I_2 **c** methane, CH_4
 d chlorine, Cl_2 **e** butane, C_4H_{10} **f** ethanol, C_2H_5OH
 g ammonium sulfate $(NH_4)_2SO_4$

5.4 Some calculations about masses and %

Two laws of chemistry

If you know the *actual* amounts of two substances that react, you can:

- predict other amounts that will react
- say how much product will form.

You just need to remember these two laws of chemistry:

1 **Elements always react in the same ratio, to form a given compound.**
   For example, when carbon burns in oxygen to form carbon dioxide:
   6 g of carbon combines with **16 g** of oxygen, so
   12 g of carbon will combine with **32 g** of oxygen, and so on.

2 **The total mass does not change, during a chemical reaction.**
   So total mass of reactants = total mass of products.
   So **6 g** of carbon and **16 g** of oxygen give **22 g** of carbon dioxide.
   12 g of carbon and **32 g** of oxygen give **44 g** of carbon dioxide.

▲ A model of the carbon dioxide molecule. The amounts of carbon and oxygen that react to give this compound are always in the same ratio.

Calculating quantities

Calculating quantities is quite easy, using the laws above.

Example 64 g of copper reacts with 16 g of oxygen to give the black compound copper(II) oxide.

a What mass of copper will react with 32 g of oxygen?
   64 g of copper reacts with 16 g of oxygen, so
   2×64 g or **128 g** of copper will react with 32 g of oxygen.

b What mass of oxygen will react with 32 g of copper?
   16 g of oxygen reacts with 64 g of copper, so
   $\frac{16}{2}$ or **8 g** of oxygen will react with 32 g of copper.

c What mass of copper(II) oxide will be formed, in **b**?
   40 g of copper(II) oxide will be formed. $(32 + 8 = 40)$

d How much copper and oxygen will give 8 g of copper(II) oxide?
   64 g of copper and 16 g of oxygen give 80 g of copper(II) oxide, so
   $\frac{64}{10}$ of copper and $\frac{16}{10}$ g of oxygen will give 8 g of copper(II) oxide, so
   6.4 g of copper and **1.6 g of oxygen** are needed.

Percentages: a reminder

Calculations in chemistry often involve percentages. Remember:

- **The full amount of anything is 100%.**
- **To change a fraction to a %, just multiply it by 100.**

Example 1 Change the fractions $\frac{1}{2}$ and $\frac{18}{25}$ to percentages.

$\frac{1}{2} \times 100 = \textbf{50%}$ $\qquad$ $\frac{18}{25} \times 100 = \textbf{72%}$

Example 2 Give 19 % as a fraction.

$19\% = \frac{19}{100}$

▲ The French scientist Antoine Lavoisier (1743–1794) was the first to state that the total mass does not change, during a reaction. He was executed during the French Revolution, when he was 51.

Calculating the percentage composition of a compound

The **percentage composition** of a compound tells you how much of each element it contains, *as a percentage of the total mass*. This is how to work it out:

1 Write down the formula of the compound.
2 Using A_r values, work out its molecular or formula mass (M_r).
3 Write the mass of the element as a fraction of the M_r.
4 Multiply the fraction by 100, to give a percentage.

Example   Calculate the percentage of oxygen in sulfur dioxide.

1 The formula of sulfur dioxide is SO_2.
2 The M_r of the compound is 64, as shown on the right.
3 Mass of oxygen as a fraction of the total $= \dfrac{32}{64}$

4 Mass of oxygen as a percentage of the total $= \dfrac{32}{64} \times 100 = 50\%$
So the compound is **50% oxygen**.
This means it is also 50% sulfur ($100\% - 50\% = 50\%$).

> **M_r for sulfur dioxide, SO_2**
> A_r: S = 32, O = 16.
> So the M_r is:
> 1 S   = 32
> 2 O = 2 × 16   = 32
>   Total   = 64

Calculating % purity

A **pure** substance has nothing else mixed with it.
But substances often contain unwanted substances, or **impurities**.
Purity is usually given as a percentage. This is how to work it out:

$$\% \text{ purity of a substance} = \frac{\text{mass of pure substance in it}}{\text{total mass}} \times 100\%$$

Example   Impure copper is refined (purified), to obtain pure copper for use in computers. 20 tonnes of copper gave 18 tonnes of pure copper, on refining.

a What was the % purity of the copper before refining?

$$\% \text{ purity of the copper} = \frac{18 \text{ tonnes}}{20 \text{ tonnes}} \times 100\% = \textbf{90\%}$$

So the copper was **90% pure**.

b How much pure copper will 50 tonnes of the impure copper give?
The impure copper is 90% pure.
90% is $\dfrac{90}{100}$.
So 50 tonnes of it will give $\dfrac{90}{100} \times 50$ tonnes or **45 tonnes** of pure copper.

▲ Copper of high purity is needed for circuit boards like this one, in computers.

1 Magnesium burns in chlorine to give magnesium chloride, $MgCl_2$. In an experiment, 24 g of magnesium was found to react with 71 g of chlorine.
   a How much magnesium chloride was obtained in the experiment?
   b How much chlorine will react with 12 g of magnesium?
   c How much magnesium chloride will form, in **b**?

2 Methane has the formula CH_4. Work out the % of carbon and hydrogen in it. (A_r: C = 12, H = 1)
3 In an experiment, a sample of lead(II) bromide was made. It weighed 15 g. But the sample was found to be impure. In fact it contained only 13.5 g of lead(II) bromide.
   a Calculate the % purity of the sample.
   b What mass of impurity was present in the sample?

Checkup on Chapter 5

Revision checklist

Core syllabus content

Make sure you can …

☐ name a simple compound, when you are given the names of the two elements that form it

☐ work out the formula of a compound from a drawing of its structure

☐ work out the formula of a simple compound when you know the two elements in it, by balancing their valencies

☐ work out the formula of an ionic compound by balancing the charges of the ions, so that the total charge is zero

☐ write the equation for a reaction:
  – as a word equation
  – as a symbol equation

☐ balance a symbol equation

☐ say what the state symbols mean: (s), (l), (g), (aq)

☐ define *aqueous*

☐ explain that the carbon-12 atom is taken as the standard, for working out masses of atoms

☐ explain what the *relative atomic mass* of an element is, and why it takes isotopes into account

☐ say what these two symbols mean: A_r M_r

☐ explain the difference between *relative formula mass* and *relative molecular mass* (both known as M_r for short)

☐ work out M_r values, given the A_r values

☐ predict other amounts of reactants that will react, when you are given some actual amounts

☐ calculate:
  – the mass of a product, when you are given the masses of the reactants that combine to form it
  – the percentage of an element in a compound, using the formula and A_r values
  – the percentage purity of a substance, when you are given the total mass of the impure substance, and the amount of the pure substance in it

Questions

Core syllabus content

If you are not sure about symbols for the elements, you can check the Periodic Table on page 27.

1 Write the formulae for these compounds:
  a water **b** carbon monoxide
  c carbon dioxide **d** sulfur dioxide
  e sulfur trioxide **f** sodium chloride
  g magnesium chloride **h** hydrogen chloride
  i methane **j** ammonia

2 You can work out the formula of a compound from the ratio of the different atoms in it. Sodium carbonate has the formula Na_2CO_3 because it contains 2 atoms of sodium for every 1 atom of carbon and 3 atoms of oxygen.
  Deduce the formula for each compound **a** to **h**:

	Compound	Ratio in which the atoms are combined in it
a	lead oxide	1 of lead, 2 of oxygen
b	lead oxide	3 of lead, 4 of oxygen
c	potassium nitrate	1 of potassium, 1 of nitrogen, 3 of oxygen
d	nitrogen oxide	2 of nitrogen, 1 of oxygen
e	nitrogen oxide	2 of nitrogen, 4 of oxygen
f	sodium hydrogen carbonate	1 of sodium, 1 of hydrogen, 1 of carbon, 3 of oxygen
g	sodium sulfate	2 of sodium, 1 of sulfur, 4 of oxygen
h	sodium thiosulfate	2 of sodium, 2 of sulfur, 3 of oxygen

3 For each compound, write down the ratio of atoms present:
  a copper(II) oxide, CuO
  b copper(I) oxide, Cu_2O
  c aluminium chloride, $AlCl_3$
  d nitric acid, HNO_3
  e calcium hydroxide, $Ca(OH)_2$
  f ethanoic acid, CH_3COOH
  g ammonium nitrate, NH_4NO_3
  h ammonium sulfate, $(NH_4)_2SO_4$
  i sodium phosphate, $Na_3(PO_4)_2$
  j hydrated iron(II) sulfate, $FeSO_4.7H_2O$
  k hydrated cobalt(II) chloride, $CoCl_2.6H_2O$

4 Write the chemical formulae for the compounds with the structures shown below:

a $H-Br$

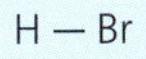

b $Cl-P-Cl$
    $\quad\quad\;\; |$
    $\quad\quad\;\; Cl$

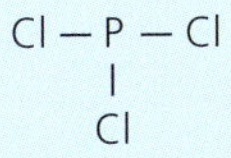

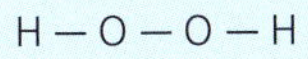

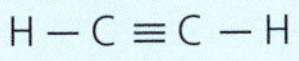

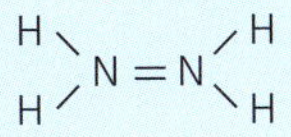

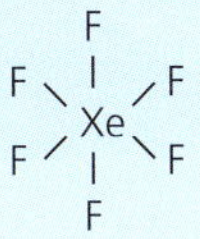

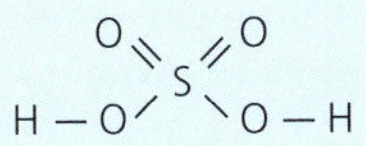

c $H-O-O-H$

d $H-C\equiv C-H$

e
$\begin{array}{c} H \\ \;\;\diagdown \\ \quad N=N \\ \;\;\diagup \quad\;\; \diagdown \\ H \quad\quad\quad H \end{array}$ with H

f
$\begin{array}{c} F \\ F\diagdown | \diagup F \\ \quad Xe \\ F \diagup | \diagdown F \\ F \end{array}$

g
$\begin{array}{c} O \diagdown \quad \diagup O \\ \quad S \\ H-O \diagup \quad \diagdown O-H \end{array}$

h
$\begin{array}{c} O \\ || \\ S \\ H-O \diagup \;\; \diagdown O-H \end{array}$

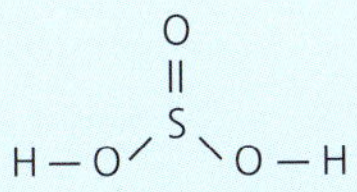

5 This shows the structure of an ionic compound.
  a Name the compound.
  b What is the simplest formula for it?

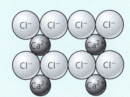

6 This shows the structure of a molecular compound. 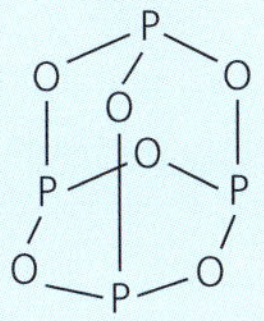
  a Name the compound.
  b What is the simplest formula for it?

7 Write these as word equations:
  a $Zn + 2HCl \longrightarrow ZnCl_2 + H_2$
  b $Na_2CO_3 + H_2SO_4 \longrightarrow Na_2SO_4 + CO_2 + H_2O$
  c $2Mg + CO_2 \longrightarrow 2MgO + C$
  d $ZnO + C \longrightarrow Zn + CO$
  e $Cl_2 + 2NaBr \longrightarrow 2NaCl + Br_2$
  f $CuO + 2HNO_3 \longrightarrow Cu(NO_3)_2 + H_2O$

8 Balance these equations:
  a $N_2 + ...\, O_2 \longrightarrow ...\, NO_2$
  b $K_2CO_3 + ...\, HCl \longrightarrow ...\, KCl + CO_2 + H_2O$
  c $C_3H_8 + ...\, O_2 \longrightarrow ...\, CO_2 + 4H_2O$
  d $Fe_2O_3 + ...CO \longrightarrow ...Fe + ...CO_2$
  e $Ca(OH)_2 + ...\, HCl \longrightarrow CaCl_2 + ...\, H_2O$
  f $2Al + ...\, HCl \longrightarrow 2AlCl_3 + ...\, H_2$

9 Copy and complete these equations:
  a $MgSO_4 + ... \longrightarrow MgSO_4.6H_2O$
  b $...\, C + ... \longrightarrow 2CO$
  c $2CuO + C \longrightarrow 2Cu + ...$
  d $C_2H_6 \longrightarrow .... + H_2$
  e $ZnO + C \longrightarrow Zn + ...$
  f $NiCO_3 \longrightarrow NiO + ...$
  g $CO_2 + ... \longrightarrow CH_4 + O_2$
  h $NaOH + HNO_3 \longrightarrow NaNO_3 + ...$
  i $C_2H_6 \longrightarrow C_2H_4 + ...$

10 Calculate M_r for these compounds.
   (A_r values are given at the top of page 325.)
   a water, H_2O
   b ammonia, NH_3
   c ethanol, CH_3CH_2OH
   d sulfur trioxide, SO_3
   e sulfuric acid, H_2SO_4
   f hydrogen chloride, HCl
   g phosphorus(V) oxide, P_2O_5

11 Calculate the relative formula mass for these ionic compounds. (A_r values are given on page 325.)
   a magnesium oxide, MgO
   b lead sulfide, PbS
   c calcium fluoride, CaF_2
   d sodium chloride, $NaCl$
   e silver nitrate, $AgNO_3$
   f ammonium sulfate, $(NH_4)_2SO_4$
   g potassium carbonate, K_2CO_3
   h hydrated iron(II) sulfate, $FeSO_4.7H_2O$

12 Iron reacts with excess sulfuric acid to give iron(II) sulfate. The equation for the reaction is:
   $Fe + H_2SO_4 \longrightarrow FeSO_4 + H_2$
   5 g of iron gives 13.6 g of iron(II) sulfate.
   a Using excess acid, how much iron(II) sulfate can be obtained from:
      i 10 g of iron?   **ii** 1 g of iron?
   b How much iron will be needed to make 136 g of iron(II) sulfate?
   c A 10 g sample of impure iron(II) sulfate contains 8 g of iron(II) sulfate. Calculate the percentage purity of the iron(II) sulfate.

13 Aluminium is extracted from the ore bauxite, which is impure aluminium oxide.
   1 tonne (1000 kg) of the ore was found to have this composition:
      aluminium oxide 825 kg
      iron(III) oxide 100 kg
      sand 75 kg
   a What percentage of this ore is impurities?
   b 1 tonne of the ore gives 437 kg of aluminium.
      i How much aluminium will be obtained from 5 tonnes of the ore?
      ii What mass of sand is in this 5 tonnes?
   c What will the percentage of aluminium oxide in the ore be, if all the iron(III) oxide is removed, leaving only the aluminium oxide and sand?

6.1 The mole

What is a mole?

As you saw on page 66, the masses of atoms are found by comparing them with the carbon-12 atom:

carbon-12   magnesium

This is an atom of carbon-12. It is chosen as the standard atom. Its A_r is taken as 12. Then other atoms are compared with it.

This is a magnesium atom. It is twice as heavy as a carbon-12 atom, so its A_r is 24. So it follows that ...

… 24 g of magnesium contains the *same number of atoms* as 12 g of carbon-12. 24 g of magnesium is called **a mole** of magnesium atoms.

A mole of a substance is the amount that contains the same number of units as the number of carbon atoms in 12 grams of carbon-12.
These units can be atoms, or molecules, or ions, as you will see.

The Avogadro constant

Thanks to the work of the Italian scientist Avogadro, we know that 12 g of carbon-12 contains **602 000 000 000 000 000 000 000** carbon atoms!

This huge number is called **the Avogadro constant**.
It is written in a short way as **6.02×10^{23}**. (The 10^{23} tells you to move the decimal point 23 places to the right, to get the full number.)

So 1 mole of magnesium atoms contains 6.02×10^{23} magnesium atoms.

More examples of moles

Sodium is made of single sodium atoms. Its symbol is Na. Its A_r is **23**.	Iodine is made of iodine molecules. Its formula is I_2. Its M_r is **254**. 	Water is made of water molecules. Its formula is H_2O. Its M_r is **18**.

This is **23 grams** of sodium. It contains 6.02×10^{23} sodium atoms, or **1 mole of sodium atoms**.	This is **254 grams** of iodine. It contains 6.02×10^{23} iodine molecules, or **1 mole of iodine molecules**.	The beaker contains **18 grams** of water, or 6.02×10^{23} water molecules, or **1 mole of water molecules**.

So you can see that:
One mole of a substance is obtained by weighing out the A_r or M_r of the substance, in grams.

Finding the mass of a mole

You can find the mass of one mole of any substance by these steps:

1 Write down the symbol or formula of the substance.
2 Find its A_r or M_r.
3 Express that mass in grams (g).

This table shows three more examples:

Substance	Symbol or formula	A_r	M_r	Mass of 1 mole
helium	He	He = 4	exists as single atoms	**4 grams**
oxygen	O_2	O = 16	2 × 16 = 32	**32 grams**
ethanol	C_2H_5OH	C = 12 H = 1 O = 16	2 × 12 = 24 6 × 1 = 6 1 × 16 = 16 — 46	**46 grams**

Some calculations on the mole

These equations will help you:

Mass of a given number of moles = mass of 1 mole × number of moles

$$\text{Number of moles in a given mass} = \frac{\text{mass}}{\text{mass of 1 mole}}$$

Example 1 Calculate the mass of 0.5 moles of bromine atoms.
The A_r of bromine is 80, so 1 mole of bromine atoms has a mass of 80 g.
So 0.5 moles of bromine atoms has a mass of 0.5 × 80 g, or **40** g.

Example 2 Calculate the mass of 0.5 moles of bromine molecules.
A bromine *molecule* contains 2 atoms, so its M_r is 160.
So 0.5 moles of bromine molecules has a mass of 0.5 × 160 g, or **80** g.

Example 3 How many moles of oxygen molecules are in 64 g of oxygen?
The M_r of oxygen is 32, so 32 g of it is 1 mole.
Therefore 64 g is $\frac{64}{32}$ moles, or **2 moles** of oxygen molecules.

Use the calculation triangle

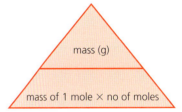

▲ Cover the one you want to find – and you will see how to calculate it.

$A_r = 80$
$M_r = 160$

$A_r = 16$
$M_r = 32$

Q

1 How many atoms are in 1 mole of atoms?
2 How many molecules are in 1 mole of molecules?
3 What name is given to the number 6.02×10^{23}?
4 Find the mass of 1 mole of:
  a hydrogen atoms   b iodine atoms
  c chlorine atoms   d chlorine molecules
5 Find the mass of 2 moles of:
  a oxygen atoms   b oxygen molecules

6 Find the mass of 3 moles of ethanol, C_2H_5OH.
7 How many moles of molecules are there in:
  a 18 grams of hydrogen, H_2?
  b 54 grams of water?
8 Sodium chloride is made up of Na^+ and Cl^- ions.
  a How many sodium ions are there in 58.5 g of sodium chloride? (A_r: Na = 23; Cl = 35.5.)
  b What is the mass of 1 mole of chloride ions?

6.2 Calculations from equations, using the mole

What an equation tells you

When carbon burns in oxygen, the reaction can be shown as:

 + →

1 atom of carbon   1 molecule of oxygen   1 molecule of carbon dioxide

or in a short way, using the symbol equation:

$$C\ (s) + O_2\ (g) \longrightarrow CO_2\ (g)$$

This equation tells you that:

| 1 carbon atom | reacts with | 1 molecule of oxygen | to give | 1 molecule of carbon dioxide |

Now suppose there is 1 *mole* of carbon atoms. Then we can say that:

| 1 mole of carbon atoms | reacts with | 1 mole of oxygen molecules | to give | 1 mole of carbon dioxide molecules |

So from the equation, we can tell how many moles react.
But moles can be changed to grams, using A_r and M_r.
The A_r values are: C = 12, O = 16.
So the M_r values are: O_2 = 32, CO_2 = (12 + 32) = 44, and we can write:

| 12 g of carbon | reacts with | 32 g of oxygen | to give | 44 g of carbon dioxide |

Since substances always react in the same ratio, this also means that:

| 6 g of carbon | reacts with | 16 g of oxygen | to give | 22 g of carbon dioxide |

and so on.
So we have gained a great deal of information from the equation.
In fact you can obtain the same information from any equation.

From the equation for a reaction you can tell:

- **how many moles of each substance take part**
- **how many grams of each substance take part.**

Reminder: the total mass does not change

Look what happens to the total mass, during the reaction above:

mass of carbon and oxygen at the start: 12 g + 32 g = **44 g**
mass of carbon dioxide at the end: **44 g**

The total mass has not changed, during the reaction. This is because no atoms have disappeared. They have just been rearranged.

That is one of the two laws of chemistry that you met on page 68:
The total mass does not change, during a chemical reaction.

▲ Iron and sulfur reacting: the total mass is the same before and after.

Calculating masses from equations

These are the steps to follow:

1 Write the balanced equation for the reaction. (It gives *moles*.)

2 Write down the A_r or M_r for each substance that takes part.

3 Using A_r or M_r, change the moles in the equation to *grams*.

4 Once you know the theoretical masses from the equation, you can then find any *actual* mass.

Example Hydrogen burns in oxygen to form water. What mass of oxygen is needed for 1 g of hydrogen, and what mass of water is obtained?

1 The equation for the reaction is: $2H_2\,(g) + O_2\,(g) \longrightarrow 2H_2O\,(l)$

2 A_r: H = 1, O = 16. M_r: H_2 = 2, O_2 = 32, H_2O = 18.

3 So, for the equation, the amounts in grams are:

$$2H_2\,(g) \quad + \quad O_2\,(g) \quad \longrightarrow \quad 2H_2O\,(l)$$

$$2 \times 2\text{ g} \qquad 32\text{ g} \qquad 2 \times 18\text{ g} \qquad \text{or}$$

$$4\text{ g} \qquad 32\text{ g} \qquad 36\text{ g}$$

4 But you start with only 1 g of hydrogen, so the *actual* masses are:

$$1\text{ g} \qquad 32/4\text{ g} \qquad 36/4\text{ g} \qquad \text{or}$$

$$1\text{ g} \qquad 8\text{ g} \qquad 9\text{ g}$$

So **1 g** of hydrogen needs **8 g** of oxygen to burn, and gives **9 g** of water.

▲ These models show how the atoms are rearranged, when hydrogen burns in oxygen. Which colour is oxygen?

Working out equations, from masses

If you know the actual masses that react, you can work out the equation for the reaction. Just change the masses to moles.

Example Iron reacts with a solution of copper(II) sulfate ($CuSO_4$) to give copper and a solution of iron sulfate. The formula for the iron sulfate could be either $FeSO_4$ or $Fe_2(SO_4)_3$. 1.4 g of iron gave 1.6 g of copper. Write the correct equation for the reaction.

1 A_r: Fe = 56, Cu = 64.

2 Change the masses to moles of atoms:

$\dfrac{1.4}{56}$ moles of iron atoms gave $\dfrac{1.6}{64}$ moles of copper atoms, or

0.025 moles of iron atoms gave 0.025 moles of copper atoms, so

1 mole of iron atoms gave 1 mole of copper atoms.

3 So the equation for the reaction must be:

$$Fe \quad + \quad CuSO_4 \quad \longrightarrow \quad Cu \quad + \quad FeSO_4$$

4 Add the state symbols to complete it:

$$Fe\,(s) \quad + \quad CuSO_4\,(aq) \quad \longrightarrow \quad Cu\,(s) \quad + \quad FeSO_4\,(aq)$$

▲ Iron wool reacting with copper(II) sulfate solution. Iron is more reactive than copper so displaces the copper (deep pink) from solution.

Q

1 The reaction between magnesium and oxygen is:

$$2Mg\,(s) + O_2\,(g) \rightarrow 2MgO\,(s)$$

   a Write a word equation for the reaction.

   b How many moles of magnesium atoms react with 1 mole of oxygen molecules?

   c The A_r values are: Mg = 24, O = 16.
   How many grams of oxygen react with:

     i 48 g of magnesium? ii 12 g of magnesium?

2 Copper(II) carbonate breaks down on heating, like this:

$$CuCO_3\,(s) \rightarrow CuO\,(s) + CO_2\,(g)$$

   a Write a word equation for the reaction.

   b Find the mass of 1 mole of each substance taking part in the reaction. (A_r: Cu = 64, C = 12, O = 16.)

   c When 31 g of copper(II) carbonate is used:

     i how many grams of carbon dioxide form?

     ii what mass of solid remains after heating?

6.3 Reactions involving gases

A closer look at some gases

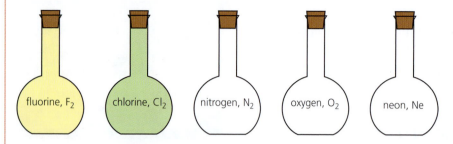

Imagine five very large flasks, each with a volume of 24 dm³. Each is filled with a different gas. Each gas is at room temperature and pressure, or **rtp**.

(We take **room temperature and pressure** as the standard conditions for comparing gases; rtp is 20 °C and 1 atmosphere.)

If you weighed the gas in the five flasks, you would discover something amazing. There is exactly 1 mole of each gas!

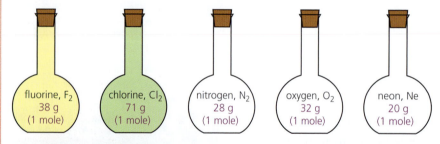

So we can conclude that:

1 mole of every gas occupies the same volume, at the same temperature and pressure. At room temperature and pressure, this volume is 24 dm³.

This was discovered by Avogadro, in 1811. So it is often called **Avogadro's Law**. It does not matter whether a gas exists as atoms or molecules, or whether its atoms are large or small. The law still holds.

The volume occupied by 1 mole of a gas is called its **molar volume**.
The molar volume of a gas is 24 dm³ at rtp.

> **Remember**
> 24 dm³ = 24 litres
>   = 24 000 cm³
>
> Imagine a ball about 36 cm in diameter. Its volume is about 24 dm³.

Another way to look at it

Look at these two gas jars.
A is full of nitrogen dioxide, NO_2.
B is full of oxygen, O_2.

The two gas jars have identical volumes, and the gases are at the same temperature and pressure.

You cannot see the gas molecules – let alone count them. But, from Avogadro's Law, you can say that the two jars contain the same number of molecules.

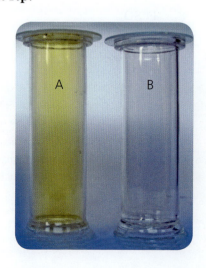

Calculating gas volumes from moles and grams

Avogadro's Law makes it easy to work out the volumes of gases.

Example 1   What volume does 0.25 moles of a gas occupy at rtp?
1 mole occupies 24 dm³ so
0.25 moles occupies 0.25 × 24 dm³ = 6 dm³
so 0.25 moles of any gas occupies **6 dm³** (or **6000 cm³**) at rtp.

Example 2   What volume does 22 g of carbon dioxide occupy at rtp?
M_r of carbon dioxide = 44, so
44 g = 1 mole, so
22 g = 0.5 mole
so the volume occupied = 0.5 × 24 dm³ = **12 dm³**.

Calculating gas volumes from equations

From the equation for a reaction, you can tell how many *moles* of a gas take part. Then you can use Avogadro's Law to work out its *volume*. In these examples, all volumes are measured at rtp.

Example 1   What volume of hydrogen will react with 24 dm³ of oxygen to form water?

1   The equation for the reaction is: $2H_2 (g) + O_2 (g) \longrightarrow 2H_2O (l)$
2   So 2 volumes of hydrogen react with 1 of oxygen, or
    2 × 24 dm³   react with 24 dm³, so
    48 dm³ of hydrogen will react.

Example 2   When sulfur burns in air it forms sulfur dioxide. What volume of this gas is produced when 1 g of sulfur burns? (A_r: S = 32.)

1   The equation for the reaction is: $S (s) + O_2 (g) \longrightarrow SO_2 (g)$
2   32 g of sulfur atoms = 1 mole of sulfur atoms, so
    $1 g = \dfrac{1}{32}$ mole or 0.03125 moles of sulfur atoms.
3   1 mole of sulfur atoms gives 1 mole of sulfur dioxide molecules so
    0.03125 moles   give 0.03125 moles.
4   1 mole of sulfur dioxide molecules has a volume of 24 dm³ at rtp so
    0.03125 moles has a volume of 0.03125 × 24 dm³ at rtp, or 0.75 dm³.
    So **0.75 dm³** (or **750 cm³**) of sulfur dioxide are produced.

Use the calculation triangle

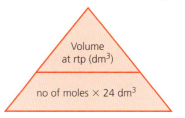

▲ Cover the one you want to find – and you will see how to calculate it.

▲ Sulfur dioxide is one of the gases given out in volcanic eruptions. These scientists are collecting gas samples on the slopes of an active volcano.

Q

(A_r: O = 16, N = 14, H = 1, C = 12.)
1   What does *rtp* mean? What values does it have?
2   What does *molar volume* mean, for a gas?
3   What is the molar volume of neon gas at rtp?
4   For any gas, calculate the volume at rtp of:
    a 7 moles   **b** 0.5 moles   **c** 0.001 moles
5   Calculate the volume at rtp of:
    a 16 g of oxygen (O_2)   **b** 1.7 g of ammonia (NH_3)

6   You burn 6 grams of carbon in plenty of air:
    $C (s) + O_2 (g) \rightarrow CO_2 (g)$
    a What volume of gas will form (at rtp)?
    b What volume of oxygen will be used up?
7   If you burn the carbon in limited air, the reaction is
    different: $2C (s) + O_2 (g) \rightarrow 2CO (g)$
    a What volume of gas will form this time?
    b What volume of oxygen will be used up?

6.4 The concentration of a solution

What does 'concentration' mean?

A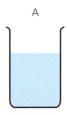

Solution A contains 2.5 grams of copper(II) sulfate in 1 dm³ of water. So its concentration is **2.5 g/dm³**.

B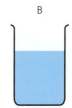

Solution B contains 25 grams of copper(II) sulfate in 1 dm³ of water. So its concentration is **25 g/dm³**.

C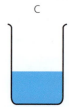

Solution C contains 125 grams of copper(II) sulfate in 0.5 dm³ of water. So its concentration is **250 g/dm³**.

The concentration of a solution is the amount of solute, in grams or moles, that is dissolved in 1 dm³ of solution.

Finding the concentration in moles

Example   Find the concentrations of A and C above, in moles per dm³.

First, change the mass of the solute to moles.
The formula mass of copper(II) sulfate is 250, as shown on the right.
So 1 mole of the compound has a mass of 250 g.

Solution A has 2.5 g of the compound in 1 dm³ of solution.

2.5 g = 0.01 moles

so its concentration is **0.01 mol/dm³**.

Note the unit of concentration: **mol/dm³**. This is often shortened to **M**, so the concentration of solution A can be written as **0.01 M**.

Solution C has 250 g of the compound in 1 dm³ of solution.
250 g = 1 mole
so its concentration is **1 mol/dm³**, or **1 M** for short.

A solution that contains 1 mole of solute per dm³ of solution is often called a **molar solution**. So C is a molar solution.

> In general, to find the concentration of a solution in moles per dm³:
>
> $$\text{concentration (mol/dm}^3) = \frac{\text{amount of solute (mol)}}{\text{volume of solution (dm}^3)}$$

Use the equation above to check that the last column in this table is correct:

Amount of solute (mol)	Volume of solution (dm³)	Concentration of solution (mol/dm³)
1.0	1.0	1.0
0.2	0.1	2.0
0.5	0.2	2.5
1.5	0.3	5.0

> **!**
>
> M_r **for copper(II) sulfate**
> Its formula is $CuSO_4.5H_2O$.
> This has 1 Cu, 1 S, 9 O, and 10 H.
> So the formula mass is:
> 1 Cu = 1 × 64 =   64
> 1 S = 1 × 32 =   32
> 9 O = 9 × 16 = 144
> 10 H = 10 × 1 =   10
>   Total = 250

> **!**
>
> **Remember**
> - 1 dm³ = 1 litre
>   = 1000 cm³
>   = 1000 ml
> - All these mean the same thing:
>   moles per dm³
>   mol/dm³
>   mol dm⁻³
>   moles per litre

Finding the amount of solute in a solution

If you know the concentration of a solution, and its volume:

- you can work out how much solute it contains, in moles.
  Just rearrange the equation from the last page:
  amount of solute (mol) = concentration (mol/dm³) × volume (dm³)

- you can then convert moles to grams, by multiplying the number of moles by M_r.

Sample calculations

The table shows four solutions, with different volumes and concentrations. Check that you understand the calculations that give the masses of solute in the bottom row.

Use the calculation triangle

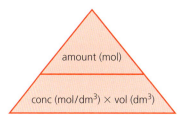

▲ Cover the one you want to find – and you will see how to calculate it. To draw this triangle, remember that **a**lligators **c**hew **v**isitors!

solution	sodium hydroxide NaOH	sodium thiosulfate $Na_2S_2O_3$	lead nitrate $Pb(NO_3)_2$	silver nitrate $AgNO_3$
	2 dm³	250 cm³	100 cm³	25 cm³
concentration (mol/dm³)	1	2	0.1	0.05
amount of solute present (moles)	$1 \times 2 = 2$	$2 \times \dfrac{250}{1000} = 0.5$	$0.1 \times \dfrac{100}{1000} = 0.01$	$0.05 \times \dfrac{25}{1000} = 0.00125$
M_r	40	158	331	170
mass of solute present (g)	80	79	3.31	0.2125

1 How many moles of solute are in:
  a 500 cm³ of solution, of concentration 2 mol/dm³?
  b 2 litres of solution, of concentration 0.5 mol/dm³?
2 What is the concentration of a solution containing:
  a 4 moles in 2 dm³ of solution?
  b 0.3 moles in 200 cm³ of solution?
3 Different solutions of salt X are made up. What volume of:
  a a 4 mol/dm³ solution contains 2 moles of X?
  b a 6 mol/dm³ solution contains 0.03 moles of X?

4 The M_r of sodium hydroxide is 40. How many grams of sodium hydroxide are there in:
  a 500 cm³ of a molar solution?
  b 25 cm³ of a 0.5 M solution?
5 What is the concentration in moles per litre of:
  a a sodium carbonate solution containing 53 g of the salt (Na_2CO_3) in 1 litre?
  b a copper(II) sulfate solution containing 62.5 g of the salt ($CuSO_4.5H_2O$) in 1 litre?

6.5 Finding the empirical formula

What a formula tells you about moles and masses

The formula of carbon dioxide is **CO$_2$**. Some molecules of it are shown on the right. You can see that:

A_r: C = 12, O = 16

| 1 carbon atom | combines with | 2 oxygen atoms | so |

| 1 mole of carbon atoms | combines with | 2 moles of oxygen atoms |

Moles can be changed to grams, using A_r and M_r. So we can write:

| 12 g of carbon | combines with | 32 g of oxygen |

In the same way:
6 g of carbon combines with 16 g of oxygen
24 kg of carbon combines with 64 kg of oxygen, and so on.
The masses of substances that combine are *always in the same ratio*.

Therefore, from the formula of a compound, you can tell:

- **how many moles of the different atoms combine**
- **how many grams of the different elements combine.**

Finding the empirical formula

From the formula of a compound you can tell what masses of the elements combine. But you can also do things the other way round.
If you know what masses combine, you can work out the formula.
These are the steps:

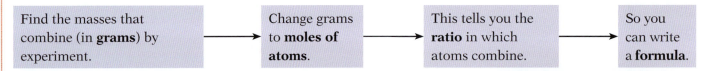

| Find the masses that combine (in **grams**) by experiment. | → | Change grams to **moles of atoms**. | → | This tells you the **ratio** in which atoms combine. | → | So you can write a **formula**. |

A formula found in this way is called the **empirical formula.**
The empirical formula shows the simplest ratio in which atoms combine.

Example 1 32 grams of sulfur combine with 32 grams of oxygen to form an oxide of sulfur. What is its empirical formula?

Draw up a table like this:

Elements that combine	sulfur	oxygen
Masses that combine	32 g	32 g
Relative atomic masses (A_r)	32	16
Moles of atoms that combine	32/32 = 1	32/16 = 2
Ratio in which atoms combine	1:2	
Empirical formula	**SO$_2$**	

So the empirical formula of the oxide that forms is **SO$_2$**.

▲ Sulfur combines with oxygen when it burns.

Example 2 An experiment shows that compound Y is 80% carbon and 20% hydrogen. What is its empirical formula?

Y is 80% carbon and 20% hydrogen. So 100 g of Y contains 80 g of carbon and 20 g of hydrogen. Draw up a table like this:

Elements that combine	carbon	hydrogen
Masses that combine	80 g	20 g
Relative atomic masses (A_r)	12	1
Moles of atoms that combine	80/12 = 6.67	20/1 = 20
Ratio in which atoms combine	6.67 : 20 or 1:3 in its simplest form	
Empirical formula	CH_3	

▲ Empirical formulae are found by experiment – and that usually involves weighing.

So the empirical formula of Y is **CH_3**.

But we can tell right away that the *molecular* formula for Y must be different. (A carbon atom does not bond to only 3 hydrogen atoms.) You will learn how to find the molecular formula from the empirical formula in the next unit.

An experiment to find the empirical formula

To work out the empirical formula, you need to know the masses of elements that combine. *The only way to do this is by experiment*.

For example, magnesium combines with oxygen to form magnesium oxide. The masses that combine can be found like this:

1 Weigh a crucible and lid, empty. Then add a coil of magnesium ribbon and weigh it again, to find the mass of the magnesium.
2 Heat the crucible. Raise the lid carefully at intervals to let oxygen in. The magnesium burns brightly.
3 When burning is complete, let the crucible cool (still with its lid on). Then weigh it again. The increase in mass is due to oxygen.

The results showed that 2.4 g of magnesium combined with 1.6 g of oxygen. Draw up a table again:

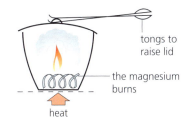

tongs to raise lid

the magnesium burns

heat

Elements that combine	magnesium	oxygen
Masses that combine	2.4 g	1.6 g
Relative atomic masses (A_r)	24	16
Moles of atoms that combine	2.4/24 = 0.1	1.6/16 = 0.1
Ratio in which atoms combine	1:1	
Empirical formula	**MgO**	

Mg   O

$A_r = 24$ $A_r = 16$

So the empirical formula for the oxide is **MgO**.

Q

1 a How many atoms of hydrogen combine with one carbon atom to form methane, CH_4?
   b How many grams of hydrogen combine with 12 grams of carbon to form methane?
2 What does the word *empirical* mean? (Check the glossary?)

3 56 g of iron combine with 32 g of sulfur to form iron sulfide. Find the empirical formula for iron sulfide. (A_r: Fe = 56, S = 32.)
4 An oxide of sulfur is 40% sulfur and 60% oxygen. What is its empirical formula?

6.6 From empirical to final formula

The formula of an ionic compound

You saw in the last unit that the empirical formula shows the *simplest ratio* in which atoms combine.

The diagram on the right shows the structure of sodium chloride. The sodium and chlorine atoms are in the ratio 1:1 in this compound. So its empirical formula is NaCl.

The formula of an ionic compound is the same as its empirical formula.

In the experiment on page 81, the empirical formula for magnesium oxide was found to be MgO. So the formula for magnesium oxide is also **MgO**.

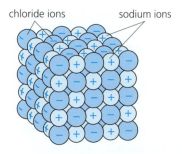

chloride ions   sodium ions

▲ The structure of sodium chloride.

The formula of a molecular compound

The gas ethane is one of the alkane family of compounds. An ethane molecule is drawn on the right. It contains only hydrogen and carbon atoms, so ethane is a **hydrocarbon**.

From the drawing you can see that the ratio of carbon to hydrogen atoms in ethane is 2:6. The simplest ratio is therefore 1:3.
So the *empirical* formula of ethane is CH_3. (It is compound Y on page 81.) But its *molecular* formula is C_2H_6.

The molecular formula shows the *actual* numbers of atoms that combine to form a molecule.

The molecular formula is more useful than the empirical formula, because it gives you more information.

For some molecular compounds, both formulae are the same. For others they are different. Compare them for the alkanes in the table on the right. What do you notice?

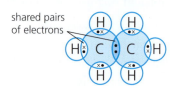

shared pairs of electrons

▲ An ethane molecule.

Alkane	Molecular formula	Empirical formula
methane	CH_4	CH_4
ethane	C_2H_6	CH_3
propane	C_3H_8	C_3H_8
butane	C_4H_{10}	C_2H_5
pentane	C_5H_{12}	C_5H_{12}
hexane	C_6H_{14}	C_3H_7

How to find the molecular formula

To find the molecular formula for an unknown compound, you need to know these:

- the **relative molecular mass** of the compound (M_r). This can be found using a mass spectrometer.

- its **empirical formula**. This is found by experiment, as on page 81.

- its **empirical mass**. This is the mass calculated using the empirical formula and A_r values.

Once you know those, you can work out the molecular formula by following these steps:

To find the molecular formula:

i   Calculate $\dfrac{M_r}{\text{empirical mass}}$ for the compound. This gives a number, *n*.

ii Multiply the numbers in the empirical formula by *n*.

Let's look at two examples.

▲ A mass spectrometer, for finding relative molecular mass. It compares the mass of a molecule with the mass of a carbon-12 atom, using an electric field.

Calculating the molecular formula

Example 1   A molecular compound has the empirical formula HO.
Its relative molecular mass is 34. What is its molecular formula?
(A_r: H = 1, O = 16.)

For the empirical formula HO, the empirical mass = 17. But M_r = 34.

So $\dfrac{M_r}{\text{empirical mass}} = \dfrac{34}{17} = 2$

So the molecular formula is 2 × HO, or **H_2O_2**.
So the compound is hydrogen peroxide.
Note how you write the 2 *after* the symbols, when you multiply.

Example 2   Octane is a hydrocarbon – it contains only carbon and
hydrogen. It is 84.2% carbon and 15.8% hydrogen by mass. Its M_r is 114.
What is its molecular formula?

1   **First find the empirical formula for the compound.**
   From the %, we can say that in 100 g of octane, 84.2 g is carbon
   and 15.8 g is hydrogen.
   So 84.2 g of carbon combines with 15.8 g of hydrogen.
   Changing masses to moles:
   $\dfrac{84.2}{12}$ moles of carbon atoms combine with $\dfrac{15.8}{1}$ moles of hydrogen
   atoms, or
   7.02 moles of carbon atoms combine with 15.8 moles of hydrogen
   atoms, so
   1 mole of carbon atoms combines with $\dfrac{15.8}{7.02}$ or **2.25 moles** of
   hydrogen atoms.
   So the atoms combine in the ratio of 1 : 2.25 or **4 : 9**.
   (Give the ratio as *whole* numbers, since only whole atoms combine.)
   The empirical formula of octane is therefore **C_4H_9**.

2   **Then use M_r to find the molecular formula.**
   For the empirical formula (C_4H_9), the empirical mass = 57.
   But M_r = 114.

   So $\dfrac{M_r}{\text{empirical mass}} = \dfrac{114}{57} = 2$

So the molecular formula of octane is 2 × C_4H_9 or **C_8H_{18}**.

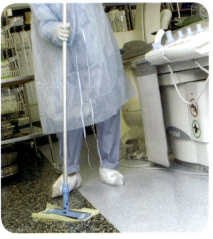

▲ Using hydrogen peroxide solution to
clean a hospital floor. Hydrogen peroxide
acts as a bleach, and kills germs.

▲ Octane is one of the main ingredients
in gasoline (petrol). When it burns in the
engine, it gives out lots of energy to
move that car.

Q

1   In the ionic compound magnesium chloride, magnesium and
   chlorine atoms combine in the ratio 1:2. What is the formula
   of magnesium chloride?
2   In the ionic compound aluminium fluoride, aluminium and
   fluorine atoms combine in the ratio 1:3. What is the formula
   of aluminium fluoride?
3   What is the difference between an empirical formula and a
   molecular formula? Can they ever be the same?

4   What is the empirical formula of benzene, C_6H_6?
5   A compound has the empirical formula CH_2.
   Its M_r is 28. What is its molecular formula?
6   A hydrocarbon is 84% carbon, by mass. Its relative molecular
   mass is 100. Find:
   a   its empirical formula   b   its molecular formula
7   An oxide of phosphorus has an M_r value of 220. It is 56.4%
   phosphorus. Find its molecular formula. (A_r of phosphorus, 31.)

6.7 Finding % yield and % purity

Yield and purity

The **yield** is the amount of product you obtain from a reaction. Suppose you own a factory that makes paint or fertilisers. You will want the highest yield possible, for the lowest cost!

Now imagine your factory makes medical drugs, or flavouring for foods. The yield will still be important – but the **purity** of the product may be even more important. Impurities could harm people.

In this unit you'll learn how to calculate the % yield from a reaction, and remind yourself how to calculate the % purity of the product obtained.

Finding the % yield

You can work out % yield like this:

$$\% \text{ yield} = \frac{\text{actual mass obtained}}{\text{calculated mass}} \times 100\%$$

Example   The medical drug aspirin is made from salicyclic acid. 1 mole of salicylic acid gives 1 mole of aspirin:

$$C_7H_6O_3 \xrightarrow{\text{chemicals}} C_9H_8O_4$$

salicylic acid   aspirin

In a trial, 100.0 grams of salicylic acid gave 121.2 grams of aspirin. What was the % yield?

1   A_r: C = 12, H = 1, O = 16.

   So M_r: salicyclic acid = 138, aspirin = 180.

2   138 g of salicyclic acid = 1 mole

   so 100 g = $\frac{100}{138}$ mole = 0.725 moles

3   1 mole of salicylic acid gives 1 mole of aspirin
   so 0.725 moles give 0.725 moles of aspirin
   or 0.725 × 180 g = 130.5 g

   So 130.5 g is the **calculated mass** for the reaction.

4   But the **actual mass** obtained in the trial was 121.2 g.

   So % yield = $\frac{121.2}{130.5}$ g × 100 = **92.9%**

This is a high yield – so it is worth continuing with those trials.

Finding the % purity

When you make something in a chemical reaction, and separate it from the final mixture, it will not be pure. It will have impurities mixed with it – for example small amounts of unreacted substances, or another product. You can work out the % purity of the product you obtained like this:

$$\% \text{ purity of a product} = \frac{\text{mass of the pure product}}{\text{mass of the impure product obtained}} \times 100\%$$

▲ Everything is carefully controlled in a chemical factory, to give a high yield – and as quickly as possible.

▲ For some products, a very high level of purity is essential – for example when you are creating new medical drugs.

Below are examples of how to work out the % purity.

Example 1 Aspirin is itself an acid. (Its full name is acetylsalicylic acid.) It is neutralised by sodium hydroxide in this reaction:

$$C_9H_8O_4 \ (aq) + NaOH \ (aq) \longrightarrow C_9H_7O_4Na \ (aq) + H_2O \ (l)$$

Some aspirin was prepared in the lab. Through titration, it was found that 4.00 g of the aspirin were neutralised by 17.5 cm³ of 1M sodium hydroxide solution. How pure was the aspirin sample?

1 M_r of $C_9H_8O_4$ = 180   (A_r: C = 12, H = 1, O = 16)

2 17.5 cm³ of 1M sodium hydroxide contain $\frac{17.5}{1000}$ moles or 0.0175 moles of NaOH

3 1 mole of NaOH reacts with 1 mole of $C_9H_8O_4$ so 0.0175 moles react with 0.0175 moles.

4 0.0175 moles of $C_9H_8O_4$ = 0.0175 × 180 g or 3.15 g of aspirin.

5 But the mass of the aspirin sample was 4 g.

   So % purity of the aspirin = $\frac{3.15}{4} \times 100\%$ or **78.75%**.

This is far from acceptable for medical use. The aspirin could be purified by crystallisation. Repeated crystallisation might be needed.

Example 2 Chalk is almost pure calcium carbonate.
10 g of chalk was reacted with an excess of dilute hydrochloric acid. 2280 cm³ of carbon dioxide gas was collected at room temperature and pressure (rtp). What was the purity of the sample?

You can work out its purity from the volume of carbon dioxide given off. The equation for the reaction is:

$$CaCO_3 \ (s) + 2HCl \ (aq) \longrightarrow CaCl_2 \ (aq) + H_2O \ (l) + CO_2 \ (g)$$

1 M_r of $CaCO_3$ = 100   (A_r: Ca = 40, C = 12, O = 16.)

2 1 mole of $CaCO_3$ gives 1 mole of CO_2 and

   1 mole of gas has a volume of 24 000 cm³ at rtp.

3 So 24 000 cm³ of gas is produced by 100 g of calcium carbonate and

   2280 cm³ is produced by $\frac{2280}{24\,000} \times 100$ g or **9.5 g**.

   So there is 9.5 g of calcium carbonate in the 10 g of chalk.

   So the % purity of the chalk = $\frac{9.5}{10}$ g × 100 = **95%**.

▲ White chalk cliffs on the Danish island of Mon. Chalk forms in the ocean floor, over many millions of years, from the hard parts of tiny marine organisms.

1 Define the term: **a** % yield   **b** % purity
2 100 g of aspirin was obtained from 100 g of salicylic acid. What was the % yield?
3 17 kg of aluminium was produced from 51 kg of aluminium oxide (Al_2O_3) by electrolysis. What was the percentage yield? (A_r: Al = 27, O = 16.)

4 Some seawater is evaporated. The sea salt obtained is found to be 86% sodium chloride. How much sodium chloride could be obtained from 200 g of this salt?
5 A 5.0 g sample of dry ice (solid carbon dioxide) turned into 2400 cm³ of carbon dioxide gas at rtp. What was the percentage purity of the dry ice? (M_r of CO_2 = 44.)

Checkup on Chapter 6

Revision checklist

Extended syllabus content

Make sure you can …

☐ explain what a *mole* of atoms or molecules or ions is, and give examples

☐ say what the *Avogadro constant* is

☐ do these calculations, using A_r and M_r:
  – find the mass of 1 mole of a substance
  – change moles to masses
  – change masses to moles

☐ use the idea of the mole to:
  – calculate the masses of reactants or products, from the equation for a reaction
  – work out the equation for a reaction, given the masses of the reactants and products

☐ define *molar volume* and *rtp*

☐ calculate the volume that a gas will occupy at rtp, from its mass, or number of moles

☐ calculate the volume of gas produced in a reaction, given the equation and the mass of one substance

☐ explain what *concentration of a solution* means and give examples, using grams and moles

☐ state the units used for concentration

☐ explain what a *molar solution* is

☐ work out:
  – the concentration of a solution, when you know the amount of solute dissolved in it
  – the amount of solute dissolved in a solution, when you know its concentration

☐ explain what the *empirical formula* of a substance is

☐ work out the empirical formula, from the masses that react

☐ work out the correct formula, using the empirical formula and M_r

☐ define *% yield*

☐ calculate the % yield for a reaction, from the equation and the actual mass of product obtained

☐ define *% purity*

☐ calculate the % purity of a product, given the mass of the impure product, and the mass of pure product it contains

Questions

Extended syllabus content

1 Iron is obtained by reducing iron(III) oxide using the gas carbon monoxide. The reaction is:
$$Fe_2O_3 \ (s) + 3CO \ (g) \longrightarrow 2Fe \ (s) + 3CO_2 \ (g)$$
  a Write a word equation for the reaction.
  b What is the formula mass of iron(III) oxide? (A_r: Fe = 56, O = 16.)
  c How many moles of Fe_2O_3 are there in 320 kg of iron(III) oxide? (1 kg = 1000 g.)
  d How many moles of Fe are obtained from 1 mole of Fe_2O_3?
  e From **c** and **d**, find how many moles of iron atoms are obtained from 320 kg of iron(III) oxide.
  f How much iron (in kg) is obtained from 320 kg of iron(III) oxide?

2 With strong heating, calcium carbonate undergoes thermal decomposition:
$$CaCO_3 \ (s) \longrightarrow CaO \ (s) + CO_2 \ (g)$$
  a Write a word equation for the change.
  b How many moles of $CaCO_3$ are in 50 g of calcium carbonate? (A_r: Ca = 40, C = 12, O = 16.)
  c **i** What mass of calcium oxide is obtained from the thermal decomposition of 50 g of calcium carbonate, assuming a 40% yield ?
  ii What mass of carbon dioxide will be given off at the same time?
  iii What volume will this gas occupy at rtp?

3 Nitroglycerine is used as an explosive. The equation for the explosion reaction is:
$$4C_3H_5(NO_3)_3 \ (l) \longrightarrow$$
$$12CO_2 \ (g) + 10H_2O \ (l) + 6N_2 \ (g) + O_2 \ (g)$$
  a How many moles does the equation show for:
  i nitroglycerine?
  ii *gas* molecules produced?
  b How many moles of gas molecules are obtained from 1 mole of nitroglycerine?
  c What is the total volume of gas (at rtp) obtained from 1 mole of nitroglycerine?
  d What is the mass of 1 mole of nitroglycerine? (A_r: H = 1, C = 12, N = 14, O = 16.)
  e What will be the total volume of gas (at rtp) from exploding 1 kg of nitroglycerine?
  f Using your answers above, try to explain *why* nitroglycerine is used as an explosive.

4 Nitrogen monoxide reacts with oxygen like this:
$$2NO\ (g) + O_2\ (g) \longrightarrow 2NO_2\ (g)$$
   a How many moles of oxygen molecules react with 1 mole of nitrogen monoxide molecules?
   b What volume of oxygen will react with 50 cm³ of nitrogen monoxide?
   c Using the volumes in **b**, what is:
      i the total volume of the two reactants?
      ii the volume of nitrogen dioxide formed?

5 2 g (an excess) of iron is added to 50 cm³ of 0.5 M sulfuric acid. When the reaction is over, the reaction mixture is filtered. The mass of the unreacted iron is found to be 0.6 g. (A_r: Fe = 56.)
   a What mass of iron took part in the reaction?
   b How many moles of iron atoms took part?
   c How many moles of sulfuric acid reacted?
   d Write the equation for the reaction, and deduce the charge on the iron ion that formed.
   e What volume of hydrogen (calculated at rtp) bubbled off during the reaction?

6 27 g of aluminium burns in chlorine to form 133.5 g of aluminium chloride. (A_r: Al = 27, Cl = 35.5.)
   a What mass of chlorine is present in 133.5 g of aluminium chloride?
   b How many moles of chlorine atoms is this?
   c How many moles of aluminium atoms are present in 27 g of aluminium?
   d Use your answers for parts **b** and **c** to find the simplest formula of aluminium chloride.
   e 1 dm³ of an aqueous solution is made using 13.35 g of aluminium chloride. What is its concentration in moles per dm³?

7 You have to prepare some 2 M solutions, with 10 g of solute in each. What volume of solution will you prepare, for each solute below?
   (A_r: H = 1, Li = 7, N = 14, O = 16, Mg = 24, S = 32.)
   a lithium sulfate, Li_2SO_4
   b magnesium sulfate, $MgSO_4$
   c ammonium nitrate, NH_4NO_3

8 Phosphorus forms two oxides, which have the empirical formulae P_2O_3 and P_2O_5.
   a Which oxide contains the higher percentage of phosphorus? (A_r: P = 31, O = 16.)
   b What mass of phosphorus will combine with 1 mole of oxygen molecules (O_2) to form P_2O_3?
   c What is the molecular formula of the oxide that has a formula mass of 284?
   d Suggest a molecular formula for the other oxide.

9 Zinc and phosphorus react to give zinc phosphide. 9.75 g of zinc combines with 3.1 g of phosphorus.
   a Find the empirical formula for the compound. (A_r: Zn = 65, P = 31.)
   b Calculate the percentage of phosphorus in it.

10 110 g of manganese was extracted from 174 g of manganese oxide. (A_r: Mn = 55, O = 16.)
   a What mass of oxygen is there in 174 g of manganese oxide?
   b How many moles of oxygen atoms is this?
   c How many moles of manganese atoms are there in 110 g of manganese?
   d Give the empirical formula of manganese oxide.
   e What mass of manganese can obtained from 1000 g of manganese oxide?

11 Find the molecular formulae for these compounds.
   (A_r: H = 1, C = 12, N = 14, O = 16.)

Compound	M_r	Empirical formula	Molecular formula
a hydrazine	32	NH_2	
b cyanogen	52	CN	
c nitrogen oxide	92	NO_2	
d glucose	180	CH_2O	

12 Hydrocarbons A and B both contain 85.7% carbon. Their molar masses are 42 and 84 g respectively.
   a Which elements does a hydrocarbon contain?
   b Calculate the empirical formulae of A and B.
   c Calculate the molecular formulae of A and B.

13 Mercury(II) oxide breaks down on heating:
$$2HgO\ (s) \longrightarrow 2Hg\ (l) + O_2\ (g)$$
   a Calculate the mass of 1 mole of mercury(II) oxide. (A_r: O = 16, Hg = 201)
   b How much mercury and oxygen *could* be obtained from 21.7 g of mercury(II) oxide?
   c Only 19.0 g of mercury was collected. Calculate the % yield of mercury for this experiment.

14 A 5-g sample of impure magnesium carbonate is reacted with an excess of hydrochloric acid:
$$MgCO_3\ (s) + 2HCl\ (aq) \longrightarrow$$
$$MgCl_2\ (aq) + H_2O\ (l) + CO_2\ (g)$$
   1250 cm³ of carbon dioxide is collected at rtp.
   a How many moles of CO_2 are produced?
   b What mass of pure magnesium carbonate would give this volume of carbon dioxide?
   (A_r: C = 12, O = 16, Mg = 24.)
   c Calculate the % purity of the 5-g sample.

7.1 Oxidation and reduction

Different groups of reactions

Thousands of different reactions go on around us, in labs, and factories, and homes. We can divide them into different groups. For example two of the groups are **neutralisation reactions** and **precipitation reactions**.

One big group is the **redox reactions**, in which **oxidation** and **reduction** occur. We focus on those in this chapter.

Oxidation: oxygen is gained

Magnesium burns in air with a dazzling white flame. A white ash is formed. The reaction is:

magnesium + oxygen ⟶ magnesium oxide
$$2Mg\ (s) \quad + \quad O_2\ (g) \longrightarrow \quad 2MgO\ (s)$$

The magnesium has gained oxygen. We say it has been oxidised.

A gain of oxygen is called *oxidation*. The substance has been oxidised.

▲ Magnesium burning in oxygen.

Reduction: oxygen is lost

Now look what happens when hydrogen is passed over heated copper(II) oxide. The black compound turns pink:

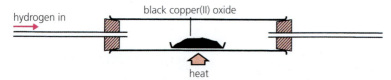

hydrogen in

black copper(II) oxide

heat

This reaction is taking place:

copper(II) oxide + hydrogen ⟶ copper + water
$$CuO\ (s) \quad + \quad H_2\ (g) \quad \longrightarrow Cu\ (s)\ +\ H_2O\ (l)$$

This time the heated substance is *losing* oxygen. It is being **reduced**.

A loss of oxygen is called *reduction*. The substance is reduced.

▲ Iron occurs naturally in the earth as iron(III) oxide, Fe_2O_3. This is **reduced** to iron in the blast furnace. Here, molten iron runs out from the bottom of the furnace.

▲ And here, iron is being **oxidised** to iron(III) oxide again! We call this process **rusting**. It is ruining the bikes. The formula for rust is $Fe_2O_3.2H_2O$.

Oxidation and reduction take place together

Look again at the reaction between copper(II) oxide and hydrogen.

Copper(II) oxide loses oxygen, and hydrogen gains oxygen:

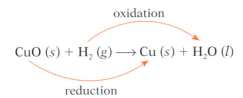

$$CuO\ (s) + H_2\ (g) \longrightarrow Cu\ (s) + H_2O\ (l)$$

So the copper(II) oxide is reduced, and the hydrogen is oxidised.

**Oxidation and reduction *always* take place together.
So the reaction is called a *redox reaction*.**

Two more examples of redox reactions

The reaction between calcium and oxygen   Calcium burns in air with a red flame, to form the white compound calcium oxide. It is easy to see that calcium has been oxidised. But oxidation and reduction *always* take place together, which means oxygen has been reduced:

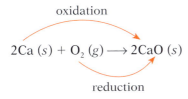

$$2Ca\ (s) + O_2\ (g) \longrightarrow 2CaO\ (s)$$

The reaction between hydrogen and oxygen   Hydrogen reacts explosively with oxygen, to form water. Hydrogen is oxidised, and oxygen is reduced:

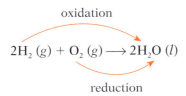

$$2H_2\ (g) + O_2\ (g) \longrightarrow 2H_2O\ (l)$$

▲ A redox reaction that cooks our food. The gas reacts with the oxygen in air, giving out heat.

▲ Roaming around on redox. The burning of petrol is a redox reaction. So is the 'burning' of glucose in our cells. It reacts with oxygen to give us energy, in a process called **respiration**.

Those burning reactions
- Another name for burning is **combustion**.
- Combustion is a redox reaction.
- For example, when an element burns in oxygen, it is oxidised to its oxide.

Q

1 Copy and complete the statements:
   a Oxidation means …
   b Reduction means …
   c Oxidation and reduction always …
2 Magnesium reacts with sulfur dioxide like this:

   $$2Mg\ (s) + SO_2\ (g) \rightarrow 2MgO\ (s) + S\ (s)$$

   Copy the equation, and use labelled arrows to show which substance is oxidised, and which is reduced.

3 Explain where the term *redox* comes from.
4 Many people cook with natural gas, which is mainly methane, CH_4. The equation for its combustion is:

   $$CH_4\ (g) + 2O_2\ (g) \rightarrow CO_2\ (g) + 2H_2O\ (l)$$

   Show that this is a redox reaction.
5 Write down the equation for the reaction between magnesium and oxygen. Use labelled arrows to show which element is oxidised, and which is reduced.

7.2 Redox and electron transfer

Another definition for oxidation and reduction

When magnesium burns in oxygen, magnesium oxide is formed:

$$2Mg\ (s) + O_2\ (g) \longrightarrow 2MgO\ (s)$$

The magnesium has clearly been oxidised. Oxidation and reduction *always* take place together, so the oxygen must have been reduced. But how? Let's see what is happening to the electrons:

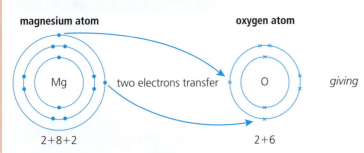
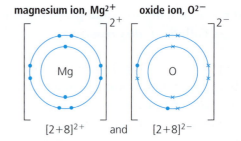

During the reaction, each magnesium atom loses two electrons and each oxygen atom gains two. This leads us to a new definition:

If a substance loses electrons during a reaction, it has been oxidised.

If it gains electrons, it has been reduced.

The reaction is a redox reaction.

> **Remember OILRIG!**
> **O**xidation **I**s **L**oss of electrons.
> **R**eduction **I**s **G**ain of electrons.

Writing half-equations to show the electron transfer

You can use **half-equations** to show the electron transfer in a reaction. One half-equation shows electron loss, and the other shows electron gain.

This is how to write the half-equations for the reaction above:

1 **Write down each reactant, with the electrons it gains or loses.**
   magnesium: $Mg \longrightarrow Mg^{2+} + 2e^-$
   oxygen: $O + 2e^- \longrightarrow O^{2-}$

2 **Check that each substance is in its correct form (ion, atom or molecule) on each side of the arrow. If it is not, correct it.**
   Oxygen is not in its correct form on the left above. It exists as molecules, so you must change O to O_2. That means you must also double the number of electrons and oxide ions:
   oxygen: $O_2 + 4e^- \longrightarrow 2O^{2-}$

3 **The number of electrons must be the same in both equations. If it is not, multiply one (or both) equations by a number, to balance them.**
   So we must multiply the magnesium half-equation by 2.
   magnesium: $2Mg \longrightarrow 2Mg^{2+} + 4e^-$
   oxygen: $O_2 + 4e^- \longrightarrow 2O^{2-}$
   The equations are now balanced, each with 4 electrons.

> **Two ways to show oxidation**
> You can show oxidation (the loss of electrons) in two ways:
> $Mg \rightarrow Mg^{2+} + 2e^-$
> or
> $Mg - 2e^- \rightarrow Mg^{2+}$
> Both are correct!

Redox without oxygen

Our definition of redox reactions is now much broader:
Any reaction in which electron transfer takes place is a redox reaction.
So the reaction does not have to include oxygen! Look at these examples:

1 **The reaction between sodium and chlorine**
The equation is:
$$2Na\ (s) + Cl_2\ (g) \longrightarrow 2NaCl\ (s)$$

The sodium atoms give electrons to the chlorine atoms, forming ions as shown on the right. So sodium is oxidised, and chlorine is reduced.

So the reaction is a redox reaction. Look at the half-equations:
sodium: $2Na \longrightarrow 2Na^+ + 2e^-$ \qquad (oxidation)
chorine: $Cl_2 + 2e^- \longrightarrow 2Cl^-$ \qquad (reduction)

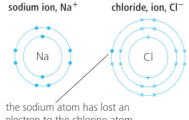

sodium ion, Na$^+$ \qquad chloride, ion, Cl$^-$

the sodium atom has lost an electron to the chlorine atom

2 **The reaction between chlorine and potassium bromide**
When chlorine gas is bubbled through a colourless solution of potassium bromide, the solution goes orange due to this reaction:
$$Cl_2\ (g) + 2KBr\ (aq) \longrightarrow 2KCl\ (aq) + Br_2\ (aq)$$
\qquad\qquad colourless \qquad\qquad\qquad orange

Bromine has been **displaced**. The half-equations for the reaction are:
chlorine: $Cl_2 + 2e^- \longrightarrow 2Cl^-$ \qquad (reduction)
bromide ion: $2Br^- \longrightarrow Br_2 + 2e^-$ \qquad (oxidation)

From half-equations to the ionic equation

Adding the balanced half-equations gives the **ionic equation** for the reaction.
An ionic equation shows the ions that take part in the reaction.

For example, for the reaction between chlorine and potassium bromide:

$$Cl_2 + 2e^- \longrightarrow 2Cl^-$$
$$2Br^- \longrightarrow Br_2 + 2e^-$$
$$\overline{Cl_2 + \cancel{2e^-} + 2Br^- \longrightarrow 2Cl^- + Br_2 + \cancel{2e^-}}$$

The electrons cancel, giving the ionic equation for the reaction:
$$Cl_2 + 2Br^- \longrightarrow 2Cl^- + Br_2$$

Redox: a summary

Oxidation is gain of oxygen, or loss of electrons.
Reduction is loss of oxygen, or gain of electrons.
Oxidation and reduction always take place together, in a **redox reaction**.

▲ Bromine being displaced by chlorine, from a colourless solution of potassium bromide. The solution goes orange.

Q

1 Give a *full* definition for: **a** oxidation \qquad **b** reduction
2 What does a *half-equation* show?
3 Potassium and chlorine react to form potassium chloride.
  a It is a redox reaction. Explain why.
  b See if you can write the balanced half-equations for it.

4 Bromine displaces iodine from a solution of potassium iodide.
  a Write the balanced half-equations for this reaction.
  b Add the half-equations, to give the ionic equation for the reaction.

7.3 Redox and changes in oxidation state

What does oxidation state mean?

Oxidation state tells you how many electrons each atom of an element has gained, lost, or shared, in forming a compound.

As you will see, oxidation states can help you to identify redox reactions.

The rules for oxidation states

1 Each atom in a formula has an oxidation state.

2 The oxidation state is usually given as a Roman numeral. Note these Roman numerals:

number	0	1	2	3	4	5	6	7
Roman numeral	0	I	II	III	IV	V	VI	VII

3 Where an element is not combined with other elements, its atoms are in oxidation state 0.

4 Many elements have the same oxidation state in most or all of their compounds. Look at these:

Element	Usual oxidation state in compounds
hydrogen	+I
sodium and the other Group I metals	+I
calcium and the other Group II metals	+II
aluminium	+III
chlorine and the other Group VII non-metals, in compounds without oxygen	−I
oxygen (except in peroxides)	−II

5 But atoms of transition elements can have variable oxidation states in their compounds. Look at these:

Element	Common oxidation states in compounds
iron	+II and +III
copper	+I and +II
manganese	+II, +IV, and +VII
chromium	+III and +VI

So for these elements, the oxidation state is included in the compound's name. For example iron(III) chloride, copper(II) oxide.

6 Note that in any formula, the oxidation states must add up to zero. Look at the formula for magnesium chloride, for example:

$$MgCl_2$$

$$+II \qquad 2 \times -I \qquad \text{Total} = \text{zero}$$

So you could use oxidation states to check that formulae are correct.

▲ The element sodium: oxidation state 0 (zero).

▲ Copper in its three oxidation states:
A – copper metal, 0
B – copper(I) chloride, +I
C – copper(II) chloride, +II

Oxidation states change during redox reactions

Look at the equation for the reaction between sodium and chlorine:

$$2Na\ (s)\ +\ Cl_2\ (g)\ \longrightarrow\ 2NaCl\ (s)$$
$$0 \qquad\quad 0 \qquad\qquad\quad +I\ -I$$

The oxidation states are also shown, using the rules on page 92. Notice how they have changed during the reaction.

Each sodium atom loses an electron during the reaction, to form an Na^+ ion. So sodium is oxidised, and its oxidation state rises from 0 to $+I$. Each chlorine atom gains an electron, to form a Cl^- ion. So chlorine is reduced, and its oxidation state falls from 0 to $-I$.

If oxidation states change during a reaction, it is a redox reaction.

> A rise in oxidation number means oxidation has occurred.
>
> $-IV \quad -III \quad -II \quad -I \quad 0 \quad +I \quad +II \quad +III \quad +IV$
>
> A fall in oxidation number means reduction has occurred.

▲ Sodium burning in chlorine, to form sodium chloride.

Using oxidation states to identify redox reactions

Example 1   Iron reacts with sulfur to form iron(II) sulfide:

$$Fe\ (s)\ +\ S\ (s)\ \longrightarrow\ FeS\ (s)$$
$$0 \qquad\quad 0 \qquad\qquad +II\ -II$$

The oxidation states are shown, using the rules on page 92.
There has been a change in oxidation states. So this is a redox reaction.

Example 2   When chlorine is bubbled through a solution of iron(II) chloride, iron(III) choride is formed. The equation and oxidation states are:

$$2FeCl_2\ (aq)\ +\ Cl_2\ (aq)\ \longrightarrow\ 2FeCl_3\ (aq)$$
$$+II\ \ -I \qquad\qquad 0 \qquad\qquad +III\ \ -I$$

There has been a change in oxidation states. So this is a redox reaction.

Example 3   When ammonia and hydrogen chloride gases mix, they react to form ammonium chloride. The equation and oxidation states are:

$$NH_3\ (g)\ +\ HCl\ (g)\ \longrightarrow\ NH_4Cl\ (s)$$
$$-III\ +I \qquad\quad +I\ -I \qquad\qquad -III\ +I\ -I$$

There has been no change in oxidation states. So this is *not* a redox reaction.

▲ Iron filings reacting with sulfur. You need heat to start the reaction off – but then it gives out heat.

1 a Write a word equation for this reaction:

   $2H_2\ (g) + O_2\ (g) \rightarrow 2H_2O\ (l)$

   b Now copy out the chemical equation from **a**. Below each symbol write the oxidation state of the atoms.

   c Is the reaction a redox reaction? Give evidence.

   d Say which substance is oxidised, and which reduced.

2 Repeat the steps in question **1** for each of these equations:

   i $2KBr\ (s) \rightarrow 2K\ (s) + Br_2\ (l)$

   ii $2KI\ (aq) + Cl_2\ (g) \rightarrow 2KCl\ (aq) + I_2\ (aq)$

3 a Read point **6** on page 92.

   b Using the idea in point **6**, work out the oxidation state of the carbon atoms in carbon dioxide, CO_2.

   c Carbon burns in oxygen to form carbon dioxide. Write a chemical equation for the reaction.

   d Now using oxidation states, show that this is a redox reaction, and say which substance is oxidised, and which is reduced.

4 *Every reaction between two elements is a redox reaction.*
   Do you agree with this statement? Explain.

7.4 Oxidising and reducing agents

What are oxidising and reducing agents?

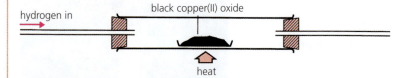

hydrogen in

black copper(II) oxide

heat

When hydrogen reacts with heated copper(II) oxide, the reaction is:

copper(II) oxide + hydrogen $\longrightarrow$ copper + water
$$CuO\ (s)\quad +\quad H_2\ (g)\ \longrightarrow\ Cu\ (s)\ +\ H_2O\ (l)$$

The copper(II) oxide is **reduced** to copper by reaction with hydrogen. So hydrogen acts as a **reducing agent**.

The hydrogen is itself **oxidised** to water, in the reaction. So copper(II) oxide acts as an **oxidising agent**.

An oxidising agent oxidises another substance during a redox reaction – and is itself reduced.

A reducing agent reduces another substance during a redox reaction – and is itself oxidised.

Oxidising and reducing agents in the lab

Some substances have a strong drive to gain electrons. So they are strong oxidising agents. They readily oxidise other substances by taking electrons from them. Examples are oxygen and chlorine.

Some substances are strong reducing agents, readily giving up electrons. For example hydrogen, carbon monoxide, and reactive metals like sodium.

Some oxidising and reducing agents show a colour change when they react. This makes them useful in lab tests. Let's look at two examples.

1 Potassium manganate(VII): an oxidising agent

Manganese is a transition element. Like other transition elements, it can exist in different oxidation states. (Look back at point **5** on page 92.)

Potassium manganate(VII) is a purple compound. Its formula is $KMnO_4$. In this, manganese is in oxidation state $+VII$. But it is much more stable in oxidation state $+II$. So it is strongly driven to reduce its oxidation state to $+II$, by gaining electrons.

That is why potassium manganate(VII) acts as a powerful oxidising agent. It takes electrons from other substances, in the presence of a little acid. It is itself reduced in the reaction – with a colour change:

$$MnO_4^-\ (aq)\ \xrightarrow{\text{reduction}}\ Mn^{2+}\ (aq)$$

manganate(VII) ion manganese(II) ion
(purple) (colourless)

This colour change means that potassium manganate(VII) can be used to test for the presence of a reducing agent. If a reducing agent is present, the purple colour will fade.

> **Oxidants and reductants** !
> - Oxidising agents are also called **oxidants**.
> - Reducing agents are called **reductants**.

> **Remember OILRIG!** !
> **O**xidation **I**s **L**oss of electrons.
> **R**eduction **I**s **G**ain of electrons.

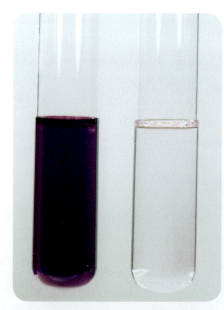

▲ If a reducing agent is present, the strong purple colour of potassium manganate will fade, as seen in the right-hand tube.

2 Potassium iodide: a reducing agent

When potassium iodide solution is added to hydrogen peroxide, in the presence of sulfuric acid, this redox reaction takes place:

$$H_2O_2 \, (aq) \; + \; 2KI \, (aq) \; + \; H_2SO_4 \, (aq) \longrightarrow I_2 \, (aq) \; + \; K_2SO_4 \, (aq) \; + \; 2H_2O \, (l)$$

hydrogen   potassium   iodine   potassium
peroxide   iodide   sulfate

You can see that the hydrogen peroxide loses oxygen: it is reduced. The potassium iodide acts as a reducing agent. At the same time the potassium iodide is oxidised to iodine. This causes a colour change:

$$2I^- \, (aq) \xrightarrow{\;\;\text{oxidation}\;\;} I_2 \, (aq)$$

    colourless   red-brown

So potassium iodide is used to test for the presence of an oxidising agent.

Oxidising agents outside the lab

Strong oxidising agents have many uses outside the lab.

- They kill bacteria and moulds, so they are widely used in household cleaning products.
- They break down coloured compounds by oxidising them, so they are used in bleaches for clothing and hair.
- The oxidising agent potassium dichromate(VI) is used in breathalysers, to test drivers for alcohol. It oxidises the ethanol in alcohol, and at the same time its colour changes from orange to green.

▲ The test-tube shows the red-brown colour you get when potassium iodide is oxidised by an oxidising agent.

▲ Yes. My hair met hydrogen peroxide. (An oxidising agent, formula H_2O_2.)

▲ Has he been drinking? Yes or no? Alcohol on the breath will turn orange potassium dichromate(VI) green.

▲ Gotcha! Many household cleaners contain oxidising agents to kill bacteria.

Q

1. What is:
   a an oxidising agent?   b a reducing agent?
2. Identify the oxidising and reducing agents in these reactions, by looking at the gain and loss of oxygen:
   a $2Mg \, (s) \; + \; O_2 \, (g) \rightarrow 2MgO \, (s)$
   b $Fe_2O_3 \, (s) \; + \; 3CO \, (g) \rightarrow 2Fe \, (l) \; + \; 3CO_2 \, (g)$

3. Now identify the oxidising and reducing agents in these:
   a $2Fe \; + \; 3Cl_2 \rightarrow 2FeCl_3$
   b $Fe \; + \; CuSO_4 \rightarrow FeSO_4 \; + \; Cu$
4. Explain why:
   a potassium manganate(VII) is a powerful oxidising agent
   b potassium iodide is used to test for oxidising agents.

Checkup on Chapter 7

Revision checklist

Core syllabus content

Make sure you can …
- ☐ define *oxidation* as a gain of oxygen
- ☐ define *reduction* as a loss of oxygen
- ☐ explain that oxidation and reduction always occur together, and give an example
- ☐ explain what a *redox reaction* is
- ☐ say what is being oxidised, and what is being reduced, in reactions involving oxygen

Extended syllabus content

Make sure you can also …
- ☐ define oxidation and reduction in terms of electron transfer
- ☐ explain these terms:
  half-equation *ionic equation*
- ☐ write balanced half-equations for a redox reaction, to show the electron transfer
- ☐ give the ionic equation for a reaction, by adding the balanced half-equations
- ☐ explain the term *oxidation state*
- ☐ give the usual oxidation state for these elements, in their compounds:

  hydrogen oxygen aluminium
  sodium and other Group I metals
  calcium and other Group II metals
  chlorine and other Group VII non-metals

- ☐ tell the oxidation state from a compound's name, for elements with variable oxidation states
- ☐ work out the oxidation state for each element in a compound (they must add up to zero)
- ☐ give the oxidation state for each element present, in the equation for a reaction
- ☐ identify a redox reaction from changes in oxidation states, in the equation for a reaction
- ☐ define these terms:
  oxidising agent *reducing agent*
- ☐ explain why some substances are:
  strong oxidising agents *strong reducing agents*
  and give examples
- ☐ explain why potassium manganate(VII) is used in the lab to test for the presence of reducing agents
- ☐ explain why potassium iodide is used in the lab to test for the presence of oxidising agents

Questions

Core syllabus content

1 If a substance gains oxygen in a reaction, it has been oxidised. If it loses oxygen, it has been reduced. Oxidation and reduction always take place together, so if one substance is oxidised, another is reduced.

  a First, see if you can write a word equation for each redox reaction **A** to **F** below.

  b Then, using the ideas above, say which substance is being oxidised, and which is being reduced, in each reaction.

  A $Ca\ (s) + O_2\ (g) \longrightarrow 2\ CaO\ (s)$
  B $2CO\ (g) + O_2\ (g) \longrightarrow 2CO_2\ (g)$
  C $CH_4\ (g) + 2O_2\ (g) \longrightarrow CO_2\ (g) + 2H_2O\ (l)$
  D $2CuO\ (s) + C\ (s) \longrightarrow 2Cu\ (s) + CO_2\ (g)$
  E $4Fe\ (s) + 3O_2\ (g) \longrightarrow 2Fe_2O_3\ (s)$
  F $Fe_2O_3\ (s) + 3CO\ (g) \longrightarrow 2Fe\ (s) + 3CO_2\ (g)$

2 a Is this a redox reaction? Give your evidence.
  A $2Mg\ (s) + CO_2\ (g) \longrightarrow 2MgO\ (s) + C\ (s)$
  B $SiO_2\ (s) + C\ (s) \longrightarrow Si\ (s) + CO_2\ (g)$
  C $NaOH\ (aq) + HCl\ (aq) \longrightarrow NaCl\ (aq) + H_2O\ (l)$
  D $Fe\ (s) + CuO\ (s) \longrightarrow FeO\ (s) + Cu\ (s)$
  E $C\ (s) + PbO\ (s) \longrightarrow CO\ (g) + Pb\ (s)$

  b For each redox reaction you identify, state:
  i what is being oxidised
  ii what is being reduced.

Extended syllabus content

3 All reactions in which electron transfer take place are redox reactions. This diagram shows the electron transfer during one redox reaction.

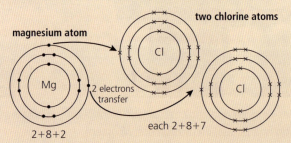

  a What is the product of this reaction?
  b Write a balanced equation for the full reaction.
  c i Which element is being oxidised?
  ii Write a half-equation for the oxidation.
  d i Which element is being reduced?
  ii Write a half-equation for the reduction of this element.

4 Redox reactions involve electron transfer.
  a Fluorine, from Group VII, reacts with lithium, from Group I, to form a poisonous white compound. What is its name?
  b Write a balanced equation for the reaction.
  c Draw a diagram to show the electron transfer that takes place during the reaction.
  d **i** Which element is oxidised in the reaction?
    ii Write a half-equation for this oxidation.
  e Write a half-equation for the reduction of the other element.

5 Chlorine gas is bubbled into a solution containing sodium bromide. The equation for the reaction is:
$Cl_2 (g) + 2NaBr (aq) \rightarrow Br_2 (aq) + 2NaCl (aq)$
  a Chlorine takes the place of bromine, in the metal compound. What is this type of reaction called?
  b The compounds of Group I metals are white, and give colourless solutions. What would you see as the above reaction proceeds?
  c **i** Write a half-equation for the reaction of the chlorine.
    ii Is the chlorine oxidised, or reduced, in this reaction? Explain.
  d Write a half-equation for the reaction of the bromide ion.
  e Reactive elements have a strong tendency to exist as ions. Which is more reactive, chlorine or bromine? Explain why you think so.
  f **i** Which halide ion could be used to convert bromine back to the bromide ion?
    ii Write the ionic equation for this reaction.

6 Iodine is extracted from seaweed using acidified hydrogen peroxide, in a redox reaction. The ionic equation for the reaction is:
$2I^- (aq) + H_2O_2 (aq) + 2H^+ (aq) \rightarrow$
$$I_2 (aq) + 2H_2O (l)$$
  a In which oxidation state is the iodine in seaweed?
  b There is a colour change in this reaction. Why?
  c **i** Is the iodide ion oxidised, or reduced?
    ii Write the half-equation for this change.
  d In hydrogen peroxide, the oxidation state of the hydrogen is +I.
    i What is the oxidation state of the oxygen in hydrogen peroxide?
    ii How does the oxidation state of oxygen change during the reaction?
    iii Copy and complete this half-equation for hydrogen peroxide:
    $H_2O_2 (aq) + 2H^+ (aq) + \ldots\ldots \rightarrow 2H_2O (l)$

7 The oxidation states in a formula add up to zero.
  a Give the oxidation state of the underlined atom in each formula below:
    i aluminium oxide, $\underline{Al}_2O_3$
    ii ammonia, $\underline{N}H_3$
    iii $H_2\underline{C}O_3$ (aq), carbonic acid
    iv phosphorus trichloride, $\underline{P}Cl_3$
    v copper(I) chloride, $\underline{Cu}Cl$
    vi copper(II) chloride, $\underline{Cu}Cl_2$
  b Now comment on the compounds in **v** and **vi**.

8 The oxidising agent potassium manganate(VII) can be used to analyse the % of iron(II) present in iron tablets. Below is an **ionic equation**, showing the ions that take part in the reaction:
$MnO_4^- (aq) + 8H^+ (aq) + 5Fe^{2+} (aq) \rightarrow$
$$Mn^{2+} (aq) + 5Fe^{3+} (aq) + 4H_2O (l)$$
  a What does the H^+ in the equation tell you about this reaction? (Hint: check page 146.)
  b Describe the colour change.
  c Which is the reducing reagent in this reaction?
  d How could you tell when all the iron(II) had reacted?
  e Write the half-equation for the iron(II) ions.

9 Potassium chromate(VI) is yellow. In acid it forms orange potassium dichromate(VI). These are the ions that give those colours:

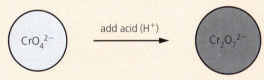

  a What is the oxidation state of chromium in:
    i the yellow compound?
    ii the orange compound?
  b This reaction of chromium ions is not a redox reaction. Explain why.

10 When solutions of silver nitrate and potassium chloride are mixed, a white precipitate forms. The ionic equation for the reaction is:
$Ag^+ (aq) + Cl^- (aq) \rightarrow AgCl (s)$
  a **i** What is the name of the white precipitate?
    ii Is it a soluble or insoluble compound?
  b Is the precipitation of silver chloride a redox reaction or not? Explain your answer.
  c When left in light, silver chloride decomposes to form silver and chlorine gas.
    Write an equation for the reaction and show clearly that this is a redox reaction.

8.1 Conductors and insulators

Batteries and electric current

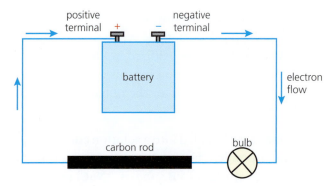

The photograph above shows a battery, a bulb and a rod of graphite joined or **connected** to each other by copper wires. (Graphite is a form of carbon.) This arrangement is called an **electric circuit**.

The bulb is lit: this shows that electricity must be flowing in the circuit. **Electricity is a stream of electrons.**

The diagram shows how the electrons move through the circuit. The battery acts like an electron pump. Electrons leave it through the **negative terminal**. They travel through the wire, bulb, and rod, and enter the battery again through the **positive terminal**.

When the electrons stream through the fine wire in the bulb, they cause it to heat up. It gets white-hot and gives out light.

Conductors and insulators

In the circuit above, the graphite and copper wire allow electricity to pass through. So they are called conductors.

But if you connect a piece of plastic or ceramic into the circuit, the bulb will not light. Plastic and ceramic do not let electricity pass through them. They are **non-conductors** or **insulators**.

▲ Copper carries the current into the styling iron. Then it flows through wire made of nichrome (a nickel-chromium alloy) which heats up. Meanwhile, the plastic protects you.

Some uses for conductors and insulators

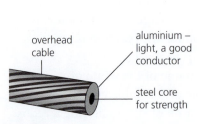

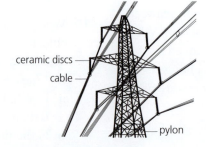

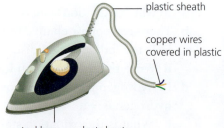

The cables that carry electricity around the country are made of aluminium and steel. Both are conductors. (Aluminium is a better conductor than steel.)

At pylons, ceramic discs support the bare cables. Since it is an insulator, the ceramic prevents the current from running down the pylon. (Dangerous!)

Copper is used for wiring, at home. It is a very good conductor. But the wires are sheathed in plastic, and plug cases are made of plastic (an insulator), for safety.

Testing substances to see if they conduct

You can test any substance to see if it conducts, by connecting it into a circuit like the one on page 98. For example:

Bunsen burner

Tin. A strip of tin is connected into the circuit, in place of the graphite rod. The bulb lights, so tin must be a conductor.

Ethanol. The liquid is connected into the circuit by placing graphite rods in it. The bulb does not light, so ethanol is a non-conductor.

Lead bromide. It does not conduct when solid. But if you melt it, it conducts, and gives off a choking brown vapour.

The results   These are the results from a range of tests:

1 **The only solids that conduct are the metals and graphite.**
    These conduct because of their free electrons (pages 57 and 58).
    The electrons get pumped out of one end of the solid by the battery, while more electrons flow in the other end.
    For the same reason, *molten* metals conduct. (It is hard to test molten graphite, because at room pressure graphite sublimes.)

2 **Molecular substances are non-conductors.**
    This is because they contain no free electrons, or other charged particles, that can flow through them.
    Ethanol (above) is molecular. So are petrol, paraffin, sulfur, sugar, and plastic. These never conduct, whether solid or molten.

3 **Ionic substances don't conduct when solid, but do conduct when melted or dissolved in water. They *break down at the same time*.**
    An ionic substance contains no free electrons. But it does contain **ions**, which have a charge. The ions become free to move when the substance is melted or dissolved, so they conduct the electricity.
    Lead bromide (above) is ionic. It conducts when it melts. The vapour that forms is bromine. So electricity has caused the lead bromide to decompose (break down). This process is called **electrolysis**.

    Electrolysis is the breaking down of an ionic compound, when molten or in aqueous solution, by the passage of electricity.

    A liquid that conducts electricity is called an **electrolyte**.
    Molten lead bromide is an electrolyte. So is a solution of sodium chloride. But ethanol is a non-electrolyte.

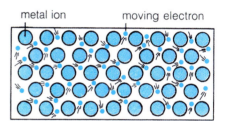

metal ion   moving electron

▲ Metals conduct, thanks to their free electrons, which form a current.

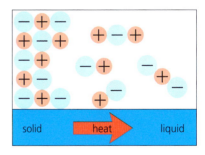

solid   heat   liquid

▲ An ionic solid conducts when it melts, because the ions become free to move.

Q

1 What is a *conductor* of electricity?
2 Draw a circuit you could use to see if mercury conducts.
3 Explain why metals are able to conduct electricity.
4 What does *electrolysis* mean?

5 Cooking oil is a mixture of molecular substances. Do you think it will conduct electricity? Explain.
6 What is: **a** an electrolyte?   **b** a non-electrolyte? Give *three* examples of each.

8.2 The principles of electrolysis

Electrolysis: breaking down by electricity

Any liquid that contains ions will conduct electricity.
This is because the ions are free to move. But at the same time, decomposition takes place.

So you can use electricity to break down a substance. The process is called **electrolysis**.

The electrolysis of molten lead bromide

The diagram on the right shows the apparatus.

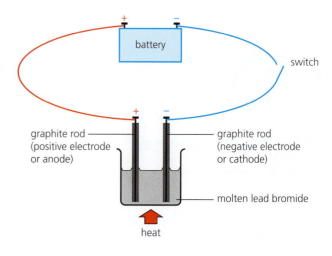

- The graphite rods are called **electrodes**.
- The electrode attached to the positive terminal of the battery is also positive. It is called the **anode**.
- The negative electrode is called the **cathode**.

The molten lead bromide contains lead ions (Pb^{2+}) and bromide ions (Br^-). This shows what happens when the switch is closed:

5 Electrons flow from the anode, along the wire, to the positive terminal of the battery.

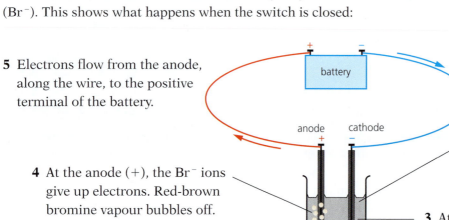

1 Electrons flow along the wire, from the negative terminal of the battery to the cathode.

2 In the liquid, the **ions** carry the current. They move to the electrode of opposite charge.

3 At the cathode ($-$), the Pb^{2+} ions accept electrons. Lead begins to appear below the cathode.

4 At the anode ($+$), the Br^- ions give up electrons. Red-brown bromine vapour bubbles off.

The result is that the lead bromide has **decomposed**:

lead bromide $\longrightarrow$ lead + bromine
$PbBr_2\,(l) \longrightarrow Pb\,(l) + Br_2\,(g)$

Note that:

- Electrons carry the current through the wires and electrodes. But the ions carry it through the liquid.
- The graphite electrodes are inert. They carry the current into the liquid, but remain unchanged. (Electrodes made of platinum are also inert.)

> **Which electrode is positive?**
> Remember **PA**!
>
> **P**ositive **A**node.

The electrolysis of other molten compounds

The pattern is the same for all molten ionic compounds of two elements: **Electrolysis breaks the molten ionic compound down to its elements, giving the metal at the cathode, and the non-metal at the anode.**

So it is a very important process. We depend on it to obtain reactive metals such as lithium, sodium, potassium, magnesium, and aluminium, from compounds dug from the earth.

> **Obtaining aluminium**
> Find out how electrolysis is used to extract aluminium, on page 197.

The electrolysis of aqueous solutions

Electrolysis can also be carried out on solutions of ionic compounds in water, because the ions in solutions are free to move. But the result may be different than for the molten compound. Compare these:

Electrolyte	At the cathode (−) you get ...	At the anode (+) you get ...
molten sodium chloride	sodium	chlorine
a concentrated solution of sodium chloride	hydrogen	chlorine

Why the difference? *Because the water itself produces ions*. Although water is molecular, a tiny % of its molecules is split up into ions:

some water molecules $\longrightarrow$ hydrogen ions + hydroxide ions

$$H_2O\ (l) \longrightarrow H^+\ (aq) + OH^-\ (aq)$$

These ions also take part in the electrolysis, so the products may change.

The rules for the electrolysis of a solution

At the cathode (−), either a metal or hydrogen forms.

1 The more reactive an element, the more it 'likes' to exist as ions. So if a metal is more reactive than hydrogen, its ions stay in solution and hydrogen bubbles off. (Look at the list on the right.)
2 But if the metal is less reactive than hydrogen, the metal forms.

At the anode (+), a non-metal other than hydrogen forms.

1 If it is a concentrated solution of a **halide** (a compound containing Cl^-, Br^- or I^- ions), then chlorine, bromine, or iodine form.
2 But if the halide solution is dilute, or there is no halide, oxygen forms.

Order of reactivity

potassium
sodium
calcium
magnesium
aluminium
zinc
iron
lead
hydrogen
copper
silver

increasing reactivity

Look at these examples. Do they follow the rules?

Electrolyte	At the cathode (−) you get...	At the anode (+) you get ...
a concentrated solution of potassium bromide, KBr	hydrogen	bromine
a concentrated solution of silver nitrate, AgNO$_3$	silver	oxygen
concentrated hydrochloric acid, HCl	hydrogen (H$^+$ is the only positive ion present)	chlorine
a dilute solution of sodium chloride, NaCl	hydrogen	oxygen
dilute sulfuric acid, H$_2$SO$_4$	hydrogen	oxygen

Look again at the last two examples. The water has been decomposed!

Q

1 a Which type of compounds can be electrolysed? Why?
   b What form must they be in?
2 What does electrolysis of these molten compounds give?
   a sodium chloride, NaCl   b aluminium oxide, Al$_2$O$_3$
   c calcium fluoride, CaF$_2$ d lead sulfide, PbS

3 Name the products at each electrode, when these aqueous solutions are electrolysed using inert electrodes:
   a a concentrated solution of magnesium chloride, MgCl$_2$
   b concentrated hydrochloric acid, HCl
   c a dilute solution of copper(II) sulfate, CuSO$_4$

8.3　The reactions at the electrodes

What happens to ions in the molten lead bromide?

In molten lead bromide, the ions are free to move. This shows what happens to them, when the switch in the circuit is closed:

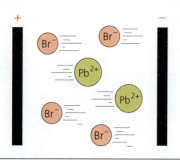 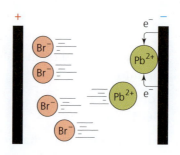 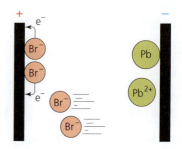

First, the ions move.	**At the cathode (−):**	**At the anode (+):**
Opposite charges attract. So the positive lead ions (Pb^{2+}) move to the cathode (−). The negative bromide ions (Br^-) move to the anode (+). The moving ions carry the current.	the lead ions each receive two electrons and become lead atoms. The **half-equation** is: $Pb^{2+} (l) + 2e^- \longrightarrow Pb (l)$ Lead collects on the electrode and eventually drops off it.	the bromide ions each give up an electron, and become atoms. These then pair up to form molecules. The **half-equation** is: $2Br^- (l) \longrightarrow Br_2 (g) + 2e^-$ The bromine gas bubbles off.
The free ions move.	Ions gain electrons: **reduction.**	Ions lose electrons: **oxidation.**

Remember **OILRIG:**
Oxidation **I**s **L**oss of electrons,
Reduction **I**s **G**ain of electrons.

Overall, electrolysis is a redox reaction.
Reduction takes place at the cathode
and oxidation at the anode.

The reactions for other molten compounds follow the same pattern.

For a concentrated solution of sodium chloride

This time, ions from water are also present:

> **Remember RAC!**
> **R**eduction **A**t **C**athode.

anode   cathode

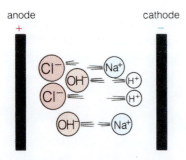 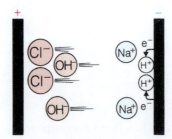 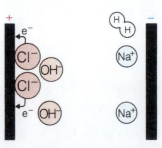

The solution contains Na^+ ions and Cl^- ions from the salt, and H^+ and OH^- ions from water.
The positive ions go to the cathode and the negative ions to the anode.

At the cathode, the H^+ ions accept electrons, since hydrogen is less reactive than sodium:

$2H^+ (aq) + 2e^- \longrightarrow H_2 (g)$

The hydrogen gas bubbles off.

At the anode, the Cl^- ions give up electrons more readily than the OH^- ions do.

$2Cl^- (aq) \longrightarrow Cl_2 (aq) + 2e^-$

The chlorine gas bubbles off.

When the hydrogen and chlorine bubble off, Na^+ and OH^- ions are left behind – so a solution of sodium hydroxide is formed.

For a dilute solution of sodium chloride

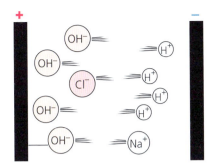

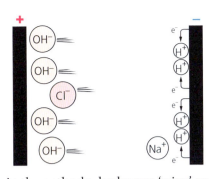

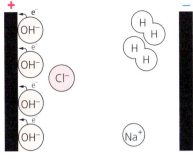

The same ions are present as before. But now the proportion of Na^+ and Cl^- ions is lower, since this is a dilute solution.
So the result will be different.

At the cathode, hydrogen 'wins' as before, and bubbles off:

$$4H^+ (aq) + 4e^- \longrightarrow 2H_2 (g)$$

(4 electrons are shown, to balance the half-equation at the anode.)

At the anode, OH^- ions give up electrons, since not many Cl^- ions are present. Oxygen bubbles off:

$$4OH^- (aq) \longrightarrow$$
$$O_2 (g) + 2H_2O (l) + 4e^-$$

When the hydrogen and oxygen bubble off, the Na^+ and Cl^- ions are left behind. So we *still* have a solution of sodium chloride!
The overall result is that water has been decomposed.

Half-equations for electrode reactions

A **half-equation** shows the electron transfer at an electrode.
This table shows how to write half-equations for both electrodes.
(But you'll be asked only for those at the cathode!)

The steps	Example: the electrolysis of molten magnesium chloride
1 First, name the ions present, and the products.	Magnesium ions and chloride ions are present. Magnesium and chlorine form.
2 Write each half-equation correctly. • Give the ion its correct charge. • Remember, positive ions go to the cathode, and negative ions to the anode. • Write the correct symbol for the element that forms. For example, Cl_2 for chlorine (not Cl). • The number of electrons in the equation should be the same as the total charge on the ion(s) in it.	Ions: Mg^{2+} and Cl^- **At the cathode:** $Mg^{2+} + 2e^- \longrightarrow Mg$ **At the anode:** $2Cl^- \longrightarrow Cl_2 + 2e^-$ (two Cl^- ions, so a total charge of $2-$) Note that it is also correct to write the anode reaction as: $2Cl^- - 2e^- \longrightarrow Cl_2$
3 You could then add the state symbols.	$Mg^{2+} (l) + 2e^- \longrightarrow Mg (s)$ $2Cl^- (l) \longrightarrow Cl_2 (g) + 2e^-$

Q

1 At which electrode does reduction always take place?
2 Give the half-equation for the reaction at the anode, during the electrolysis of these molten compounds:
   a potassium chloride   b calcium oxide

3 Give the two half-equations for the electrolysis of:
   a a concentrated solution of hydrochloric acid, HCl
   b a dilute solution of sodium nitrate, $NaNO_3$
   c a dilute solution of copper(II) chloride, $CuCl_2$

8.4 The electrolysis of brine

What is brine?

Brine is a concentrated solution of sodium chloride, or common salt. It can be obtained by pumping water into salt mines to dissolve the salt, or by evaporating seawater.

Brine might not sound very exciting – but from it, we get chemicals needed for thousands of products we use every day. When it undergoes electrolysis, the overall reaction is:

$$2NaCl\ (aq) + 2H_2O\ (l) \xrightarrow{electrolysis} 2NaOH\ (aq)\ +\ Cl_2\ (g)\ +\ H_2\ (g)$$

brine   sodium hydroxide   chlorine   hydrogen

The electrolysis

The diagram below shows one type of cell used for this electrolysis. (You won't be asked to draw it!) The anode is made of titanium, and the cathode of nickel. The membrane down the middle lets ions through, but keeps the gases apart.

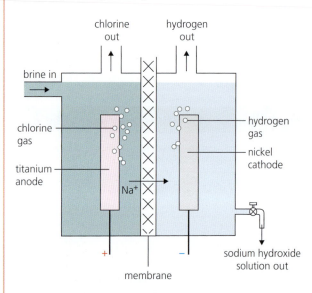

▲ Inside a salt mine. Many countries have underground salt beds. They were deposited millions of years ago, when the sea drained away from the land.

The ions present are Na$^+$ and Cl$^-$ from the salt, and H$^+$ and OH$^-$ from the water. The reactions at the electrodes are exactly as shown at the bottom of page 102. (Look back at them.)

At the cathode Hydrogen is discharged in preference to sodium:

$$2H^+\ (aq) + 2e^- \longrightarrow H_2\ (g)$$

As usual at the cathode, this is a reduction.

At the anode Chlorine is discharged in preference to oxygen:

$$2Cl^-\ (aq) \longrightarrow Cl_2\ (g) + 2e^-$$

As usual at the anode, this is an oxidation.

The two gases bubble off. Na$^+$ and OH$^-$ ions are left behind, giving a solution of sodium hydroxide. Some of the solution is evaporated to a give a more concentrated solution, and some is evaporated to dryness, giving solid sodium hydroxide.

▲ Chlorine has many uses. One is to kill germs in water. Behind the scenes at a swimming pool, this man makes sure there is chlorine in the water.

What the products are used for

The electrolysis of brine is an important process, because the products are so useful. Look at these:

Chlorine, a poisonous yellow-green gas

Used for making ...

- the plastic PVC (nearly 1/3 of it used for this)
- solvents for degreasing and drycleaning
- medical drugs (a large % of these involve chlorine)
- weedkillers and pesticides (most of these involve chlorine)
- paints and dyestuffs
- bleaches
- hydrogen chloride and hydrochloric acid

It is also used as a sterilising agent, to kill bacteria in water supplies and swimming pools.

Sodium hydroxide solution, alkaline and corrosive

Used in making ...

- soaps
- detergents
- viscose (rayon) and other textiles
- paper (like the paper in this book)
- ceramics (tiles, furnace bricks, and so on)
- dyes
- medical drugs

Hydrogen, a colourless flammable gas

Used ...

- in making nylon
- to make hydrogen peroxide
- to 'harden' vegetable oils to make margarine
- as a fuel in hydrogen fuel cells

Of the three chemicals, chlorine is the most widely used.
Around 50 million tonnes of it are produced each year, around the world.

▲ All three products from the electrolysis of brine must be transported with care. Why?

▲ Some hydrogen goes to hydrogen filling stations, for cars with hydrogen fuel cells instead of petrol engines (page 117).

Q

1 What is brine? Where is it obtained from?

2 Write a word equation for the electrolysis of brine.

3 a Draw a rough sketch of a cell for the electrolysis of brine.
   b Mark in where the oxidation and reduction reactions take place, and write the half-equations for them.

4 What is the membrane for, in the cell on page 104?

5 The electrolysis of brine is a very important process.
   a Explain why.
   b Give three uses for each of the products.

6 Your job is to keep a brine electrolysis plant running safely and smoothly. Try to think of three or four safety precautions you might need to take.

8.5 Two more uses of electrolysis

When electrodes are not inert

A solution of copper(II) sulfate contains blue Cu^{2+} ions, SO_4^{2-} ions, and H^+ and OH^- ions from water. Electrolysis of the solution will give different results, *depending on the electrodes*. Compare these:

A Using inert electrodes (carbon or platinum)

At the cathode   Copper ions are discharged:

$$2Cu^{2+}(aq) + 4e^- \longrightarrow 2Cu(s)$$

The copper coats the electrode.

At the anode   Oxygen bubbles off:

$$4OH^-(aq) \longrightarrow 2H_2O(l) + O_2(g) + 4e^-$$

So copper and oxygen are produced. This fits the rules on page 101. The blue colour of the solution fades as the copper ions are discharged.

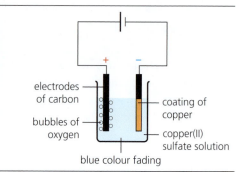

electrodes of carbon

bubbles of oxygen

coating of copper

copper(II) sulfate solution

blue colour fading

B Using copper electrodes

At the cathode   Again, copper is formed, and coats the electrode:

$$Cu^{2+}(aq) + 2e^- \longrightarrow Cu(s)$$

At the anode The anode dissolves, giving copper ions in solution:

$$Cu(s) \longrightarrow Cu^{2+}(aq) + 2e^-$$

So this time, the electrodes are not inert. The anode dissolves, giving copper ions. These move to the cathode, to form copper. So *copper moves from the anode to the cathode*. The colour of the solution does not fade.

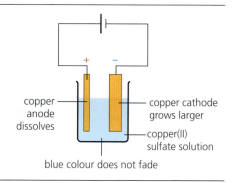

copper anode dissolves

copper cathode grows larger

copper(II) sulfate solution

blue colour does not fade

The idea in B leads to two important uses of electrolysis: for **refining** (or purifying) copper, and for **electroplating**.

Refining copper

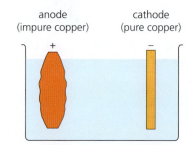

anode (impure copper)   cathode (pure copper)

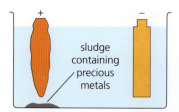

sludge containing precious metals

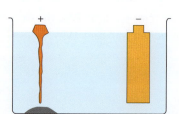

The anode is made of impure copper. The cathode is pure copper. The electrolyte is dilute copper(II) sulfate solution.

The copper in the anode dissolves. But the impurities do not dissolve. They just drop to the floor of the cell as a sludge.

A layer of pure copper builds up on the cathode. When the anode is almost gone, the anode and cathode are replaced.

The copper deposited on the cathode is over 99.9% pure.
The sludge may contain valuable metals such as platinum, gold, silver, and selenium. These are recovered and sold.

▲ The purer it is, the better copper is at conducting electricity. Highly refined copper is used for the electrics in cars. A car like this will contain more than 1 km of copper wiring.

▲ A steel tap plated with chromium, to make it look bright and shiny. Chromium does not stick well to steel. So the steel is first electroplated with copper or nickel, and then chromium.

Electroplating

Electroplating means using electricity to coat one metal with another, to make it look better, or to prevent corrosion. For example, steel car bumpers are coated with chromium. Steel cans are coated with tin to make tins for food. And cheap metal jewellery is often coated with silver.

The drawing on the right shows how to electroplate a steel jug with silver. The jug is used as the cathode. The anode is made of silver. The electrolyte is a solution of a soluble silver compound, such as silver nitrate.

At the anode   The silver dissolves, forming silver ions in solution:

$$Ag\,(s) \longrightarrow Ag^+\,(aq) + e^-$$

At the cathode   The silver ions are attracted to the cathode. There they receive electrons, forming a coat of silver on the jug:

$$Ag^+\,(aq) + e^- \longrightarrow Ag\,(s)$$

When the layer of silver is thick enough, the jug is removed.

▲ Silverplating: electroplating with silver. When the electrodes are connected to a power source, electroplating begins.

> **To electroplate**
> In general, to electroplate an object with metal X, the set-up is:
> **cathode**   –   object to be electroplated
> **anode**   –   metal X
> **electrolyte** –   a solution of a soluble compound of X.

Q

1 Copper(II) ions are blue. When copper(II) sulfate solution is electrolysed, the blue solution:
   a loses its colour when carbon electrodes are used
   b keeps its colour when copper electrodes are used.
   Explain each of these observations.

2 If you want to purify a metal by electrolysis, will you make it the anode or the cathode? Why?

3 Describe the process of refining copper.

4 What does *electroplating* mean?

5 Steel cutlery is often electroplated with nickel. Why?

6 You plan to electroplate steel cutlery with nickel.
   a What will you use as the anode?
   b What will you use as the cathode?
   c Suggest a suitable electrolyte.

Checkup on Chapter 8

Revision checklist

Core syllabus content

Make sure you can …

☐ define the terms *conductor* and *insulator*

☐ give examples of how we make use of conductors and insulators

☐ explain what these terms mean:

electrolysis   *electrolyte*   *electrode*

inert electrode   *anode*   *cathode*

☐ explain why an ionic compound must be melted, or dissolved in water, for electrolysis

☐ predict what will be obtained at each electrode, in the electrolysis of a molten ionic compound

☐ say what *halides* are

☐ say why the products of electrolysis may be different, when a compound is dissolved in water, rather than melted

☐ give the general rules for the products at the anode and cathode, in the electrolysis of a solution

☐ name the product at each electrode, for the electrolysis of these solutions:

– concentrated hydrochloric acid

– a concentrated solution of sodium chloride

– dilute sulfuric acid

☐ explain what *electroplating* is, and why it is used

☐ describe how electroplating is carried out

Extended syllabus content

Make sure you can also …

☐ predict the products, for the electrolysis of halides in dilute and concentrated solutions

☐ describe the reactions at the electrodes, for the reactions you met in this chapter, and write half-equations for the reactions at the cathode

☐ describe the electrolysis of brine, and name the three products, and give some uses for them

☐ describe the differences, when the electrolysis of copper(II) sulfate is carried out:

– using inert electrodes (carbon or platinum)

– using copper electrodes

☐ describe how electrolysis is used to refine impure copper, and say why this is important

Questions

Core syllabus content

1 Electrolysis of molten lead bromide is carried out:

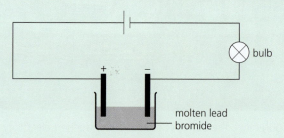

a The bulb will not light until the lead bromide has melted. Why not?

b What will be *seen* at the anode?

c Name the substance in **b**.

d What will be formed at the cathode?

2 Six substances A to F were dissolved in water, and connected in turn into the circuit below.

A represents an ammeter, which is used to measure current. The table shows the results.

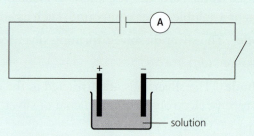

Substance	Current (amperes)	At cathode (−)	At anode (+)
A	0.8	copper	chlorine
B	1.0	hydrogen	chlorine
C	0.0	——	——
D	0.8	hydrogen	chlorine
E	1.2	hydrogen	oxygen
F	0.7	silver	oxygen

a Which solution conducts best?

b Which solution is a non-electrolyte?

c Which solution could be:

i silver nitrate?   ii copper(II) chloride?

iii sugar?   iv dilute sulfuric acid?

d i Two of the solutions give the same products at the electrodes. Which two?

ii Name two chemicals which would give those products, when connected into the circuit as concentrated solutions.

Extended syllabus content

3 The electrolysis below produces gases A and B.

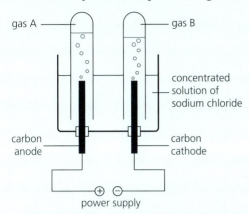

a Why does the solution conduct electricity?

b Identify each gas, and describe a test you could carry out to confirm its identity.

c Name one product manufactured from:
  i gas A   **ii** gas B

d **i** Write half-equations to show how the two gases are produced.
  ii The overall reaction is a *redox* reaction. Explain why.

e The solution remaining after the electrolysis will turn litmus paper blue.
  i What is the name of this solution?
  ii State one chemical property for it.

4 **a** List the ions that are present in concentrated solutions of:
  i sodium chloride   **ii** copper(II) chloride

b Explain why and how the ions move, when each solution is electrolysed using platinum electrodes.

c Write the half-equation for the reaction at:
  i the anode   **ii** the cathode
  during the electrolysis of each solution.

d Explain why the anode reactions for both solutions are the same.

e **i** The anode reactions will be different if the solutions are made very dilute. Explain why.
  ii Write the half-equations for the new anode reactions.

f Explain why copper is obtained at the cathode, but sodium is not.

g Name another solution that will give the same products as the concentrated solution of sodium chloride does, on electrolysis.

h Which solution in **a** could be the electrolyte in an electroplating experiment?

5 Molten lithium chloride contains lithium ions (Li^+) and chloride ions (Cl^-).

a Copy the following diagram and use arrows to show which way:
  i the ions move when the switch is closed
  ii the electrons flow in the wires

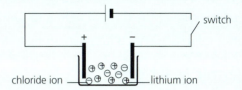

b **i** Write equations for the reaction at each electrode, and the overall reaction.
  ii Describe each of the reactions using the terms *reduction, oxidation* and *redox*.

6 This question is about the electrolysis of a dilute aqueous solution of lithium chloride.

a Give the names and symbols of the ions present.

b Say what will be formed, and write a half-equation for the reaction:
  i at the anode   **ii** at the cathode

c Name another compound that will give the same products at the electrodes.

d How will the products change, if a concentrated solution of lithium chloride is used?

7 An experiment is needed, to see if an iron object can be electroplated with chromium.

a Suggest a solution to use as the electrolyte.

b **i** Draw a labelled diagram of the apparatus that could be used for the electroplating.
  ii Show how the *electrons* will travel from one electrode to the other.

c Write half-equations for the reactions at each electrode.

d At which electrode does oxidation take place?

e The concentration of the solution does not change. Why not?

8 Nickel(II) sulfate ($NiSO_4$) is green. A solution of this salt is electrolysed using nickel electrodes.

a Write a half-equation for the reaction at each electrode.

b At which electrode does reduction take place? Explain your answer

c What happens to the size of the anode?

d The colour of the solution does not change, during the electrolysis. Explain why.

e Suggest one industrial use for this electrolysis.

9.1 Energy changes in reactions

Energy changes in reactions

During a chemical reaction, there is always an energy change.

Energy is given out or taken in. The energy is usually in the form of heat. (But some may be in the form of light and sound.)

So reactions can be divided into two groups: exothermic and endothermic.

Exothermic reactions

Exothermic reactions give out energy. So there is a temperature rise.
Here are three examples:

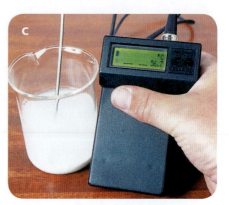

To start off the reaction between iron and sulfur, you must heat the mixture. But soon it glows red hot – *without* the Bunsen burner!

Mixing silver nitrate and sodium chloride solutions gives a white precipitate of silver chloride – and a temperature rise.

When you add water to lime (calcium oxide) heat is given out, so the temperature rises. Here the rise is being measured.

These reactions can be described as:

reactants ⟶ products + energy

The total energy is the same on each side of the arrow, in a reaction. So in exothermic reactions, the products have lower energy than the reactants. This is shown on the **energy level diagram** on the right.

The energy change

Energy is measured in kilojoule (kJ). For reaction **A** above:

$Fe (s) + S (s) \longrightarrow FeS (s)$ the energy change = -100 kJ

So 100 kJ of energy is given out when the amounts of reactants in the equation (56 g of iron and 32 g of sulfur, or 1 mole of each) react together. **The minus sign shows that energy is given out.**

▲ An energy level diagram for an exothermic reaction. The products have lower energy than the reactants.

Other examples of exothermic reactions

All these are exothermic:

- the neutralisation of acids by alkalis.

- the combustion of fuels. We burn fuels to obtain heat for cooking, heating homes, and so on. The more energy they give out, the better!

- respiration in your body cells. It provides the energy to keep your heart and lungs working, and for warmth and movement.

Endothermic reactions

Endothermic reactions take in energy from their surroundings.

Here are three examples:

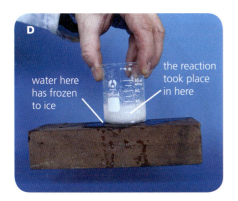

When barium hydroxide reacts with ammonium chloride, the temperature falls so sharply that water *under* the beaker will freeze!

Sherbet is citric acid plus the base sodium hydrogen carbonate. The neutralisation that occurs takes in heat – so your tongue cools.

The crucible contains calcium carbonate. If you keep on heating, it will all decompose to calcium oxide and carbon dioxide.

These reactions can be described as:

reactants + energy $\longrightarrow$ products

The energy is transferred from the surroundings: in **D** from the air and wet wood, in **E** from your tongue, and in **F** from the Bunsen burner. Since energy is taken in, the products must have higher energy than the reactants. This is shown on the energy level diagram on the right.

The energy change

For reaction **F** above:

$CaCO_3$ (s) $\longrightarrow$ CaO (s) + CO_2 (g)   the energy change = + 178 kJ

So 178 kJ of energy is needed to make 100 g (or 1 mole) of $CaCO_3$ decompose. **The plus sign shows that energy is taken in.**

Other examples of endothermic reactions

Reactions **D** and **E** above are **spontaneous**. They start off on their own. But many endothermic reactions are like **F**, where energy must be put in to start the reaction *and* keep it going. For example:

- reactions that take place in cooking
- photosynthesis. This is the process in which plants convert carbon dioxide and water to glucose. It depends on the energy from sunlight.

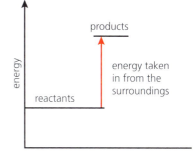

▲ An energy level diagram for an endothermic reaction. The products have higher energy than the reactants.

Remember!
Exo means *out* (think of Exit)
Endo means *in*

Q

1 Is it exothermic or endothermic?
   a the burning of a candle
   b the reaction between sodium and water
   c the change from raw egg to fried egg
2 Which unit is used to measure energy changes?

3 2Na (s) + Cl_2 (g) $\rightarrow$ 2NaCl (s)
   The energy change for this reaction is − 822.4 kJ.
   What can you conclude about the reaction?
4 Draw an energy level diagram for:
   a an endothermic reaction   **b** an exothermic reaction

9.3 Energy from fuels

What is a fuel?

A fuel is any substance we use to provide energy.
We convert the chemical energy in the fuel into another form of energy.
We burn most fuels, to obtain their energy in the form of heat.

The fossil fuels

The **fossil fuels** − **coal, petroleum (oil),** and **natural gas** (methane) −
are the main fuels used around the world. We burn them to release heat.

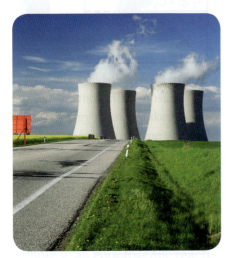

We burn fossil fuels in power
stations, to heat water to make
steam. A jet of steam drives the
turbines that generate electricity.

We burn them in factories to heat
furnaces, and in homes for cooking
and heating. (Kerosene, from
petroleum, is also used in lamps.)

Petrol and diesel (from petroleum)
are burned in engines, to give the
hot gas that moves the pistons.
These then make the wheels turn.

The world uses up enormous quantities of the fossil fuels. For example,
nearly 12 million tonnes of petroleum *every day*!

So what makes a good fuel?

These are the main questions to ask about a fuel:

- **How much heat does it give out?** We want as much heat as possible,
  per tonne of fuel.

- **Does it cause pollution?** If it causes a lot of pollution, we may be
  better off without it!

- **Is it easily available?** We need a steady and reliable supply.

- **Is it easy and safe to store and transport?** Most fuels catch fire quite
  easily, so safety is always an issue.

- **How much does it cost?** The cheaper the better.

The fossil fuels give out a lot of heat. But they cause pollution, with coal
the worst culprit. The pollutants include carbon dioxide, which is linked
to global warming, and other gases that cause acid rain. (See page 210.)

What about availability? We are using up the fossil fuels fast. Some
experts say we could run out of petroleum and gas within 50 years. But
there is probably enough coal to last several hundred years.

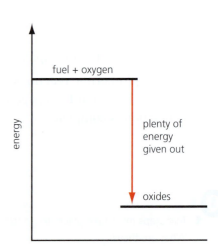

▲ The burning of fuel is an exothermic
reaction. The more heat given out the
better − as long as the fuel is safe to use.

Endothermic reactions

Endothermic reactions take in energy from their surroundings.
Here are three examples:

When barium hydroxide reacts with ammonium chloride, the temperature falls so sharply that water *under* the beaker will freeze!

Sherbet is citric acid plus the base sodium hydrogen carbonate. The neutralisation that occurs takes in heat – so your tongue cools.

The crucible contains calcium carbonate. If you keep on heating, it will all decompose to calcium oxide and carbon dioxide.

These reactions can be described as:

reactants + energy $\longrightarrow$ products

The energy is transferred from the surroundings: in **D** from the air and wet wood, in **E** from your tongue, and in **F** from the Bunsen burner. Since energy is taken in, the products must have higher energy than the reactants. This is shown on the energy level diagram on the right.

The energy change

For reaction **F** above:

$CaCO_3\ (s) \longrightarrow CaO\ (s) + CO_2\ (g)$ the energy change = + 178 kJ

So 178 kJ of energy is needed to make 100 g (or 1 mole) of $CaCO_3$ decompose. **The plus sign shows that energy is taken in.**

Other examples of endothermic reactions

Reactions **D** and **E** above are **spontaneous**. They start off on their own. But many endothermic reactions are like **F**, where energy must be put in to start the reaction *and* keep it going. For example:

- reactions that take place in cooking
- photosynthesis. This is the process in which plants convert carbon dioxide and water to glucose. It depends on the energy from sunlight.

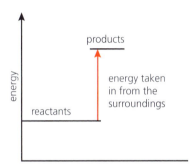

▲ An energy level diagram for an endothermic reaction. The products have higher energy than the reactants.

Remember!
Exo means **out** (think of Exit)
Endo means **in**

Q

1 Is it exothermic or endothermic?
  a the burning of a candle
  b the reaction between sodium and water
  c the change from raw egg to fried egg
2 Which unit is used to measure energy changes?

3 $2Na\ (s) + Cl_2\ (g) \longrightarrow 2NaCl\ (s)$
The energy change for this reaction is − 822.4 kJ.
What can you conclude about the reaction?
4 Draw an energy level diagram for:
  a an endothermic reaction **b** an exothermic reaction

9.2 Explaining energy changes

Making and breaking bonds

In a chemical reaction, bonds must first be broken. Then new bonds form. **Breaking bonds takes in energy. Making bonds releases energy.**

Example 1: an exothermic reaction

Hydrogen reacts with chlorine in sunshine, to form hydrogen chloride:

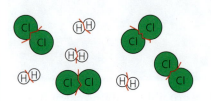 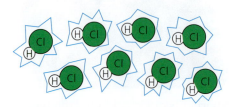

1 First, the bonds in the hydrogen and chlorine molecules must be broken. Energy must be taken in, for this. (Energy from sunshine will do!)

2 Now new bonds form between hydrogen and chlorine atoms, giving molecules of hydrogen chloride. This step releases energy.

But the energy taken in for step 1 is less than the energy given out in step 2. So this reaction gives out energy, overall. It is **exothermic**.

If the energy taken in to break bonds is *less than* the energy released in making bonds, the reaction is exothermic.

▲ Hydrogen burning in chlorine in the lab. Bonds break and new bonds form, giving hydrogen chloride.

Example 2: an endothermic reaction

If you heat ammonia strongly, it breaks down to nitrogen and hydrogen. Here we use lines to show the bonds. (Note the triple bond in nitrogen.)

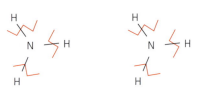 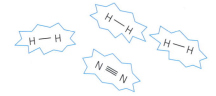

1 First, the bonds in ammonia must be broken. Energy must be taken in, for this. (You supply it by heating.)

2 Now the hydrogen atoms bond together. So do the nitrogen atoms. This releases energy.

This time, the energy taken in for step **1** is greater than the energy given out in step **2**. So the reaction takes in energy, overall. It is **endothermic**.

If the energy taken in to break bonds is *greater than* the energy released in making bonds, the reaction is endothermic.

Bond energies

The energy needed to make or break bonds is called the **bond energy** Look at the list on the right. 242 kJ must be supplied to break the bonds in a mole of chlorine molecules, to give chlorine atoms. If these atoms join again to form molecules, 242 kJ of energy are given out.

The bond energy is the energy needed to break bonds, or released when these bonds form. It is given in kJ/mole.

Bond energy (kJ / mole)	
H−H	436
Cl−Cl	242
H−Cl	431
C−C	346
C=C	612
C−O	358
C−H	413
O=O	498
O−H	464
N≡N	946
N−H	391

Calculating the energy changes in reactions

So let's calculate the energy change for those reactions on page 112.

1 The exothermic reaction between hydrogen and chlorine

$$H—H + Cl—Cl \longrightarrow 2\ H—Cl$$

Energy in to break each mole of bonds:

$1 \times$ H—H	436 kJ
$1 \times$ Cl—Cl	242 kJ
Total energy in	678 kJ

Energy out from the two moles of bonds forming:

$2 \times$ H—Cl	$2 \times 431 = 862$ kJ

Energy in − energy out = 678 kJ − 862 kJ = −184 kJ
So the reaction gives out **184 kJ** of energy, overall.
Its energy level diagram is shown on the right.

2 The endothermic decomposition of ammonia

$$2 \ \overset{\displaystyle H}{\underset{\displaystyle H}{N—H}} \longrightarrow N \equiv N + 3\ H—H$$

Energy in to break the two moles of bonds:

$6 \times$ N—H	$6 \times 391 = 2346$ kJ

Energy out from the four moles of bonds forming:

$1 \times$ N≡N	946 kJ
$3 \times$ H—H	$3 \times 436 = 1308$ kJ
Total energy out	2254 kJ

Energy in − energy out = 2346 kJ − 2254 kJ = +92 kJ
So the reaction takes in **92 kJ** of energy, overall.
Look at its energy level diagram.

Starting a reaction off

To start a reaction, bonds must be broken. As you saw, this needs energy.

- For some reactions, not much energy is needed. Just mix the reactants at room temperature. (For example, reactions **B** and **C** on page 110.)

- Some exothermic reactions need heat from a Bunsen burner just to start bonds breaking. Then the energy given out by the reaction breaks further bonds. (For example, reaction **A** on page 110.)

- But for endothermic reactions like the decomposition of calcium carbonate (reaction **F** on page 111), you must continue heating until the reaction is complete.

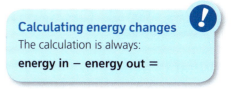
Calculating energy changes
The calculation is always:
energy in − energy out =

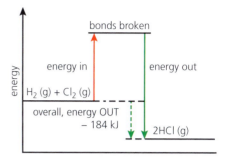

▲ For the hydrogen / chlorine reaction.

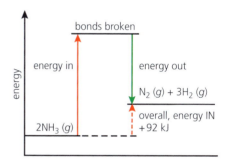

▲ For the decomposition of ammonia.

▲ One way to start bonds breaking!

Q

1 Two steps must take place, to go from reactants to products. What are they?

2 Some reactions are endothermic. Explain why, using the ideas of bond breaking and bond making.

3 Hydrogen reacts with oxygen. Draw the equation for the reaction as above, with lines to show the bonds.

4 Now see if you can calculate the energy change for the reaction in **3**, using the bond energy table on page 112.

9.3 Energy from fuels

What is a fuel?

A fuel is any substance we use to provide energy.
We convert the chemical energy in the fuel into another form of energy.
We burn most fuels, to obtain their energy in the form of heat.

The fossil fuels

The **fossil fuels** − **coal, petroleum (oil),** and **natural gas** (methane) −
are the main fuels used around the world. We burn them to release heat.

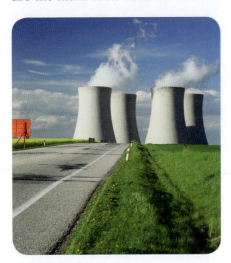

We burn fossil fuels in power stations, to heat water to make steam. A jet of steam drives the turbines that generate electricity.

We burn them in factories to heat furnaces, and in homes for cooking and heating. (Kerosene, from petroleum, is also used in lamps.)

Petrol and diesel (from petroleum) are burned in engines, to give the hot gas that moves the pistons. These then make the wheels turn.

The world uses up enormous quantities of the fossil fuels. For example, nearly 12 million tonnes of petroleum *every day*!

So what makes a good fuel?

These are the main questions to ask about a fuel:

- **How much heat does it give out?** We want as much heat as possible, per tonne of fuel.

- **Does it cause pollution?** If it causes a lot of pollution, we may be better off without it!

- **Is it easily available?** We need a steady and reliable supply.

- **Is it easy and safe to store and transport?** Most fuels catch fire quite easily, so safety is always an issue.

- **How much does it cost?** The cheaper the better.

The fossil fuels give out a lot of heat. But they cause pollution, with coal the worst culprit. The pollutants include carbon dioxide, which is linked to global warming, and other gases that cause acid rain. (See page 210.)

What about availability? We are using up the fossil fuels fast. Some experts say we could run out of petroleum and gas within 50 years. But there is probably enough coal to last several hundred years.

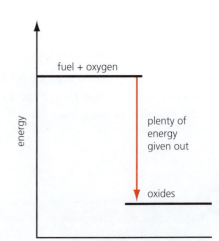

▲ The burning of fuel is an exothermic reaction. The more heat given out the better − as long as the fuel is safe to use.

Two fuels growing in importance

Because of fears about global warming, there is a push to use new fuels. Like these two:

Ethanol This is an alcohol, with the formula C_2H_5OH. It can be made from any plant material. For example, it is made from sugar cane in Brazil, and from corn (maize) in the USA. It is used in car engines, on its own or mixed with petrol. See pages 252−253 for more.

Hydrogen This gas burns explosively in oxygen, giving out a lot of energy − so it is used to fuel space rockets. It is also used in **fuel cells** (without burning) to give energy in the form of electricity. See page 117 for more.

▲ Filling up with a mixture of 85% ethanol, 15% gasoline.

Different amounts of heat

Some fuels give out a lot more heat than others. Compare these:

Fuel	Equation for burning in oxygen	Heat given out per gram of fuel/kJ
natural gas (methane)	$CH_4\,(g) + 2O_2\,(g) \rightarrow CO_2\,(g) + 2H_2O\,(l)$	−55
ethanol	$C_2H_5OH\,(l) + 3O_2\,(g) \rightarrow 2CO_2\,(g) + 3H_2O\,(l)$	−86
hydrogen	$2H_2\,(g) + O_2\,(g) \rightarrow 2H_2O\,(g)$	−143

Nuclear fuels

Nuclear fuels are *not* burned. They contain unstable atoms called **radioisotopes** (page 30). Over time, these break down naturally into new atoms, giving out radiation and a lot of energy.

But you can also *force* radioisotopes to break down, by shooting neutrons at them. That is what happens in a **nuclear power station**. The energy given out is used to heat water, to make steam. Jets of steam are then used to drive the turbines for generating electricity.

The radioisotope uranium-235 is often used. When it decays, the new atoms that form are also unstable, and break down further.

Nuclear fuel has two big advantages:

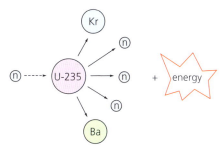
▲ When hit by a neutron, a U-235 atom breaks down to other atoms, giving out a huge amount of energy.

- It gives out huge amounts of energy. A pellet of nuclear fuel the size of a pea can give as much energy as a tonne of coal.

- No carbon dioxide or other polluting gases are formed.

But it is not all good news. An explosion in a nuclear power station could spread radioactive material over a huge area, carried in the wind. The waste material produced in a nuclear power station is also radioactive, and may remain very dangerous for hundreds of years. Finding a place to store it safely is a major problem.

▲ The radiation hazard warning sign.

1 a Sketch an energy level diagram that you think shows:
  i a good fuel   **ii** a very poor fuel
  b What else do you need to think about, to decide whether a substance would make a good fuel?

2 Look at the table above. From *all* the information given, which of the three fuels do you think is best? Explain.

3 The fuel butane (C_4H_{10}) burns to give the same products as methane. Write a balanced equation for its combustion.

9.4 Giving out energy as electricity

Electricity: a form of energy

Electricity is a current of electrons. Like heat, it is a form of energy.
When you burn a fuel, chemical energy is converted to heat.
But a reaction can also give out energy as electricity.

Electricity from a redox reaction

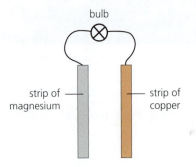

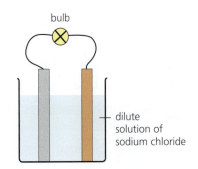

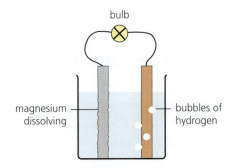

Connect a strip of magnesium, a strip of copper, and a light bulb, like this. (Note: no battery!) Nothing happens.

Now stand the strips in a dilute solution of sodium chloride. Something amazing happens: the bulb lights! A current is flowing.

At the same time bubbles of hydrogen start to form on the copper strip, and the magnesium strip begins dissolving.

So what is going on?

1 Magnesium is more reactive than copper. (See the list on the right.) That means it has a stronger drive to form ions. So the magnesium atoms give up electrons, and go into solution as ions:

$$Mg\,(s) \longrightarrow Mg^{2+}\,(aq) + 2e^-$$ (magnesium is oxidised)

2 The electrons flow along the wire to the copper strip, as a current.

3 The solution contains Na^+ and Cl^- ions from sodium chloride, and some H^+ and OH^- ions from water. Hydrogen is less reactive than sodium, so the H^+ ions accept electrons from the copper strip:

$$2H^+\,(aq) + 2e^- \longrightarrow H_2\,(g)$$ (hydrogen ions are reduced)

So a redox reaction is giving out energy in the form of a current.

A simple cell

The metal strips, wire, and beaker of solution above form a **simple cell**. Electrons flow from the magnesium strip, so it is called the **negative pole**. The copper strip is the **positive pole**. The solution is the **electrolyte**.

A simple cell consists of two metals and an electrolyte. The more reactive metal is the negative pole of the cell. Electrons flow from it.

Other metals can also be used, as long as they differ in reactivity. And any solution can be used, as long as it contains ions.

You could connect a voltmeter into the circuit, to measure the voltage. The bigger the difference in reactivity of the metals, the larger the voltage, and the more brightly the bulb will light. Find out more on page 186.

Order of reactivity

This shows the order of reactivity of some metals compared to hydrogen:

potassium
sodium
calcium
magnesium
aluminium
zinc
iron
hydrogen
copper
silver

increasing reactivity

Remember!

• In electrolysis, a current brings about a reaction.
• In simple cells, reactions produce a current.

The hydrogen fuel cell

In the hydrogen fuel cell, hydrogen and oxygen combine without burning. It is a redox reaction. The energy is given out as an electric current.

Like the simple cell, the fuel cell has a negative pole that gives out electrons, a positive pole that accepts them, and an electrolyte.

Both poles are made of carbon.

The negative pole is surrounded by hydrogen, and the positive pole by oxygen (in air). The electrolyte contains OH^- ions.

(You won't be asked to draw a fuel cell.)

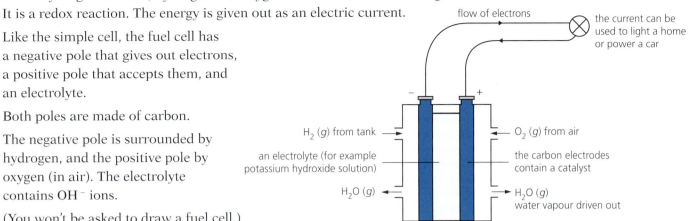

flow of electrons

the current can be used to light a home or power a car

H_2 (g) from tank

O_2 (g) from air

an electrolyte (for example potassium hydroxide solution)

the carbon electrodes contain a catalyst

H_2O (g)

H_2O (g)
water vapour driven out

At the negative pole	At the positive pole
Hydrogen loses electrons to the OH^- ions. It is oxidised: $2H_2\ (g) + 4OH^-\ (aq) \rightarrow 4H_2O\ (l) + 4e^-$ A current of electrons flows through the wire to the positive pole. You can make use of it on the way. For example, pass it through light bulbs to light your home.	The electrons are accepted by oxygen molecules. Oxygen is reduced to OH^- ions: $O_2\ (g) + 2H_2O\ (l) + 4e^- \rightarrow 4OH^-\ (aq)$ But the concentration of OH^- ions in the electrolyte does not increase. Why not?

Adding the two half-equations gives the full equation for the redox reaction:

$$2H_2\ (g) + O_2\ (g) \rightarrow 2H_2O\ (l)$$

So the overall reaction is that hydrogen and oxygen combine to form water.

Advantages of the hydrogen fuel cell

- Only water is formed. No pollutants!
- The reaction gives out plenty of energy. 1 kg of hydrogen gives about 2.5 times as much energy as 1 kg of natural gas (methane).
- We will not run out of hydrogen. It can be made by the electrolysis of water with a little acid added. Solar power could provide cheap electricity for this. Certain algae also produce hydrogen. And scientists have shown that it can be made from waste plant material, using bacteria.

But there is a drawback. Hydrogen is very flammable. A spark or lit match will cause a mixture of hydrogen and air to explode. So it must be stored safely.

▲ This car has a hydrogen fuel cell instead of a petrol engine.

 Q

1 Can you get electricity from a non-redox reaction? Explain.
2 In a simple cell, which metal gives up electrons to produce the current: the more reactive or less reactive one?
3 A wire connects strips of magnesium and copper, standing in an electrolyte. Bubbles appear at the copper strip. Why?

4 You connect two strips of iron using wire, and stand them in an electrolyte. Will a current flow? Explain your answer.
5 a In the hydrogen fuel cell, what is the *fuel*?
 b How are the electrons transferred in this cell?
 c What is the purpose of the electrolyte?

The batteries in your life

Batteries and you

We depend a lot on batteries. Cars and buses will not start without them. Torches need them. So do mobile phones, laptops, cameras, iPods …

The diagram on the right shows a simple model of a battery (or cell). All batteries contain two solid substances of different reactivity, and an electrolyte. The more reactive substance gives up electrons more readily. These flow out of the battery as an electric current.

Since the more reactive substance provides the electrons, it is called the **negative pole**, or **negative electrode**, or **negative terminal**.

The simple cell, shown on page 116, is the simplest battery of all. But it is not very practical. You could not use it in a torch, for example, and it does not have enough voltage to start a car. You need other types of battery.

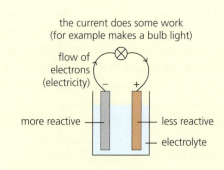

the current does some work
(for example makes a bulb light)

flow of electrons (electricity)

more reactive

less reactive

electrolyte

A torch battery

Torch batteries are 'dry', and easy to carry around:

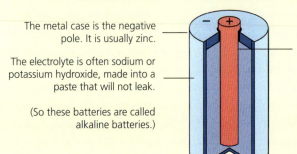

The metal case is the negative pole. It is usually zinc.

The electrolyte is often sodium or potassium hydroxide, made into a paste that will not leak.

(So these batteries are called alkaline batteries.)

The positive pole is down the middle.

It is often manganese(IV) oxide, packed around a carbon rod. The Mn^{4+} ions accept electrons to become Mn^{3+} ions.

The battery 'dies' when the reactions stop.

▲ Gotcha! Thanks to redox reactions in the torch battery.

A car battery

A car battery consists of plates of lead and lead(IV) oxide, standing in a solution of sulfuric acid, as shown on the right. This is what happens:

1 The lead plate reacts with the sulfuric acid, giving lead(II) sulfate:

$$Pb\ (s) + H_2SO_4\ (aq) \longrightarrow PbSO_4\ (s) + 2H^+ + 2e^-$$

The lead(II) sulfate coats the plate.

2 The electrons go off through the wire as an electric current. It gets the car's starter motor working.

3 The electrons flow back through the wire to the lead(IV) oxide plate. This also reacts with the acid to form lead(II) sulfate, which coats the plate:

$$PbO_2\ (s) + H_2SO_4\ (aq) + 2H^+ + 2e^- \longrightarrow$$
$$PbSO_4\ (s) + 2H_2O\ (l)$$

In fact the car battery usually has six sets of plates linked together, giving a total voltage of 12 volts.

car's starter motor

electron flow

lead

lead(IV) oxide

sulfuric acid

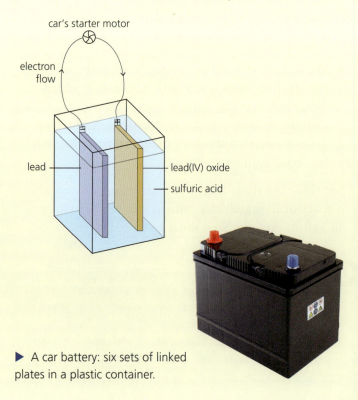

▶ A car battery: six sets of linked plates in a plastic container.

Recharging the car battery

While the car battery is running, the plates are being coated with lead(II) sulfate, and the sulfuric acid is being used up. So if it runs for long enough, the battery will stop working, or 'go flat'.

But it needs to run for only a short time, to start the car. And then something clever happens: electricity generated by the motor causes the reactions to reverse. The lead(II) sulfate on the plates is converted back to lead and lead(IV) oxide, ready for next time.

▲ Meanwhile, the battery is recharging.

A button battery

You probably have a button battery in your watch. Button batteries often use lithium as the negative terminal. Here is a cross-section through one:

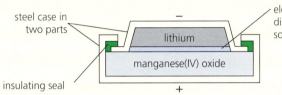

steel case in two parts

lithium

manganese(IV) oxide

electrolyte (a lithium salt dissolved in an organic solvent)

insulating seal

Lithium is a good choice because it is highly reactive: it gives up electrons easily. These flow out through the top of the steel case, to the connection in your watch. They flow back through the lower part of the case, and Mn^{4+} ions accept them, to become Mn^{3+} ions.

▲ Button batteries come in different sizes, for different uses.

A lithium-ion battery

Lithium-ion batteries are **rechargeable**. So they are used in laptops, mobile phones, and iPods.

The battery consists of thin sheets of lithium cobalt oxide ($LiCoO_2$), and graphite (carbon). The electrolyte is a solution of a lithium salt in an organic solvent. This is how the battery works:

When it is charging	When you use it
flow of current — charger — lithium cobalt oxide / graphite — Li⁺ ions	flow of current — phone — lithium cobalt oxide / graphite — Li⁺ ions
When your phone is charging, the graphite becomes negative, and attracts lithium ions from the lithium cobalt oxide.	When you use it, the lithium ions flow back to lithium cobalt oxide, and electrons flow from the graphite to power your phone.

▲ Lithium-ion batteries.

So your calls and texts depends on those lithium ions moving. Remember that, next time you use your mobile!

▲ Keeping in touch, via lithium ions.

9.5 Reversible reactions

When you heat copper(II) sulfate crystals …

The blue crystals above are *hydrated* copper(II) sulfate. On heating, they turn to a white powder. This is *anhydrous* copper(II) sulfate:

$CuSO_4.5H_2O$ (s) →
$CuSO_4$ (s) + $5H_2O$ (l)

The reaction is easy to reverse: add water! The anhydrous copper(II) sulfate gets hot and turns blue. The reaction is:

$CuSO_4$ (s) + $5H_2O$ (l) →
$CuSO_4.5H_2O$ (s)

> **Water of crystallisation** !
> - The water in blue copper(II) sulfate crystals is called **water of crystallisation.**
> - *Hydrated* means it has water molecules built into its structure.
> - *Anhydrous* means no water is present.

So the reaction can go in either direction: it is **reversible**.
The reaction we start with (**1** above) is called the **forward** reaction.
Reaction **2** is the **back** reaction.

We use the symbol ⇌ instead of a single arrow, to show that a reaction is reversible. So the equation for the reaction above is:

$CuSO_4.5H_2O$ (s) ⇌ $CuSO_4$ (s) + $5H_2O$ (l)

What about the energy change?

Reaction **1** above requires heat − it is **endothermic**. In **2**, the white powder gets hot and spits when you drip water on it − so that reaction is **exothermic**. It gives out *the same amount of heat* as reaction **1** took in.

A reversible reaction is endothermic in one direction, and exothermic in the other. The same amount of energy is transferred each time.

> **Two tests for water** !
> Water will turn:
> - white anhydrous copper(II) sulfate blue
> - blue cobalt(II) chloride paper pink.
>
> Both compounds add on water of crystallisation, giving the colour change. To reverse, just heat!

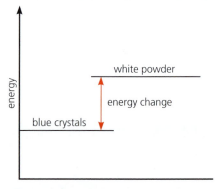

▲ In a reversible reaction, the energy change is the same in both directions.

Some important reversible reactions

Many important reactions are reversible. Here are some examples:

Reaction	Comments
N_2 (g) + $3H_2$ (g) ⇌ $2NH_3$ (g) nitrogen hydrogen ammonia	This is a very important reaction, because ammonia is used to make nitric acid and fertilisers.
$2SO_2$ (g) + O_2 (g) ⇌ $2SO_3$ (g) sulfur dioxide oxygen sulfur trioxide	This is a key step in the manufacture of sulfuric acid.
$CaCO_3$ (s) ⇌ CaO (s) + CO_2 (g) calcium carbonate calcium oxide carbon dioxide	This is a **thermal decomposition**: it needs heat. Calcium oxide (called lime, or quicklime) has many uses (page 236).

Reversible reactions and equilibrium

As you saw in the last table, the reaction between nitrogen and hydrogen to make ammonia is **reversible**:

$$N_2\,(g) + 3H_2\,(g) \rightleftharpoons 2NH_3\,(g)$$

So let's see what happens during the reaction:

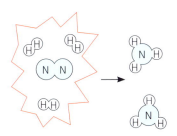 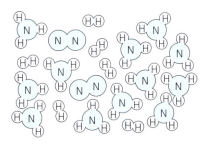 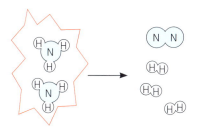

Three molecules of hydrogen react with one of nitrogen to form two of ammonia. So if you put the correct mixture of nitrogen and hydrogen into a closed container …

… will it all turn into ammonia? No! Once a certain amount of ammonia is formed, the system reaches a state of **dynamic equilibrium**. From then on …

… every time two ammonia molecules form, another two break down into nitrogen and hydrogen. So the level of ammonia remains unchanged.

Equilibrium means there is *no overall change*. The amounts of nitrogen, hydrogen, and ammonia remain steady. But **dynamic** means there is continual change: ammonia molecules continually break down, while new ones form.

In a closed system, a reversible reaction reaches a state of dynamic equilibrium, where the forward and back reactions take place at the same rate. So there is no overall change.

The term **dynamic equilibrium** is usually shortened to **equilibrium**.

A challenge for industry

Imagine you run a factory that makes ammonia. You want the yield of ammonia to be as high as possible.

But the reaction between nitrogen and hydrogen is *never complete*. Once equilibium is reached, a molecule of ammonia breaks down every time a new one forms.

This is a problem. What can you do to increase the yield of ammonia? You will find out in the next unit.

▲ Worried about the yield?

Q

1 What is a *reversible* reaction?

2 Write a word equation for the reaction between solid copper(II) sulfate and water.

3 How would you turn hydrated copper(II) sulfate into anhydrous copper(II) sulfate?

4 What will you observe if you place pink cobalt(II) chloride paper in a warm oven?

5 Explain the term *dynamic equilibrium*.

6 Nitrogen and hydrogen are mixed, to make ammonia.
  a Soon, two reactions are going on in the mixture. Give the equations for them.
  b For a time, the rate of the forward reaction is greater than the rate of the back reaction. Has equilibrium been reached? Explain.

9.6 Shifting the equilibrium

The challenge

Reversible reactions present a challenge to industry, because they *never complete*. Let's look at that reaction between nitrogen and hydrogen again:

$$N_2 (g) + 3H_2 (g) \rightleftharpoons 2NH_3 (g)$$

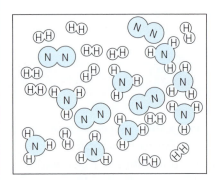

This represents the reaction mixture at equilibrium. The amount of ammonia in it will not increase …

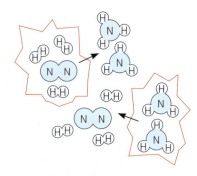

… because every time a new molecule of ammonia forms, another breaks down.

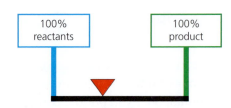

Here the red triangle represents the equilibrium mixture. It is only part way along the scale. Why?

What can be done?

You want as much ammonia as possible. So how can you increase the yield? This idea, called Le Chatelier's principle, will help you:

When a reversible reaction is in equilibrium and you make a change, the system acts to oppose the change, and restore equilibrium. A new equilibrium mixture forms.

A reversible reaction *always* reaches equilibrium, in a closed system. But by changing conditions, you can *shift equilibrium, so* that the mixture contains more product. Let's look at four changes you could make.

▲ A change in reaction conditions has led to a new equilibrium mixture, with more ammonia. *Equilibrium has shifted to the right, to favour the product.*

1 Change the temperature

Will raising the temperature help you obtain more ammonia? Let's see.

$$N_2 + 3H_2 \xrightarrow{\text{heat out}} 2NH_3$$
$$2NH_3 \xrightarrow{\text{heat in}} N_2 + 3H_2$$

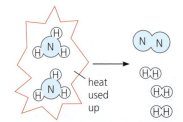

heat used up

The forward reaction is exothermic – it gives out heat. The back reaction is endothermic – it takes it in. Heating speeds up *both* reactions …

… but if you heat the equilibrium mixture, it acts to oppose the change. More ammonia breaks down in order to use up the heat you add.

So the reaction reaches equilibrium faster – but the new equilibrium mixture has *less* ammonia. So you are worse off than before.

What if you lower the temperature? The system acts to oppose the change: more ammonia forms, giving out heat. Great! But if the temperature is *too* low, the reaction takes too long to reach equilibrium. Time is money, in a factory. So it is best to choose a moderate temperature.

2 Change the pressure

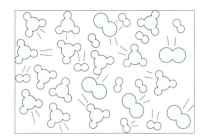 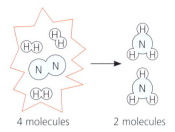

4 molecules   2 molecules

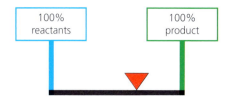

Pressure is caused by the gas molecules colliding with the walls of the container. *The more molecules present, the higher the pressure.*

When you increase the pressure, the equilibrium mixture acts to oppose this. More ammonia forms, so there are fewer molecules.

So the amount of ammonia in the mixture has *increased*. Equilibrium has shifted to the right. Well done. You are on the right track.

3 Remove the ammonia

The equilibrium mixture is a balance between nitrogen, hydrogen, and ammonia. Suppose you cool the mixture. Ammonia condenses first, so you can run it off as a liquid. Then warm the remaining nitrogen and hydrogen again. More ammonia will form, to restore the balance.

4 Add a catalyst

Iron is a catalyst for this reaction.
A catalyst speeds up the forward and back reactions equally.
So the reaction reaches equilibrium faster, which saves you time. But *the amount of ammonia does not change*.

Choosing the optimum conditions

So to get the best yield of ammonia, it is best to:

- use high pressure, and remove ammonia, to improve the yield
- use a moderate temperature, and a catalyst, to get a decent rate.

Page 223 shows how these ideas are applied in an ammonia factory.

A note about rate

By now, you should realise that:

- a change in temperature *always* shifts equilibrium: it changes the yield.
- a change in pressure will shift equilibrium *only if* the number of molecules is different on each side of the equation.

But how do these changes affect the *rate*? Raising the temperature or pressure increases the rate of *both the forward and back reactions*, so equilibrium is reached faster. (A temperature rise gives the molecules more energy. An increase in pressure forces them closer. So in both cases, the number of successful collisions increases.)

A summary, for reversible reactions of gases

- Forward reaction *exothermic*: temperature ↑ means yield ↓.
- Forward reaction *endothermic*: temperature ↑ means yield ↑.
- *Fewer molecules* on the right-hand side of the equation: pressure ↑ means yield ↑.

What about solutions?

Many reversible reactions take place in solution:

reactants (*aq*) ⇌ products (*aq*)

You can shift the equilibrium:

- by adding more of a reactant (increasing its concentration). So more product will form to oppose this change.
- by changing the temperature. A rise in temperature will favour the endothermic reaction.

Q

1 The reaction between nitrogen and hydrogen is *reversible*. This causes a problem for the ammonia factory. Why?

2 What is Le Chatelier's principle? Write it down.

3 In manufacturing ammonia, explain why:
   a high pressure is used   b ammonia is removed

4 Sulfur dioxide (SO_2) and oxygen react exothermically to form sulfur trioxide (SO_3). The reaction is reversible.
   a Write the symbol equation for this reaction.
   b What happens to the yield of sulfur trioxide if you:
      i increase the pressure?   ii raise the temperature?

Checkup on Chapter 9

Revision checklist

Core syllabus content

Make sure you can …

☐ explain what these terms mean:
   exothermic reaction   *endothermic reaction*
☐ give examples of exothermic and endothermic reactions and draw energy level diagrams for them
☐ state the unit used for measuring energy
☐ say what the + and − signs mean, in energy values
☐ explain what the purpose of a fuel is
☐ name the *fossil fuels*, and say how we use them
☐ explain what *nuclear fuels* are, and where we use them, and name one
☐ give advantages and disadvantages of nuclear fuels
☐ say how hydrogen and ethanol are used as fuels
☐ explain what a *reversible reaction* is, with examples
☐ write the symbol for a reversible reaction
☐ describe how to change hydrated copper(II) sulfate to the anhydrous compound, and back again
☐ explain why anhydrous copper(II) sulfate can be used to test for the presence of water

Extended syllabus content

Make sure you can also …

☐ use the idea of bond making and bond breaking to explain why a reaction is exo- or endothermic
☐ define *bond energy*
☐ calculate the energy change in a reaction, given the equation, and bond energy values
☐ describe a *simple cell*, and explain that the current comes from a redox reaction
☐ predict which metal will be the negative pole in a simple cell
☐ give half-equations for reactions that take place in a simple cell (like the one on page 116)
☐ describe the hydrogen fuel cell, and give the overall reaction that takes place in it
☐ explain that a reversible reaction never completes, in a closed container – it reaches equilibrium
☐ give ways to obtain more product in a gaseous reversible reaction
☐ predict the effect of a change in temperature and pressure, for a given reversible reaction
☐ say how a catalyst will affect a reversible reaction
☐ predict the effect of a change in conditions for a reversible reaction in solution

Questions

Core syllabus content

1 Look at this reaction:
   $NaOH\ (aq) + HCl\ (aq) \longrightarrow NaCl\ (aq) + H_2O\ (l)$
   a Which type of reaction is it?
   b It is *exothermic*. What does that mean?
   c What will happen to the temperature of the solution, as the chemicals react?
   d Draw an energy diagram for the reaction.

2 Water at 25 °C was used to dissolve two compounds. The temperature of each solution was measured immediately afterwards.

Compound	Temperature of solution/°C
ammonium nitrate	21
calcium chloride	45

   a List the apparatus needed for this experiment.
   b Calculate the temperature change on dissolving each compound.
   c i   Which compound dissolved exothermically?
      ii How did you decide this?
      iii What can you say about the energy level of its ions in the solution, compared with in the solid compound?
   d For each solution, estimate the temperature of the solution if:
      i   the amount of water is halved, but the same mass of compound is used
      ii the mass of the compound is halved, but the volume of water is unchanged
      iii both the mass of the compound, and the volume of water, are halved.

3 Hydrated copper(II) sulfate crystals were heated:

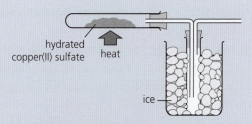

   a What is the ice for?
   b What colour change will occur in the test-tube?
   c The reaction is *reversible*. What does that mean?
   d How would you show that it is reversible?
   e Write the equation for the reversible reaction.

Extended syllabus content

4 The fuel *natural* gas is mostly methane.
Its combusion in oxygen is exothermic:

$$CH_4(g) + 2O_2(g) \longrightarrow CO_2(g) + 2H_2O(l)$$

a Explain *why* this reaction is exothermic, in terms of bond breaking and bond making.

b i Copy and complete this energy diagram for the reaction, indicating:

    A the overall energy change

    B the energy needed to break bonds

    C the energy given out by new bonds forming.

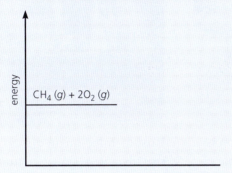

  ii Methane will not burn in air until a spark or flame is applied. Why not?

c When 1 mole of methane burns in oxygen, the energy change is -890 kJ.

  i What does the $-$ sign tell you?

  ii Which word describes a reaction with this type of energy change?

d How much energy is given out when 1 gram of methane burns? (A_r: C = 12, H = 1.)

5 Strips of copper foil and magnesium ribbon were cleaned with sandpaper and then connected as shown below. The bulb lit up.

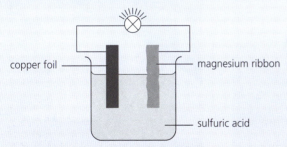

a Why were the metals cleaned?

b Name the electrolyte used.

c Explain why the bulb lit up.

d Which metal releases electrons into the circuit?

e In this arrangement, energy is being changed from one form to another. Explain.

f What is this type of arrangement called?

g Give reasons why the set-up shown above would not be used as a torch battery.

6 The gas hydrazine, N_2H_4, burns in oxygen like this:

$$\underset{H}{\overset{H}{N}}-\underset{H}{\overset{H}{N}}(g) + O=O(g) \longrightarrow N\equiv N(g) + 2O\overset{H}{\underset{}{}}(g)$$

a Count and list the bonds broken in this reaction.

b Count and list the new bonds formed.

c Calculate the total energy:

  i required to break the bonds

  ii released when the new bonds form.

  (The bond energies in kJ/mole are: N−H 391; N−N 158; N≡N 945; O−H 464; O=O 498.)

d Calculate the energy change in the reaction.

e Is the reaction exothermic, or endothermic?

f Where is energy transferred from, and to?

g Comment on the suitability of hydrazine as a fuel.

7 Hydrogen and bromine react reversibly:

$$H_2(g) + Br_2(g) \rightleftharpoons 2HBr(g)$$

a Which of these will favour the formation of more hydrogen bromide?

  i add more hydrogen   **ii** remove bromine

  iii remove the hydrogen bromide as it forms

b Explain why increasing the pressure will have no effect on the amount of product formed.

c However, the pressure *is* likely to be increased, when the reaction is carried out in industry. Suggest a reason for this.

8 Ammonia is made from nitrogen and hydrogen. The energy change in the reaction is -92 kJ/mole. The reaction is reversible, and reaches equilibrium.

a Write the equation for the reaction.

b Is the forward reaction endothermic, or exothermic? Give your evidence.

c Explain why the *yield* of ammonia:

  i rises if you increase the pressure

  ii falls if you increase the temperature

d What effect does increasing:

  i the pressure   **ii** the temperature

  have on the *rate* at which ammonia is made?

e Why is the reaction carried out at 450 °C rather than at a lower temperature?

9 The dichromate ion $Cr_2O_7{}^{2-}$ and chromate ion $CrO_4{}^{2-}$ exist in equilibrium, like this:

$$Cr_2O_7{}^{2-}(aq) + H_2O(l) \rightleftharpoons 2CrO_4{}^{2-}(aq) + 2H^+(aq)$$

orange   yellow

a What would you see if you added dilute acid to a solution containing chromate ions?

b How would you reverse the change?

10.1 Rates of reaction

Fast and slow

Some reactions are **fast** and some are **slow**. Look at these examples:

The precipitation of silver chloride, when you mix solutions of silver nitrate and sodium chloride. This is a very fast reaction.

Concrete setting. This reaction is quite slow. It will take a couple of days for the concrete to fully harden.

Rust forming on an old car. This is usually a very slow reaction. It will take years for the car to rust completely away.

But it is not always enough to know just that a reaction is fast or slow. In factories where they make products from chemicals, they need to know *exactly* how fast a reaction is going, and how long it will take to complete. In other words, they need to know the **rate** of the reaction.

What is rate?

Rate is a measure of how fast or slow something is. Here are some examples.

This plane has just flown 800 kilometres in 1 hour. It flew at a **rate** of 800 km per hour.

This petrol pump can pump out petrol at a **rate** of 50 litres per minute.

This machine can print newspapers at a **rate** of 10 copies per second.

From these examples you can see that:

Rate is a measure of the change that happens in a single unit of time.

Any suitable unit of time can be used – a second, a minute, an hour, even a day.

Rate of a chemical reaction

When zinc is added to dilute sulfuric acid, they react together. The zinc disappears slowly, and a gas bubbles off.

As time goes by, the gas bubbles off more and more slowly. This is a sign that the reaction is slowing down.

Finally, no more bubbles appear. The reaction is over, because all the acid has been used up. Some zinc remains behind.

The gas that bubbles off is hydrogen. The equation for the reaction is:

zinc + sulfuric acid $\longrightarrow$ zinc sulfate + hydrogen

$$Zn\ (s) +\ H_2SO_4\ (aq)\ \longrightarrow ZnSO_4\ (aq) +\ H_2\ (g)$$

Both zinc and sulfuric acid get used up in the reaction. At the same time, zinc sulfate and hydrogen form.

You could measure the rate of the reaction, by measuring:

- the amount of zinc used up per minute *or*
- the amount of sulfuric acid used up per minute *or*
- the amount of zinc sulfate produced per minute *or*
- the amount of hydrogen produced per minute.

For this reaction, it is easiest to measure the amount of hydrogen produced per minute, since it is the only gas that forms. It can be collected as it bubbles off, and its volume can be measured.

In general, to find the rate of a reaction, you should measure:
the amount of a reactant used up per unit of time *or*
the amount of a product produced per unit of time.

Q

1 Here are some reactions that take place in the home. Put them in order of decreasing rate (the fastest one first).
  a raw egg changing to hard-boiled egg
  b fruit going rotten
  c cooking gas burning
  d bread baking
  e a metal tin rusting

2 Which of these rates of travel is slowest?
  5 kilometres per second
  20 kilometres per minute
  60 kilometres per hour

3 Suppose you had to measure the rate at which zinc is used up in the reaction above. Which of these units would be suitable? Explain your choice.
  a litres per minute
  b grams per minute
  c centimetres per minute

4 Iron reacts with sulfuric acid like this:

$$Fe\ (s) + H_2SO_4\ (aq) \rightarrow FeSO_4\ (aq) + H_2\ (g)$$

  a Write a word equation for this reaction.
  b Write down four different ways in which the rate of the reaction could be measured.

10.2 Measuring the rate of a reaction

A reaction that produces a gas

The rate of a reaction is found by measuring the amount of a **reactant** used up per unit of time, or the amount of a **product** produced per unit of time. Look at this reaction:

magnesium + hydrochloric acid $\longrightarrow$ magnesium chloride + hydrogen

$$Mg\ (s)\quad +\quad 2HCl\ (aq)\quad \longrightarrow\quad MgCl_2\ (aq)\quad +\quad H_2\ (g)$$

Here hydrogen is the easiest substance to measure, because it is the only gas in the reaction. It bubbles off and can be collected in a **gas syringe**, where its volume is measured.

The experiment

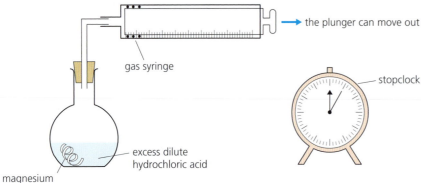

the plunger can move out

gas syringe

stopclock

excess dilute hydrochloric acid

magnesium

▲ Testing an explosive substance. The rate of a fast reaction like this, giving a mix of gases, is not easy to measure.

Clean the magnesium with sandpaper. Put dilute hydrochloric acid in the flask. Drop the magnesium into the flask, and insert the stopper and syringe immediately. Start the clock at the same time.

Hydrogen begins to bubble off. It rises up the flask and into the gas syringe, pushing the plunger out:

   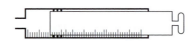

At the start, no gas has yet been produced or collected. So the plunger is all the way in.

Now the plunger has been pushed out to the 20 cm³ mark. 20 cm³ of gas have been collected.

The volume of gas in the syringe is noted at intervals – for example every half a minute. How will you know when the reaction is complete?

Typical results

Time / minutes	0	$\frac{1}{2}$	1	$1\frac{1}{2}$	2	$2\frac{1}{2}$	3	$3\frac{1}{2}$	4	$4\frac{1}{2}$	5	$5\frac{1}{2}$	6	$6\frac{1}{2}$
Volume of hydrogen / cm³	0	8	14	20	25	29	33	36	38	39	40	40	40	40

This table shows some typical results for the experiment.

You can tell quite a lot from this table. For example, you can see that the reaction lasted about five minutes. But a graph of the results is even more helpful. The graph is shown on the next page.

A graph of the results

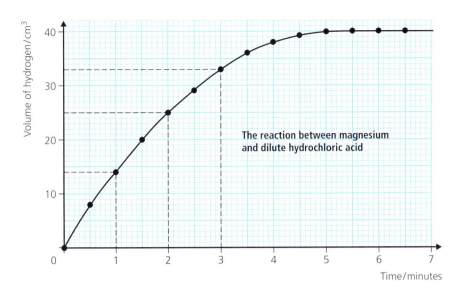

The reaction between magnesium and dilute hydrochloric acid

Notice these things about the results:

1 In the first minute, 14 cm³ of hydrogen are produced.
So the rate for the first minute is 14 cm³ of hydrogen per minute.
In the second minute, only 11 cm³ are produced. (25 − 14 = 11)
So the rate for the second minute is 11 cm³ of hydrogen per minute.
The rate for the third minute is 8 cm³ of hydrogen per minute.
So the rate decreases as time goes on.
The rate changes all through the reaction. It is greatest at the start, but decreases as the reaction proceeds.

2 The reaction is fastest in the first minute, and the curve is steepest then. It gets less steep as the reaction gets slower.
The faster the reaction, the steeper the curve.

3 After 5 minutes, no more hydrogen is produced, so the volume no longer changes. The reaction is over, and the curve goes flat.
When the reaction is over, the curve goes flat.

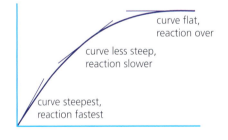

4 Altogether, 40 cm³ of hydrogen are produced in 5 minutes.

$$\text{The } average \text{ rate for the reaction} = \frac{\text{total volume of hydrogen}}{\text{total time for the reaction}}$$

$$= \frac{40 \text{ cm}^3}{5 \text{ minutes}}$$

$$= \textbf{8 cm}^3 \textbf{ of hydrogen per minute.}$$

Note that this method can be used for *any* reaction where one product is a gas.

1 For the experiment in this unit, explain why:
  a the magnesium ribbon is cleaned first
  b the clock is started the moment the reactants are mixed
  c the stopper is replaced immediately
2 From the graph at the top of this page, how can you tell when the reaction is over?

3 Look again at the graph at the top of the page.
  a How much hydrogen is produced in the first:
    i 2.5 minutes?   **ii** 4.5 minutes?
  b How long did it take to get 20 cm³ of hydrogen?
  c What is the rate of the reaction during:
    i the fourth minute?   **ii** the sixth minute?

10.3 Changing the rate of a reaction (part I)

Ways to change the rate of a reaction

There are several ways to speed up or slow down a reaction. For example you could change the concentration of a reactant, or the temperature. The *rate* will change – but the *amount of product* you obtain will not.

1 By changing concentration

Here you will see how rate changes with the **concentration** of a reactant.

The method   Repeat the experiment from page 128 twice (A and B below). Keep everything the same each time *except* the concentration of the acid. In B it is *twice as concentrated* as in A.

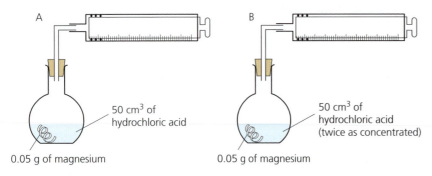

A
50 cm³ of hydrochloric acid

0.05 g of magnesium

B
50 cm³ of hydrochloric acid (twice as concentrated)

0.05 g of magnesium

The results   Here are both sets of results, shown on the same graph.

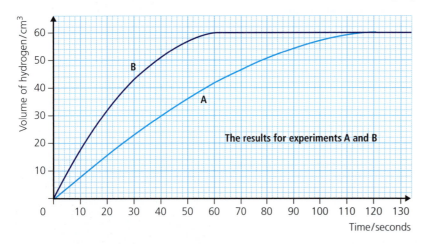

The results for experiments A and B

Notice these things about the results:

1 Curve B is steeper than curve A. So the reaction was faster for B.
2 In B, the reaction lasts for 60 seconds. In A it lasts for 120 seconds.
3 Both reactions produced 60 cm³ of hydrogen. Do you agree?
4 So in B the average rate was 1 cm³ of hydrogen per second. (60 ÷ 60)
   In A it was 0.5 cm³ of hydrogen per second. (60 ÷ 120)
   The average rate in B was twice the average rate in A.
   So in this example, doubling the concentration doubled the rate.

These results show that:
A reaction goes faster when the concentration of a reactant is increased.
This means you can also slow down a reaction, by reducing concentration.

▲ Bleach reacts with coloured substances. The more concentrated the solution of bleach, the faster this stain will disappear.

2 By changing temperature

Here you will see how rate changes with the **temperature** of the reactants.

The method Dilute hydrochloric acid and sodium thiosulfate solution react to give a fine yellow precipitate of sulfur. You can follow the rate of the reaction like this:

1 Mark a cross on a piece of paper.
2 Place a beaker containing sodium thiosulfate solution on top of the paper, so that you can see the cross through it, from above.
3 Quickly add hydrochloric acid, start a clock at the same time, and measure the temperature of the mixture.
4 The cross grows fainter as the precipitate forms. Stop the clock the moment you can no longer see the cross. Note the time.
5 Now repeat steps 1–4 several times, changing *only* the temperature. You do this by heating the sodium thiosulfate solution to different temperatures, before adding the acid.

▲ The low temperature in the fridge slows down reactions that make food rot.

View from above the beaker:

The cross grows fainter with time

The results This table shows some typical results:

Temperature/°C	20	30	40	50	60
Time for cross to disappear/seconds	200	125	50	33	24

The higher the temperature, the faster the cross disappears ➤

The cross disappears when enough sulfur has formed to hide it. This took 200 seconds at 20 °C, but only 50 seconds at 40 °C. So the reaction is *four times faster* at 40 °C than at 20 °C.

A reaction goes faster when the temperature is raised. When the temperature increases by 10 °C, the rate generally doubles.

That it why food cooks much faster in pressure cookers than in ordinary saucepans. (The temperature in a pressure cooker can reach 125 °C.) And if you want to slow a reaction down, of course, you can lower the temperature.

▲ Oh dear. Oven too hot? Reactions faster than expected?

 Q

1 Look at the graph on the opposite page.
  a How much hydrogen was obtained after 2 minutes in:
    i experiment A? ii experiment B?
  b How can you tell which reaction was faster, from the shape of the curves?
2 Explain why experiments A and B both gave the same amount of hydrogen.

3 Copy and complete: A reaction goes …… when the concentration of a …… is increased. It also goes …… when the …… is raised.
4 Raising the temperature speeds up a reaction. Try to give two (new) examples of how this is used in everyday life.
5 What happens to the rate of a reaction when the temperature is *lowered*? How do we make use of this?

131

10.4 Changing the rate of a reaction (part II)

3 By changing surface area

In many reactions, one reactant is a solid. The reaction between hydrochloric acid and calcium carbonate (marble chips) is an example. Carbon dioxide gas is produced:

$$CaCO_3 \ (s) + 2HCl \ (aq) \longrightarrow CaCl_2 \ (aq) + H_2O \ (l) + CO_2 \ (g)$$

The rate can be measured using the apparatus on the right.

The method Place the marble in the flask and add the acid. Quickly plug the flask with cotton wool to stop any liquid splashing out. Then weigh it, starting the clock at the same time. Note the mass at regular intervals until the reaction is complete.

Carbon dioxide is a heavy gas. It escapes through the cotton wool, which means that the flask gets lighter as the reaction proceeds. So by weighing the flask at regular intervals, you can follow the rate of reaction.

The experiment is repeated twice. Everything is kept exactly the same each time, except the *surface area* of the marble chips.

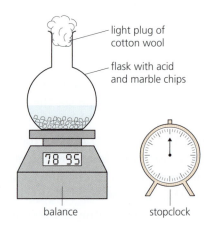
light plug of cotton wool
flask with acid and marble chips
78 95
balance
stopclock

For experiment 1, large chips are used. Their surface area is the total area of exposed surface.

For experiment 2, the same *mass* of marble is used – but the chips are small so the surface area is greater.

The results The results of the two experiments are plotted here:

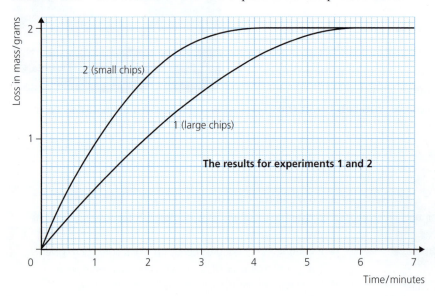

2 (small chips)

1 (large chips)

The results for experiments 1 and 2

Loss in mass/grams

Time/minutes

So what can you conclude about surface area? Did it affect the rate of the reaction?

How to draw the graph

First you have to find the *loss in mass* at different times:

loss in mass at a given time = mass at start − mass at that time

Then you plot the values for loss in mass against time.

Notice these things about the results:

1 Curve 2 is steeper than curve 1. This shows that the reaction is faster for the small chips.

2 In both experiments, the final loss in mass is 2.0 grams. In other words, 2.0 grams of carbon dioxide are produced each time.

3 For the small chips, the reaction is complete in 4 minutes. For the large chips, it takes 6 minutes.

These results show that:

The rate of a reaction increases when the surface area of a solid reactant is increased.

Explosion!

As you have seen, you can increase the rate of a reaction by increasing:

● the concentration of a reactant

● the temperature

● the surface area of a solid reactant

In some situations, an increase in any of these can lead to a dangerously fast reaction. You get an **explosion**. Here are examples.

In flour mills Flour particles are tiny, so flour has a very large surface area. It can also catch fire. In a flour mill, if there is a lot of flour dust in the air, a spark from a machine could be enough to cause an explosion.

For the same reason, explosions are a risk in wood mills, from wood dust, and in silos where wheat and other grains are stored. And in factories that make custard powder, and dried milk. The dust from all these will burn.

In coal mines In coal mines, methane (CH_4) and other flammable gases collect in the air. At certain concentrations they form an explosive mix with the air. A spark is enough to set off an explosion.

▲ In the old days, miners used candles to see their way underground – which caused many explosions. Now they use sealed lamps powered by batteries.

◀ A fire at a grain silo in Ghent, Belgium, after wheat dust exploded. Several people were injured.

Q

1 This question is about the graph on the opposite page. For each experiment find:
   a the mass of carbon dioxide produced in the first minute
   b the average rate of production of the gas, for the complete reaction.

2 a Which has the largest surface area: 1 g of large marble chips, or 1 g of small marble chips?
   b Which 1 g sample will disappear first when reacted with excess hydrochloric acid? Why?

3 Explain why fine flour dust in the air is a hazard, in flour mills.

10.5 Explaining rates

The collision theory

Magnesium and dilute hydrochloric acid react together like this:

magnesium + hydrochloric acid ⟶ magnesium chloride + hydrogen

$$Mg\,(s) \quad + \quad 2HCl\,(aq) \quad \longrightarrow \quad MgCl_2\,(aq) \quad + \quad H_2\,(g)$$

In order for the magnesium and acid particles to react together:

- **the particles must collide with each other, and**

- **the collision must have enough energy to be successful. In other words, enough energy to break bonds to allow reaction to occur.**

This is called the **collision theory**. It is shown by the drawings below.

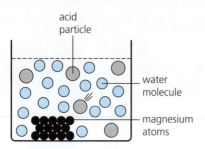

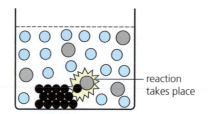

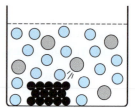

The particles in the liquid move non-stop. To react, an acid particle must collide with a magnesium atom, and bonds must break.

This collision has enough energy to break bonds. So it is successful. The particles react and new bonds form, giving magnesium chloride and hydrogen.

But this collision did not have enough energy. It was not successful. No bonds were broken. The acid particle just bounced away again.

If there are lots of successful collisions in a given minute, then a lot of hydrogen is produced in that minute. In other words, the rate of reaction is high. If there are not many, the rate of reaction is low.

The rate of a reaction depends on how many successful collisions there are in a given unit of time.

Changing the rate of a reaction

Why rate increases with concentration   If the concentration of the acid is increased, the reaction goes faster. It is easy to see why:

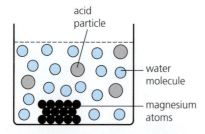

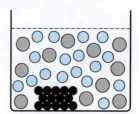

In dilute acid, there are not so many acid particles. So there is less chance of an acid particle hitting a magnesium atom.

Here the acid is more concentrated – there are more acid particles. So there is now more chance of a successful collision.

The more successful collisions there are, the faster the reaction.

> **Reactions between gases**
> - When you increase the pressure on two reacting gases, it means you squeeze more gas molecules into a given space.
> - So there is a greater chance of successful collisions.
> - So if pressure ↑ then rate ↑ for a gaseous reaction.

That idea also explains why the reaction between magnesium and hydrochloric acid slows down over time:

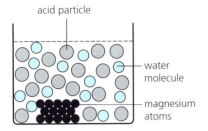

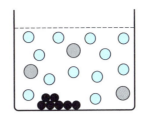

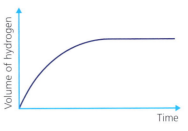

At the start, there are plenty of magnesium atoms and acid particles. But they get used up in successful collisions.

After a time, there are fewer magnesium atoms, and the acid is less concentrated. So there is less chance of successful collisions.

As a result, the slope of the reaction curve decreases with time, as shown above. It goes flat when the reaction is over.

Why rate increases with temperature   On heating, *all* the particles take in heat energy.

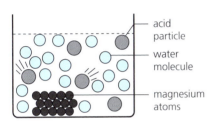

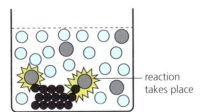

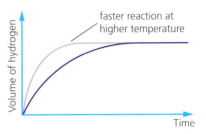

This makes the acid particles move faster – so they collide more often with magnesium particles.

The collisions have more energy too – so more are successful. So the reaction rate increases.

In fact, as you saw earlier, the rate generally doubles for an increase in temperature of 10 °C.

So the reaction rate increases with temperature for *two* reasons: there are more collisions, and more of them have enough energy to be successful.

Why rate increases with surface area   The reaction between the magnesium and acid is much faster when the metal is powdered:

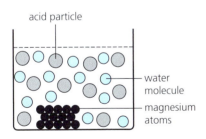

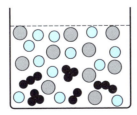

The acid particles can collide only with the magnesium atoms in the outer layer of the metal ribbon.

In the powdered metal, many more atoms are exposed. So the chance of a collision increases.

Q

1 Copy and complete: Two particles can react together only if they ...... and the ...... has enough ...... to be ....

2 What is meant by:
   a a successful collision?
   b an unsuccessful collision?

3 Reaction between magnesium and acid speeds up when:
   a the concentration of the acid is doubled. Why?
   b the temperature is raised. Why? (Give *two* reasons.)
   c the acid is stirred. Why?
   d the metal is ground to a powder. Why?

10.6 Catalysts

What is a catalyst?

You saw that a reaction can be speeded up by increasing the temperature, or the concentration of a reactant, or the surface area of a solid reactant.

There is another way to increase the rate of some reactions: use a **catalyst**.

A catalyst is a substance that speeds up a chemical reaction, but remains chemically unchanged itself.

Example: the decomposition of hydrogen peroxide

Hydrogen peroxide is a colourless liquid that breaks down very slowly to water and oxygen:

hydrogen peroxide $\longrightarrow$ water + oxygen
$$2H_2O_2 \, (l) \longrightarrow 2H_2O \, (l) + O_2 \, (g)$$

You can show how a catalyst affects the reaction, like this:

1 Pour some hydrogen peroxide into three measuring cylinders. The first one is the control.

2 Add manganese(IV) oxide to the second, and raw liver to the third.

3 Now use a glowing wooden splint to test the cylinders for oxygen. The splint will burst into flame if there is enough oxygen present.

▲ Many different substances can act as catalysts. They are usually made into shapes that offer a very large surface area.

The results

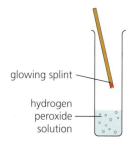

glowing splint

hydrogen peroxide solution

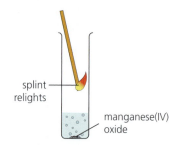

splint relights

manganese(IV) oxide

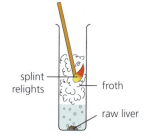

splint relights

froth

raw liver

Since hydrogen peroxide breaks down very slowly, there is not enough oxygen to relight the splint.

Manganese(IV) oxide makes the reaction go thousands of times faster. The splint bursts into flame.

Raw liver also speeds it up. The liquid froths as the oxygen bubbles off – and the splint relights.

So manganese(IV) oxide acts as a catalyst for the reaction. If you add more manganese(IV) oxide, the reaction will go even faster.

Something in the raw liver acts as a catalyst too. That 'something' is an **enzyme** called catalase.

What are enzymes?

Enzymes are proteins made by cells, to act as biological catalysts.
Enzymes are found in every living thing. You have thousands of different enzymes inside you. For example catalase speeds up the decomposition of hydrogen peroxide in your cells, before it can harm you. Amylase in your saliva speeds up the breakdown of the starch in your food.

Without enzymes, most of the reactions that take place in your body would be far too slow at body temperature. You would die.

▲ Enzyme molecules are large and complex, as this model shows.

How do catalysts work?

For a reaction to take place, the reacting particles must collide with enough energy for bonds to break and reaction to occur.

When a catalyst is present, the reactants are able to react in a way that requires less energy.

This means that more collisions now have enough energy to be successful. So the reaction speeds up. But the catalyst itself is unchanged.

Note that a catalyst must be chosen to suit the particular reaction. It may not work for other reactions.

Catalysts in the chemical industry

In industry, many reactions need heat. Fuel can be a very big expense.

With a catalyst, a reaction goes faster at a given temperature. So you get the product faster, saving time. Even better, it may go fast enough *at a lower temperature* – which means a lower fuel bill.

So catalysts are very important in the chemical industry. They are often **transition elements** or their **oxides**. Two examples are:

- **iron** used in the manufacture of ammonia
- **vanadium(IV) oxide** used in the manufacture of sulfuric acid.

▲ A catalyst of platinum and rhodium, in the form of a gauze, is being fitted into a tank. It will catalyse the production of nitric acid from ammonia and oxygen.

Making use of enzymes

There are thousands of different enzymes, made by living things. We are finding many uses for them.

For example some bacteria make enzymes that catalyse the breakdown of fat, starch, and proteins. The bacteria can be grown in tanks, in factories. The enzymes are removed, and used in **biological detergents**. In the wash, they help to break down grease, food stains, and blood stains on clothing.

Enzymes work best in conditions like those in the living cells that made them.

- If the temperature gets too high, the enzyme is destroyed or **denatured**. It loses its shape.
- An enzyme also works best in a specific pH range. You can denature it by adding acid or alkali.

▲ Now add a biological detergent? But do not use them for wool or silk, as they cause the proteins in these to break down.

> **More means faster**
> The more catalyst you add, the faster the reaction goes.

Q

1 What is a *catalyst*?
2 Which of these does a catalyst *not* change?
   a the speed of a reaction
   b the products that form
   c the total amount of each product formed
3 Explain what an *enzyme* is, and give an example.

4 Why do our bodies need enzymes?
5 Catalysts are very important in industry. Explain why.
6 Give two examples of catalysts used in the chemical industry.
7 A box of biological detergent had this instruction on the back: *Do not use in a wash above 60 °C.* Suggest a reason.

More about enzymes

Mainly from microbes

Enzymes are proteins made by living things, to act as catalysts for their own reactions. So we can obtain enzymes from plants, and animals, and microbes such as bacteria and fungi. In fact we get most from microbes.

Traditional uses for enzymes

We humans have used enzymes for thousands of years. For example …

- **In making bread**   Bread dough contains yeast (a fungus), and sugar. When the dough is left in a warm place, the yeast cells feed on the sugar to obtain energy. Enzymes in the yeast catalyse the reaction, which is called **fermentation**:

$$C_6H_{12}O_6 \, (aq) \xrightarrow[\text{by enzymes}]{\text{catalysed}} 2C_2H_5OH \, (aq) \; + \; 2CO_2 \, (g) \; + \; \text{energy}$$
$$\text{glucose} \qquad\qquad \text{ethanol} \qquad \text{carbon dioxide}$$

  The carbon dioxide gas makes the dough rise. Later, in the hot oven, the gas expands even more, while the bread sets. So you end up with spongy bread. The heat kills the yeast off.

- **In making yoghurt**   To make yoghurt, special bacteria are added to milk. They feed on the lactose (sugar) in it, to obtain energy. Their enzymes catalyse its conversion to lactic acid and other substances, which turn the milk into yoghurt.

Making enzymes the modern way

In making bread and yoghurt, the microbes that make the enzymes are present. But in most modern uses for enzymes, they are not. Instead:

- bacteria and other microbes are grown in tanks, in a rich broth of nutrients; so they multiply fast
- then they are killed off, and their enzymes are separated and purified
- the enzymes are sold to factories.

▲ Yeast cells. Cells are living things, but the enzymes they make are not. Enzymes are just chemicals (proteins).

▲ The holes in bread are where carbon dioxide gas expanded.

▲ Anyone home? The tank contains bacteria, busy making enzymes. For example it could be the enzyme amylase, which catalyses the conversion of starch to sugar.

▲ The amylase is sold to a company that uses it to make a sweet syrup from corn (maize) flour. The syrup is used in biscuits, cakes, soft drinks, and sauces.

Modern uses of enzymes

Enzymes have many different uses. Here are some common ones:

- **In making soft-centred chocolates**   How do they get the runny centres into chocolates? By using the enzyme **invertase**.

  First they make a paste containing sugars, water, flavouring, colouring, and invertase. Then they dip blobs of it into melted chocolate, which hardens. Inside, the invertase catalyses the breakdown of the sugars to more soluble ones, so the paste goes runny.

  Other enzymes are used in a similar way to 'soften' food, to make tinned food for infants.

- **In making stone-washed denim**   Once, denim was given a worn look by scrubbing it with pumice stone. Now an enzyme does the job.

- **In making biological detergents**   As you saw on page 137, these contain enzymes to catalyse the breakdown of grease and stains.

- **In DNA testing**   Suppose a criminal leaves tiny traces of skin or blood at a crime scene. The enzyme **polymerase** is used to 'grow' the DNA in them, to give enough to identify the criminal.

▲ Thanks to invertase …

▲ Thanks to polymerase …

How do they work?

This shows how an enzyme molecule catalyses the breakdown of a reactant molecule:

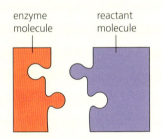

First, the two molecules must fit together like jigsaw pieces. (So the reactant molecule must be the right shape, for the enzyme.)

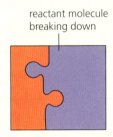

The 'complex' that forms makes it easy for the reactant molecule to break down. You do not need to provide energy by heating.

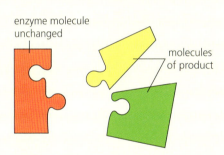

When decomposition is complete the molecules of the product move away. Another molecule of the reactant takes their place.

Enzymes are a much more complex shape than the drawing suggests. Even so, this model gives you a good idea of how they work.

The search for extremophiles

Most of the enzymes we use work around 40 °C, and at a pH not far from 7. In other words, in conditions like those in the cells that made them.

But around the world, scientists are searching high and low for microbes that live in very harsh conditions. For example deep under the ice in Antarctica, or at hot vents in the ocean floor, or in acidic lakes around volcanoes. They call these microbes **extremophiles**.

Why do scientists want them? Because the enzymes made by these microbes will work in the same harsh conditions. So they may find a great many uses in industry.

▲ Many bacteria live around hot vents in the ocean floor – in water at up to 400 °C.

10.7 Photochemical reactions

Some reactions need light

Some chemical reactions obtain the energy they need from light. They are called **photochemical reactions**. Examples are photosynthesis, and the reactions that occur in film photography.

Photosynthesis

- **Photosynthesis** is the reaction between carbon dioxide and water, in the presence of chlorophyll and sunlight, to produce glucose:

$$6CO_2\,(g) + 6H_2O\,(l) \xrightarrow[\text{chlorophyll}]{\text{light}} C_6H_{12}O_6\,(aq) + 6O_2\,(g)$$

carbon dioxide   water   glucose   oxygen

- It takes place in plant leaves. Carbon dioxide enters the leaves through tiny holes called **stomata**.
- **Chlorophyll**, the green pigment in leaves, is a **catalyst** for the reaction.
- The water is taken in from the soil, through the plant's roots.
- Sunlight provides the energy for this endothermic reaction.
- The plant then uses the glucose for energy, and to build the cellulose and other substances it needs for growth.

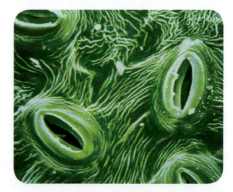

▲ The stomata of a leaf, magnified by about 700. Carbon dioxide passes in through them, and oxygen passes out.

Changing the rate of the photosynthesis reaction

Could you change the rate by changing the strength of the light? Let's see.

The method   Pondweed is a suitable plant to use for the experiment.

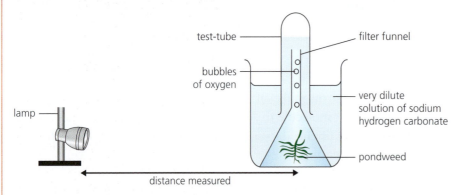

test-tube — filter funnel

bubbles of oxygen

very dilute solution of sodium hydrogen carbonate

lamp —

pondweed

distance measured

1. Put some pondweed in a beaker containing a very dilute solution of sodium hydrogen carbonate, $NaHCO_3$. (This compound decomposes, giving off carbon dioxide.) Place a funnel over it.
2. Place a test-tube full of the solution over the funnel, as shown.
3. Place the lamp 50 cm from the beaker. (Look at the arrow above.)
4. Let the pondweed adjust to the conditions for 1 minute. Then count the bubbles of oxygen it gives off, over 1 minute. Repeat twice more to get an average value per minute. Record your results.
5. Repeat step **4**, with the lamp placed at 40, 30, 20, and 10 cm from the beaker.

You can then plot a graph for your results.

▲ The plant on the right is unhealthy because it did not get enough light – so it made glucose too slowly.

The results   This graph shows that the number of bubbles per minute *increases* as the lamp is brought closer to the plant.

The closer it is, the greater the strength or **intensity** of the light that reaches the plant. So we can say that the rate of photosynthesis increases as the intensity of the light increases.

That makes sense. Light provides the energy for the reaction. The stronger it is, the more energy it provides. So more molecules of carbon dioxide and water gain enough energy to react.

A photochemical reaction can be speeded up by increasing the intensity of the light.

This is true of all photochemical reactions.

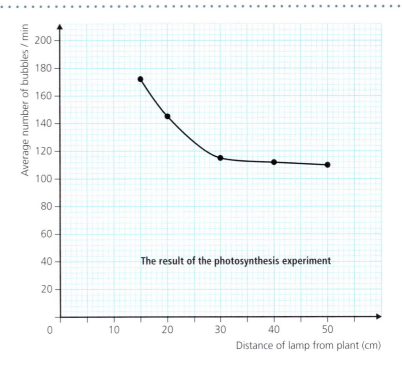

The result of the photosynthesis experiment

The reactions in film photography

Black-and-white film photography relies on a photochemical reaction. The film is covered with a coating of gel that contains tiny grains of silver bromide. Light causes this to break down:

$$2AgBr\,(s) \longrightarrow 2Ag\,(s) + Br_2\,(l)$$

It is both a photochemical reaction *and* a redox reaction.

The silver ions are **reduced**: $2Ag^+ + 2e^- \longrightarrow 2Ag$ (electron gain)

The bromide ions are **oxidised**: $2Br^- \longrightarrow Br_2 + 2e^-$ (electron loss)

So how is a photo produced?

1   When you click to take the photo, the camera shutter opens briefly. Light enters and strikes the film. The silver bromide decomposes, giving tiny dark particles of silver. Where brighter light strikes (from brighter parts of the scene), decomposition is faster, giving more silver.

2   Next the film is **developed**: unreacted silver bromide is washed away, leaving clear areas on the film. The silver remains, giving darker areas.

3   Then the film is printed. In this step, light is shone through the film onto photographic paper, which is also coated with silver bromide. The light passes through the clear areas of the film easily, causing the silver bromide to decompose. But the darker areas block light.

4   The unreacted silver bromide is washed from the paper. This leaves a black-and-white image of the original scene, made of silver particles.

▲ A 'negative' portrait in silver particles. What will the printed photo show?

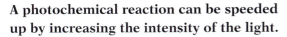

Q

1   What is a *photochemical reaction*? Give two examples.

2   **a**   Write down the equation for photosynthesis.
  b   What is the purpose of the chlorophyll?
  c   Stronger light speeds up photosynthesis. Why?

3   **a**   Why is silver bromide used in photographic film?
  b   Its decomposition is a redox reaction. Explain why.

4   *The more intense the light, the faster the photochemical reaction.* Explain how this idea is used in photography with film.

Checkup on Chapter 10

Revision checklist

Core syllabus content

Make sure you can …

☐ explain what *the rate of a reaction* means

☐ describe a way to measure the rate of a reaction that produces a gas, using a gas syringe

☐ describe a way to measure the rate of a reaction that produces carbon dioxide (a heavy gas), using a balance

☐ give the correct units for the rate of a given reaction (for example cm³ per minute, or grams per minute)

☐ work out, from the graph for a reaction:
  – how long the reaction lasted
  – how much product was obtained
  – the average rate of the reaction
  – the rate in any given minute

☐ give three ways to increase the rate of a reaction

☐ say which of two reactions was faster, by comparing the slope of their curves on a graph

☐ explain why there is a risk of explosions in flour mills and coal mines

☐ explain these terms: *catalyst* *enzyme*

☐ explain why enzymes are important in our bodies

☐ explain why catalysts are important in industry

☐ give examples of the use of catalysts, including enzymes

Extended syllabus content

Make sure you can also …

☐ describe the collision theory

☐ use the collision theory to explain *why* the rate of a reaction increases with concentration, temperature (*two reasons!*) and surface area

☐ explain how catalysts work

☐ say how enzymes can be destroyed

☐ define *photochemical reaction* and give examples

☐ say what *photosynthesis* is, and give the word and chemical equations for it

☐ name the catalyst for photosynthesis

☐ explain why a photochemical reaction can be speeded up by increasing the intensity of the light

☐ give the equation for the photochemical reaction that takes place on black-and-white film and photographic paper

☐ show that this reaction is also a redox reaction

Questions

Core syllabus content

1 The rate of the reaction between magnesium and dilute hydrochloric acid can be measured using this apparatus:

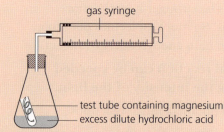

test tube containing magnesium
excess dilute hydrochloric acid

a What is the purpose of:
  i the test-tube?   **ii** the gas syringe?

b How would you get the reaction to start?

2 Some magnesium and an *excess* of dilute hydrochloric acid were reacted together. The volume of hydrogen produced was recorded every minute, as shown in the table:

Time / min	0	1	2	3	4	5	6	7
Volume of hydrogen / cm³	0	14	23	31	38	40	40	40

a What does an *excess* of acid mean?

b Plot a graph of the results.

c What is the *rate of reaction* (in cm³ of hydrogen per minute) during:
  i the first minute?
  ii the second minute?
  iii the third minute?

d Why does the rate change during the reaction?

e How much hydrogen was produced in total?

f How long does the reaction last?

g What is the *average rate* of the reaction?

h How could you slow down the reaction, while keeping the amounts of reactants unchanged?

3 Suggest a reason for each observation below.

a Hydrogen peroxide decomposes much faster in the presence of the enzyme catalase.

b The reaction between manganese carbonate and dilute hydrochloric acid speeds up when some concentrated hydrochloric acid is added.

c Powdered magnesium is used in fireworks, rather than magnesium ribbon.

4 In two separate experiments, two metals A and B were reacted with an excess of dilute hydrochloric acid. The volume of hydrogen was measured every 10 seconds. These graphs show the results:

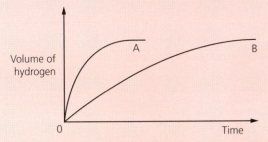

a **i** Which piece of apparatus can be used to measure the volume of hydrogen produced?

ii What other measuring equipment is needed?

b Which metal, A or B, reacts faster with hydrochloric acid? Give your evidence.

c Sketch and label the curves that will be obtained for metal B if:

i more concentrated acid is used (curve X)

ii the reaction is carried out at a lower temperature (curve Y).

5 Copper(II) oxide catalyses the decomposition of hydrogen peroxide. 0.5 g of the oxide was added to a flask containing 100 cm³ of hydrogen peroxide solution. A gas was released. It was collected, and its volume noted every 10 seconds.
This table shows the results:

Time / s	0	10	20	30	40	50	60	70	80	90
Volume / cm³	0	18	30	40	48	53	57	58	58	58

a What is a catalyst?

b Draw a diagram of suitable apparatus for this experiment.

c Name the gas that is formed.

d Write a balanced equation for the decomposition of hydrogen peroxide.

e Plot a graph of the volume of gas (vertical axis) against time (horizontal axis).

f Describe how rate changes during the reaction.

g What happens to the concentration of hydrogen peroxide as the reaction proceeds?

h What chemicals are present in the flask after 90 seconds?

i What mass of copper(II) oxide would be left in the flask at the end of the reaction?

j Sketch on your graph the curve that might be obtained for 1.0 g of copper(II) oxide.

k Name one other substance that catalyses this decomposition.

Extended syllabus content

6 Marble chips (lumps of calcium carbonate) react with hydrochloric acid as follows:
$$CaCO_3\ (s) + 2HCl\ (aq) \longrightarrow$$
$$CaCl_2\ (aq) + CO_2\ (g) + H_2O\ (l)$$

a What gas is released during this reaction?

b Describe a laboratory method that could be used to investigate the rate of the reaction.

c How will this affect the rate of the reaction?

i increasing the temperature

ii adding water to the acid

d Explain each of the effects in **c** in terms of collisions between reacting particles.

e If the lumps of marble are crushed first, will the reaction rate change? Explain your answer.

7 Zinc and iodine solution react like this:
$$Zn\ (s) + I_2\ (aq) \longrightarrow ZnI_2\ (aq)$$
The rate of reaction can be followed by measuring the mass of zinc metal at regular intervals, until all the iodine has been used up.

a What will happen to the mass of the zinc, as the reaction proceeds?

b Which reactant is in excess? Explain your choice.

c The reaction rate slows down with time. Why?

d Sketch a graph showing the mass of zinc on the y axis, and time on the x axis.

e How will the graph change if the temperature of the iodine solution is increased by 10 °C?

f Explain your answer to **e** using the idea of collisions between particles.

8 Some pondweed is placed as shown:

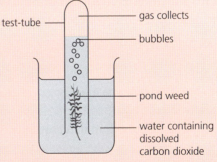

a **i** Name the gas that collects in the test tube.

ii What other product is produced?

b This experiment must be carried out in the light. Why?

c Using the apparatus above, suggest a method by which the rate of reaction could be found.

d What would be the effect of bringing a lamp close to the beaker? Explain your answer.

11.1 Acids and alkalis

Acids

One important group of chemicals is called **acids**:

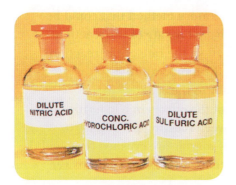

You have probably seen these acids in the lab. They are all solutions of pure compounds in water. They can be dilute or concentrated.

They must be handled carefully, especially the concentrated solutions, because they are **corrosive**. They can eat away metals, skin, and cloth.

But some acids are not so corrosive, even when concentrated. These are called **weak** acids. Ethanoic acid is one example. It is found in vinegar.

You can tell if something is an acid, by its effect on **litmus**. Litmus is a purple dye. It can be used as a solution, or on paper. **Acids turn litmus red**.

Some common acids

The main acids you will meet in chemistry are:

hydrochloric acid HCl (*aq*)
sulfuric acid H_2SO_4 (*aq*)
nitric acid HNO_3 (*aq*)
ethanoic acid CH_3COOH (*aq*)

But there are many others. For example, lemon and lime juice contain **citric acid**, ant stings contain **methanoic acid**, and fizzy drinks contain **carbonic acid**, formed when carbon dioxide dissolves in water.

Alkalis

There is another group of chemicals that also affect litmus, but in a different way. They are the **alkalis**.
Alkalis turn litmus blue.
Like acids, they must be handled carefully. They too can burn skin.

Some common alkalis

The pure alkalis are solids – except for ammonia, which is a gas. They are used in the lab as aqueous solutions. The main ones you will meet are:

sodium hydroxide $NaOH$ (*aq*)
potassium hydroxide KOH (*aq*)
calcium hydroxide $Ca(OH)_2$ (*aq*)
ammonia NH_3 (*aq*)

Remember:
aci**d** turns litmus re**d**

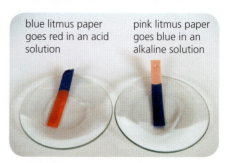

blue litmus paper goes red in an acid solution

pink litmus paper goes blue in an alkaline solution

▲ Testing with litmus paper.

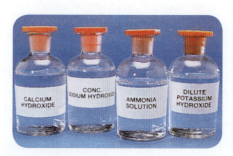

▲ Common laboratory alkalis. The solution of calcium hydroxide is called limewater.

Indicators

Litmus is called an **indicator**, because it indicates whether something is an acid or an alkali.

This table shows litmus and two other indicators. All three give a colour change from acid to alkali. That is why they are used.

Indicator	Colour in acid	Colour in alkali
litmus	red	blue
methyl orange	red	yellow
phenolphthalein	colourless	pink

Neutral substances

Many substances are not acids or alkalis. They are **neutral**. Examples are pure water, and aqueous solutions of sodium chloride and sugar.

The pH scale

You can say how acidic or alkaline a solution is using a scale of numbers called the **pH scale**. The numbers go from 0 to 14:

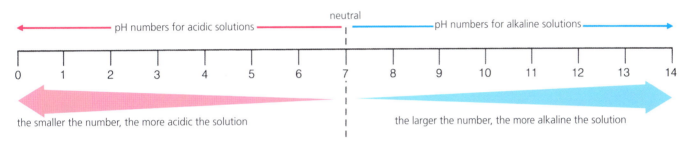

On this scale:

An acidic solution has a pH number less than 7.
An alkaline solution has a pH number greater than 7.
A neutral solution has a pH number of exactly 7.

Universal indicator paper

You can find the pH of any solution by using **universal indicator**. This is a mixture of dyes. Like litmus, it can be used as a solution, or a paper strip. Its colour changes with pH, as shown here:

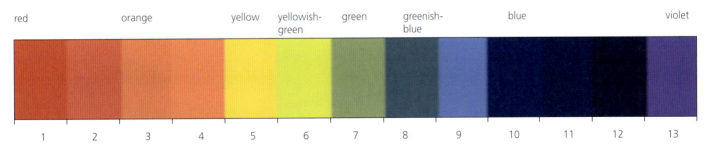

Q

1 What does *corrosive* mean?
2 How would you test a substance, to see if it is an acid?
3 Write down the formula for:
   sulfuric acid   nitric acid
   calcium hydroxide   ammonia solution
4 Phenolpthalein is an *indicator*. What does that mean?
5 What does this pH value tell you about the solution?
   a 9   **b** 4   **c** 7   **d** 1   **e** 10   **f** 3
6 What colour is universal indicator, in an aqueous solution of sugar? Why?

11.2 A closer look at acids and alkalis

Acids produce hydrogen ions

Hydrogen chloride is a gas, made of molecules. It dissolves in water to give hydrochloric acid. But this is not molecular. In water, the molecules break up or **dissociate** into ions:

$$HCl\ (aq) \longrightarrow H^+\ (aq) + Cl^-\ (aq)$$

So hydrochloric acid contains hydrogen ions. All other solutions of acids do too. The hydrogen ions give them their 'acidity'.
Solutions of acids contain hydrogen ions.

Remember!
Acidic solutions contain hydrogen ions:

H^+

It is what makes them 'acidic'.

Comparing acids

Since solutions of acids contain ions, they conduct electricity. We can measure how well they conduct using a conductivity meter. We can also check their pH using a pH meter.

Samples of acids of the same concentration were tested. This table gives the results. (The unit of conductivity is the siemens, or S.)

Acid	For a 0.1 M solution ...		
	conductivity (µS/cm)	pH	
hydrochloric acid	25	1.0	
sulfuric acid	40	0.7	strong acids
nitric acid	25	1.0	
methanoic acid	2	2.4	
ethanoic acid	0.5	2.9	weak acids
citric acid	4	2.1	

▲ The conductivity meter measures the current passing through the liquid, carried by ions. (Lemon juice contains hydrogen ions and citrate ions.)

So the acids fall into two groups. The first group shows high conductivity, and low pH. These are **strong acids**. The second group does not conduct nearly so well, and has a higher pH. These are **weak acids**.

The difference between strong and weak acids

In a solution of hydrochloric acid, all the molecules of hydrogen chloride have become ions:

$$HCl\ (aq) \xrightarrow{100\%} H^+\ (aq) + Cl^-\ (aq)$$

But in weak acids, only some of the molecules have become ions. For example, for ethanoic acid:

$$CH_3COOH\ (aq) \xrightarrow{\text{much less than } 100\%} H^+\ (aq) + CH_3COO^-\ (aq)$$

In solutions of strong acids, all the molecules become ions.
In solutions of weak acids, only some do.

So strong acids conduct better because there are more *ions* present. They have a lower pH because there are more *hydrogen ions* present.
The higher the concentration of hydrogen ions, the lower the pH.

▲ Strong and weak: the car battery contains sulfuric acid, and the oranges contain citric acid.

Alkalis produce hydroxide ions

Now let's turn to alkalis, with sodium hydroxide as our example. It is an ionic solid. When it dissolves, all the ions separate:

$$NaOH\ (aq) \longrightarrow Na^+\ (aq) + OH^-\ (aq)$$

So sodium hydroxide solution contains hydroxide ions. The same is true of all alkaline solutions.
Solutions of alkalis contain hydroxide ions.

Remember!
Alkaline solutions contain hydroxide ions:

OH⁻

It is what makes them alkaline.

Comparing alkalis

We can compare the conductivity and pH of alkalis too. Look at these results:

Alkali	For a 0.1 M solution ...		
	conductivity (µS/cm)	pH	
sodium hydroxide	20	13.0	strong alkali
potassium hydroxide	15	13.0	
ammonia solution	0.5	11.1	weak alkali

The first two alkalis show high conductivity, and high pH. They are **strong alkalis**. But the ammonia solution shows much lower conductivity, and a lower pH. It is a **weak alkali**.

Why ammonia solution is different

In sodium hydroxide solution, all the sodium hydroxide exists as ions:

$$NaOH\ (aq) \xrightarrow{100\%} Na^+\ (aq) + OH^-\ (aq)$$

The same is true for potassium hydroxide. But ammonia gas is molecular. When it dissolves in water, this is what happens:

$$NH_3\ (aq) + H_2O\ (l) \xrightarrow{\text{much less than 100\%}} NH_4^+\ (aq) + OH^-\ (aq)$$

Only some of the ammonia molecules form ions. So there are fewer hydroxide ions present than in a sodium hydroxide solution of the same concentration.

The sodium hydroxide solution is a better conductor than the ammonia solution because it contains more *ions*. And it has a higher pH because it contains more *hydroxide ions*.

The higher the concentration of hydroxide ions, the higher the pH.

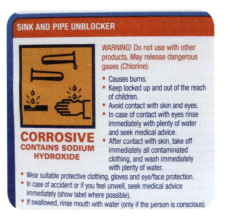

▲ Alkalis react with grease. So the strong alkali sodium hydroxide is used to clear blocked sinks and pipes in homes. What does the drawing tell you?

Q

1 Write an equation to show what happens when hydrogen chloride dissolves in water.
2 All acids have something in common. What is it?
3 For the table on page 146, explain why ethanoic acid has:
   a lower conductivity
   b a higher pH
   than hydrochloric acid.

4 What do all alkaline solutions have in common?
5 Write an equation to show what happens when ammonia gas dissolves in water.
6 For the table above, explain why the ammonia solution has:
   a lower conductivity
   b a lower pH
   than the potassium hydroxide solution.

11.3 The reactions of acids and bases

When acids react

When acids react with metals, bases and carbonates, a **salt** is produced. Salts are ionic compounds. Sodium chloride, NaCl, is an example.

The name of the salt depends on the acid you start with:

hydrochloric acid   gives   chlorides
sulfuric acid   gives   sulfates
nitric acid   gives   nitrates

Typical acid reactions

1 **With metals:**   **acid + metal ⟶ salt + hydrogen**

For example:

magnesium + sulfuric acid ⟶ magnesium sulfate + hydrogen

$$Mg\,(s) \quad + H_2SO_4\,(aq) \quad \longrightarrow \quad MgSO_4\,(aq) \quad + \quad H_2\,(g)$$

So the metal drives the hydrogen out of the acid, and takes its place: it **displaces** hydrogen. A solution of the salt magnesium sulfate is formed.

2 **With bases:**   **acid + base ⟶ salt + water**

Bases are compounds that react with acid to give *only* a salt and water. **Metal oxides and hydroxides** are bases. **Alkalis** are soluble bases.

Example for an acid and alkali:

hydrochloric acid + sodium hydroxide ⟶ sodium chloride + water

$$HCl\,(aq) \quad + \quad NaOH\,(aq) \quad \longrightarrow \quad NaCl\,(aq) \quad + H_2O\,(l)$$

Example for an acid and insoluble base:

sulfuric acid   +   copper(II) oxide ⟶ copper(II) sulfate +   water

$$H_2SO_4\,(aq) \quad + \quad CuO\,(s) \quad \longrightarrow \quad CuSO_4\,(aq) \quad + H_2O\,(l)$$

3 **With carbonates:**

acid + carbonate ⟶ salt + water + carbon dioxide

For example:

calcium   +   hydrochloric ⟶   calcium   +   water   +   carbon
carbonate   acid   chloride   dioxide

$$CaCO_3\,(s) \quad + \quad 2HCl\,(aq) \quad \longrightarrow \quad CaCl_2\,(aq) + H_2O\,(l) \quad + \quad CO_2\,(g)$$

Reactions of bases

1 Bases react with acids, as you saw above, giving only a salt and water. That is what identifies a base.

2 Bases such as sodium, potassium, and calcium hydroxides react with ammonium salts, driving out ammonia gas. For example:

calcium   + ammonium ⟶   calcium   +   water   + ammonia
hydroxide   chloride   chloride

$$Ca(OH)_2\,(s) + 2NH_4Cl\,(s) \longrightarrow CaCl_2\,(s) + 2H_2O\,(l) \quad + \quad 2NH_3\,(g)$$

This reaction is used for making ammonia in the laboratory.

▲ Magnesium reacting with dilute sulfuric acid. Hydrogen bubbles off.

▲ In A, black copper(II) oxide is reacting with dilute sulfuric acid. The solution turns blue as copper(II) sulfate forms. B shows how the final solution will look.

▲ Calcium carbonate reacting with dilute hydrochloric acid. What is that gas?

Neutralisation

Neutralisation is a reaction with acid that gives *water* as well as a salt.
So the reactions of bases and carbonates with acids are neutralisations.
We say the acid is **neutralised**.
But the reactions of acids with metals are not neutralisations. Why not?

Making use of neutralisation

We often make use of neutralisation outside the lab.
For example, to reduce acidity in soil.

Soil forms when rock is broken up over many years
by the action of rain and the weather. It may be acidic
because of the type of rock it came from. But rotting
vegetation, and heavy use of fertilisers, can also make
it acidic.

Most crops grow best when the pH of the soil is close
to 7. If the soil is too acidic, crops grow badly or not at
all. That could be a disaster for farmers.

So to reduce its acidity, the soil is treated with crushed
limestone (which is calcium carbonate), or **lime**
(calcium oxide) or **slaked lime** (calcium hydroxide).
A neutralisation reaction takes place.

▲ The soil is too acidic, so the farmer is spreading lime. It is more soluble than limestone. Is that an advantage – or not?

Acids and redox reactions

Look again at the three groups of acid reactions.

The reactions of acids with metals are **redox reactions**, because electrons
are transferred. For example when magnesium reacts with hydrochloric
acid, magnesium ions form. The magnesium is oxidised:

$$Mg\ (s) \longrightarrow Mg^{2+}\ (aq) + 2e^- \quad \text{(oxidation is loss of electrons)}$$

But in neutralisation reactions, no electrons are transferred. You can
check this by looking at the oxidation states in the equation. For example,
for the reaction between hydrochloric acid and sodium hydroxide:

$$HCl\ (aq) + NaOH\ (aq) \longrightarrow NaCl\ (aq) + H_2O\ (l)$$

$$+I-I \qquad +I-II+I \qquad +I-I \qquad +I-II$$

No element changes its oxidation state. So this is not a redox reaction.

In the next unit, you can find out what does go on during neutralisation.

▲ Bee stings are acidic. To neutralise the
sting, rub on some baking soda (sodium
hydrogen carbonate) or calamine lotion
(which contains zinc carbonate).

Q

1 Write a word equation for the reaction of dilute sulfuric
   acid with: **a** zinc   **b** sodium carbonate
2 Which reaction in question **1** is *not* a neutralisation?
3 Salts are ionic compounds. Name the salt that forms
   when calcium oxide reacts with hydrochloric acid, and
   say which ions it contains.
4 Zinc oxide is a base. Suggest a way to make zinc nitrate from
   it. Write a word equation for the reaction.

5 In what ways are the reactions of hydrochloric acid with
   calcium oxide and calcium carbonate:
   a similar?
   b different?
6 **a** Lime can help to control acidity in soil. Why?
   b Name one product that will form when it is used.
7 Zinc reacts with hydrochloric acid to form zinc chloride, $ZnCl_2$,
   and hydrogen. Show that this is a redox reaction.

11.4 A closer look at neutralisation

The neutralisation of an acid by an alkali (a soluble base)

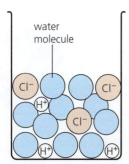

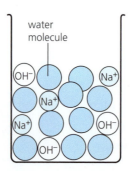

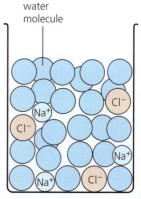

This is a solution of hydrochloric acid. It contains H⁺ and Cl⁻ ions. It will turn litmus red.

This is a solution of sodium hydroxide. It contains Na⁺ and OH⁻ ions. It will turn litmus blue.

When you mix the two solutions, the OH⁻ ions and H⁺ ions join to form **water molecules**. You end up with a neutral solution of sodium chloride, with no effect on litmus.

The overall equation for this neutralisation reaction is:

$$HCl\ (aq) + NaOH\ (aq) \longrightarrow NaCl\ (aq) + H_2O\ (l)$$

The ionic equation for the reaction

The best way to show what is going on in a neutralisation reaction is to write an **ionic equation** for it.

The ionic equation shows just the ions that take part in the reaction.

This is how to write the ionic equation for the reaction above:

1 **First, write down all the ions present in the equation.**

The drawings above will help you to do that:

$$H^+\ (aq) + Cl^-\ (aq) + Na^+\ (aq) + OH^-\ (aq) \longrightarrow$$
$$Cl^-\ (aq) + Na^+\ (aq) + H_2O\ (l)$$

2 **Now cross out any ions that appear, unchanged, on both sides of the equation.**

$$H^+\ (aq) + \cancel{Cl^-\ (aq)} + \cancel{Na^+\ (aq)} + OH^+\ (aq) \longrightarrow$$
$$\cancel{Cl^-\ (aq)} + \cancel{Na^+\ (aq)} + H_2O\ (l)$$

The crossed-out ions are present in the solution, but do not take part in the reaction. So they are called **spectator ions**.

3 **What's left is the ionic equation for the reaction.**

$$H^+\ (aq) + OH^-\ (aq) \longrightarrow H_2O\ (l)$$

So an H⁺ ion combines with an OH⁻ ion to produce a water molecule. This is all that happens during neutralisation.

During neutralisation, H⁺ ions combine with OH⁻ ions to form water molecules.

But an H⁺ ion is just a **proton**, as the drawing on the right shows. So, in effect, the acid **donates** (gives) **protons** to the hydroxide ions. The hydroxide ions accept these protons, to form water molecules.

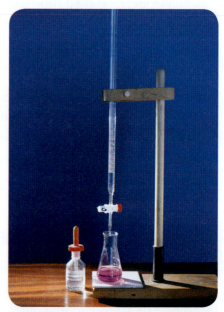

▲ A titration: sodium hydroxide solution was added to hydrochloric acid, from the burette. Neutralisation is complete: the phenolphthalein has turned pink.

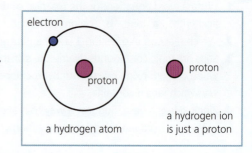

a hydrogen atom

a hydrogen ion is just a proton

The neutralisation of an acid by an insoluble base

Magnesium oxide is insoluble. It does not produce hydroxide ions.
So how does it neutralise an acid? Like this:

 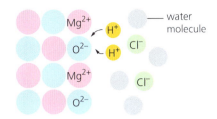 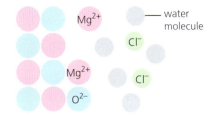

Magnesium oxide is a lattice of magnesium and oxygen ions. It is insoluble in water. But when you add dilute hydrochloric acid …

… the acid donates protons to the oxide ions. The oxide ions accept them, forming water molecules. So the lattice breaks down.

The magnesium ions join the chloride ions in solution. If you evaporate the water you will obtain the salt magnesium chloride.

The equation for this neutralisation reaction is:

$$2HCl\ (aq) + MgO\ (s) \longrightarrow MgCl_2\ (aq) + H_2O\ (l)$$

The ionic equation for it is:

$$2H^+\ (aq) + O^{2-}\ (s) \longrightarrow H_2O\ (l)$$

Proton donors and acceptors

Now compare the ionic equations for the two neutralisations in this unit:

$$H^+\ (aq) + OH^-\ (aq) \longrightarrow H_2O\ (l)$$
$$2H^+\ (aq) + O^{2-}\ (s) \longrightarrow H_2O\ (l)$$

In both:

- the protons are donated by the acids
- ions in the bases accept them, forming water molecules.

So this gives us a new definition for acids and bases:
Acids are proton donors, and bases are proton acceptors.

▲ Help is at hand. Indigestion is due to excess hydrochloric acid in the stomach. Milk of Magnesia contains magnesium hydroxide, which will neutralise it.

1 a What is an *ionic equation*?
  b Hydrochloric acid is neutralised by a solution of potassium hydroxide.
     What do you expect the ionic equation for this neutralisation reaction to be? Write it down.
2 What are *spectator ions*? Explain in your own words.
3 An H^+ ion is just a proton. Explain why. (Do a drawing?)
4 a Acids act as *proton donors*. What does that mean?
  b Bases act as *proton acceptors*. Explain what that means.
5 Neutralisation is *not* a redox reaction. Explain why, using the word *proton* in your answer.

6 How to write an ionic equation:
  i Write down all the ions present in the full equation.
  ii Cross out any that are the same on both sides of the equation.
  iii What is left is the ionic equation. Rewrite it neatly.
    a Follow steps i – iii for the reaction between magnesium oxide and hydrochloric acid above.
    b Does your ionic equation match the one shown above? If so, well done!
7 Hydrochloric acid is neutralised by a solution of sodium carbonate. Write the ionic equation for this reaction.

11.5 Oxides

What are oxides?

Oxides are compounds containing oxygen and another element.
You have seen already that metal oxides act as bases. Here we look more
closely at different types of oxides, and their behaviour.

Basic oxides

Look how these metals react with oxygen:

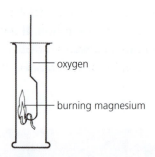

oxygen

burning magnesium

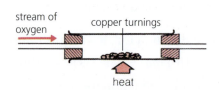

stream of oxygen

copper turnings

heat

Magnesium ribbon is lit over a
Bunsen flame, and plunged into
a jar of oxygen. It burns with
a brilliant white flame, leaving
a white ash, **magnesium oxide**:

$2Mg\ (s) + O_2\ (g) \rightarrow 2MgO\ (s)$

Hot iron wool is plunged into a
gas jar of oxygen. It glows bright
orange, and throws out a shower
of sparks. A black solid is left in
the gas jar. It is **iron(III) oxide**:

$4Fe\ (s) + 3O_2\ (g) \rightarrow 2Fe_2O_3\ (s)$

Copper is too unreactive to catch
fire in oxygen. But when it is
heated in a stream of the gas, its
surface turns black. The black
substance is **copper(II) oxide**:

$2Cu\ (s) + O_2\ (g) \rightarrow 2CuO\ (s)$

The more reactive the metal, the more vigorously it reacts.

The copper(II) oxide produced in the last reaction above is insoluble
in water. But it does dissolve in dilute acid:

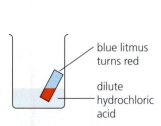

blue litmus
turns red

dilute
hydrochloric
acid

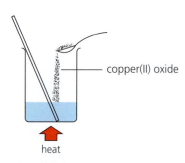

copper(II) oxide

heat

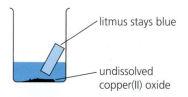

litmus stays blue

undissolved
copper(II) oxide

This is dilute hydrochloric acid.
It turns blue litmus paper red, like
all acids do.

Copper(II) oxide dissolves in it,
when it is warmed. But after a
time, no more will dissolve.

The resulting liquid has no effect
on blue litmus. So the oxide has
neutralised the acid.

Copper(II) oxide is called a **basic oxide** since it can neutralise an acid:

base + acid $\longrightarrow$ salt + water
$CuO\ (s) + 2HCl\ (aq) \longrightarrow CuCl_2\ (aq) + H_2O\ (l)$

Iron(III) oxide and magnesium oxide behave in the same way – they too
can neutralise acid, so they are basic oxides.

In general, metals react with oxygen to form basic oxides.
Basic oxides belong to the larger group of compounds called bases.

Acidic oxides

Now look how these non-metals react with oxygen:

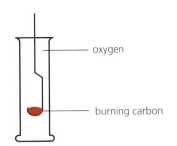
oxygen
burning carbon

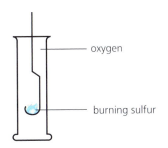
oxygen
burning sulfur

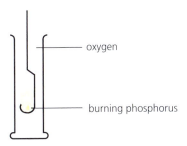
oxygen
burning phosphorus

Powdered carbon is heated over a Bunsen burner until red-hot, then plunged into a jar of oxygen. It glows bright red, and the gas **carbon dioxide** is formed:

$C\ (s) + O_2\ (g) \longrightarrow CO_2\ (g)$

Sulfur catches fire over a Bunsen burner, and burns with a blue flame. In pure oxygen it burns even brighter. The gas **sulfur dioxide** is formed:

$S\ (s) + O_2\ (g) \longrightarrow SO_2\ (g)$

Phosphorus bursts into flame in air or oxygen, without heating. (So it is stored under water!) A white solid, **phosphorus (V) oxide**, is formed:

$P_4\ (s) + 5O_2\ (g) \longrightarrow P_4O_{10}\ (s)$

Carbon dioxide is slightly soluble in water. The solution will turn litmus red: it is acidic. The weak acid carbonic acid has formed:

$$CO_2\ (g) + H_2O\ (l) \longrightarrow H_2CO_3\ (aq)$$

Sulfur dioxide and phosphorus(V) oxide also dissolve in water to form acids. So they are all called **acidic oxides**.

In general, non-metals react with oxygen to form acidic oxides.

Amphoteric oxides

Aluminium is a metal, so you would expect aluminium oxide to be a base. In fact it is both acidic *and* basic. It acts as a base with hydrochloric acid:

$$Al_2O_3\ (s) + 6HCl\ (aq) \longrightarrow 2AlCl_3\ (aq) + 3H_2O\ (l)$$

But it acts as an acidic oxide with sodium hydroxide, giving a compound called sodium aluminate:

$$Al_2O_3\ (s) + 6NaOH\ (aq) \longrightarrow 2Na_3AlO_3\ (aq) + 3H_2O\ (l)$$

So aluminium oxide is called an **amphoteric oxide**.
An amphoteric oxide will react with both acids and alkalis.
Zinc oxide is also amphoteric.

acid   alkali   ZINC OXIDE

▲ Zinc oxide: an amphoteric oxide. It will react with both acid and alkali.

Neutral oxides

Some oxides of non-metals are neither acidic nor basic: they are **neutral**.
Neutral oxides do not react with acids or bases.

The gases carbon monoxide, CO, and dinitrogen oxide, N_2O are neutral. (Other nitrogen oxides are acidic.)

▲ No pain. The neutral oxide dinitrogen oxide (N_2O) is used as an anaesthetic by dentists. It is also called **laughing gas**.

Q

1 How would you show that magnesium oxide is a base?
2 Copy and complete: Metals usually form .... oxides while non-metals form .... oxides.
3 See if you can arrange carbon, phosphorus and sulfur in order of reactivity, using their reaction with oxygen.
4 What colour change would you see, on adding litmus solution to a solution of phosphorus pentoxide?
5 What is an *amphoteric* oxide? Give two examples.
6 Dinitrogen oxide is a *neutral* oxide. It is quite soluble in water. How could you prove it is neutral?

153

11.6 Making salts

You can make salts by reacting acids with metals, or insoluble bases, or soluble bases (alkalis), or carbonates.

Starting with a metal

Zinc sulfate can be made by reacting dilute sulfuric acid with zinc:

$$Zn\ (s) + H_2SO_4\ (aq) \longrightarrow ZnSO_4\ (aq) + H_2\ (g)$$

These are the steps:

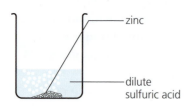
zinc
dilute
sulfuric acid

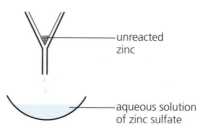
unreacted
zinc
aqueous solution
of zinc sulfate

crystals
form

1 Add the zinc to the acid in a beaker. It starts to dissolve, and hydrogen bubbles off. Bubbling stops when all the acid is used up.

2 Some zinc is still left. (The zinc was *in excess*.) Remove it by filtering. This leaves an aqueous solution of zinc sulfate.

3 Heat the solution to evaporate some water, to obtain a saturated solution. Leave this to cool. Crystals of zinc sulfate appear.

This method is fine for making salts of magnesium, aluminium, zinc, and iron. But you could not use it with sodium, potassium, or calcium, because these metals react violently with acids.

At the other extreme, the reaction of lead with acids is too slow, and copper, silver and gold do not react at all. (There is more about the reactivity of metals with acids in Unit 13.2.)

Starting with an insoluble base

Copper will not react with dilute sulfuric acid. So to make copper(II) sulfate, you must start with a base such as copper(II) oxide, which is insoluble. The reaction that takes place is:

$$CuO\ (s) + H_2SO_4\ (aq) \longrightarrow CuSO_4\ (aq) + H_2O\ (l)$$

The method is quite like the one above:

▲ Crystals of copper(II) sulfate. They are *hydrated*: they contain water molecules in the crystal structure. Their full formula is $CuSO_4.5H_2O$.

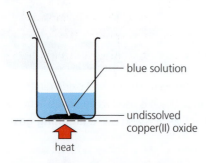
blue solution
undissolved
copper(II) oxide
heat

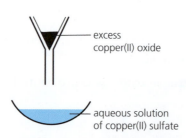
excess
copper(II) oxide
aqueous solution
of copper(II) sulfate

blue crystals
form

1 Add some copper(II) oxide to dilute sulfuric acid. It dissolves on warming, and the solution turns blue. Add more until no more will dissolve …

2 … which means all the acid has now been used up. Remove the excess solid by filtering. This leaves a blue solution of copper(II) sulfate in water.

3 Heat the solution to obtain a saturated solution. Then leave it to cool. Crystals of copper(II) sulfate form. They look like the crystals in the photo above.

You could also use copper(II) carbonate as the starting compound here.

Starting with an alkali (soluble base)

It is dangerous to add sodium to acid. So to make sodium salts, start with sodium hydroxide. You can make sodium chloride like this:

$$NaOH (aq) + HCl (aq) \longrightarrow NaCl (aq) + H_2O (l)$$

Both reactants are soluble, and no gas bubbles off. So how can you tell when the reaction is complete? By carrying out a **titration**.

In a titration, one reactant is slowly added to the other in the presence of an indicator. The indicator changes colour when the reaction is complete. So you know how much reactant is needed for a complete reaction. Now you can mix the correct amounts, *without* the indicator.

▲ The phenolphthalein says 'alkaline'.

The steps in making sodium chloride

You could use **phenolphthalein** as the indicator. It is pink in alkaline solution, but colourless in neutral and acid solutions. These are the steps:

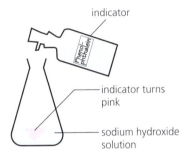 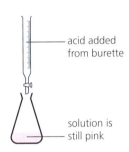 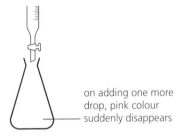

1 Put 25 cm³ of sodium hydroxide solution into a flask, using a pipette (for accuracy). Add two drops of phenolphthalein.

2 Add the acid from a burette, just a little at a time. Swirl the flask carefully, to help the acid and alkali mix.

3 The indicator suddenly turns colourless. So the alkali has all been used up. The solution is now neutral. Add no more acid!

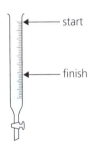 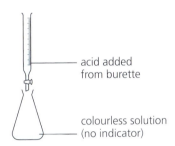 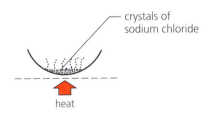

4 Find how much acid you added, using the scale on the burette. This tells you how much acid is needed to neutralise 25 cm³ of the alkali.

5 Now repeat *without* the indicator. (It would be an impurity.) Put 25 cm³ of alkali in the flask. Add the correct amount of acid to neutralise it.

6 Finally, heat the solution from the flask to evaporate the water. White crystals of sodium chloride will be left behind.

You could use the same method for making potassium salts from potassium hydroxide, and ammonium salts from ammonia solution.

Q

1 What will you start with, to make the salt zinc chloride?

2 You would not make lead salts by reacting lead with acids.
 a Why not? **b** Suggest a way to make lead nitrate.

3 Look at step **2** at the top of page 154. The zinc was *in excess*. What does that mean? (Check the glossary?)

4 What is the purpose of a titration?

5 For carrying out a titration, a burette and pipette are used rather than measuring cylinders. Why?

6 You are asked to make the salt ammonium nitrate. Which reactants will you use?

11.7 Making insoluble salts by precipitation

Not all salts are soluble

The salts we looked at so far have all been soluble. You could obtain them as crystals, by evaporating solutions. But not all salts are soluble.
This table shows the 'rules' for the solubility of salts:

Soluble		Insoluble
All sodium, potassium, and ammonium salts		
All nitrates		
Chlorides …	*except*	silver and lead chloride
Sulfates …	*except*	calcium, barium, and lead sulfate
Sodium, potassium, and ammonium carbonates …		but all other carbonates are insoluble

Making insoluble salts by precipitation

Insoluble salts can be made by **precipitation**.
Barium sulfate is an insoluble salt. You can make it by mixing solutions of barium chloride and magnesium sulfate:

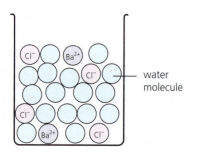

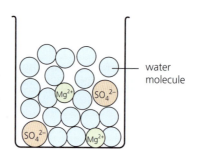

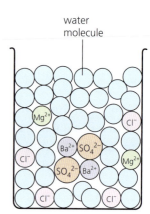

water molecule

A solution of barium chloride, $BaCl_2$, contains barium ions and chloride ions, as shown here.

A solution of magnesium sulfate, $MgSO_4$, contains magnesium ions and sulfate ions.

When you mix the two solutions, the barium and sulfate ions bond together. Barium sulfate forms as a precipitate.

The equation for the reaction is:

$$BaCl_2\ (aq) + MgSO_4\ (aq) \longrightarrow BaSO_4\ (s) + MgCl_2\ (aq)$$

The ionic equation is:

$$Ba^{2+}\ (aq) + SO_4^{2-}\ (aq) \longrightarrow BaSO_4\ (s)$$

This does not show the magnesium and chloride ions, because they are **spectator ions**. They are present, but do not take part in the reaction.

The steps in making barium sulfate

1 Make up solutions of barium chloride and magnesium sulfate.

2 Mix them. A white precipitate of barium sulfate forms at once.

3 Filter the mixture. The precipitate is trapped in the filter paper.

4 Rinse the precipitate by running distilled water through it.

5 Then place it in a warm oven to dry.

▲ The precipitation of barium sulfate.

Choosing the starting compounds

Barium sulfate can also be made from barium nitrate and sodium sulfate, since both of these are soluble. As long as barium ions and sulfate ions are present, barium sulfate will precipitate.

To precipitate an insoluble salt, you must mix a solution that contains its positive ions with one that contains its negative ions.

Some uses of precipitation

Precipitation has some important uses in industry. For example:

- It is used to make coloured pigments for paint.
- It is sometimes used to remove harmful substances dissolved in water, when cleaning up waste water.
- It is used in making film, for photography. For this, solutions of silver nitrate and potassium bromide are mixed with gelatine. A precipitate of tiny crystals of insoluble silver bromide forms. The mixture is then coated onto clear film, giving **photographic film**.

Later, when light strikes the film, the silver bromide will break down:

$$2AgBr\ (s) \longrightarrow 2Ag\ (s) + Br_2\ (l)$$

You can find out more about the photographic process on page 141.

▲ The paint we use for home decoration contains insoluble pigments like these – usually made by precipitation.

▲ Putting film in a camera. Most of the film is inside the yellow cartridge, at the top, protected from light.

▲ Steady on! More and more movies are being shot using digital cameras. But some directors prefer to use film. Film for movies is coated with silver halides in gelatine – just like camera film. Chemicals mixed with the halides provide the colour.

Digital cameras

Today digital cameras are more popular than cameras that use film.

In a digital camera, the light strikes a surface that generates a current. This is converted to an image by a little computer inside the camera.

1 Explain what *precipitation* means, in your own words.
2 Name four salts you could not make by precipitation.
3 Choose two starting compounds you could use to make these insoluble salts:
   a calcium sulfate   b magnesium carbonate
   c zinc carbonate   d lead chloride

4 Write a balanced equation for each reaction you chose in **3**.
5 a What is a spectator ion?
   b Identify the spectator ions in your reactions for **3**.
6 Write the ionic equations for your reactions for **3**.
7 Why is precipitation necessary, in making photographic film?

11.8 Finding concentrations by titration

How to find a concentration by titration

On page 155, the volume of acid needed to neutralise an alkali was found by adding the acid a little at a time, until the indicator showed that the reaction was complete. This method is called **titration**.

You can find the *concentration* of an acid using the same method. You use a solution of alkali of known concentration (a **standard solution**) and titrate the acid against it.

An example

You are asked to find the concentration of a solution of hydrochloric acid, using a 1 M solution of sodium carbonate as the standard solution.

First, titrate the acid against your standard solution.

- Measure 25 cm³ of the sodium carbonate solution into a conical flask, using a pipette. Add a few drops of methyl orange indicator. The indicator goes yellow.

- Pour the acid into a 50 cm³ burette. Record the level.

- Drip the acid slowly into the conical flask. Keep swirling the flask. Stop adding acid when a single drop finally turns the indicator red. Record the new level of acid in the burette.

- Calculate the volume of acid used. For example:
  Starting level: 1.0 cm³
  Final level: 28.8 cm³
  Volume used: 27.8 cm³

So 27.8 cm³ of the acid neutralised 25 cm³ of the alkaline solution.

You can now calculate the concentration of the acid.

Step 1 Calculate the number of moles of sodium carbonate used.

1000 cm³ of 1 M solution contains 1 mole so

25 cm³ contains $\frac{25}{1000} \times 1$ mole or 0.025 mole.

Step 2 From the equation, find the molar ratio of acid to alkali.

$$2HCl \ (aq) + Na_2CO_3 \ (aq) \longrightarrow 2NaCl \ (aq) + H_2O \ (l) + CO_2 \ (g)$$
2 moles   1 mole
The ratio is 2 moles of acid to 1 of alkali.

Step 3 Work out the number of moles of acid neutralised.

1 mole of alkali neutralises 2 moles of acid so
0.025 mole of alkali neutralises 2×0.025 moles of acid.
0.05 moles of acid were neutralised.

Step 4 Calculate the concentration of the acid.

The volume of acid used was 27.8 cm³ or 0.0278 dm³.

$$concentration = \frac{number \ of \ moles}{volume \ in \ dm^3} = \frac{0.05}{0.0278} = 1.8 \ mol/dm^3$$

So the concentration of the hydrochloric acid is **1.8 M.**

Remember!
- Concentration is usually given as moles per dm³ or mol/dm³
- 1000 cm³ = 1 dm³
- To convert cm³ to dm³ move the decimal point 3 places left. So 250 cm³ = 0.25 dm³

Use the calculation triangle

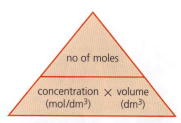

▲ Cover 'concentration' with your finger to see how to calculate it.

▲ You can find how much alkali is needed to neutralise acid by doing a titration using indicator, as here …

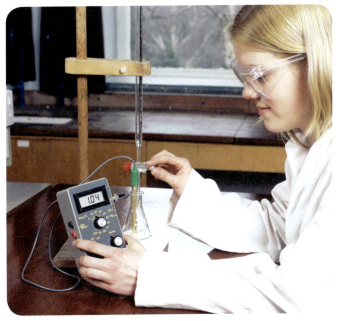

▲ … or you could use a pH meter, to measure the pH of the solution. How will you know when neutralisation is complete?

Another sample calculation

Vinegar is mainly a solution of the weak acid ethanoic acid. 25 cm³ of vinegar were neutralised by 20 cm³ of 1 M sodium hydroxide solution. What is the concentration of ethanoic acid in the vinegar?

Step 1 Calculate the number of moles of sodium hydroxide used.

1000 cm³ of 1 M solution contains 1 mole so

20 cm³ contains $\dfrac{20}{1000} \times 1$ mole or 0.02 mole.

Step 2 From the equation, find the molar ratio of acid to alkali.

$CH_3COOH\ (aq) + NaOH\ (aq) \longrightarrow CH_3COONa\ (aq) + H_2O\ (l)$
1 mole   1 mole
The ratio is 1 mole of acid to 1 mole of alkali.

Step 3 Work out the number of moles of acid neutralised.

1 mole of alkali neutralises 1 mole of acid so
0.02 mole of alkali neutralise 0.02 mole of acid.

Step 4 Calculate the concentration of the acid. (25 cm³ = 0.025 dm³)

$$\text{concentration} = \frac{\text{number of moles}}{\text{volume in dm}^3} = \frac{0.02}{0.025} = 0.8 \text{ mol/dm}^3$$

So the concentration of ethanoic acid in the vinegar is 0.8 M.

Note: ethanoic acid is only partly dissociated into ions at any given time. (It is a weak acid.) But as the neutralisation proceeds, it continues to dissociate until it has all reacted.

▲ The ethanoic acid in vinegar – the bottle on the left – gives salad dressing its tasty tang.

Q

1 What is a *standard solution*?
2 What volume of 2 M hydrochloric acid will neutralise 25 cm³ of 2 M sodium carbonate?

3 20 cm³ of 1 M sulfuric acid were neutralised by 25 cm³ of ammonia solution. Calculate the concentration of the ammonia solution. (See the equation on page 225.)

Checkup on Chapter 11

Revision checklist

Core syllabus content

Make sure you can …

☐ name the common laboratory acids and alkalis, and give their formulae

☐ describe the effect of acids and alkalis on litmus, and methyl orange

☐ explain what the pH scale is, and what pH numbers tell you

☐ describe what *universal indicator* is, and how its colour changes across the pH range

☐ define a *base*, and say that alkalis are soluble bases

☐ say what is formed when acids react with:
  metals   bases   carbonates

☐ explain what a *neutralisation* reaction is, and identify one from its equation

☐ say what gas is given off when strong bases are heated with ammonium compounds

☐ say why it is important to control acidity in soil, and how this is done

☐ explain what *basic oxides* and *acidic oxides* are, and give examples

☐ choose suitable reactants for making a salt

☐ describe methods for making a salt, starting with:
  – a metal or insoluble base
  – an alkaline solution

☐ explain how and why an indicator is used, in a titration

Extended syllabus content

Make sure you can also …

☐ define *strong acids* and *weak acids*, with examples

☐ define *strong alkalis* and *weak alkalis*, with examples

☐ explain why the reaction between an acid and a metal is a redox reaction

☐ explain what happens in a neutralisation reaction, and give the ionic equation

☐ give a definition for acids and bases using the idea of *proton transfer*

☐ say what *amphoteric oxides* and *neutral oxides* are, and give examples

☐ choose suitable reactants for making an insoluble salt by precipitation

☐ say what *spectator ions* are, and identify the spectator ions in a precipitation reaction

☐ calculate the concentration of a solution of acid or alkali, using data from a titration

Questions

Core syllabus content

1 Rewrite the following, choosing the correct word from each pair in brackets.

Acids are compounds that dissolve in water giving hydrogen ions. Sulfuric acid is an example. It can be neutralised by (acids/bases) to form salts called (nitrates/sulfates).

Many (metals/non-metals) react with acids to give (hydrogen/carbon dioxide). Acids react with (chlorides/carbonates) to give (hydrogen/carbon dioxide).

Since they contain ions, solutions of acids are (good/poor) conductors of electricity. They also affect indicators. Litmus turns (red/blue) in acids while phenolphthalein turns (pink/colourless). The level of acidity of an acid is shown by its (concentration/pH number). The (higher/lower) the number, the more acidic the solution.

2 **A** and **B** are white powders. **A** is insoluble in water, but **B** dissolves. Its solution has a pH of 3.
A mixture of **A** and **B** bubbles or effervesces in water, giving off a gas. A clear solution forms.

a Which of the two powders is an acid?

b The other powder is a carbonate. Which gas bubbles off in the reaction?

c Although **A** is insoluble in water, a clear solution forms when the mixture of **A** and **B** is added to water. Explain why.

3 Oxygen reacts with other elements to form oxides. Three examples are: calcium oxide, phosphorus pentoxide, and copper(II) oxide.

a Which of these is:
  i   an insoluble base?
  ii a soluble base?
  iii an acidic oxide?

b When the soluble base is dissolved in water, the solution changes the colour of litmus paper. What colour change will you see?

c Name the gas given off when the soluble base is heated with ammonium chloride.

d i   Write a word equation for the reaction between the insoluble base and sulfuric acid.
  ii What is this type of reaction called?

e Name another acidic oxide.

Method of preparation	Reactants	Salt formed	Other products
a acid + alkali	calcium hydroxide and nitric acid	calcium nitrate	water
b acid + metal	zinc and hydrochloric acid	....	....
c acid + alkali	.... and potassium hydroxide	potassium sulfate	water only
d acid + carbonate	.... and	sodium chloride	water and ....
e acid + metal	.... and	iron(II) sulfate	....
f acid + ...	nitric acid and sodium hydroxide	....	....
g acid + insoluble base	.... and copper(II) oxide	copper(II) sulfate	....
h acid + ...	.... and	copper(II) sulfate	carbon dioxide and ....

4 The table above is about the preparation of salts.
  i Copy it and fill in the missing details.
  ii Write balanced equations for the eight reactions.

5 The drawings show the preparation of copper(II) ethanoate, a salt of ethanoic acid.

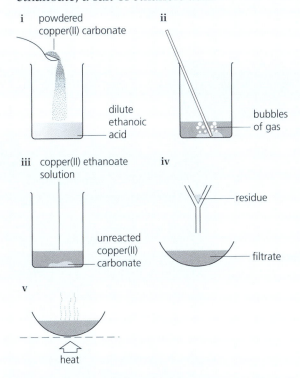

a Which gas is given off in stage **ii**?
b i Write a word equation for the reaction in ii.
  ii How can you tell when it is over?
c Which reactant above is:
  i present in excess? What is your evidence?
  ii completely used up in the reaction?
d Copper(II) carbonate is used in powder form, rather than as lumps. Suggest a reason.
e Name the residue in stage **iv**.
f Write a list of instructions for carrying out this preparation in the laboratory.
g Suggest another copper compound to use instead of copper(II) carbonate, to make the salt.

Extended syllabus content

6 Magnesium sulfate ($MgSO_4$) is the chemical name for Epsom salts. It can be made in the laboratory by neutralising the base magnesium oxide (MgO).
  a Which acid should be used to make Epsom salts?
  b Write a balanced equation for the reaction.
  c i The acid is fully dissociated in water. Which term describes this type of acid?
    ii Which ion causes the 'acidity' of the acid?
  d i What is a base?
    ii Write an ionic equation that shows the oxide ion (O^{2-}) acting as a base.

7 a i From the list on page 156, write down two starting compounds that could be used to make the insoluble compound silver chloride.
    ii What is this type of reaction called?
  b i Write the ionic equation for the reaction.
    ii List the spectator ions for the reaction.

8 **Washing soda** is crystals of hydrated sodium carbonate, $Na_2CO_3.xH_2O$.
  The value of x can be found by titration.
  In the experiment, 2 g of hydrated sodium carbonate neutralised 14 cm³ of a standard 1 M solution of hydrochloric acid.
  a What does *hydrated* mean?
  b Write a balanced equation for the reaction that took place during the titration.
  c How many moles of HCl were neutralised?
  d How many moles of sodium carbonate, Na_2CO_3, were in 2 g of the hydrated salt?
  e What mass of sodium carbonate, Na_2CO_3, is this? (M_r: Na = 23, C = 12, O = 16)
  f What mass of the hydrated sodium carbonate was water?
  g How many moles of water is this?
  h How many moles of water are there in 1 mole of $Na_2CO_3.xH_2O$?
  i Write the full formula for washing soda.

161

12.1 An overview of the Periodic Table

What is the Periodic Table?

I	II												III	IV	V	VI	VII	VIII
								1 H hydrogen 1										2 He helium 4
3 Li lithium 7	4 Be beryllium 9												5 B boron 11	6 C carbon 12	7 N nitrogen 14	8 O oxygen 16	9 F fluorine 19	10 Ne neon 20
11 Na sodium 23	12 Mg magnesium 24												13 Al aluminium 27	14 Si silicon 28	15 P phosphorus 31	16 S sulfur 32	17 Cl chlorine 35.5	18 Ar argon 40
19 K potassium 39	20 Ca calcium 40	21 Sc scandium 45	22 Ti titanium 48	23 V vanadium 51	24 Cr chromium 52	25 Mn manganese 55	26 Fe iron 56	27 Co cobalt 59	28 Ni nickel 59	29 Cu copper 64	30 Zn zinc 65		31 Ga gallium 70	32 Ge germanium 73	33 As arsenic 75	34 Se selenium 79	35 Br bromine 80	36 Kr krypton 84
37 Rb rubidium 85	38 Sr strontium 88	39 Y yttrium 89	40 Zr zirconium 91	41 Nb niobium 93	42 Mo molybdenum 96	43 Tc technetium –	44 Ru ruthenium 101	45 Rh rhodium 103	46 Pd palladium 106	47 Ag silver 108	48 Cd cadmium 112		49 In indium 115	50 Sn tin 117	51 Sb antimony 122	52 Te tellurium 128	53 I iodine 127	54 Xe xenon 131
55 Cs caesium 133	56 Ba barium 137	57–71 lanthanoids	72 Hf hafnium 178	73 Ta tantalum 181	74 W tungsten 184	75 Re rhenium 186	76 Os osmium 190	77 Ir iridium 192	78 Pt platinum 195	79 Au gold 197	80 Hg mercury 201		81 Tl thallium 204	82 Pb lead 207	83 Bi bismuth 209	84 Po polonium –	85 At astatine –	86 Rn radon –
87 Fr francium –	88 Ra radium –	89–103 actinoids	104 Rf rutherfordium –	105 Db dubnium –	106 Sg seaborgium –	107 Bh bohrium –	108 Hs hassium –	109 Mt meitnerium –	110 Ds darmstadtium –	111 Rg roentgenium –	112 Cr copernicium –		114 Fl flerovium –		116 Lv livermorium –			

lanthanoids	57 La lanthanum 139	58 Ce cerium 140	59 Pr praseodymium 141	60 Nd neodymium 144	61 Pm promethium –	62 Sm samarium 150	63 Eu europium 152	64 Gd gadolinium 157	65 Tb terbium 159	66 Dy dysprosium 163	67 Ho holmium 165	68 Er erbium 167	69 Tm thulium 169	70 Yb ytterbium 173	71 Lu lutetium 175
actinoids	89 Ac actinium –	90 Th thorium 232	91 Pa protactinium 231	92 U uranium 238	93 Np neptunium –	94 Pu plutonium –	95 Am americium –	96 Cm curium –	97 Bk berkelium –	98 Cf californium –	99 Es einsteinium –	100 Fm fermium –	101 Md mendelevium –	102 No nobelium –	103 Lr lawrencium –

You met the Periodic Table briefly in Chapter 3. Let's review its key points.

- The Periodic Table is a way of classifying the elements.

- It shows them in order of their proton number.
  Lithium has 3 protons, beryllium has 4, boron has 5, and so on.
  (The proton number is the *upper* number beside each symbol.)

- When arranged by proton number, the elements show **periodicity**:
  elements with similar properties appear at regular intervals.
  The similar elements are arranged in columns.

- Look at the columns numbered I to VIII. The elements in these form
  families called **groups**.

- The rows are called **periods**. They are numbered 0 to 7.

- The heavy zig-zag line above separates **metals** from **non-metals**, with
  the non-metals to the right (except for hydrogen).

More about the groups

- The group number is the same as the number of outer-shell electrons
  in the atoms, except for helium in Group VIII. In Group I the atoms
  have one outer-shell electron, in Group II they have two, and so on.

- The outer-shell electrons are also called **valency electrons**. And they
  are very important: they dictate how an element behaves.

- So all the elements in a group have similar reactions, because they
  have the same number of valency electrons.

- The atoms of the Group VIII elements have a very stable arrangement
  of electrons in their outer shells. This makes them **unreactive**.

The small numbers
The two numbers beside a symbol
tell you about the particles in the
nucleus of its atoms:

    proton number
              symbol
  nucleon number

- The proton number is the number
  of protons.
- The nucleon number is the total
  number of particles in the nucleus
  (protons + neutrons)

These numbers are for the main
isotope of each element.

Groups with special names
Group I: **the alkali metals**
Group II: **the alkaline earth metals**
Group VII: **the halogens**
Group VIII: **the noble gases**

More about the periods

The period number tells you the number of electron shells in the atoms.
So in the elements of Period 2, the atoms have two electron shells.
In Period 3 they have three, and so on.

The metals and non-metals

Look again at the table. The metals are to the left of the zig-zag line. There are
far more metals than non-metals. In fact over 80% of the elements are metals.
Metals and non-metals have very different properties. See Unit 3.5 for more.

Hydrogen

Find hydrogen in the table. It sits alone. That is because it has one outer
electron, and forms a positive ion (H^+) like the Group I metals – but
unlike them it is a gas, and usually reacts like a non-metal.

The transition elements

The **transition elements**, in the block in the middle of the Periodic Table,
are all metals. There is more about these in Unit 12.5.

Artificial elements

Some of the elements in the Periodic Table are artificial: they have been
created in the lab. Most of these are in the lowest block. They include
neptunium (Np) to lawrencium (Lr) in the bottom row. These artificial
elements are radioactive, and their atoms break down very quickly.
(That is why they are not found in nature.)

Patterns and trends in the Periodic Table

As you saw, the elements in a group behave in a similar way. But they also
show **trends**. For example as you go down Group I, the elements become
more reactive. Down Group VII, they become *less* reactive.

Across a period there is another trend: a change from metal to non-metal.
For example in Period 3, only sodium, magnesium, and aluminium are
metals. The rest are non-metals.

So if you know where an element is, in the Periodic Table, you can use the
patterns and trends to predict how it will behave.

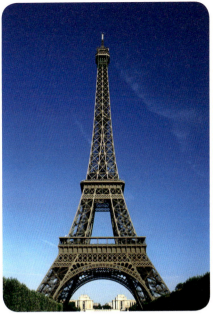

▲ A world-famous structure, made
from iron. Find iron in the Periodic Table.
Which block is it in?

▲ Aluminium is used for drinks cans.
How many valency electrons?

Q

1 Use the Periodic Table to find the names of:
   a three metals in common use around you
   b two non-metals that you breathe in.
2 Using *only* the Periodic Table to help you, write down
   everything you can about: a nitrogen   b magnesium
3 Only two groups in the table are completely non-metal.
   Which two?

4 Name three elements that are likely to react in a similar
   way to: a sodium   b fluorine
5 Which is likely to be more reactive, oxygen or krypton?
   Why?
6 Which element is named after:
   a Europe?   b Dmitri Mendeleev?   c America?
7 Chemists consider the Periodic Table very useful. Why?

12.2 Group I: the alkali metals

What are they?

The **alkali metals** are in Group I in the Periodic Table: lithium, sodium, potassium, rubidium, caesium and francium. Only the first three of these are safe to keep in the school lab. The rest are violently reactive.

Their physical properties

The alkali metals are *not* typical metals.

- Like all metals, they are good conductors of heat and electricity.
- But they are softer than most other metals. You can cut them with a knife.
- They are 'lighter' than most other metals – they have **low density**. So they float on water – while reacting with it.
- They have low melting and boiling points, compared with most metals.

▲ A piece of sodium, cut with a knife.

The trends in their physical properties

Like any family, the alkali metals are all a little different. Look at this table:

Metal	This metal is silvery and …	Density in g/cm³	Melts at /°C
lithium, Li	soft	0.53	181
sodium, Na	a little softer	0.97	98
potassium, K	softer still	0.86	63
rubidium, Rb	even softer	1.53	39
caesium, Cs	the softest	1.88	29

softness increases *density increases* *melting points decrease*

So there is an overall increase or decrease for each property, as you go down the table. This kind of pattern is called a trend.

Their chemical properties

Let's compare the reactions of lithium, sodium, and potassium, in the lab.

1 Reaction with water

All three react violently with water, giving hydrogen and a hydroxide.

Experiment	What you see
metal trough of water and indicator	• lithium floats and fizzes • sodium shoots across the water • potassium melts with the heat of the reaction, and the hydrogen catches fire

increasing reactivity

Note the trend in reactivity. For sodium the reaction is:

sodium + water → sodium hydroxide + hydrogen

Sodium hydroxide is an alkali, so the indicator changes colour.

The alkali metals react vigorously with water. Hydrogen bubbles off, leaving solutions of their hydroxides, which are alkalis.

2 Reaction with chlorine

If you heat the three metals, and plunge them into gas jars of chlorine, they burst into flame. They burn brightly, forming chlorides. For example:

sodium + chlorine → sodium chloride

3 Reaction with oxygen

The three metals also burst into flame when you heat them and plunge them into gas jars of oxygen. They burn fiercely to form **oxides**. These dissolve in water to give alkaline solutions.

The same trend in reactivity is shown in all three reactions. Each time, lithium is the least reactive of the three elements, and potassium the most:

Reactivity increases as you go down Group I.

Why do they react in a similar way?

All the alkali metals react in a similar way. Why? Because they have the same number of valency (outer-shell) electrons:

| 2,1 | 2,8,1 | 2,8,8,1 |

Atoms with the same number of valency electrons react in a similar way.

Why are they so reactive?

The alkali metals are the most reactive of all the metals.

Why? Because they need to lose only one electron, to gain a stable outer shell. So they have a strong drive to react with other elements and compounds, in order to give up this electron. They become **ions**. The compounds they form are **ionic**. For example sodium chloride is made up of the ions Na^+ and Cl^-.

The alkali metals form ionic compounds, in which the metal ion has a charge of 1+. The compounds are white solids. They dissolve in water to give colourless solutions.

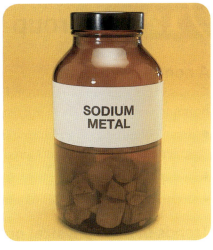

▲ Lithium, sodium, and potassium are stored under oil in the lab, to prevent reaction with oxygen and water.

Why does reactivity *increase* down Group I?

In reactions, the Group I atoms lose their outer electron, to gain a stable outer shell.

The more shells there are, the further the outer electron is from the positive nucleus – so the easier it is to lose.

And the easier it is to lose an electron, the more reactive the metal will be!

And the winner is …

Lithium is the lightest of all metals.

Q

1 a What is the other name for the Group I elements?
 b Why are they called that?
2 Which best describes the Group I metals:
 a soft or hard? **b** reactive or unreactive?
3 The Group I metals show a *trend* in melting points.
 a What does that mean?
 b Describe two other physical trends for the group.
 c One measurement in the table on page 164 does *not* fit the trend. See if you can spot it.

4 a What forms when potassium reacts with chlorine?
 b What colour is this compound?
 c What will you *see* when you dissolve it in water?
 d Will the solution conduct electricity? Explain.
5 Which holds its outer electron more strongly: a lithium atom, or a sodium atom? Explain why you think so.
6 Rubidium is below potassium, in Group I. Predict how it will react with: **a** water **b** chlorine and describe the products that form.

12.3 Group VII: the halogens

A non-metal group

Group VII is a group of non-metal elements. It includes fluorine, chlorine, bromine, and iodine. These are usually called the **halogens**. They all:

- form coloured gases. Fluorine is a pale yellow gas and chlorine is a green gas. Bromine forms a red vapour, and iodine a purple vapour.

- are poisonous.

- form diatomic molecules (containing two atoms). For example, Cl_2.

Trends in their physical properties

As usual, the group shows trends in physical properties. Look at these:

Halogen	At room temperature the element is ...			Boiling point / °C	
fluorine, F_2	a yellow gas			−188	
chlorine, Cl_2	a green gas	colour gets deeper	density increases	−35	boiling points increase
bromine, Br_2	a red liquid			59	
iodine, I_2	a black solid			184	

Trends in their chemical properties

The halogens are among the most reactive elements in the Periodic Table. They react with metals to form compounds called **halides**. For example:

Halogen	Reaction with iron wool		The product	Its appearance
fluorine	Iron wool bursts into flame as fluorine passes over it – without any heating!		iron(III) fluoride, FeF_3	pale green solid
chlorine	Hot iron wool glows brightly when chlorine passes over it.	reactivity decreases	iron(III) chloride, $FeCl_3$	yellow solid
bromine	Hot iron wool glows, but less brightly, when bromine vapour passes over it.		iron(III) bromide, $FeBr_3$	red-brown solid
iodine	Hot iron wool shows a faint red glow when iodine vapour passes over it.		iron(III) iodide, FeI_3	black solid

So they all react in a similar way. But note the trend in reactivity:
Reactivity decreases as you go down Group VII.

Why do they react in a similar way?

The halogens react in a similar way because their atoms all have 7 valency (outer-shell) electrons. Compare the fluorine and chlorine atoms:

2,7   2,8,7

Atoms with the same number of valency electrons react in a similar way.

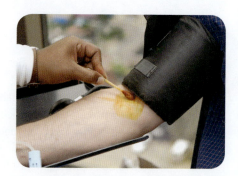

▲ Iodine is a disinfectant. His skin is being wiped with a solution of iodine in ethanol, before he gives blood.

Why are they so reactive?

The halogen atoms need just one more electron to reach a stable outer shell of 8 electrons. So they have a strong drive to react with other elements or compounds, to gain this electron. That is why they are so reactive.

When halogen atoms react with metal atoms they accept electrons, forming halide ions. So the products are ionic. For example the reaction between iron and chlorine gives iron(III) chloride, made up of Fe^{3+} and Cl^- ions.

But with non-metal atoms such as hydrogen and carbon, they share electrons, forming molecules with covalent bonds. For example hydrogen and chlorine atoms share electrons, to form molecules of hydrogen chloride, HCl.

How the halogens react with halides

1   When chlorine water (a solution of chlorine) is added to a colourless solution of potassium bromide, the solution turns orange, as shown in the photo. This reaction is taking place:

$$Cl_2\,(aq) \quad + \quad 2KBr\,(aq) \quad \longrightarrow \quad 2KCl\,(aq) \quad + \quad Br_2\,(aq)$$
colourless   orange

Bromine has been pushed out of its compound, or **displaced**.

2   And when chlorine water is added to a colourless solution of potassium iodide, the solution turns red-brown, because of this reaction:

$$Cl_2\,(aq) \quad + \quad 2KI\,(aq) \quad \longrightarrow \quad 2KCl\,(aq) \quad + \quad I_2\,(aq)$$
colourless   red-brown

This time iodine has been displaced.

But what happens if you use bromine or iodine instead of chlorine? This table gives the results:

▲ Chlorine displacing bromine from aqueous potassium bromide.

If the solution contains ...	when you add chlorine ...	when you add bromine ...	when you add iodine ...
chloride ions (Cl^-)		there is no change	there is no change
bromide ions (Br^-)	bromine is displaced		there is no change
iodide ions (I^-)	iodine is displaced	iodine is displaced	

You know already that chlorine is more reactive than bromine, and bromine is more reactive than iodine. So now you can see that:

A halogen will displace a less reactive halogen from a solution of its halide.

Q

1   What do the halogens look like? Describe them.

2   **a** Describe the trend in reactivity in Group VII.
     b Is this trend the same as for Group I? (Check back!)

3   **a** Describe any similarities you notice in the *products* that form when the halogens react with iron wool.
     b Which type of bonding do they have?

4   What makes the halogens so reactive?

5   **a** Write a word equation for the reaction of bromine with potassium iodide. What do you expect to *see*?
     b Now explain *why* the reaction in **a** occurs.

6   The fifth element in Group VII is called astatine. It is a very rare element. Do you expect it to be:
     a a gas, a liquid, or a solid? Give your reason.
     b coloured or colourless?   **c** harmful or harmless?

12.4 Group VIII: the noble gases

The noble gases

This group of non-metals contains the elements helium, neon, argon, krypton, and xenon. These elements are all:

- non-metals

- colourless gases, which occur naturally in air

- **monatomic** – they exist as single atoms

- unreactive. This is their most striking property. They do not normally react with anything. That is why they are called **noble**.

Why are they unreactive?

As you have seen, atoms react in order to gain a stable outer shell of electrons. But the atoms of the noble gases already have a stable outer shell – with 8 electrons, except for helium which has 2 (since it has only one shell):

He	Ne	Ar
a helium atom	a neon atom	an argon atom

So the atoms have no need to react in order to gain or lose electrons. **The noble gases are unreactive, and monatomic, because their atoms already have a stable outer electron shell.**

Trends in their physical properties

Like all groups, the Group VIII elements do show trends. Look at this table.

Noble gas	% in air
helium	tiny traces
argon	just under 1 %
neon	0.002 %
krypton	0.0001 %
xenon	less than 0.0001 %

Helium is the second most abundant element in the universe, after hydrogen. But it is so light that it escapes from our atmosphere.

Where we get them
We obtain helium from natural gas, in which it is an impurity.
We get the other noble gases from the air, in the fractional distillation of liquid air (page 208).

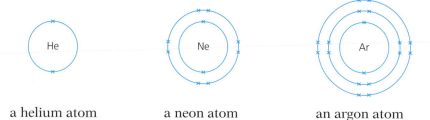

Noble gas	Its atoms		A balloon full of this gas …		Boiling point / °C	
helium	$^{4}_{2}$He		rises quickly into the air		−269	
neon	$^{20}_{10}$Ne	the atoms increase in size and mass	rises slowly	the density of the gases increases	−246	the boiling points increase
argon	$^{40}_{18}$Ar		falls slowly		−186	
krypton	$^{84}_{36}$Kr		falls quickly		−152	
xenon	$^{131}_{54}$Xe		falls very quickly		−107	

The gases grow denser (or 'heavier') down the group, because the mass of the atoms increases. The increase in boiling points is a sign of increasing attraction between atoms. It gets harder to separate them to form a gas.

Compare these physical trends with those for the Group VII non-metals on page 166. What do you notice?

▲ Colourful signs in Tokyo, thanks to neon.

▲ Cool blue headlamps, thanks to xenon.

Uses of the noble gases

The noble gases are unreactive or inert, which makes them safe to use. They also glow when a current is passed through them at low pressure. These properties lead to many uses.

- Helium is used to fill balloons and airships, because it is much lighter than air – and will not catch fire.

- Argon is used to provide an inert atmosphere. For example it is used:
  - as a filler in tungsten light bulbs. (If air were used, the oxygen in it would make the tungsten filament burn away.)
  - to protect metals that are being welded. It won't react with the hot metals (unlike the oxygen in air).

- Neon is used in advertising signs. It glows red, but the colour can be changed by mixing it with other gases.

- Krypton is used in lasers – for example for eye surgery – and in car headlamps.

- Xenon gives a light like bright daylight, but with a blue tinge. It is used in lighthouse lamps, lights for hospital operating rooms, and car headlamps.

▲ There is an easy way to blow up balloons: buy a canister of helium.

Q

1 Why do the members of Group VIII have similar properties?

2 Explain why the noble gases are unreactive.

3 a What are the trends in density and boiling point for the noble gases?

  b Are these trends the same as for:
    i Group I?   ii Group VII?   (Check back!)

4 The noble gases are widely used. Explain why, and give one use for each.

5 The sixth element in Group VIII is radon (Rn). Would you expect it to be:

  a a gas, a liquid, or a solid, at room temperature?

  b heavier, or lighter, than xenon?

  c chemically reactive?

12.5 The transition elements

What are they?

The transition elements are the block of 30 elements in the middle of the Periodic Table. They are all metals, and include most of the metals we use every day – such as iron, tin, copper, and silver.

The transition elements

Their physical properties

Here are three of the transition elements:

Iron: the most widely used metal; grey with a metallic lustre (shine).

Copper: reddish with a metallic lustre.

Nickel: silvery with a metallic lustre.

Here is some data for them, with sodium for comparison:

Element	Symbol	Density in g/cm³	Melting point/°C
iron	Fe	7.9	1535
copper	Cu	8.9	1083
nickel	Ni	8.9	1455
sodium	Na	0.97	98

The transition elements share these physical properties:

- **hard, tough, and strong**. They are not soft like the Group I metals.
- **high melting points**. Look at the values in the table. But mercury is an exception. It is a liquid at room temperature. (It melts at $-39\,°C$.)
- **malleable** (can be hammered into different shapes) and **ductile** (can be drawn out into wires).
- **good conductors of heat and electricity**. Of all the metals, silver is the best conductor of electricity, and copper is next.
- **high density**. They are heavy. 1 cm³ cube of iron weighs 7.9 grams – over 8 times more than 1 cm³ cube of sodium.

Some transition elements

iron	copper
nickel	zinc
silver	gold
platinum	mercury
chromium	cadmium

Their chemical properties

1. **They are much less reactive than the Group I metals.**
   For example copper and nickel do not react with water, or catch fire in air – unlike sodium. In general, the transition elements do not corrode readily in the atmosphere. But iron is an exception – it rusts easily. We spend a fortune every year on rust prevention.

2. **They show no clear trend in reactivity, unlike the Group I metals.**
   But those next to each other in the Periodic Table do tend to be similar.

3. **Most transition elements form coloured compounds.** In contrast, the Group I metals form white compounds.

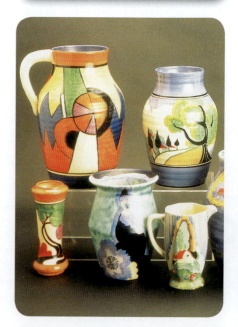

▲ Because they are coloured, compounds of the transition elements are used in pottery glazes.

4 Most can form ions with different charges. Compare these:

Metal	Charge on ions	Examples
Group I metals	always 1+	sodium: Na^+
Group II metals	always 2+	magnesium: Mg^{2+}
Group III metals	always 3+	aluminium: Al^{3+}
Transition elements	variable	copper: Cu^+, Cu^{2+} iron: Fe^{2+}, Fe^{3+}

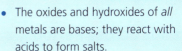

Salts of transition elements
- The oxides and hydroxides of *all* metals are bases; they react with acids to form salts.
- So you can make salts of the transition elements by starting with their oxides or hydroxides, and reacting these with acids.

5 They can form more than one compound with another element. That's because they can form ions with different charges. Look:

copper(I) oxide, Cu_2O copper(II) oxide, CuO

iron(II) oxide, FeO iron(III) oxide, Fe_2O_3

The Roman numeral tells you how many electrons each metal atom has lost, to form the compound. This number is called its **oxidation state**. **The transition elements have variable oxidation states.**

6 Most transition elements can form complex ions. For example, if you add ammonia to a solution containing copper(II) ions, a pale blue precipitate of copper(II) hydroxide forms. It dissolves again if you add more ammonia, giving a deep blue solution.

It dissolves because each copper ion attracts four ammonia molecules and two water molecules, forming a large soluble **complex ion**. This ion gives the solution its deep blue colour.

Testing for copper ions
The reaction in point 6 is used in the test for copper(II) ions (page 281.)
The formula of the complex ion that forms is $[Cu(H_2O)_2(NH_3)_4]^{2+}$.

Uses of the transition elements
- The hard, strong transition elements are used in structures such as bridges, buildings, and cars. Iron is the most widely used – usually in the form of **alloys** called **steels**. (In alloys, small amounts of other substances are mixed with a metal, to improve its properties.)

- Many transition elements are used in making alloys. For example, chromium and nickel are mixed with iron to make **stainless steel**.

- Transition elements are used as conductors of heat and electricity. For example, steel is used for radiators, and copper for electric wiring.

- Many transition elements and their compounds acts as **catalysts**. Catalysts speed up reactions, while remaining unchanged themselves. For example, iron is used as a catalyst in making ammonia (page 123).

▲ Iron rods give the building strength.

Q

1 Name five transition elements.
2 Which best describes the transition elements, overall:
   a soft or hard?   b high density or low density?
   c high melting point or low melting point?
   d reactive or unreactive, with water?
3 What is unusual about mercury?

4 Most paints contain compounds of transition elements. Why do you think this is?
5 A certain metal compound has (IV) in its name. To which block in the Periodic Table does the metal belong?
6 Suggest reasons why copper is used in hot water pipes, while iron is not.

12.6 Across the Periodic Table

Trends across Period 3

As you saw, there are trends within groups in the Periodic Table. There are also trends across a period. Look at this table for Period 3:

Group	I	II	III	IV	V	VI	VII	VIII
Element	sodium	magnesium	aluminium	silicon	phosphorus	sulfur	chlorine	argon
Valency electrons	1	2	3	4	5	6	7	8
Element is a . . .	metal	metal	metal	metalloid	non-metal	non-metal	non-metal	non-metal
Reactivity	high →			low →			high	unreactive
Melting point/°C	98	649	660	1410	590	119	−101	−189
Boiling point/°C	883	1107	2467	2355	(ignites)	445	−35	−186
Oxide is . . .	basic		amphoteric				acidic	–
Typical compound	NaCl	$MgCl_2$	$AlCl_3$	$SiCl_4$	PH_3	H_2S	HCl	–
Valency shown in that compound	1	2	3	4	3	2	1	–

Notice these trends across the period:

1 The number of valency (outer-shell) electrons increases by 1 each time. It is the same as the group number, for Groups I to VII.

2 The elements go from metal to non-metal. Silicon is in between. It is like a metal in some ways and a non-metal in others. It is called a **metalloid**.

3 Melting and boiling points rise to the middle of the period, then fall to very low values on the right. (Only chlorine and argon are gases at room temperature.)

4 The oxides of the metals are **basic** – they react with acids to form salts. Those of the non-metals are **acidic** – they react with alkalis to form salts. But aluminium oxide is in between: it reacts with both acids and alkalis to form salts. So it is called an **amphoteric oxide**. (See page 153 for more.)

The elements in Period 2 show similar trends.

The change from metal to non-metal

The change from metal to non-metal is not clear-cut. Silicon is called a metalloid because it is like metal in some ways, and a non-metal in others.

In fact there are metalloids in all the periods of the table. They lie along the zig-zag line that separates metals from non-metals. Look on the right.

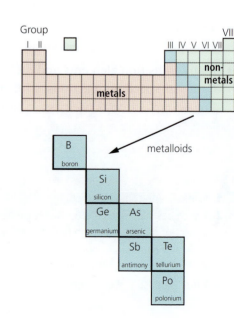

Metals conduct electricity. Metalloids can too, under certain conditions. So they are called **semi-conductors**. This leads to their use in computer chips and PV cells for solar power. Silicon is used the most.

Valency

Look at the last two rows in the table. One shows a typical compound of each element. The other shows the **valency** of the element in that compound.

The valency of an element is the number of electrons its atoms lose, gain or share, to form a compound.

Sodium always loses 1 electron to form a compound. So it has a valency of 1. Chlorine shares or gains 1, so it also has a valency of 1. Valency rises to 4 in the middle of the period, then falls again. It is zero for the noble gases.

Note that *valency* is not the same as *the number of valency electrons*. But:

- the valency does match the number of valency electrons, up to Group IV
- the valency matches the charge on the ion, where an element forms ions.

What about reactivity?

As you know, metal atoms lose their outer electrons when they react, while non-metal atoms accept or share electrons.

Reactivity across Period 3 changes roughly like this:

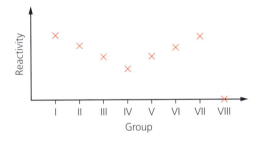

Note that:

- reactivity *decreases* across the metals. Aluminium is a lot less reactive than sodium, for example. Why? Because the more electrons a metal atom needs to lose, the more difficult it is. (The electrons must have enough energy to overcome the pull of the nucleus.)

- reactivity *increases* across the non-metals (apart from Group VIII). So chlorine is more reactive than sulfur. Why? Because the fewer electrons a non-metal atom needs to gain, the easier it is to attract them.

▲ Silicon occurs naturally in sand as silicon (IV) oxide (silicon dioxide or silica). To extract it the silicon (IV) oxide is heated with carbon (coke).

▲ Silicon is the main element used in solar cells, to generate electricity from sunlight. It has to be 99.9999% pure!

Q

1 a Describe how the number of valency electrons changes with group number, across the Periodic Table.

b Describe the change in character from metal to non-metal, across Period 3.

2 How does the reactivity of the metals change as you move across a period? Why?

3 What does *valency of an element* mean? Give two examples.

4 What is a *metalloid*? Give three examples.

5 What is a *semi-conductor*? Name one.

6 a A challenge! Make a table like the one opposite, but for *Period 2*. For each element the table should show:

i the group number

ii the name of the element

iii the number of valency electrons it has

iv a typical compound

v the valency shown in that compound.

b Now try to predict melting and boiling points for the elements in the period. (The earlier units may help!)

How the Periodic Table developed

Life before the Periodic Table

Imagine you find a box of jigsaw pieces. You really want to build that jigsaw. But the lid has only scraps of the picture. Many of the pieces are missing. And the image on some pieces is not complete. How frustrating!

That's how chemists felt, about 150 years ago. They had found more and more new elements. For example 24 metals, including lithium, sodium, potassium, calcium, and magnesium, were discovered between 1800 and 1845. They could tell that these fitted a pattern of some kind. They could see fragments of the pattern – but could not work out what the overall pattern was.

And then the Periodic Table was published in 1869, and everything began to make sense.

A really clever summary

The Periodic Table is *the* summary of chemistry. It names the elements that make up our world. It shows the families they belong to, and how these relate to each other. It even tells you about the numbers of protons, electrons, and electron shells in their atoms.

Today we take the Periodic Table for granted. But it took hundreds of years, and hard work by hundreds of chemists, to develop. There were some good tries along the way, like the 'Law of Octaves'.

The Law of Octaves

By 1863, 56 elements were known. John Newlands, an English chemist, noted that there were many pairs of similar elements. In each pair, the **atomic weights** (or relative atomic masses) differed by a multiple of 8. So he produced a table with the elements in order of increasing atomic weight, and put forward the **Law of Octaves**: *an element behaves like the eighth one following it in the table*.

This was the first table to show a repeating or **periodic** pattern of properties. But it had many inconsistencies. For example it had copper and sodium in the same group – even though they behave very differently. So it was rejected by other chemists.

▲ 250 years ago, nobody knew of aluminium. Today, planes are about 80% aluminium by mass.

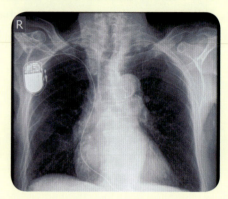

▲ Lithium was discovered in 1817. Lithium batteries are used in pacemakers, to keep the heartbeat steady.

Examples of octaves

element	atomic weight
potassium	39
sodium	− 23
	16 or 2 × 8
calcium	40
magnesium	− 24
	16 or 2 × 8

Now we use **relative atomic mass** instead of **atomic weight**.

◄ Newlands knew of all these, in 1865. How many of them can you name? Find Di (for didymium). This 'element' was later found to be a mixture.

Newland's Table of Octaves, presented to the Chemical Society in London in 1865

H	Li	Be	B	C	N	O
F	Na	Mg	Al	Si	P	S
Cl	K	Ca	Cr	Ti	Mn	Fe
Co, Ni	Cu	Zn	Y	In	As	Se
Br	Rb	Sr	Ce, La	Zr	Di, Mo	Ro, Ru
Pd	Ag	Cd	U	Sn	Sb	Te
I	Cs	Ba, V	Ta	W	Nb	Au
Pt, Ir	Tl	Pb	Th	Hg	Bi	Os

▲ Dmitri Mendeleev (1834–1907). Element 101 in the Periodic Table – the artificial element Mendelevium (Md) – is named after him. So is a crater on the moon.

▲ New elements are still being added to the Periodic Table. This is the team that created the artificial element 112, which was officially named copernicium (Cn) in 2010.

The Periodic Table arrives

Dmitri Ivanovich Mendeleev was born in Russia in 1834, the youngest of at least 14 children. By the age of 32, he was a Professor of Chemistry.

Mendeleev had gathered a huge amount of data about the elements. He wanted to find a pattern that made sense of it, to help his students. So he made a card for each of the known elements (by then 63). He played around with the cards on a table, first putting the elements in order of atomic weight, and then into groups with similar behaviour. The result was the Periodic Table. It was published in 1869.

Mendeleev took a big risk: he left gaps for elements not yet discovered. He even named three: **eka-aluminium**, **eka-boron**, and **eka-silicon**, and predicted their properties. And soon three new elements were found, that matched his predictions – gallium, scandium, and germanium. This helped to convince other chemists, and his table was accepted.

Atomic structure and the Periodic Table

Mendeleev had put the elements in order of *atomic weight*. But he was puzzled, because he then had to swop some to get them into the right groups. For example potassium ($A_r = 39$) is lighter than argon ($A_r = 40$), so should come before argon. But a reactive metal like potassium clearly belongs to Group I, not Group VIII. So he switched those two around.

In 1911 the proton was discovered. It soon became clear that the **proton number** was the key factor in deciding an element's position in the table. So Mendeleev was right to swop those elements. But it was still not clear why the groups were so different. Then scientists discovered that:

- the number of electrons equals the number of protons, in an atom
- the electrons are arranged in shells
- the outer-shell electrons dictate reactions. So elements with the same number of outer-shell electrons react in the same way.

By 1932, 63 years after it appeared, Mendeleev's table finally made sense. Today's Periodic Table contains many more elements. But his table, nearly 150 years old, is still the blueprint for it.

▲ Mendeleev knew of aluminium, titanium, and molybdenum, which are all used in today's racing bikes.

▲ Mendeleev would recognise all the elements in these health tablets too.

Checkup on Chapter 12

Revision checklist

Core syllabus content

Make sure you can …

- ☐ state the link between the Periodic Table and proton number
- ☐ point out where in the Periodic Table these are:
  *Group I   Group VII   Group VIII
  hydrogen   the transition elements*
- ☐ define *valency electrons*
- ☐ state the link between:
  – group number and the number of valency electrons
  – period number and the number of electron shells
- ☐ describe the change from metal to non-metal, across a period
- ☐ say why elements in a group react in a similar way
- ☐ give the other name for Group I, and name at least three elements in this group
- ☐ describe the trends in softness, melting point, density, and reactivity, for the Group I elements
- ☐ give at least two typical reactions for Group I elements, and describe the products
- ☐ explain why the Group I elements are so reactive
- ☐ give the other name for Group VII
- ☐ name at least four Group VII elements and say what they look like at room temperature
- ☐ describe the trend in reactivity for Group VII
- ☐ explain why the Group VII elements are so reactive
- ☐ describe how halogens react with solutions of other halides, and explain the pattern
- ☐ give the other name for Group VIII, and name five elements in this group
- ☐ explain why the Group VIII elements are unreactive
- ☐ give one use for each Group VIII element you name
- ☐ give three physical properties and three chemical properties of the transition elements
- ☐ explain why compounds of transition elements often have Roman numerals in their names
- ☐ give some uses of transition elements, including as catalysts

Extended syllabus content

Make sure you can also …

- ☐ say that the transition elements have *variable oxidation states*, and explain what that means
- ☐ give more detail about the trends across a period, including the change from metal to non-metal

Questions

Core syllabus content

1 This extract from the Periodic Table shows the symbols for the first 20 elements.

					H												He
Li	Be											B	C	N	O	F	Ne
Na	Mg											Al	Si	P	S	Cl	Ar
K	Ca																

Look at the row from lithium (Li) to neon (Ne).

a What is this row of the Periodic Table called?

b Which element in it is the least reactive? Why?

Look at the column of elements from lithium (Li) to potassium (K).

c What is this column of the table called?

d Of the three elements shown in this column, which one is the most reactive?

2 Rubidium is an alkali metal. It lies below potassium in Group I. Here is data for Group I:

Element	Proton number	Melting point /°C	Boiling point /°C	Chemical reactivity
lithium	3	180	1330	quite reactive
sodium	11	98	890	reactive
potassium	19	64	760	very reactive
rubidium	37	?	?	?
caesium	55	29	690	violently reactive

a Describe the trends in melting point, boiling point, and reactivity, as you go down the group.

b Now predict the missing data for rubidium.

c In a rubidium atom:
  i how many electron shells are there?
  ii how many electrons are there?
  iii how many valency electrons are there?

3 Identify these non-metal elements:

a a colourless gas, used in balloons and airships

b a poisonous green gas

c a colourless gas that glows with a red light in advertising signs

d a red liquid

e a yellow gas which is so reactive that it is not allowed in school labs

f a black solid that forms a purple vapour when you heat it gently.

4 This diagram shows some of the elements in Group VII of the Periodic Table.

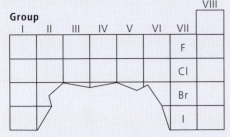

a What are the elements in this group called?

b Chlorine reacts explosively with hydrogen. The word equation for the reaction is:

hydrogen + chlorine → hydrogen chloride

The reaction requires sunlight, but not heat.

   i How would you expect fluorine to react with hydrogen?

   ii Write the word equation for the reaction.

c i How might bromine react with hydrogen?

   ii Write the word equation for that reaction.

5 The Periodic Table is the result of hard work by many scientists, in many countries, over hundreds of years. They helped to develop it by discovering, and investigating, new elements.

The Russian chemist Mendeleev was the first person to produce a table like the one we use today. He put all the elements he knew of into his table. But he realised that gaps should be left for elements not yet discovered. He even predicted the properties of some of these.

Mendeleev published his Periodic Table in 1869. The extract on the right below shows Groups I and VII from his table. Use the modern Periodic Table (page 324) to help you answer these questions.

a What does Period 2 mean?

b i How does Group I in the modern Periodic Table differ from Group I in Mendeleev's table?

   ii The arrangement in the modern table is more appropriate for Group I. Explain why.

   iii What do we call the Group I elements today?

c i What do we call the Group VII elements?

   ii The element with the symbol Mn is out of place in Group VII. Why?

   iii Where is the element Mn in today's table?

d Mendeleev left gaps in several places in his table. Why did he do this?

e There was no group to the right of Group VII, in Mendeleev's table. Suggest a reason for this omission.

Extended syllabus content

6 This question is about **elements** from these families: alkali metals, alkaline earth metals (Group II), transition elements, halogens, noble gases.

   A is a soft, silvery metal that reacts violently with water.

   B is a gas at room temperature. It reacts violently with other elements, without heating.

   C is an unreactive gas that sinks in air.

   D is a hard solid at room temperature, and shows variable oxidation states in its compounds.

   E conducts electricity, and reacts slowly with water. Its atoms each give up two electrons.

   F is a reactive liquid; it does not conduct electricity; it shows a valency of 1 in its compounds.

   G is a hard solid that conducts electricity, can be beaten into shape, and rusts easily.

   a For each element above, say which of the listed families it belongs to.

   b i Comment on the position of elements **A**, **B**, and **C** within their families.

   ii Describe the valence (outer) shell of electrons for each of the elements **A**, **B**, and **C**.

   c Explain why the arrangement of electrons in their atoms makes some elements very reactive, and others unreactive.

   d Name elements that fit descriptions **A** to **G**.

   e Which of **A** to **G** may be useful as catalysts?

7 The elements of Group VIII are called the noble gases. They are all monatomic gases.

   a Name four of the noble gases.

   b i What is meant by *monatomic*?

   ii Explain *why* the noble gases, unlike all other gaseous elements, are monatomic.

   When elements react, they become like noble gases.

   c i Explain what the above statement means.

   ii What can you conclude about the reactivity of Group VII *ions*?

An extract from Mendeleev's Periodic Table

	Group I	Group VII
Period 1	H	
Period 2	Li	F
Period 3	Na	Cl
Period 4	K	Mn
	Cu	Br
Period 5	Rb	
	Ag	I

13.1 Metals: a review

So far …
You have met quite a lot of information about metals already. We review it in this unit, before going on to look more closely at their reactivity.

Metals and the Periodic Table
The metals lie to the left of the zig-zag line in the Periodic Table. There are far more metals than non-metals. Here is a reminder of some of them:

Group I – the alkali metals (lithium, sodium, potassium …)

Group III has aluminium, the most abundant metal in the Earth's crust, in this position …

Group

I   II   III IV V VI VII   VIII

transition elements   non-metals

metals

Group II – the alkaline earth metals (beryllium, magnesium, calcium …)

the transition elements – all are metals, and they include most of the metals in everyday use, like iron, copper, tin, zinc, lead, silver, gold …

The general properties of metals
The general properties of metals are given below. (But they don't all have *all* of these properties!)

Physical properties
1   They are **strong**. If you press on them, or drop them, or try to tear them, they won't break – and it is hard to cut them.
2   They are **malleable**. They can be hammered into shape without breaking.
3   They are **ductile**: they can be drawn out into wires.
4   They are **sonorous**: they make a ringing noise when you strike them.
5   They are shiny when polished.
6   They are good conductors of electricity and heat.
7   They have high melting and boiling points. (They are all solid at room temperature, except mercury.)
8   They have high **density** – they feel 'heavy'. (Look at the blue panel.)

Chemical properties
1   They react with dilute acids to form **salts**. Hydrogen gas is given off.
2   They react with oxygen to form **oxides**, and these oxides are **bases** – they neutralise acids, forming salts and water.
3   Metals form positive ions when they react.
     For metals in the numbered groups in the Periodic Table, the charge on the ion is the same as the group number. But transition elements can form ions with different charges. For example Cu^+ and Cu^{2+}.

What is density?
It tells you 'how heavy'.

$$density = \frac{mass\ (in\ grams)}{volume\ (in\ cm^3)}$$

Compare these:

1 cm³ of iron, mass 7.86 g
density 7.86 g/cm³

1 cm³ of lead, mass 11.34 g
density 11.34 g/cm³

▲ Dropping anchor … helped by the density of the iron.

▲ We are over 96% non-metal (mainly oxygen, carbon, nitrogen, hydrogen) but we also contain metals: calcium, potassium, sodium, magnesium, copper, zinc, iron, and more …

▲ Trucks at a copper mine in the USA. Metals are big business. World trade in metals is worth over 2000 billion US dollars a year, and the metals industry employs around 70 million people.

All metals are different!

The properties on the opposite page are typical of metals.
But all metals are different. They do not share *all* of those properties.
For example, all *do* conduct electricity, and their oxides act as bases.
But compare these:

Iron is malleable and strong. Good for gates like these! But it rusts easily in damp air. And unlike most other metals, it is **magnetic**. It melts at 1530 °C.

Sodium is so soft that you can cut it with a knife. It floats on water – and reacts immediately with it, forming a solution. So no good for gates. It melts at only 98 °C.

Gold is unreactive. It is malleable, ductile, and looks attractive. It is also quite rare. So it is used for jewellery and precious objects. It melts at 1064 °C.

So of those three metals, sodium is clearly the most reactive, and gold the least. In the next two units we will look at reactions you can do in the lab, to compare the reactivities of metals.

Q

1 Not all metals share the typical metal properties. See if you can name a metal (not shown in the photos) that is *not*:
   a hard and strong   b malleable at room temperature
2 10 cm³ of aluminium weighs 28 g.
   10 cm³ of tin weighs 74 g.
   a Which is more dense, aluminium or tin?
   b How many times more dense is it than the other metal?

3 Suggest reasons for this use of a metal:
   a silver for jewellery   b copper for electrical wiring
4 For some uses, a highly *sonorous* metal is needed.
   See if you can give two examples.
5 Try to think of *two* reasons why:
   a mercury is used in thermometers
   b aluminium is used for soft-drinks cans.

13.2 Comparing metals for reactivity

What does *reactive* mean?

A *reactive* metal has a strong drive to give up electrons and form ions, with stable outer electron shells. So it reacts readily to form compounds. Let's look at some reactions.

1 The reaction of metals with water

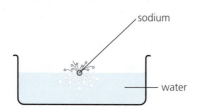

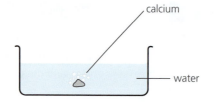

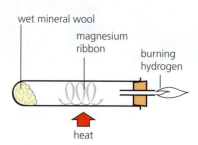

Sodium reacts violently with cold water, whizzing over the surface. Hydrogen gas and a clear solution of sodium hydroxide are formed.

The reaction between calcium and cold water is slower. Hydrogen bubbles off, and a cloudy solution of calcium hydroxide forms.

Magnesium reacts very slowly with cold water, but vigorously on heating in steam: it glows brightly. Hydrogen and solid magnesium oxide form.

You can tell from their behaviour that sodium is the most reactive of the three metals, and magnesium the least.

Compare the equations for the three reactions, below. What pattern do you notice?

$$2Na \ (s) \ + \ 2H_2O \ (l) \ \longrightarrow \ 2NaOH \ (aq) \ + \ H_2 \ (g)$$
$$Ca \ (s) \ + \ 2H_2O \ (l) \ \longrightarrow \ Ca(OH)_2 \ (aq) \ + \ H_2 \ (g)$$
$$Mg \ (s) \ + \ H_2O \ (g) \ \longrightarrow \ MgO \ (s) \ + \ H_2 \ (g)$$

Now compare the reactions of those metals with the others in this table:

Metal	Reaction	Order of reactivity	Products
potassium	very violent with cold water; catches fire	most reactive	hydrogen and a solution of potassium hydroxide, KOH
sodium	violent with cold water		hydrogen and a solution of sodium hydroxide, NaOH
calcium	less violent with cold water		hydrogen and calcium hydroxide, $Ca(OH)_2$, which is only slightly soluble
magnesium	very slow with cold water, but vigorous with steam		hydrogen and solid magnesium oxide, MgO
zinc	quite slow with steam		hydrogen and solid zinc oxide, ZnO
iron	slow with steam		hydrogen and solid iron oxide, Fe_2O_3
copper	no reaction		
silver			
gold		least reactive	

Note the order of reactivity, based on the reaction with water.
And note that only the first three metals in the list produce hydroxides.
The others produce insoluble oxides, if they react at all.

2 The reaction of metals with hydrochloric acid

It is not safe to add sodium or potassium to acid in the lab, because the reactions are explosively fast. But other metals can be tested safely. Compare these reactions with dilute hydrochloric acid:

Metal	Reaction with hydrochloric acid	Order of reactivity	Products
magnesium	vigorous	most reactive	hydrogen and a solution of magnesium chloride, $MgCl_2$
zinc	quite slow		hydrogen and a solution of zinc chloride, $ZnCl_2$
iron	slow		hydrogen and a solution of iron(II) chloride, $FeCl_2$
lead	slow, and only if the acid is concentrated		hydrogen and a solution of lead(II) chloride, $PbCl_2$
copper silver gold	no reaction, even with concentrated acid	least reactive	

The equation for the reaction with magnesium this time is:

$$Mg\,(s)\ +\ 2HCl\,(aq)\ \longrightarrow\ MgCl_2\,(aq)\ +\ H_2\,(g)$$

Now compare the order of the metals in the two tables, and the equations for the reactions. What patterns can you see?

Hydrogen is displaced

When a metal *does* react with water or hydrochloric acid, it drives hydrogen out, and takes its place. This shows that the metal is *more reactive* than hydrogen. It has a stronger drive to form a compound.

But copper and silver do not react with water or acid. So they are *less reactive* than hydrogen.

It is a redox reaction

The displacement of hydrogen is a **redox reaction**. When magnesium reacts with hydrochloric acid, its atoms lose electrons. The hydrogen ions from the acid gain electrons. The half-equations are:

magnesium: $Mg\,(s)\ \longrightarrow\ Mg^{2+}\,(aq) + 2e^-$ (oxidation)
hydrogen ions: $2H^+\,(aq) + 2e^-\ \longrightarrow\ H_2\,(g)$ (reduction)

▲ Magnesium displacing hydrogen from hydrochloric acid.

Remember **OIL RIG!**
Oxidation **I**s **L**oss of electrons.
Reduction **I**s **G**ain of electrons.

Q

1 Write a balanced equation for the reaction of potassium with water.

2 Which is more reactive? And what is your evidence?
   a potassium or sodium?   b copper or zinc?

3 Which gas is always produced if a metal reacts with water, or dilute acid?

4 Explain why the reaction of iron with hydrochloric acid is a redox reaction.

13.3 Metals in competition

When metals compete

You saw that metals can be put in order of reactivity, using their reactions with water and hydrochloric acid. Now let's see what happens when they compete with each other, and with carbon, to form a compound.

1 Competing with carbon

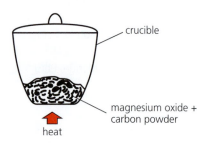

magnesium oxide + carbon powder

heat

crucible

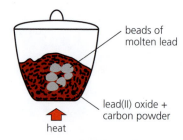

beads of molten lead

lead(II) oxide + carbon powder

heat

calcium	*more reactive than carbon*
aluminium	
carbon	
zinc	
iron	*less reactive than carbon*
copper	

Magnesium oxide is mixed with powdered carbon and heated. No reaction! So magnesium must be more reactive than carbon.

But when lead(II) oxide is used instead, it turns to lead, and carbon dioxide gas forms. So carbon is more reactive than lead.

The oxides of the metals above were also tested. Two were found to be more reactive than carbon. The other three were less reactive than carbon.

The equation for the reaction with lead(II) oxide is:

$$2PbO\ (s)\quad +\quad C\ (s)\quad \longrightarrow\quad 2Pb\ (s)\quad +\quad CO_2\ (g)$$

lead(II) oxide + carbon ⟶ lead + carbon dioxide

So carbon has reduced the lead(II) oxide to lead. The reaction is a redox reaction.

Carbon is more reactive than some metals. It will reduce their oxides to the metal.

2 Competing with other metals, for oxygen

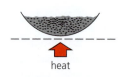

heat

zinc	
iron	the metal grabs oxygen from the oxide of the metal below it
lead	
copper	

Some powdered iron is heated with copper(II) oxide, CuO. Can the iron grab the oxygen from the copper(II) oxide?

The reaction gives out heat, once it gets going. The mixture glows. Iron(II) oxide and copper are formed. The iron has won.

Other metals were compared in the same way. This shows their order of reactivity. It is the same as in the table on page 181.

The tests confirm that iron, zinc, and lead are all more reactive than copper. The equation for the reaction with iron is:

$$Fe\ (s)\quad +\quad CuO\ (s)\quad \longrightarrow\quad FeO\ (s)\quad +\quad Cu\ (s)$$

iron + copper(II) oxide ⟶ iron(II) oxide + copper

The iron is acting as a **reducing agent**, removing oxygen.

A metal will reduce the oxide of a less reactive metal. The reduction always gives out heat – it is exothermic.

3 Competing to form ions in solution

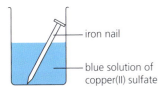
- iron nail
- blue solution of copper(II) sulfate

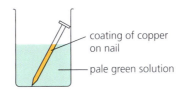
- coating of copper on nail
- pale green solution

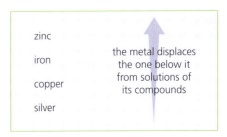
zinc
iron
copper
silver

the metal displaces the one below it from solutions of its compounds

Copper(II) sulfate solution contains blue copper(II) ions and sulfate ions. An iron nail is placed in it. Will anything happen?

Yes! Copper soon coats the nail. The solution turns green, which indicates iron(II) ions. Iron has pushed copper out of solution.

Other metals and solutions were tested too, with the results above. What do you notice about the order of the metals in this list?

Once again, iron wins against copper. It **displaces** the copper from the copper(II) sulfate solution:

$$Fe\ (s)\ +\quad CuSO_4\ (aq)\quad \longrightarrow\quad FeSO_4\ (aq)\quad +\quad Cu\ (s)$$

iron	+	copper(II) sulfate	$\longrightarrow$	iron(II) sulfate	+	copper
		(blue)		(green)		

Other metals displace less reactive metals in the same way.

A metal displaces a less reactive metal from solutions of its compounds.

Comparing the drive to form ions

All the reactions in this unit are redox reactions: electron transfer takes place. Compare the competitions between iron and copper:

	Competing for oxygen	Competing to form ions in solution
Equation	$Fe\ (s)\ +\ CuO\ (s) \longrightarrow FeO\ (s)\ +\ Cu\ (s)$	$Fe\ (s)\ +\ CuSO_4\ (aq) \longrightarrow FeSO_4\ (aq)\ +\ Cu\ (s)$
The half-equations for electron loss for electron gain	$Fe \longrightarrow Fe^{2+}\ +\ 2e^-$ $Cu^{2+}\ +\ 2e^- \longrightarrow Cu$	$Fe \longrightarrow Fe^{2+}\ +\ 2e^-$ $Cu^{2+}\ +\ 2e^- \longrightarrow Cu$
The ionic equation (add the half-equations and cancel the electrons)	$Fe\ +\ Cu^{2+} \longrightarrow Fe^{2+}\ +\ Cu$	$Fe\ +\ Cu^{2+} \longrightarrow Fe^{2+}\ +\ Cu$

In each case iron gives up electrons to form positive ions. And copper ions accept electrons. So iron has a stronger drive than copper, to form ions.
The more reactive metal forms positive ions more readily.

Q

1. In the reaction between carbon and lead(II) oxide, which substance is oxidised?
2. a What do you expect to happen when carbon powder is heated with: **i** calcium oxide? **ii** zinc oxide?
   b Give a word equation for any reaction that occurs in **a**.
3. When chromium(III) oxide is heated with powdered aluminium, chromium and aluminium oxide form. Which is more reactive, chromium or aluminium?

4. Iron displaces copper from copper(II) sulfate solution. Explain what *displaces* means, in your own words.
5. When copper wire is put into a colourless solution of silver nitrate, crystals of silver form on the wire, and the solution goes blue. Explain these changes.
6. For the reaction described in **5**:
   a write the half equations, to show the electron transfer
   b give the ionic equation for the reaction.

13.4 The reactivity series

Pulling it all together: the reactivity series

We can use the results of the experiments in the last two units to put the metals in final order, with the most reactive one first. The list is called **the reactivity series**. Here it is.

The reactivity series		
potassium, K	most reactive	
sodium, Na		
calcium, Ca		
magnesium, Mg		metals above the blue line:
aluminium, Al		carbon can't reduce their oxides
carbon		
zinc, Zn	increasing	
iron, Fe	reactivity	
lead, Pb		metals above the red line: they
		displace hydrogen from acids
hydrogen		
copper, Cu		
silver, Ag		
gold, Au	least reactive	

The non-metals carbon and hydrogen are included for reference.
The list is not complete, of course. You could test many other metals, for example tin, and nickel, and platinum, and add them in the right place.

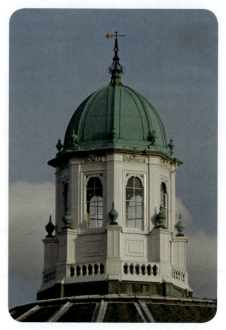

▲ Copper is used for roofing, since it is unreactive. But over time it does form a coat of blue-green copper(II) carbonate.

Things to remember about the reactivity series

1 The reactivity series is really a list of the metals in order of their drive to give up electrons, and form positive ions with stable outer shells. The stronger the drive, the more reactive the metal will be.

2 So a metal will react with a compound of a less reactive metal (for example an oxide, or a salt in solution) by pushing the less reactive metal out of the compound and taking its place.

3 The more reactive the metal, the more **stable** its compounds are. They do not break down easily.

4 The more reactive the metal, the more difficult it is to extract from its ores, since these are stable compounds. For the most reactive metals you need the toughest method of extraction: electrolysis.

5 The less reactive the metal, the less it likes to form compounds. That is why copper, silver and gold are found as elements in the Earth's crust. The other metals are *always* found as compounds.

Metals we had to wait for …
- Because they are easy to obtain from their ores, the less reactive metals have been known and used for thousands of years. For example copper has been in wide use for 6000 years, and iron for 3500 years.
- But the more reactive metals, such as sodium and potassium, had to wait until the invention of electrolysis, in 1800, for their discovery.

▲ A metal's position in the reactivity series will give you clues about its uses. Only unreactive metals are used in coins.

Comparing the stability of some metal compounds

Many compounds break down easily on heating. In other words, they undergo **thermal decomposition**.

But reactive metals have more stable compounds. Will they break down easily? Let's compare some compounds of sodium and copper:

Compound	Effect of heat on the sodium compound	Effect of heat on the copper compound
carbonate	There is no change in this white compound.	This blue-green compound readily breaks down to black copper(II) oxide and carbon dioxide: $$CuCO_3\,(s) \longrightarrow CuO\,(s) + CO_2\,(g)$$
hydroxide	There is no change in this white compound.	This pale blue compound readily breaks down to copper(II) oxide and water: $$Cu(OH)_2\,(s) \longrightarrow CuO\,(s) + H_2O\,(l)$$
nitrate	This white compound partly decomposes to the nitrite and oxygen: $$2NaNO_3\,(s) \longrightarrow 2NaNO_2\,(s) + O_2\,(g)$$ sodium nitrite	This bright blue compound readily breaks down to copper(II) oxide and the brown gas nitrogen dioxide: $$2Cu(NO_3)_2\,(s) \longrightarrow 2CuO\,(s) + 4NO_2\,(g) + O_2\,(g)$$

So the compounds of copper, the less reactive metal, break down easily. The compounds of sodium do not.

The general rules for thermal decomposition

These are the general rules:

- The lower a metal is in the reactivity series, the more readily its compounds decompose when heated.

- Carbonates, *except* those of sodium and potassium, decompose to the oxide and carbon dioxide.

- Hydroxides, *except* those of sodium and potassium, decompose to the oxide and water.

- Nitrates, *except* those of sodium and potassium, decompose to the oxide, nitrogen dioxide, and oxygen. The nitrates of sodium and potassium form nitrites and oxygen.

▲ Limestone (calcium carbonate) being heated in a lime kiln to give calcium oxide (called lime, or quicklime). The lime might be used to make limewash for buildings, or mixed with sand to make lime mortar.

Q

1 a List the metals of the reactivity series, in order.
  b Beside each, say where it occurs in the Periodic Table.
  c To which group in the Periodic Table do the most reactive metals belong?
  d Where in the table are the least reactive ones found?
2 Why is magnesium never found as the element, in nature?

3 Gold has been known and used for thousands of years longer than aluminium. Explain why.
4 Which will break down more easily on heating, magnesium nitrate or silver nitrate? Why?
5 Write a balanced equation for the thermal decomposition of lead(II) nitrate.

13.5 Making use of the reactivity series

Those differences in reactivity are useful!

We make clever use of the differences in reactivity of metals. Here are four examples.

1 The thermite process

This is used to repair rail and tram lines. Powdered aluminium and iron(III) oxide are put in a container over the damaged rail. When the mixture is lit, the aluminium reduces the iron(III) oxide to molten iron, in a very vigorous reaction. The iron runs into the cracks and gaps in the rail, and hardens:

$$Fe_2O_3(s) + 2Al(s) \longrightarrow 2Fe(l) + Al_2O_3(s)$$

2 Making simple cells

The diagram on the right shows a simple cell – two metal strips standing in an electrolyte. (You may have met one on page 116.) The bulb is lit, so a current must be flowing. Hydrogen is forming at the copper strip. The magnesium strip is dissolving. Why is all this happening?

1 Magnesium is more reactive than copper: it has a stronger drive to form positive ions. So when it is connected to the copper strip, it gives up electrons and goes into solution as ions:

$$Mg(s) \longrightarrow Mg^{2+}(aq) + 2e^- \qquad \text{(oxidation)}$$

2 Electrons flow along the wire to the copper strip, as a current. The bulb lights up as the current flows through it.

3 The solution contains Na^+ and Cl^- ions from sodium chloride, and some H^+ and OH^- ions from water. Hydrogen is less reactive than sodium, so the H^+ ions accept electrons from the copper strip:

$$2H^+(aq) + 2e^- \longrightarrow H_2(g) \qquad \text{(reduction)}$$

So the difference in reactivity has caused a redox reaction – that gives out energy in the form of electricity!

A simple cell consists of two different metals in an electrolyte. Electrons flow from the more reactive metal, so it is called the negative pole. The other metal is the positive pole.

Using other metals in simple cells

You can use other metals in place of copper and magnesium, in a simple cell.

A voltmeter measures the 'push' or voltage that makes electrons flow. This chart shows the voltage for different pairs of metals. For example 2.7 V for copper/magnesium, and 0.47 V for copper/lead. **The further apart the metals are in reactivity, the higher the voltage will be.**

Notice how the voltages in the chart add up: 0.47 V for copper/lead, 0.31 for lead/iron, and 0.78 V (0.47 + 0.31) for copper/iron.

▲ The thermite process being used to join new tram lines.

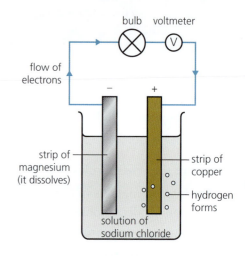

strip of magnesium (it dissolves)

strip of copper

hydrogen forms

solution of sodium chloride

bulb   voltmeter

flow of electrons

The poles in cells …

… are sometimes called **electrodes**. Don't confuse them with the rods in electrolysis!

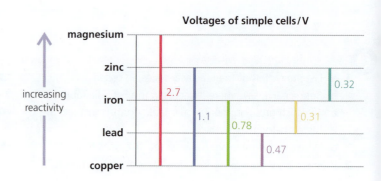

Voltages of simple cells/V

increasing reactivity

magnesium — 2.7

zinc — 1.1   0.32

iron — 0.78   0.31

lead — 0.47

copper

3   The sacrificial protection of iron

Iron is used in big structures such as oil rigs and ships. But it has one big drawback: it reacts with oxygen and water, forming iron(III) oxide or **rust**.

To prevent this, the iron can be teamed up with a more reactive metal such as zinc or magnesium. For example a block of zinc may be welded to the side of a ship. Zinc is more reactive than iron – so the zinc dissolves:

$$2Zn \ (s) \longrightarrow 2Zn^{2+} \ (aq) \ + \ 4e^- \qquad \text{(oxidation)}$$

The electrons flow to the iron, which passes them on, in this reaction:

$$O_2 \ (g) \ + \ 2H_2O \ (l) \ + \ 4e^- \longrightarrow 4OH^- \ (aq) \qquad \text{(reduction)}$$

The overall equation for the reaction is:

$$2Zn \ (s) \ + \ O_2 \ (g) \ + \ 2H_2O \ (l) \longrightarrow 2Zn(OH)_2 \ (aq)$$

So the zinc is oxidised instead of the iron. This is called **sacrificial protection**. The zinc block must be replaced before it all dissolves away.

4   Galvanising

This is another way of using zinc to protect iron. It is used for the steel in car bodies, and the corrugated iron for roofing.

- In **galvanising,** the iron or steel is coated with zinc. For car bodies, this is carried out by a form of electrolysis. For roofing, the iron is dipped in a bath of molten zinc.

- The zinc coating keeps air and moisture away. But if the coating gets damaged, the zinc will still protect the iron, by sacrificial protection.

A note about the reactivity of aluminium

Aluminium is more reactive than iron. But we can use it for things like TV aerials, and satellite dishes, and ladders, without protecting it. Why?

Because aluminium protects itself! It reacts rapidly with oxygen, forming a thin coat of aluminium oxide – so thin you cannot see it.

This sticks tight to the metal, acting as a barrier to further corrosion. So the aluminium behaves as if it were unreactive.

(You saw on page 186 that it reacts very vigorously with iron(III) oxide in the thermite process. But for this, powdered aluminium is used, and a very hot flame to start the reaction off.)

▲ Here blocks of magnesium have been welded to a ship's hull, to prevent the steel (an alloy of iron) from corroding.

▼ The aluminium ladder is protected from corrosion by its oxide layer.

Q

1 A copper rod and an iron rod stand in an electrolyte. If you connect a bulb between them, it will light dimly.
   a Why does the current flow?
   b Which acts as the positive electrode: copper or iron?
   c Suggest two metals you could use to get a brighter light.

2 From the chart on page 186, see if you can work out the voltage for a cell that uses magnesium and zinc.

3 a Steel for cars is *galvanised*. What does that mean?
   b Explain how this protects the steel.

4 Aluminium is more reactive than iron. But unlike iron, we do not need to protect it from corrosion. Why not?

5 a Write a word equation for the thermite reaction.
   b See if you can give *two* reasons why the aluminium is powdered, for this reaction.

Checkup on Chapter 13

Revision checklist

Core syllabus content

Make sure you can …
- ☐ explain these terms used about metals:
  malleable *ductile* *sonorous*
  high density *conductors*
- ☐ give at least five physical properties of metals
- ☐ give three chemical properties of metals
- ☐ explain what a *reactive* element is
- ☐ explain what the reactivity series is, and list the metals in it, in the correct order
- ☐ describe how the metals in the series react with
  – water
  – dilute acids
  and give word equations where reactions occur
- ☐ explain what *displacement of hydrogen* means
- ☐ explain why hydrogen and carbon are often shown in the reactivity series, and say where they fit in
- ☐ predict the products, when carbon is heated with the oxide of a metal below it in the series

Extended syllabus content

Make sure you can also …
- ☐ say that the reactivity of a metal is related to its drive to form positive ions
- ☐ state what the products will be, when:
  – a metal is heated with the oxide of a less reactive metal
  – a metal is placed in the solution of a compound of a less reactive metal
- ☐ explain that those reactions are redox reactions
- ☐ define *thermal decomposition*
- ☐ give the 'rules' for the effect of heat on:
  – metal carbonates
  – metal hydroxides
  – metal nitrates
  and give word equations where reactions occur
- ☐ explain what simple cells are, and
  – say why a current is produced
  – predict which metal will be the positive pole
  – decide which pair of metals will give the largest voltage, and why
- ☐ explain what these are for, and why they work, and name the metals used:
  – sacrificial protection
  – galvanising

Questions

Core syllabus content

1

Metal	Density in g/cm³
aluminium	2.7
calcium	1.6
copper	8.9
gold	19.3
iron	7.9
lead	11.4
magnesium	1.7
sodium	0.97

  a List the metals given in the table above in order of increasing density.
  b i What is meant by *density*?
  ii A block of metal has a volume of 20 cm³ and a mass of 158 g. Which metal is it?
  c Now list the metals in order of reactivity.
  d i The most reactive metal in the list has a density of .....?
  ii The least reactive one has a density of .....?
  iii Does there appear to be a link between density and reactivity? If yes, what?
  e Using low-density metals for vehicles saves money on fuel and road repairs. Explain why.
  f Which of the low-density metals above is the most suitable for vehicles? Why? Give three reasons.

2 This shows metals in order of reactivity:

  sodium   (*most reactive*)
  calcium
  magnesium
  aluminium
  zinc
  iron
  lead
  copper
  silver   (*least reactive*)

  a Which element is stored in oil?
  b Which elements will react with cold water?
  c Choose one metal that will react with steam but *not* cold water. Draw suitable apparatus for this reaction. (Show how the steam is generated.)
  d i Name the gas given off in **b** and **c**.
  ii Name another reagent that reacts with many metals to give the same gas.

3 For each description below, choose one metal that fits the description. Name the metal. Then write a word equation for the reaction that takes place.
  a A metal that displaces copper from copper(II) sulfate solution.
  b A metal that reacts gently with dilute hydrochloric acid.
  c A metal that floats on water and reacts vigorously with it.
  d A metal that reacts quickly with steam but very slowly with cold water.

4 Look again at the list of metals in **2**. Carbon can be placed between zinc and aluminium.
  a Which two of these pairs will react?
     i carbon + aluminium oxide
     ii carbon + copper(II) oxide
     iii magnesium + carbon dioxide
  b Write a word equation for the two reactions, and underline the substance that is reduced.

Extended syllabus content

5 When magnesium powder is added to copper(II) sulfate solution, a displacement reaction occurs and solid copper forms.

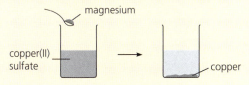

  a Write a word equation for the reaction.
  b Why does the displacement reaction occur?
  c **i** Write a half-equation to show what happens to the magnesium atoms.
     ii Which type of reaction is this?
  d **i** Write a half-equation to show what happens to the copper ions.
     ii Which type of reaction is this?
     iii Which metal shows the greater tendency to form a positive ion?
  e **i** Write the ionic equation for the displacement reaction, by adding the half-equations.
     ii Which type of reaction is it?
  f Use the reactivity series of metals to decide whether these will react together:
     i iron + copper(II) sulfate solution
     ii silver + calcium nitrate solution
     iii zinc + lead(II) nitrate solution
  g For those that react:
     i describe what you would *see*
     ii write the ionic equations for the reactions.

6 When magnesium and copper(II) oxide are heated together, this redox reaction occurs:
$$Mg\ (s)\ +\ CuO\ (s)\ \longrightarrow\ MgO\ (s)\ +\ Cu\ (s)$$
  a What does the word *redox* stand for?
  b For the above reaction, name:
     i the reducing agent **ii** the oxidising agent
  c Describe the electron transfer in the reaction.
  d Explain as fully as you can why the *reverse* reaction does not occur.
  e **i** Name one metal that would remove the oxygen from magnesium oxide.
     ii Does this metal *gain* electrons, or *lose* them, more easily than magnesium does?

7 When the pale blue compound copper(II) hydroxide is heated, thermal decomposition occurs and steam is given off.
  a **i** What does *thermal decomposition* mean?
     ii Write the chemical equation for the reaction.
     iii What colour change would you observe?
  b Name a hydroxide that does not decompose when heated.
  c In further experiments, nitrates of copper and sodium are heated.
     i Which gas is released in both experiments?
     ii One of the nitrates also releases the brown gas nitrogen dioxide. Which one?
     iii Write the equation for this reaction.
  d Relate the observations in **c** to the positions of copper and sodium in the reactivity series.

8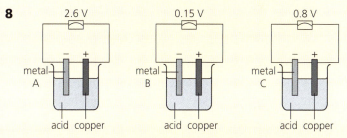

Look at the three cells above.
  a How can you tell that the three unknown metals are all more reactive than copper?
  b Place the metals in order, most reactive first.
  c What voltage will be obtained in a cell using:
     i metals A and B?   **ii** metals B and C?
  d For each cell in **c**, state which metal is the negative terminal.

9 In simple cells, chemical reactions give electricity.
  a Which other set-up also involves electricity and chemical reactions?
  b What is the key difference between it and the simple cell?

14.1 Metals in the Earth's crust

The composition of the Earth's crust

We get some metals from the sea, but most from the Earth's **crust** – the Earth's hard outer layer.

The crust is mostly made of **compounds**. But it also contains some **elements** such as copper, silver, mercury, platinum, and gold. These occur **native,** or uncombined, because they are unreactive.

If you could break all the crust down to elements, you would find it is almost half oxygen! This shows its composition:

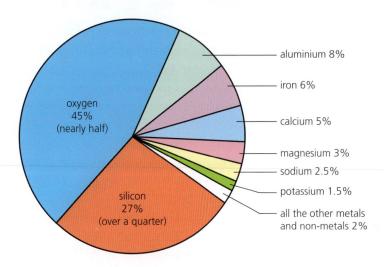

oxygen 45% (nearly half)

silicon 27% (over a quarter)

aluminium 8%
iron 6%
calcium 5%
magnesium 3%
sodium 2.5%
potassium 1.5%
all the other metals and non-metals 2%

▲ Light metals such as aluminium and titanium are used in the International Space Station, 360 km above us.

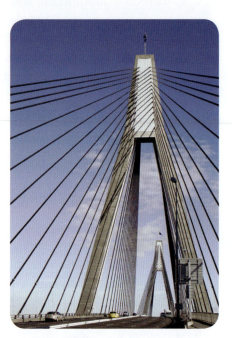

▲ We use about nine times more iron than all the other metals put together.

Note that:

- two non-metals, oxygen and silicon, make up nearly three-quarters of the crust. They occur together in compounds such as silicon dioxide (**silica** or **sand**). Oxygen is also found in compounds such as iron(III) oxide, aluminium oxide, and calcium carbonate.

- just six metals – aluminium to potassium in the pie chart – make up over one-quarter of the crust. Aluminium is the most abundant of these, and iron next. All six occur as compounds, because they are reactive.

Scarce, and precious …

All the other metals *together* make up less than 2% of the Earth's crust. Many, including lead, zinc, and copper, are considered scarce.

Gold, silver, platinum, and palladium are called **precious metals** because they are scarce, expensive, and often kept as a store of wealth.

The car industry uses a lot of metal. Cars are mainly steel, plus over 5% aluminium. But the steel is coated with zinc, and the bumpers with nickel and chromium. The electrics depend on copper, the battery uses lead, and modern exhausts contain palladium, platinum, and rhodium as catalysts.

▼ Metals on wheels.

Metal ores

The rocks in the Earth's crust are a mixture of substances. Some contain a large amount of one metal compound, or one metal, so it may be worth digging them up to extract the metal. Rocks from which metals are obtained are called ores. For example:

This is a chunk of **rock salt**, the main ore of sodium. It is mostly **sodium chloride**.

This is a piece of **bauxite**, the main ore of aluminium. It is mostly **aluminium oxide**.

Since gold is unreactive, it occurs native (uncombined). This sample is almost pure gold.

To mine or not to mine?

Before starting to mine an ore, the mining company must decide whether it is economical. It must find answers to questions like these:

1 How much ore is there?

2 How much metal will we get from it?

3 Are there any special problems about getting the ore out?

4 How much will it cost to mine the ore and extract the metal from it? (The cost will include roads, buildings, mining equipment, the extraction plant, transport, fuel, chemicals, and wages.)

5 How much will we be able to sell the metal for?

6 So will we make a profit if we go ahead?

The answers to these questions will change from year to year. For example if the selling price of a metal rises, even a low-quality or **low-grad**e ore may become worth mining.

The local people may worry that the area will be spoiled, and the air and rivers polluted. So they may object to plans for a new mine. On the positive side, they may welcome the new jobs that mining will bring.

▲ The world's biggest human-made hole: the Bingham Canyon copper mine in Utah, USA. Started in 1906, it is now over 1 km deep and 4 km wide.

Q

1 Which is the main *element* in the Earth's crust?

2 Which is the most common *metal* in the Earth's crust? Which is the second?

3 Gold occurs *native* in the Earth's crust. Explain.

4 Is it true that the most reactive metals are quite plentiful in the Earth's crust?

5 Some metals are called *precious*. Why? Name four.

6 One metal is used more than all the others put together. Which one? Why is it so popular?

7 What is a *metal ore*?

8 Name the main ore of: **a** sodium **b** aluminium What is the main compound in each ore?

14.2 Extracting metals from their ores

Extraction

You have a metal ore. And now you have to **extract** the metal from it. How will you do this? It depends on the metal's **reactivity**.

- The most unreactive metals – such as silver and gold – occur in their ores as elements. All you need to do is separate the metal from sand and other impurities. This is like removing stones from soil. It does not involve chemical reactions.

- The ores of all the other metals contain the metals as compounds. These have to be reduced, to give the metal:

  metal compound $\xrightarrow{\text{reduction}}$ metal

- The compounds of the more reactive metals are very stable, and need electrolysis to reduce them. This is a powerful method, but it costs a lot because it uses a lot of electricity.

- The compounds of the less reactive metals are less stable, and can be reduced to the metal by a chemical reaction.

> **Reduction of metal ores**
>
> Remember, you can define reduction as:
>
> - loss of oxygen
> $Fe_2O_3 \rightarrow Fe$
>
> - or gain of electrons
> $Fe^{3+} + 3e^- \rightarrow Fe$
>
> Either way, the ore is reduced to the metal.

Extraction and the reactivity series

So the method of extraction is strongly linked to the **reactivity series**, as shown below. Carbon is included for reference.

Metal	Method of extraction from ore
potassium	
sodium	
calcium	electrolysis
magnesium	
aluminium	
carbon	
zinc	
iron	heating with carbon or carbon monoxide
lead	
silver	occur naturally as elements so no chemical reactions needed
gold	

metals more reactive

ores more difficult to decompose

method of extraction more powerful

method of extraction more expensive

Using carbon to extract metals

As you saw on page 182, carbon will react with oxides of metals less reactive than itself, reducing them to the metal.

Luckily, many ores are oxides, or compounds easily converted to oxides.

The table shows that carbon can be used to extract zinc, iron, and lead. It is used in the form of coke (made from coal), which is heated with the metal oxide in a furnace. But in the process, carbon may react with a limited supply of oxygen, giving carbon monoxide gas (CO). In that case, the carbon monoxide brings about the reduction.

▲ No need to reduce gold …

Three examples of ore extraction

1 Iron ore This is mainly iron(III) oxide. It is reduced like this:

iron(III) oxide + carbon monoxide ⟶ iron + carbon dioxide

$$Fe_2O_3 (s) + 3CO (g) \longrightarrow 2Fe (l) + 3CO_2 (g)$$

We will look more closely at this extraction in Unit 14.3.

2 Aluminium ore This is mainly aluminium oxide. Aluminium is more reactive than carbon, so electrolysis is needed for this reduction:

aluminium oxide ⟶ aluminium + oxygen

$$2Al_2O_3 (l) \longrightarrow 4Al (l) + 3O_2 (g)$$

We will look more closely at this extraction in Unit 14.4.

3 Zinc blende This is mainly zinc sulfide, ZnS. First it is roasted in air, giving zinc oxide and sulfur dioxide:

zinc sulfide + oxygen ⟶ zinc oxide + sulfur dioxide

$$2ZnS (s) + 3O_2 (g) \longrightarrow 2ZnO (s) + 2SO_2 (g)$$

Then the oxide is reduced in one of these two ways:

i Using carbon monoxide This is carried out in a furnace:

zinc oxide + carbon monoxide ⟶ zinc + carbon dioxide

$$ZnO (s) + CO (g) \longrightarrow Zn (s) + CO_2 (g)$$

The final mixture contains zinc and a slag of impurities. The zinc is separated by fractional distillation. (It boils at 907 °C.)

ii Using electrolysis For this, a compound must be melted, or in solution. But zinc oxide has a very high melting point (1975 °C), and is insoluble in water!

Instead, it is dissolved in dilute sulfuric acid (made from the sulfur dioxide produced in the roasting stage). Zinc oxide is a base, so it neutralises the acid, giving a solution of zinc sulfate. This undergoes electrolysis, and zinc is deposited at the cathode:

$$Zn^{2+} (aq) + 2e^- \longrightarrow Zn (s) \quad \text{(reduction)}$$

The zinc is scraped off the cathode, and melted into bars to sell.

In fact most zinc is extracted by electrolysis, because this gives zinc of very high purity. Cadmium and lead occur as impurities in the zinc blende, and these metals are recovered and sold too.

▲ After extraction, some aluminium is made into rolls like these, ready for sale.

Zinc metal is used …
- to **galvanise** iron – coat it to stop it rusting (page 187)
- in the **sacrificial protection** of iron structures (page 187)
- to make **alloys** such as brass and bronze (page 199)
- to make batteries (page 118)

▲ An iron bucket, galvanised with zinc.

Q

1 Why is no chemical reaction needed to get gold?

2 Lead is extracted by heating its oxide with carbon:

lead oxide + carbon ⟶ lead + carbon monoxide

a Why can carbon be used for this reaction?

b One substance is *reduced*. Which one?

c Which substance brings about the reduction?

3 The reaction in question **2** is a *redox* reaction. Why?

4 Sodium is extracted from rock salt (sodium chloride).

a Electrolysis is needed for this. Explain why.

b Write a balanced equation for the reaction.

5 Zinc blende is an ore of zinc. It is mainly … ?

6 Describe the extraction of zinc by electrolysis.

14.3 Extracting iron

The blast furnace

This diagram shows the **blast furnace** used for extracting iron from its ore. It is an oven shaped like a chimney, at least 30 metres tall.

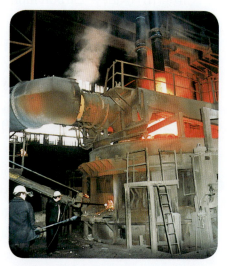

▲ Blast furnaces run non-stop 24 hours a day.

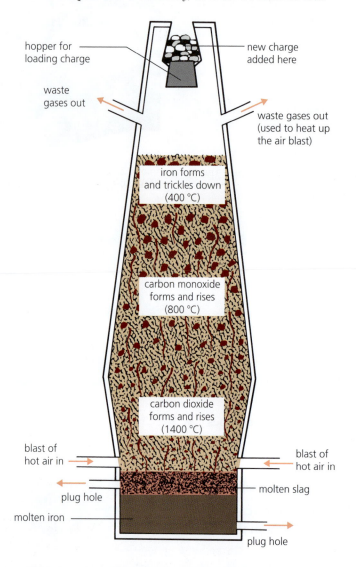

hopper for loading charge

new charge added here

waste gases out

waste gases out (used to heat up the air blast)

iron forms and trickles down (400 °C)

carbon monoxide forms and rises (800 °C)

carbon dioxide forms and rises (1400 °C)

blast of hot air in

blast of hot air in

molten slag

plug hole

molten iron

plug hole

A mixture called the **charge**, containing the iron ore, is added through the top of the furnace. Hot air is blasted in through the bottom. After a series of reactions, liquid iron collects at the bottom of the furnace.

What's in the charge?

The charge contains three things:

1 **Iron ore**. The chief ore of iron is called **hematite**. It is mainly iron(III) oxide, Fe_2O_3, mixed with sand and some other compounds.

2 **Limestone**. This common rock is mainly calcium carbonate, $CaCO_3$.

3 **Coke**. This is made from coal, and is almost pure carbon.

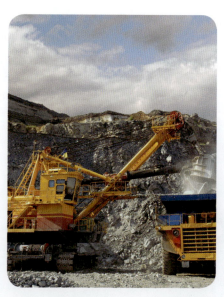

▲ Mining hematite.

The reactions in the blast furnace

Reactions, products, and waste gases	Comments
Stage 1: The coke burns, giving off heat The blast of hot air starts the coke burning. It reacts with the oxygen in the air, giving carbon dioxide: carbon + oxygen $\rightarrow$ carbon dioxide $C\ (s)\ +\ O_2\ (g)\ \rightarrow CO_2\ (g)$	This, like all combustion, is a **redox reaction**. The carbon is **oxidised** to carbon dioxide. The blast of air provides the oxygen for the reaction. The reaction is **exothermic** – it gives off heat, which helps to heat the furnace.
Stage 2: Carbon monoxide is made The carbon dioxide reacts with more coke, like this: carbon + carbon dioxide $\rightarrow$ carbon monoxide $C\ (s)\ +\ \ CO_2\ (g)\ \ \ \rightarrow\ \ \ 2CO\ (g)$	In this redox reaction, the carbon dioxide loses oxygen. It is **reduced**. The reaction is **endothermic** – it takes in heat from the furnace. That's good: stage 3 needs a lower temperature.
Stage 3: The iron(III) oxide is reduced This is where the actual extraction occurs. Carbon monoxide reacts with the iron ore, giving liquid iron: iron(III) oxide + carbon monoxide $\rightarrow$ iron + carbon dioxide $Fe_2O_3\ (s)\ +\ \ \ \ 3CO\ (g)\ \ \ \ \rightarrow 2Fe\ (l)\ +\ \ \ 3CO_2\ (g)$ The iron trickles to the bottom of the furnace.	In this redox reaction, the reduction of the iron(III) oxide to iron is brought about by carbon monoxide. The carbon monoxide is itself **oxidised** to carbon dioxide.
What is the limestone for? The limestone breaks down in the heat of the furnace: $CaCO_3 \rightarrow CaO\ (s) + CO_2\ (g)$ The calcium oxide that forms reacts with the sand, which is mainly silicon dioxide or **silica**: calcium oxide + silica $\rightarrow$ calcium silicate $CaO\ (s)\ \ \ +\ SiO_2\ (s)\ \rightarrow CaSiO_3\ (s)$ The calcium silicate forms a **slag** which runs down the furnace and floats on the iron.	The purpose of this reaction is to produce calcium oxide, which will remove the sand that was present in the ore. Calcium oxide is a basic oxide. Silica is an acidic oxide. Calcium silicate is a salt. The molten slag is drained off. When it solidifies it is sold, mostly for road building.
The waste gases: hot **carbon dioxide** and **nitrogen** come out from the top of the furnace. The heat is transferred from them to heat the incoming blast of air.	The carbon dioxide is from the reaction in stage **3**. The nitrogen is from the air blast. It has not taken part in the reactions so has not been changed.

Where next?

The iron from the blast furnace is called **pig iron**. It is impure. Carbon and sand are the main impurities.

Some is run into moulds to make **cast iron**. This is hard but brittle, because of its high carbon content – so it is used only for things like canisters for bottled gas (page 248) and drain covers.

But most of the iron is turned into **steels**. You can find how this is done in Unit 14.6.

▶ A cast-iron drain cover.

Q

1 Write the equation for the redox reaction that gives iron.
2 What is the 'blast' of the blast furnace?
3 Name the waste gases from the blast furnace.

4 The calcium carbonate in the blast furnace helps to purify the iron. Explain how, with an equation.
5 The slag and waste gases are both useful. How?

14.4 Extracting aluminium

From rocks to rockets

Aluminium is the most abundant metal in the Earth's crust. Its main ore is **bauxite**, which is aluminium oxide mixed with impurities such as sand and iron oxide. The impurities make it reddish brown.

These are the steps in obtaining aluminium:

1 First, geologists test rocks and analyse the results, to find out how much bauxite there is. If the tests are satisfactory, mining begins.

2 Bauxite is red-brown in colour. It is usually near the surface, so is easy to dig up. This is a bauxite mine in Jamaica.

3 The ore is taken to a bauxite plant, where impurities are removed. The result is white **aluminium oxide**, or **alumina**.

4 The alumina then undergoes electrolysis, to give aluminium. (It may be sent to another country with cheaper electricity, for this!)

5 The aluminium is made into sheets and blocks, and sold to other industries. It has a great many uses. For example …

6 … it is used to make drinks cans, food cartons, cooking foil, bikes, TV aerials, electricity cables, ships, trains, and aircraft.

The electrolysis

The electrolysis is carried out in a large steel tank. (See next page.) This is lined with carbon, which acts as the cathode (−). Big blocks of carbon hang in the middle of the tank, and act as the anode (+).

Alumina melts at 2045 °C. It would be impossible to keep the tank that hot. Instead, the alumina is dissolved in molten **cryolite**, or sodium aluminium fluoride, which has a much lower melting point.

The electrolysis tank

This is the tank for the electrolysis of aluminium:

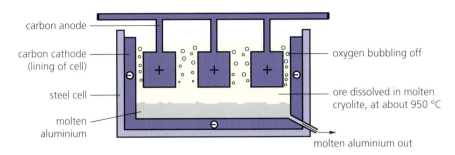

carbon anode

carbon cathode (lining of cell)

steel cell

molten aluminium

oxygen bubbling off

ore dissolved in molten cryolite, at about 950 °C

molten aluminium out

The reactions at the electrodes

Once the alumina dissolves, its aluminium and oxide ions are free to move. They move to the electrode of opposite charge.

At the cathode   The aluminium ions gain electrons:

$$4Al^{3+}\,(l) + 12e^- \longrightarrow 4Al\,(l) \qquad \text{(reduction)}$$

The aluminium drops to the bottom of the cell as molten metal. This is run off at intervals. Some will be mixed with other metals to make **alloys**. Some is run into moulds, to harden into blocks.

At the anode   The oxygen ions lose electrons:

$$6O^{2-}\,(l) \longrightarrow 3O_2\,(g) + 12e^- \qquad \text{(oxidation)}$$

The oxygen gas bubbles off, and reacts with the anode:

$$C\,(s) + O_2\,(g) \longrightarrow CO_2\,(g) \qquad \text{(oxidation of carbon)}$$

So the carbon blocks get eaten away, and need to be replaced.

The overall reaction   The alumina is broken down, giving aluminium:

aluminium oxide $\longrightarrow$ aluminium + oxygen

$$2Al_2O_3\,(l) \qquad \longrightarrow \qquad 4Al\,(l) \quad + 3O_2\,(g)$$

Some properties of aluminium

1 It is a bluish-silver, shiny metal.
2 It has a low density – it is 'light'. Iron is three times heavier.
3 It is a good conductor of heat and electricity.
4 It is malleable and ductile.
5 It resists corrosion. This is because a fine coat of aluminium oxide forms on its surface, and acts as a barrier. (See page 187.)
6 It is not very strong when pure, but it can be made much stronger by mixing it with other metals to form **alloys**. (See page 199.)
7 It is generally **non-toxic** (not harmful to health). But taking in large quantities could harm you.

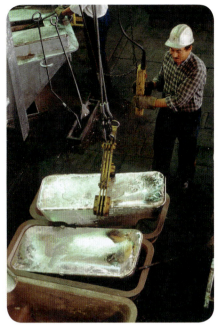

▲ Molten aluminium from the tank was run into these moulds, to make blocks.

▲ Electricity cables: aluminium (light) with a steel core (strong).

Q

1 Which compounds are used in the extraction of aluminium? Say what role each plays.
2 **a** Sketch the electrolysis cell for extracting aluminium.
   b Why do the aluminium ions move to the cathode?
   c What happens at the cathode? Give an equation.
   d The anode is replaced regularly. Why?

3 These terms all describe properties of aluminium. Say what each term means.
   a malleable   **b** ductile   **c** non-toxic
   d low density   **e** resistant to corrosion
4 List six uses of aluminium. For each, say which properties of the metal make it suitable.

14.5 Making use of metals and alloys

Properties dictate uses

Think of all the solid things you own, or use, or see around you. Some are probably made of wood, or plastic, or stone, or cloth. But which are made of metal, or contain metal?

Metals share some properties. Each also has its own special properties. How we use the metals depends on their properties. You would not use poisonous metals for food containers, for example.

Here are some examples of uses:

Aluminium foil is used for food cartons because it is non-toxic, resists corrosion, and can be rolled into thin sheets.

Copper is used for electrical wiring in homes, because it is an excellent conductor, and easily drawn into wires.

Zinc is coated onto steel car bodies, before they are painted, because zinc protects the steel from rusting, by sacrificial protection.

A summary of their uses

The three metals above have other uses too. Look at this table:

Metal	Used for ...	Properties that make it suitable
aluminium	overhead electricity cables (with a steel core for strength)	a good conductor of electricity (not as good as copper, but cheaper and much lighter); ductile, resists corrosion
	cooking foil and food cartons	non-toxic, resistant to corrosion, can be rolled into thin sheets
	drinks cans	light, non-toxic, resistant to corrosion
	coating CDs and DVDs	can be deposited as a thin film; shiny surface reflects laser beam
copper	electrical wiring	one of the best conductors of electricity, ductile
	saucepans and saucepan bases	malleable, conducts heat well, unreactive, tough
zinc	protecting steel from rusting	offers sacrificial protection to the iron in steel
	coating or **galvanising** iron and steel	resists corrosion, but offers sacrificial protection if coating cracks
	for torch batteries	gives a current when connected to a carbon pole, packed into a paste of electrolyte

Alloys: making metals more useful

The uses on page 198 are for the pure metals. But often a metal is more useful when mixed with another substance. The mixture is called an **alloy**.

For example, mixing molten zinc and copper gives the gold-coloured alloy called **brass**. When this solidifies, it is hard, strong, and shiny. It is used for door locks, keys, knobs, and musical instruments such as trumpets.

Turning a metal into an alloy changes its properties, and makes it more useful.

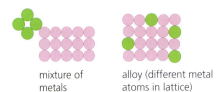

mixture of metals   alloy (different metal atoms in lattice)

▲ In an alloy, the atoms of the second metal must enter the lattice. So you need to melt the metals first, then mix them.

Why an alloy has different properties

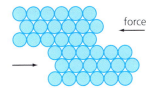
force

force

This shows the atoms in a pure metal. They are arranged in a regular lattice. (In fact they are metal ions in a sea of electrons, as you saw on page 58.)

When pressure is applied, for example by hammering the metal, the layers can slide over each other easily. That is why a metal is malleable and ductile.

But when the metal is turned into an alloy, new atoms enter the lattice. The layers can no longer slide easily. So the alloy is stronger than the original metal.

It is not only strength that changes: other properties will change too. For example the alloy may be more resistant to corrosion than the original metal was.

You can add more than one substance. You can try out different amounts of different substances, to design an alloy with exactly the properties that you need.

Some examples of alloys

There are *thousands* of different alloys. Here are a couple of examples, for the metals on the opposite page.

▲ Blow your own (brass) trumpet.

Alloy	Made from	Special properties	Uses
brass	70% copper 30% zinc	harder than copper, does not corrode	musical instruments, ornaments, door knobs and other fittings
aluminium alloy 7075 TF	90.25% aluminium 6% zinc 2.5% magnesium 1.25% copper	light but very strong does not corrode	aircraft

Look at the aluminium alloy. Aircraft need materials that are light but very strong, and will not corrode. Pure aluminium is not strong enough. So *hundreds* of aluminium alloys have been developed, for aircraft parts.

Q

1 See if you can give two *new* examples of a use for a metal, that depends on the metal being:
   a a good conductor   **b** non-toxic
   c strong   **d** resistant to corrosion

2 Bronze is 95% copper and 5% tin.
   a What is a mixture like this called?
   b What can you say about its properties?
   c See if you can give an example of a use for bronze.

14.6 Steels and steel-making

Steels: alloys of iron

Iron is very widely used – but almost never on its own.

Pure iron is no good for building things, because it is too soft, and stretches quite easily, as you can see from this photo. Even worse, it rusts very easily.

But when it has a very small amount of carbon mixed with it, the result is a **mild steel**. This is hard and strong. It is used for buildings, ships, car bodies, and machinery.

When nickel and chromium are mixed with iron, the result is **stainless steel**. This is hard and rustproof. It is used for cutlery, and equipment in chemical factories.

So mild steel and stainless steel are **alloys** of iron. Some typical mixtures are:

mild steel − 99.7% iron and 0.3% carbon

stainless steel − 70% iron, 20% chromium, and 10% nickel

There are many other types of steel too, all with different properties.

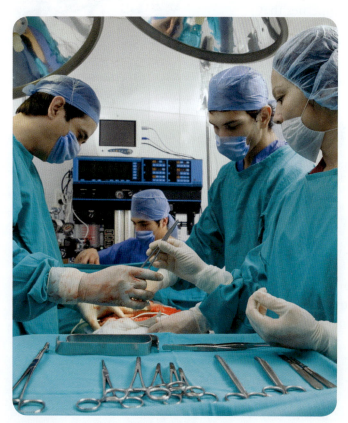

▲ Scalpel, please! Stainless steel saves lives. It is easy to sterilise, which is very important in surgery.

▲ Mild steel is used for washing machines and fridges, as well as for cars.

▲ Very often, scrap iron is added to the molten pig iron in the oxygen furnace. So the iron gets recycled as new steel.

▲ Molten steel being poured out, in a steel works. Look how it glows. Temperature: around 1400 °C.

Making steels

Steels are made using the molten iron from the blast furnace.

As you saw, the molten iron is impure. It contains about 5% carbon, from the coke used in the furnace, plus sand (which is mainly silicon dioxide) and phosphorus and sulfur, from the iron ore.

1 **First, unwanted impurities are removed from the iron.**
   - The molten iron from the blast furnace is poured into a furnace called a **basic oxygen converter**, and a jet of oxygen is turned on.
   - The oxygen reacts with the carbon and sulfur to form carbon dioxide and sulfur dioxide. These go off as gases. It reacts with phosphorus to form phosphorus(V) oxide (P_2O_5), a solid.
   - Then calcium oxide is added. It is a basic oxide. It reacts with the silicon dioxide and phosphorus(V) oxide, since they are acidic oxides. This gives a **slag** which is skimmed off.

   For some steels, *all* impurities are removed. But many steels are just iron plus a little carbon. Carbon makes steel stronger – but too much makes it brittle, and hard to shape. So the carbon content is checked continually. When it is correct, the oxygen is turned off.

2 **Then other elements may be added.**
   For example, nickel and chromium are added to give stainless steel.

   By choosing carefully what to add to the steel, and exactly how much, you can make steels which have the properties you want. Brilliant!

> **About calcium oxide**
> Calcium oxide is also called **lime** and **quicklime**.

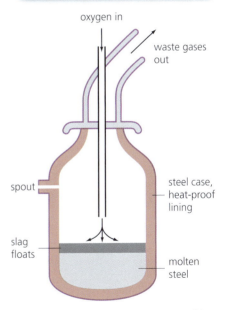

▲ The basic oxygen converter. It will be tipped to pour the molten steel out through the spout.

Q

1 What is the main metal, in steels?
2 a Why is oxygen blown through the molten iron?
   b Write a word equation for a reaction which occurs at this stage, to remove carbon.
3 What is special about calcium oxide, that makes it useful in steel-making?

4 Mild steel contains a very small % of carbon.
   a Draw diagrams to show clearly the difference between pure iron and mild steel. (Show atoms!)
   b Explain why mild steel is stronger than iron.
5 Alloys are a type of solution. Explain why.
6 Name the metals in stainless steel.

Metals, civilisation, and you

No you without metals

Without metals you probably would not exist. You certainly would not be reading this book. The world would not have nearly 7 billion people.

Our human history has been shaped by the discovery of metals. That is why two of our eras are known as the Bronze Age and the Iron Age.

> **Time past and present** !
> - **CE** stands for *common era*. It means the period since **1 AD**.
> - **BCE** stands for *before common era*. It means the same as **BC**.

Metals and civilisation

	200 000 years ago
farming starts around here	8000 BCE
gold known	6000 BCE
copper and silver known	4000 BCE
tin and lead known	3500 BCE
iron began to be used	1500 BCE
just 7 metals known by 1 CE: gold, copper, silver, lead, tin, iron, and mercury	1 CE
Industrial Revolution starts around here. 24 metals known	1750 CE / 1800 CE
65 metals known	1900 CE
over 90 metals known	2000 CE

The Stone Age

The early humans were hunters and gatherers. We had to kill and chop to get meat and fruit and firewood. We used stone and bone as tools and weapons.

Then, about 10 000 years ago, we began to farm. Farming started in the Middle East. As it spread, great civilisations grew – for example in the valley of the River Indus, in Asia, around 9000 years ago.

The Bronze Age

By 3500 BCE (over 5000 years ago) copper and tin were known – but not much used, since they are quite soft. Then someone made a discovery: mixing molten copper and tin gives a strong hard metal that can be hammered into different shapes. It was the alloy **bronze**. The Bronze Age was here!

Now a whole new range of tools could be made, for both farming and fighting.

The Iron Age

Our ancestors had no use for iron – until one day, around 2500 years ago, some got heated up with charcoal (carbon). Perhaps by accident, in a hot fire. A soggy mess of impure iron collected at the bottom of the fire.

The iron was hammered into shape with a stone. The result was a metal with vast potential. In time it led to the Industrial Revolution. It is still the most widely used metal on the Earth.

The Digital Age

We still depend on iron. But computers now touch every aspect of life. There would be no computing – and no satellites, or TV, or mobile phones – without hi-tech metals such as selenium and titanium, and old favourites such as aluminium and copper. Many new hi-tech uses are being found for these metals and their alloys.

Where next?

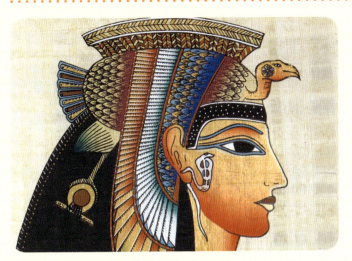

▲ The lead ore galena (lead sulfide) was used as eye make-up in ancient Egypt. Lead was probably produced by heating galena in an open fire.

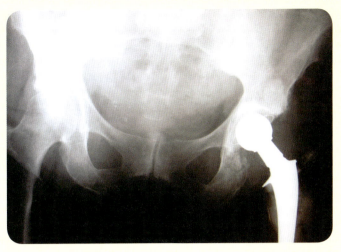

▲ A modern use of metals: this X-ray shows a replacement hip joint made from titanium. Titanium is suitable since it is strong, light, non-toxic, and does not corrode.

The earliest metals

The first metals found were the unreactive ones that exist as elements: gold, copper, and silver. But these were too soft on their own to have many uses, other than for ornaments, jewellery, and coins.

Tin, lead, and iron occur naturally as compounds, so have to be extracted by chemical reactions. It could have happened by accident, at the start. Some ore could have fallen into a fire where charcoal was burning. Or the molten metals could have appeared from clay being baked in pottery kilns, where the enclosed fires are very hot.

The invention of electrolysis

By 250 years ago, only 24 metals were known: those found naturally as elements, plus others that could be extracted easily in a furnace, using carbon. Nobody had set eyes on sodium or magnesium, for example.

Then in 1800, the first ever electrolysis was carried out (of water). The scientist Humphry Davy heard about it, and tried it out on molten compounds – with amazing results! He discovered potassium and sodium in 1807, and magnesium, calcium, strontium, and barium in 1808.

The discovery of aluminium

Aluminium is the most common metal in the Earth's crust. But it was not extracted until 1825, when aluminium chloride and potassium were heated together. (Potassium, being more reactive, displaces aluminium from its compounds.)

Only small amounts of aluminium could be extracted this way. So it became more valuable than gold! Then in 1886, the way to extract it by electrolysis, using cryolite, was developed. Aluminium had arrived.

Any more metals to find?

We know about all the metals in the Earth's crust by now. However, new metals are still being discovered. But they are artificial elements, made in labs, and radioactive. They usually break down very very fast.

▲ This gold collar is around 3000 years old. It was found in Ireland.

▲ Humphry Davy died at 52, of an illness probably caused by harmful vapours from electrolysis.

Checkup on Chapter 14

Revision checklist

Core syllabus content

Make sure you can …

☐ name the two most common metals in the Earth's crust

☐ explain what an *ore* is, and name the main ores of aluminium and iron

☐ explain what *extracting a metal* means

☐ say how the method used to extract a metal depends on the reactivity of the metal

☐ explain why electrolysis is needed to extract some metals, and name at least two of them

☐ for the extraction of iron in the blast furnace:
  – name the raw materials and explain the purpose of each
  – draw a labelled sketch of the blast furnace
  – give word equations for the reactions that take place
  – give uses for the waste products that form
  – name two impurities present in the molten iron

☐ say that aluminium is extracted from bauxite by electrolysis

☐ give at least two uses of aluminium, and state the properties that make it suitable for those uses

☐ explain what alloys are, and draw a diagram to show the structure of an alloy

☐ explain why alloys are often used instead of pure metals (they are harder and stronger)

☐ explain what brass is, and say what it is used for

☐ describe how iron from the blast furnace is turned into steels

☐ say what is in this alloy, and what it is used for:
  – mild steel
  – stainless steel

Extended syllabus content

Make sure you can also …

☐ describe the extraction of zinc from zinc blende

☐ describe the extraction of aluminium from aluminium oxide, with the help of cryolite (you will not be asked to draw the electrolysis cell)

☐ give uses for copper and zinc, and state the properties that lead to those uses

Questions

Core syllabus content

1 This table gives information about the extraction of three different metals from their main ores:

metal	formula of main compound in ore	method of extraction
iron	Fe_2O_3	heating with carbon
aluminium	Al_2O_3	electrolysis
sodium	$NaCl$	electrolysis

a Give:
  i the chemical name of each compound shown
  ii the common names for the three ores

b Arrange the three metals in order of reactivity.

c i How are the two more reactive metals extracted from their ores?
  ii Explain why this is a reduction of the ores.

d i How is the least reactive metal extracted from its ore?
  ii Explain why this is a reduction of the ore.
  iii Why can this method not be used for the more reactive metals?

e Which of the methods would you use to extract:
  i potassium? ii lead? iii magnesium?

f Gold is found native in the Earth's crust. Explain what *native* means.

g Where should gold go, in your list for b?

h Name another metal that occurs native.

2 a Draw a diagram of the blast furnace. Show clearly on your diagram:
  i where air is blasted into the furnace
  ii where the molten iron is removed
  iii where a second liquid is removed

b i Name the three raw materials used.
  ii What is the purpose of each material?

c i Name the second liquid that is removed.
  ii When it solidifies, does it have any uses? If so, name one.

d i Name a waste gas from the furnace.
  ii Does this gas have a use? If so, what?

e i Write an equation for the chemical reaction that produces the iron.
  ii Explain why this is a reduction of the iron compound.
  iii What brings about this reduction?

Extended syllabus content

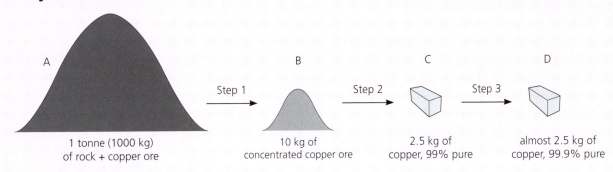

A — 1 tonne (1000 kg) of rock + copper ore

Step 1 →

B — 10 kg of concentrated copper ore

Step 2 →

C — 2.5 kg of copper, 99% pure

Step 3 →

D — almost 2.5 kg of copper, 99.9% pure

3 The diagram above shows stages in obtaining copper from a low-grade ore. The ore contains copper(II) sulfide, CuS. It may also contain small amounts of silver, gold, platinum, iron, cadmium, zinc, and arsenic.
  a What is an *ore*?
  b What is a *low-grade* ore?
  c i How much waste rock is removed per tonne in **step 1**?
  ii About what % of the ore in B is finally extracted as pure copper?
  d Why might it be economic to extract copper from a low-grade ore like this?
  e i Which type of reaction occurs in **step 2**?
  ii With what could the copper ore be reacted, to obtain the metal?
  f i Which process is carried out at **step 3,** to purify the metal?
  ii What will the main cost in this process be?
  iii As well as pure copper, this process may produce other valuable substances.
      Explain why, and where they will be found.
  g List some of the environmental problems that may arise in going from A to D.

4 Zinc and lead are obtained from ores that contain only the metal and sulfur, in the molar ratio 1:1.
  a Name the compounds in these ores.
  b Write down their formulae.
  In the extraction of the metal, the compounds are roasted in air to obtain the oxide of the metal.
  The sulfur forms sulfur dioxide.
  c i Write equations for the roasting of the ores.
  ii Which type of reaction is this?
  iii Care must be taken in disposing of the sulfur dioxide produced. Explain why.
  Then the oxide can be heated with coke (carbon) to obtain the metal and carbon monoxide.
  d i Write equations for the reactions with carbon.
  ii Which substances are reduced, in the reactions?

5 Aluminium is the most abundant metal in the Earth's crust. Iron is next.
  Iron and aluminium are extracted from their ores in large quantities. The table below summarises the two extraction processes.

Extraction	Iron	Aluminium
Chief ore	hematite	bauxite
Formula of main compound in ore	Fe_2O_3	Al_2O_3
Energy source used	burning of coke in air (exothermic reaction)	electricity
Other substances used in extraction	limestone	carbon (graphite) cryolite
Temperature at hottest part of reactor/°C	1900	1000
How the metal separates from the reaction mixture	melts and collects at the bottom	melts and collects at the bottom
Other products	carbon dioxide sulfur dioxide slag	carbon dioxide

  a In each extraction, is the metal oxide *oxidised* or *reduced*, when it is converted to the metal?
  b Explain *why* each of these substances is used:
    i limestone in the extraction of iron
    ii carbon in the extraction of aluminium
    iii cryolite in the extraction of aluminium.
  c Describe any two similarities in the extraction processes for aluminium and iron.
  d Give a use for the slag that is produced as a by-product in the extraction of iron.
  e Aluminium costs over three times as much per tonne as iron. Suggest two reasons why aluminium is more expensive than iron, even though it is more abundant in the Earth's crust.
  f Most of the iron produced is converted into steels. **i** Why? **ii** How is this carried out?
  g Both steel and aluminium are recycled.
  Suggest reasons why it is important to recycle these metals.

15.1 What is air?

The Earth's atmosphere

The **atmosphere** is the blanket of gas around the Earth.

It is about 700 km thick. We live in the lowest layer, the **troposphere**. (Look at the diagram.) The gas is at its most dense here, thanks to gravity. As you go up, it quickly thins out. In fact 90% of the atmosphere's mass is in the lowest 16 km.

Here in the troposphere, we usually call the atmosphere air.

What is in air?

This pie chart shows the gases that make up clean dry air.

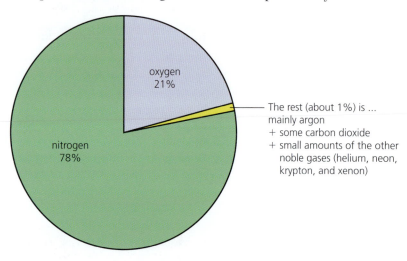

oxygen 21%

nitrogen 78%

The rest (about 1%) is ...
mainly argon
+ some carbon dioxide
+ small amounts of the other
   noble gases (helium, neon,
   krypton, and xenon)

But the air is not always clean and dry.

- There is usually some water vapour in it.

- It may also contain pollutants such as carbon monoxide and sulfur dioxide, given out by power stations, and factories, and traffic. Since air is continually on the move, the pollutants get spread around too.

| up to 700 km | −1750°C |

ionosphere
(mainly charged particles)
<1mb

80 km   −85°C

70 km

mesophere

60 km   1mb

50 km   0°C

research balloons

stratosphere

40 km

−25°C   10 mb

highest jet aircraft

30 km

natural ozone layer   −45°C

supersonic jets

20 km   −70°C   100 mb

highest cloud

jet airlines

10 km

troposphere

Mount Everest

15°C 1000 mb

sea level

▲ The Earth's atmosphere.

▲ We cannot live without oxygen. So deep-sea divers have to bring some with them ...

◀ ... and so do astronauts.

Oxygen: the gas we need most

Most of the gases in air are essential to us. For example we depend on plants for food, and they depend on carbon dioxide. And without nitrogen to dilute the oxygen, fuels would burn too fast for us.

But the gas we depend on most is oxygen. Without it, we would quickly die. We need it for the process called **respiration**, that goes on in all our cells:

glucose + oxygen $\longrightarrow$ carbon dioxide + water + energy

The energy from respiration keeps us warm, and allows us to move, and enables hundreds of reactions to go on in our bodies. (Respiration goes on in the cells of other animals too, and plants.)

Measuring the percentage of oxygen in air

The apparatus A tube of hard glass is connected to two gas syringes A and B. The tube is packed with small pieces of copper wire. At the start, syringe A contains 100 cm³ of air. B is empty.

▲ Oxygen enters through your mouth and nose, passes into your lungs, and from there diffuses into your blood.

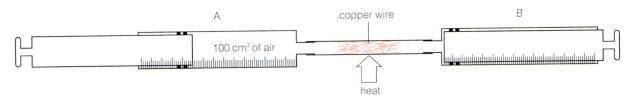

A   copper wire   B

100 cm³ of air

heat

The method These are the steps:

1 Heat the tube containing copper using a Bunsen burner. Then push in A's plunger, as shown above. This forces the air into B. When A is empty, push in B's plunger, forcing the air back to A. Repeat several times. As the air is pushed to and fro, the oxygen in it reacts with the hot copper, turning it black.

2 Stop heating the tube after about 3 minutes, and allow the apparatus to cool. Then push all the gas into one syringe and measure its volume. (It is now less than 100 cm³.)

3 Repeat steps 1 and 2 until the volume of the gas remains steady. This means all the oxygen has been used up. Note the final volume.

The results Starting volume of air: 100 cm³. Final volume of air: 79 cm³. So the volume of oxygen in 100 cm³ air is 21 cm³.

The percentage of oxygen in air is therefore $\frac{21}{100} \times 100 = \textbf{21\%}$.

▲ Fish take in the oxygen dissolved in water, through their gills.

Q

1 What percentage of air is made up of:
   a nitrogen?   b oxygen?   c nitrogen + oxygen?
2 About how much more nitrogen is there than oxygen in air, by volume?
3 What is the combined percentage of all the other gases in air?

4 Mount Everest is over 8.8 km high. Climbers carry oxygen when attempting to reach its summit. Explain why.
5 Which do you think is the most *reactive* gas in air? Why?
6 a Write down the name and formula of the black substance that forms in the experiment above.
   b Suggest a way to turn it back into copper. (Page 88!)

15.2 Making use of air

Separating gases from the air

As you saw, air is a mixture of gases. Most of them are useful to us. But first, we must separate them from each other.

How can we separate gases? There is a very clever way. First the air is cooled until it turns into a liquid. Then the liquid mixture is separated using a method you met in Chapter 2: **fractional distillation**.

The fractional distillation of liquid air

This method works because the gases in air have different boiling points. (Look at the table.) So when liquid air is warmed up, the gases boil at different temperatures, and can be collected one by one.

The boiling points of the gases in air (°C)	
carbon dioxide	−32
xenon	−108
krypton	−153
oxygen	−183
argon	−186
nitrogen	−196
neon	−246
helium	−269

increasing

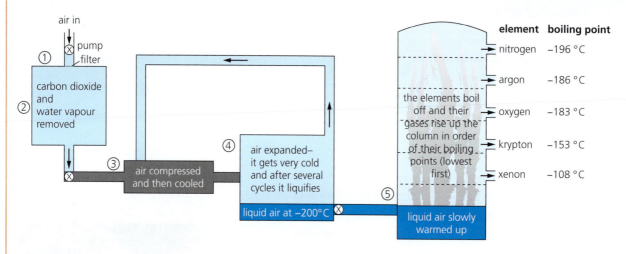

element	boiling point
nitrogen	−196 °C
argon	−186 °C
oxygen	−183 °C
krypton	−153 °C
xenon	−108 °C

the elements boil off and their gases rise up the column in order of their boiling points (lowest first)

The steps The diagram shows the steps.

1 Air is pumped into the plant, and filtered to remove dust particles.

2 Next, water vapour, carbon dioxide, and pollutants are removed (since these would freeze later and block the pipes). Like this:
   – First the air is cooled until the water vapour condenses to water.
   – Then it is passed over beds of adsorbent beads to trap the carbon dioxide, and any pollutants in it.

3 Now the air is forced into a small space, or **compressed**. That makes it hot. It is cooled down again by recycling cold air, as the diagram shows.

4 The cold, compressed air is passed through a jet, into a larger space. It expands rapidly, and this makes it very cold.

Steps 3 and 4 are repeated several times. The air gets colder each time. By −200 °C, it is liquid, except for neon and helium. These gases are removed. They can be separated from each other by adsorption on charcoal.

5 The liquid air is pumped into the fractionating column. There it is slowly warmed up. The gases boil off one by one, and are collected in tanks or cylinders. Nitrogen boils off first. Why?

▶ Liquid nitrogen is used in medical research, to keep tissue samples frozen.

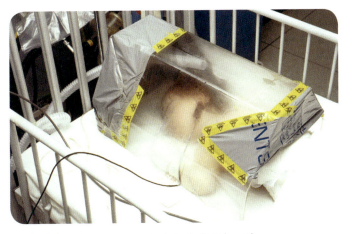

▲ An infant in an oxygen tent, to help it breathe.

▲ Easy: slicing through steel with an oxy-acetylene flame.

Some uses of oxygen

- Planes carry oxygen supplies. So do divers and astronauts.
- In hospitals, patients with breathing problems are given oxygen through an **oxygen mask**, or in an **oxygen tent**. This is a plastic tent that fits over the bed. Oxygen-rich air is pumped into it.
- In steel works, oxygen is used in converting the impure iron from the blast furnace into steels. (See page 201.)
- A mixture of oxygen and the gas **acetylene** (C_2H_2) is used as the fuel in **oxy-acetylene torches** for cutting and welding metal. When this gas mixture burns, the flame can reach 6000 °C. Steel melts at around 3150 °C, so the flame cuts through it by melting it.

Some uses of nitrogen

- Liquid nitrogen is very cold. (It boils at -196 °C.) So it is used to quick-freeze food in food factories, and to freeze liquid in cracked pipes before repairing them. It is also used in hospitals to store tissue samples.
- Nitrogen is unreactive. So it is flushed through food packaging to remove oxygen and keep the food fresh. (Oxygen helps decay.)

Some uses of the noble gases

The noble gases are unreactive or **inert**. This leads to many uses.

- Argon provides the inert atmosphere in ordinary tungsten light bulbs. (In air, the tungsten filament would quickly burn away.)
- Neon is used in advertising signs because it glows red when a current is passed through it.
- Helium is used to fill balloons, since it is very light, and safe.

For futher examples of their uses, see page 169.

▲ Frozen food on sale. In the frozen-food factory, food is dipped into liquid nitrogen to quick-freeze it.

Q

1 In the separation of air into its gases:
   a why is the air compressed and then expanded?
   b why is argon obtained *before* oxygen?
   c what do you think is the biggest expense? Explain.
2 Give two uses of oxygen gas.

3 A mixture of oxygen and acetylene burns with a much hotter flame than a mixture of air and acetylene. Why?
4 Nitrogen is used to keep food frozen during transportation. Which properties make it suitable for this?
5 Give three uses for noble gases. (Check page 169 too.)

15.3 Pollution alert!

The air: a dump for waste gases

Everyone likes clean fresh air. But every year we pump billions of tonnes of harmful gases into the air. Most come from burning **fossil fuels**.

The fossil fuels

These are **coal**, **petroleum** (or **crude oil**), and **natural gas**. Natural gas is mainly methane, CH_4. Coal and petroleum are mixtures of many compounds. Most are **hydrocarbons** – they contain only carbon and hydrogen. But some contain other elements, such as sulfur.

Fossil fuels provide us with energy for heating, and transport, and generating electricity. But there is a drawback: burning them produces harmful compounds. Look at the table below.

▲ Don't breathe in!

The main air pollutants

This table shows the main pollutants found in air, and the harm they do:

Pollutant	How is it formed?	What harm does it do?
Carbon monoxide, CO colourless gas, insoluble, no smell	Forms when the carbon compounds in fossil fuels burn in a limited supply of air. For example, inside car engines and furnaces.	Poisonous even in low concentrations. It reacts with the **haemoglobin** in blood, and prevents it from carrying oxygen around the body – so you die from oxygen starvation.
Sulfur dioxide, SO₂ an acidic gas with a sharp smell	Forms when sulfur compounds in the fossil fuels burn. Power stations are the main source of this pollutant.	Irritates the eyes and throat, and causes **respiratory** (breathing) problems. Dissolves in rain to form **acid rain**. Acid rain attacks stonework in buildings – especially limestone and marble, which are calcium carbonate. It lowers the pH in rivers and lakes, killing fish and other river life. It also kills trees and insects.
Nitrogen oxides, NO and **NO₂** acidic gases	Form when the nitrogen and oxygen in air react together, inside hot car engines and hot furnaces.	Cause respiratory problems, and dissolve in rain to give acid rain.
Lead compounds	A compound called *tetra-ethyl lead* is added to petrol in some countries, to help it burn smoothly in engines. On burning, it produces particles of other lead compounds. (So it is banned in many countries.)	Lead damages children's brains. It also damages the kidneys and nervous system in adults.

◀ Power stations are a major source of pollution – and especially those that burn coal. It can contain a lot of sulfur, and it also forms soot.

Reducing air pollution

These are some steps being taken to cut down air pollution.

- In modern power stations, the waste gas is treated with slaked lime (calcium hydroxide). This removes sulfur dioxide by reacting with it to give calcium sulfate. The process is called **flue gas desulfurisation**. See page 237 for more.

- Most countries have now banned lead in petrol. So lead pollution is much less of a problem. But it can still arise from plants where lead is extracted, and from battery factories.

- The exhausts of new cars are fitted with **catalytic converters**, in which harmful gases are converted to harmless ones. See below.

Catalytic converters for car exhausts

When petrol burns in a car engine, harmful gases are produced, including:

    oxides of nitrogen
    carbon monoxide, CO
    unburnt hydrocarbons from the petrol; these can cause cancer.

To tackle the problem, modern car exhausts contain a **catalytic converter**. In this, the harmful gases are adsorbed onto the surface of catalysts, where they react to form harmless gases. The catalysts speed up the reaction.

The converter usually has two compartments, marked A and B below:

▲ Some days the pollution is so bad that she has to wear a mask.

> In **A**, harmful compounds are **reduced**. For example:
>
> $$2NO\,(g) \longrightarrow N_2\,(g) + O_2\,(g)$$
>
> The nitrogen and oxygen from this reaction then flow into **B**.

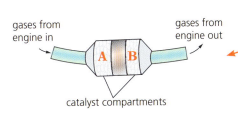

gases from engine in   A   B   gases from engine out

catalyst compartments

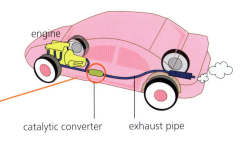

engine

catalytic converter   exhaust pipe

The catalysts are usually the transition elements platinum, palladium, and rhodium. They are coated onto a ceramic honeycomb, or ceramic beads, to give a large surface area for adsorbing the harmful gases. The harmless products flow out the exhaust pipe.

> In **B**, harmful compounds are **oxidised**, using the oxygen from **A**. For example:
>
> $$2CO\,(g) + O_2\,(g) \longrightarrow 2CO_2\,(g)$$
>
> The harmless products then flow out the exhaust pipe.

 Q

1 Look at the pollutants in the table on page 210.
  a Which come from petrol burned in car engines?
  b Which come from air? How and why do these form?

2 Natural gas or methane is a fossil fuel. In a plentiful supply of air, it burns to give carbon dioxide and water. Write a balanced equation to show this.

3 If methane burns in a poor supply of air it will give carbon monoxide and water instead. See if you can write a balanced equation to show this.

4 Catalytic converters can remove carbon monoxide.
  a Give a word equation for the reaction that takes place.
  b What is the purpose of the transition elements?

15.4 The rusting problem

What is rusting?

This car was once new and shiny. But it has been **corroded** – broken down by reaction with something in the atmosphere. In time, it will all be dust.

The corrosion of iron and steel has a special name: **rusting**. The red-brown substance that forms is called **rust**.

An experiment to investigate rusting

1 Stand three identical nails in three test-tubes.
2 Now prepare the test-tubes as below, so that:
   – test-tube 1 contains dry air
   – test-tube 2 contains water but no air
   – test-tube 3 has both air and water.
3 Leave the test-tubes to one side for several days.

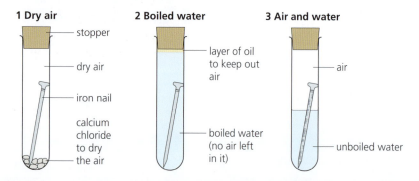

| 1 Dry air | 2 Boiled water | 3 Air and water |

- stopper
- dry air
- iron nail
- calcium chloride to dry the air
- layer of oil to keep out air
- boiled water (no air left in it)
- air
- unboiled water

Result After several days, the nails in test-tubes 1 and 2 show no signs of rusting. But the nail in test-tube 3 has rust on it. This is because:
Rusting requires oxygen and water.

In fact the iron is oxidised, in this reaction:

$$4Fe\ (s) + 3O_2\ (g) + 4H_2O\ (l) \longrightarrow 2Fe_2O_3.2H_2O\ (s)$$

iron + oxygen + water $\longrightarrow$ hydrated iron(III) oxide (rust)

▲ Rust forms as flakes. Oxygen and moisture can get behind them - so in time the iron rusts right through.

Rusting and salt

- Iron rusts faster in salty water. (Salt speeds up the oxidation.)
- So this is a problem for ships … and for cars, where salt is sprinkled on roads in winter, to melt ice. It helps the cars to rust!

▲ A rusting anchor chain. Iron rusts faster in salty water.

Stainless steel

Remember, the alloy **stainless steel** does not rust. But other steels do.

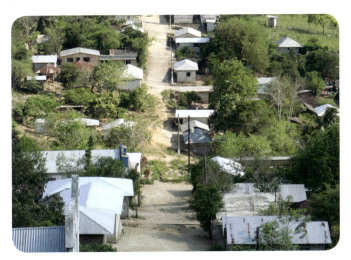

▲ Roofs of galvanised iron in a Mexican village. It is usually called **corrugated iron**.

▲ They are painting the outside of this huge ferry, to help prevent rusting.

How to prevent rusting

Iron is the most widely used metal in the world – for everything from needles to ships. But rusting destroys things. How can you prevent it? There are two approaches.

1 Cover the iron

The aim is to keep out oxygen and water. You could use:

- **paint**. Steel bridges and railings are usually painted.

- **grease**. Tools and machine parts are coated with grease or oil.

- **another metal**. Iron is coated with **zinc**, by dipping it into molten zinc, for roofing. Steel is electroplated with zinc, for car bodies. Coating with zinc has a special name: **galvanising**. For food tins, steel is coated with **tin** by electroplating.

2 Let another metal corrode instead

During rusting, iron is oxidised: it loses electrons. Magnesium is more reactive than iron, which means it has a stronger drive to lose electrons. So when a bar of magnesium is attached to the side of a steel ship, or the leg of an oil rig, it will corrode instead of the iron.

Without magnesium: $Fe \longrightarrow Fe^{2+} + 2e^-$
With magnesium: $Mg \longrightarrow Mg^{2+} + 2e^-$

The magnesium dissolves. It has been sacrificed to protect the iron. This is called **sacrificial protection**.
The magnesium bar must be replaced before it all dissolves.
Note that zinc could also be used for this. See page 187 for more.

▲ Electroplating is used to galvanise car bodies. The anode is a zinc rod, and the cathode is the steel car body.

See page 187 for more.

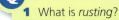

Q

1 What is *rusting*?
2 Which two substances cause iron to rust?
3 See if you can think of a way to *prove* that it is the oxygen in air, not nitrogen, that causes rusting.
4 Iron that is tin-plated does not rust. Why not?

5 You have a new bike. Suggest steps you could take to make sure it does not rust. Give a reason for each one.
6 a What does the *sacrificial protection* of iron mean?
   b Both magnesium and zinc can be used for it. Why?
   c But copper will not work. Explain why.

15.5 Water supply

Everyone needs water

We all need water.

- **At home** we need it for drinking, cooking, washing things (including ourselves), and flushing toilet waste away.
- **On farms** it is needed as a drink for animals, and to water crops.
- **In industry**, they use it as a solvent, and to wash things, and to keep hot reaction tanks cool. (Cold-water pipes are coiled around the tanks.)
- **In power stations** it is heated to make steam. The steam then drives the turbines that generate electricity.

So where does the water come from?

Much of the water we use is taken from rivers. But some is pumped up from below ground, where water that has drained down through the soil lies trapped in rocks.

This underground water is called **groundwater**. A large area of rock may hold a lot of groundwater, like a sponge. This rock is called an **aquifer**.

Is it clean?

River water is not clean – even if it looks it! It will contain particles of mud, and animal waste, and bits of dead vegetation. But worst of all are the **microbes**: bacteria and other tiny organisms that can make us ill.

Over 1 billion people around the world have no access to clean water. They depend on dirty rivers for their drinking water. And over 2 million people, mainly children, die each year from **diarrhoea** and diseases such as **cholera** and **typhoid**, caused by drinking infected water.

Providing a water supply on tap

No matter where in the world you are, the steps in providing a clean safe water supply, on tap, are the same:

> 1   Find a clean source – a river or aquifer – to pump water from.

> 2   Remove as many solid particles from the water as you can.
>   - You could make fine particles stick together and skim them off.
>   - You could filter the water through clean gravel or sand.

> 3   Add something to kill the microbes in the water. (Usually chlorine.)

> 4   Store the water in a clean covered reservoir, ready for pumping to taps.

How well you can clean the water up depends on how dirty it is, and what kind of treatment you can afford!

▲ In many places, our water supply is pumped from rivers. The water is cleaned up, the germs are killed, and then it is pumped to homes.

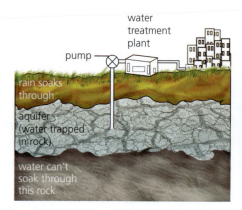

▲ Using an aquifer as a water supply.

▲ River water – now safe to drink.

A modern treatment plant

This diagram shows a modern water treatment plant. Follow the numbers to see how particles are removed and microbes killed.

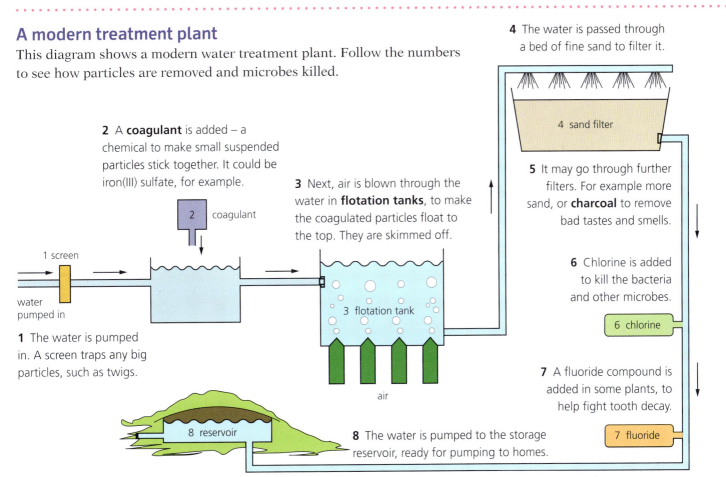

2 A **coagulant** is added – a chemical to make small suspended particles stick together. It could be iron(III) sulfate, for example.

2 coagulant

3 Next, air is blown through the water in **flotation tanks**, to make the coagulated particles float to the top. They are skimmed off.

4 The water is passed through a bed of fine sand to filter it.

4 sand filter

5 It may go through further filters. For example more sand, or **charcoal** to remove bad tastes and smells.

6 Chlorine is added to kill the bacteria and other microbes.

6 chlorine

7 A fluoride compound is added in some plants, to help fight tooth decay.

7 fluoride

1 screen

water pumped in

1 The water is pumped in. A screen traps any big particles, such as twigs.

3 flotation tank

air

8 reservoir

8 The water is pumped to the storage reservoir, ready for pumping to homes.

This treatment can remove even the tiniest particles. And chlorine can kill all the microbes. But the water may still have harmful substances dissolved in it. For example, nitrates from fertilisers, that can make babies ill.

It is possible to remove dissolved substances, using special membranes. But that is very expensive, and is not usually done. The best solution is to find the cleanest source you can, for your water supply.

Two tests for water
If a liquid contains water, it will …
- turn white anhydrous copper(II) sulfate blue
- turn blue cobalt(II) chloride paper pink.
Both colour changes can be reversed by heating.

The test for pure water
If a liquid is *pure* water, it will also …
- boil at 100 °C, and
- freeze at 0 °C.

1 What is: **a** groundwater? **b** an aquifer? **c** a microbe? (Check the glossary?)
2 What is a *coagulant* used for, in water treatment plants?
3 Why is chlorine such an important part of the treatment?
4 A fluoride compound may be added to water. Why?
5 Some water can be harmful even after treatment. Explain.

6 You need a drink of water – but there is only dirty river water. How will you make it safe for drinking?
7 You have some seawater, and cobalt(II) chloride paper.
　a How will you prove that the liquid contains water?
　b How will you prove that it is *not* pure water?
　c How might you obtain pure water from it? (Page 18?)

Living in space

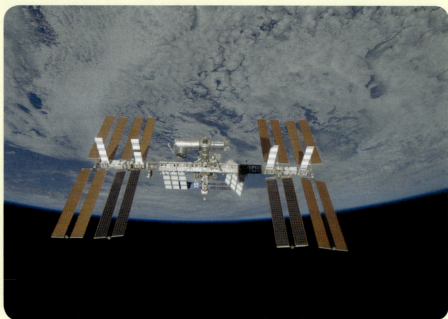

◀ The International Space Station. The 'wings' carry solar panels.

▲ Just hanging about in the lab.

The International Space Station

Right now, about 350 km above you, a large satellite is orbiting the Earth. On board are scientists: at least three. They could be asleep as you read this, or listening to music, or taking exercise. But most of the time they are doing experiments.

The satellite is the International Space Station (ISS). It is a floating lab, where scientists from different countries carry out a range of experiments. These can be more exciting than usual, since everything is weightless!

Where do they get their oxygen?

Inside the space station, the air is like that on the Earth. The main gas is nitrogen, which does not get used up. But the oxygen does – and the scientists would die without a steady supply. There is no oxygen outside the space station. So where do they get it?

- **From the electrolysis of water**   A special polymer is used as the electrolyte. The overall reaction is:

$$2H_2O\,(l) \longrightarrow 2H_2(g) + O_2\,(g)$$

  The scientists breathe the oxygen. The hydrogen is vented to space.

- **From 'oxygen candles'**   These are a back-up. They contain sodium chlorate ($NaClO_3$) mixed with iron powder. When the mixture is lit, the sodium chlorate starts breaking down to sodium chloride and oxygen. Some of the oxygen in turn reacts with the iron, giving iron oxide – and this reaction gives out the heat needed to keep the main reaction going.

- **From oxygen cylinders**   These are for emergencies only! (It costs too much to deliver oxygen cylinders from the Earth.)

The carbon dioxide that accumulates in the air is collected, and vented to space. (In future it may be kept to grow plants.)

What is the ISS for?

It is a place to study the effects of low gravity – for example on:

- the human body (and mind)
- how liquids behave
- the rate of chemical reactions

They expect to learn things that will be useful on Earth – and on long journeys to other planets.

▲ Getting to grips with the project.

Who owns the ISS?
- The ISS is jointly owned by the space agencies of the USA, Russia, Japan, and Canada, and the European Space Agency. They all send scientists to it.
- Other countries may join in before the project ends (around 2020 or later).

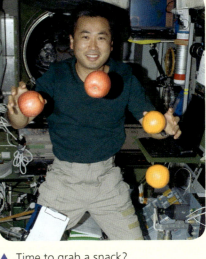

▲ Time to grab a snack?

What about water?

The scientists would die without water too. So where does the water come from, to drink and for electrolyis? Mostly from their own bodies!

All urine is collected. So is the water vapour in the air (from the scientists' breath) and any waste water from washing. It is filtered through many kinds of filter, to remove dissolved substances, and treated to kill bacteria.

In fact this water ends up much purer than our drinking water down here on the Earth! As a back-up, some containers of water are stored on board.

Note that the scientists use very little water for washing, since it is so precious. They usually have a wipe with a damp cloth.

Keeping the lights on

The other essential is electricity. Easy! Solar panels on the 'wings' of the space station convert sunlight to electricity. Some of this is used to charge batteries, for the hours when the sun is hidden. (The space station orbits the Earth once every 90 minutes, so the scientists enjoy 16 sunrises and sunsets each day!)

The electricity is used for electrolysis, and lighting, and laptops, and music players. Cooking does not take much. The food is mostly dried ready-made meals, sent up from the Earth. Add water, and warm in the oven!

What about sleep?
- The scientists sleep in sleeping bags – tied to a wall or seat to stop them floating away.
- They wear eye masks to keep the sunlight out.

A self-contained unit

This diagram sums up the life support systems, in the space station:

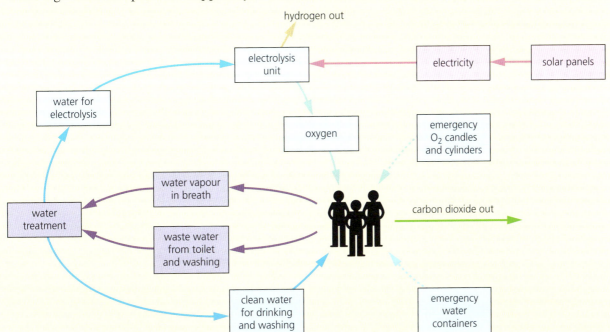

The systems developed for the space station could be very useful one day, if more of us have to move into space!

Checkup on Chapter 15

Revision checklist

Core syllabus content

Make sure you can …

☐ name the gases in clean air, and give the percentages for the two main gases present

☐ explain why oxygen is so important to us

☐ describe an experiment to find the % of oxygen in air

☐ name the fossil fuels and give examples of their use as fuels

☐ – name four common pollutants in air
　– give the source, for each
　– describe the harm they do

☐ explain what rusting is

☐ say which substances must be present for rusting to occur

☐ describe an experiment to investigate the conditions needed for rusting

☐ give examples of ways to prevent rusting, by keeping oxygen and moisture away

☐ give examples of how we use water in homes, on farms, in industry, and in power stations

☐ describe the steps in the treatment of water, to give a clean safe water supply

And check that you can …

☐ describe two tests to show that a liquid contains water

☐ say how to prove that a liquid is pure water

Extended syllabus content

Make sure you can also …

☐ describe the separation of gases from the air, using fractional distillation

☐ list the harmful gases produced in car engines

☐ – explain what catalytic converters are
　– name the metals they usually use as catalysts
　– explain how nitrogen oxides and carbon monoxide are converted to harmless gases in them

Questions

Core syllabus content

1 Copy and complete:

Dry air is a ....of different gases. 99% of it consists of the two elements .... and .... The remaining 1% contains the compound .... , and several elements called .... These elements belong to Group .... of the Periodic Table. The gas in air that we depend on most is .... This gas combines with glucose in our body cells, releasing energy. The process is called .... This gas is also used in combination with .... , in torches for welding and cutting metal.

2 Air is a *mixture* of different gases.

a Which gas makes up about 78% of the air?

b Only one gas in the mixture will allow things to burn in it. Which gas is this?

c Which noble gas is present in the greatest quantity, in air?

d i Which gas containing sulfur is a major cause of air pollution?

　ii What harmful effects does this gas have?

e Name two other gases that contribute to air pollution, and say what harm they do.

3 The rusting of iron wool gives a way to find the percentage of oxygen in air, using this apparatus:

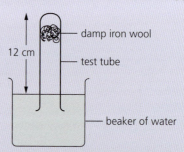

After five days the water level had risen by 2.5 cm, and the iron had rusted to iron(III) oxide.

a Write a balanced equation for the reaction between iron wool and oxygen.

b The iron wool was dampened with water before being put in the tube. Why?

c Why does the water rise up the tube?

d What result does the experiment give, for the percentage of oxygen in air?

e What sort of result would you get if you doubled the amount of iron wool?

4 Oxygen and nitrogen, the two main gases in air, are both slightly soluble in water. A sample of water was boiled, and the gases collected. The water vapour was allowed to condense and the remaining gases were measured. In a 50 cm³ sample of these gases, 18 cm³ were oxygen.
  a i What % of the dissolved air was oxygen?
  ii How does this compare to the % of oxygen in the atmosphere?
  b About what % of atmospheric air is nitrogen?
  c Which gas, nitrogen or oxygen, is more soluble in water?

5 This diagram shows one stage in the treatment of water to make it ready for piping to homes:

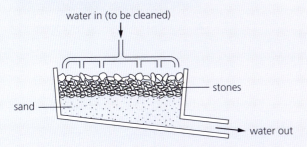

  a Name the process being carried out here.
  b The water entering this stage has already been treated with a *coagulant*, aluminium sulfate. What does a coagulant do?
  c Which kinds of impurities will the above process: **i** remove? **ii** fail to remove?
  d The next stage in treatment is *chlorination*.
  i What does this term mean?
  ii Why is this process carried out?
  e In some places the water is too acidic to be piped to homes. What could be added to reduce the acidity level?
  f At the end of treatment, another element may be added to water, for dental reasons, in the form of a salt. Which element is this?

Extended syllabus content

6 Nitrogen and oxygen are separated from air by fractional distillation. Oxygen boils at −183 °C and nitrogen at −196 °C.
  a Write the chemical formulae of these two gases.
  b What state must the air be in, before fractional distillation can be carried out?
  c Very low temperatures are required for **b**. How are these achieved?
  d Explain, using their boiling points, how the nitrogen and oxygen are separated.
  e Name one other gas obtained in the process.

7 Modern cars are fitted with catalytic converters. An outline of one is shown below:

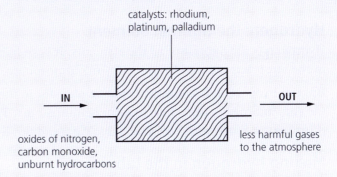

  a i Where in the car is the catalytic converter?
  ii What is the purpose of a catalytic converter?
  iii Which type of elements are rhodium, platinum, and palladium?
  b Look at the gases that enter the converter.
  i How and where are the oxides of nitrogen formed?
  ii Where do the unburnt hydrocarbons come from?
  c The gases below enter the catalytic converter. Name substances **i – iv**, to show what the gases are converted into.

gases in	converted to
oxides of nitrogen	**i** ....
carbon monoxide	**ii** ....
unburnt hydrocarbons	**iii** ....
	and **iv** ....

8 In the catalytic converters fitted to modern cars, carbon monoxide and oxides of nitrogen in the exhaust gas are converted to other substances.
  a i Why is carbon monoxide removed?
  ii Give one harmful effect of nitrogen dioxide.
  b What is meant by a *catalytic* reaction?
  c In one reaction in a catalytic converter, nitrogen monoxide (NO) reacts with carbon monoxide to form nitrogen and carbon dioxide. Write a balanced equation for this reaction.

9 Underwater steel pipelines need to be protected from corrosion. One method is to attach a block of a second metal to the pipeline.
  a i What is the key factor in choosing the second metal?
  ii Name a suitable metal.
  iii Write a half-equation to show what happens to this metal when it is attached to the pipeline.
  b What name is given to this type of protection?

16.1 Hydrogen, nitrogen, and ammonia

Hydrogen: the lightest element

Hydrogen is the lightest of all the elements. It is so light that there is none in the air: it has escaped from the Earth's atmosphere.

But out in space, it is the most common element in the universe. Inside the sun, hydrogen atoms fuse to form helium atoms. The energy given out provides the Earth with heat and light. We could not live without it.

Making hydrogen in the lab

Hydrogen is made in the lab by using a metal to drive it out or *displace* it from dilute acid. Zinc and dilute sulfuric acid are the usual choice.

The apparatus is shown on the right, below. The reaction is:

$$Zn\ (s)\ +\ H_2SO_4\ (aq)\ \longrightarrow\ ZnSO_4\ (aq)\ +\ H_2\ (g)$$

zinc   dilute sulfuric acid   zinc sulfate   hydrogen

The properties of hydrogen

1   It is the lightest of all gases – about 20 times lighter than air.
2   It is colourless, with no smell.
3   It combines with oxygen to form water.
    A mixture of the two gases will explode when lit. The reaction is:

$$2H_2\ (g)\ +\ O_2\ (g)\ \longrightarrow\ 2H_2O\ (l)$$

   It gives out so much energy that it is used to fuel space rockets. The same overall reaction takes place in hydrogen fuel cells, without burning, and gives energy in the form of electricity (page 117).
4   Hydrogen is more reactive than copper, as you can see from the panel. So it will take oxygen from copper(II) oxide, reducing it to copper. The hydrogen is itself oxidised:

$$CuO\ (s)\ +\ H_2\ (g)\ \longrightarrow\ Cu\ (s)\ +\ H_2O\ (l)$$

Nitrogen

Nitrogen is a colourless, odourless, unreactive gas, that makes up nearly 80% of the air. You breathe it in – and breathe it out again, unchanged.

But you also take in nitrogen in the **proteins** in your food. Your body uses these to build muscle, bone, skin, hair, blood, and other tissues. In fact you are nearly 3% nitrogen, by mass!

The properties of nitrogen

1   It is a colourless gas, with no smell.
2   It is only slightly soluble in water.
3   It is very unreactive compared with oxygen.
4   But it will react with hydrogen to form ammonia:

$$N_2\ (g)\ +\ 3H_2\ (g)\ \rightleftharpoons\ 2NH_3\ (g)$$

   This reversible reaction is the first step in making nitric acid, and nitrogen fertilisers. There is more about it in the next unit.

▲ Sunshine, thanks to hydrogen!

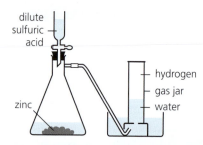

▲ One way to make hydrogen in the lab.

The hydrogen displacers

potassium
sodium
calcium
magnesium
aluminium
zinc
iron
lead
hydrogen
copper
silver
gold

increasing reactivity

metals above hydrogen in the reactivity series can displace it from acids

5 Nitrogen also combines with oxygen at high temperatures to form oxides: nitrogen monoxide (NO) and nitrogen dioxide (NO_2).

The reactions occur naturally in the air during lightning – and also inside hot car engines, and power station furnaces. The nitrogen oxides are acidic, and cause air pollution, and acid rain. (See page 210.)

Ammonia

Ammonia is a gas with the formula NH_3. It is a very important compound, because it is needed to make fertilisers, which are used everywhere. In fact ammonia is the world's second most-manufactured chemical. (Sulfuric acid is top.)

Ammonia is made in industry by reacting nitrogen with hydrogen. The details are in the next unit.

Making ammonia in the lab

You can make ammonia in the lab by heating *any* ammonium compound with a strong base. The base displaces ammonia from the compound. For example:

$$2NH_4Cl\ (s)\ +\ Ca(OH)_2\ (s)\ \longrightarrow\ CaCl_2\ (s)\ +\ 2H_2O\ (l)\ +\ 2NH_3\ (g)$$

   ammonium   calcium   calcium   water   ammonia
    chloride   hydroxide   chloride

This reaction can be used as a test for ammonium compounds. If an unknown compound gives off ammonia when heated with a strong base, it must be an ammonium compound.

The properties of ammonia

1 It is a colourless gas with a strong, choking smell.
2 It is less dense than air.
3 It reacts with hydrogen chloride gas to form a white smoke. The smoke is made of tiny particles of solid ammonium chloride:

$$NH_3\ (g)\ +\ HCl\ (g)\ \longrightarrow\ NH_4Cl\ (s)$$

This reaction can be used to test whether a gas is ammonia.
4 It is very soluble in water. (It shows the fountain effect.)
5 The solution in water is alkaline – it turns red litmus blue.
6 Since ammonia solution is alkaline, it reacts with acids to form salts. For example with nitric acid it forms ammonium nitrate:

$$NH_3\ (aq)\ +\ HNO_3\ (aq)\ \longrightarrow\ NH_4NO_3\ (aq)$$

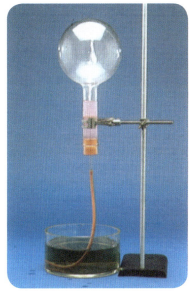

▲ The **fountain experiment**. The flask contains ammonia. It dissolves in the first drops of water that reach the top of the tube, so a fountain of water rushes up to fill the vacuum. (Any very soluble gas will show this effect.)

Q

1 Hydrogen can be made using zinc and dilute sulfuric acid. See if you can suggest a different metal and acid, for making it. (Use the reactivity series?)

2 a Hydrogen is able to react with copper(II) oxide. Why?
  b Which type of reaction is this?

3 Which is more reactive, nitrogen or oxygen?

4 Two examples of *displacement reactions* have been given in this unit. What is a displacement reaction?

5 Write a word equation for the reaction of powdered sodium hydroxide with ammonium sulfate.

16.2 Making ammonia in industry

It's a key chemical

Ammonia is a very important chemical, because it is needed to make fertilisers – and we depend on fertilisers to grow enough food. It is made from nitrogen and hydrogen. The reaction is **reversible**.

The Haber process

The process used to make ammonia is called the **Haber process**.

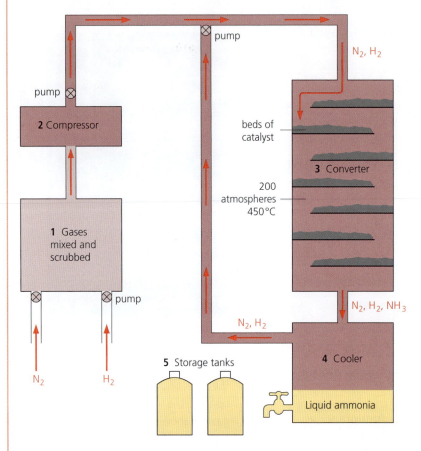

▲ Ammonia plants are often built close to oil refineries, to make use of the hydrogen from cracking.

1   The reactants are nitrogen and hydrogen. The nitrogen is obtained from air, and the hydrogen by reacting natural gas (methane) with steam, or by cracking hydrocarbons. See the details on the right. The two gases are mixed, and **scrubbed** (cleaned) to remove impurities.

2   The gas mixture is **compressed**. More and more gas is pumped in, until the pressure reaches 200 atmospheres.

3   The compressed gas flows to the **converter** – a round tank with beds of hot iron at 450 °C. The iron catalyses the reversible reaction:

$$N_2 (g) + 3H_2 (g) \rightleftharpoons 2NH_3 (g)$$

But only 15% of the mixture leaving the converter is ammonia.

4   The mixture is cooled until the ammonia condenses to a liquid. The nitrogen and hydrogen are recycled to the converter for another chance to react. Steps **3** and **4** are continually repeated.

5   The ammonia is run into tanks, and stored as a liquid under pressure.

Obtaining the reactants

Nitrogen
Air is nearly 80% nitrogen, and 20% oxygen. The oxygen is removed by burning hydrogen:
$$2H_2 (g) + O_2 (g) \longrightarrow 2H_2O (l)$$
That leaves mainly nitrogen, and a small amount of other gases. The heat given out in the reaction is used in the plant.

Hydrogen
- It is usually made by reacting **natural gas** (methane) with steam:

$$CH_4 (g) + 2H_2O (g) \xrightarrow{catalyst} CO_2 (g) + 4H_2 (g)$$

- It is also made by **cracking hydrocarbons** from petroleum. For example:

$$C_2H_6 (g) \xrightarrow{catalyst} C_2H_4 (g) + H_2 (g)$$
ethane   ethene

Improving the yield of ammonia

The reaction between nitrogen and hydrogen is reversible, and the forward reaction is exothermic: it gives out heat.

$$N_2\,(g)\ +\ 3H_2\,(g)\ \underset{\text{endothermic}}{\overset{\text{exothermic}}{\rightleftharpoons}}\ 2NH_3\,(g)$$

Since the reaction is reversible, a mixture of the two gases will *never* react completely. The yield will never be 100%.

But the yield can be improved by changing the reaction conditions, to shift equilibrium towards the product.

The graph on the right shows how the yield changes with temperature and pressure.

The chosen conditions

The temperature and pressure   As you can see, the highest yield on the graph is at **X**, at 350 °C and 400 atmospheres.

But the Haber process uses 450 °C and 200 atmospheres, at **Y** on the graph. Why? Because at 350 °C, the reaction is too slow. 450 °C gives a better rate.

And second, a pressure of 400 atmospheres needs very powerful pumps, and very strong and sturdy pipes and tanks, and a lot of electricity. 200 atmospheres is safer, and saves money.

So the conditions inside the converter do not give a high yield. But then the ammonia is removed, so that more will form. And the unreacted gases are recycled, for another chance to react. So the final yield is high.

The catalyst   Iron speeds up the reaction. But it does not change the yield!

More about the raw materials

The panel on the opposite page shows how the raw materials are obtained. Air and water are easy to find. But you need natural gas (methane), or hydrocarbons from petroleum (crude oil), to make hydrogen.

So ammonia plants are usually built close to natural gas terminals or petroleum refineries. In fact many petroleum and gas companies now make ammonia, as a way to increase their profits.

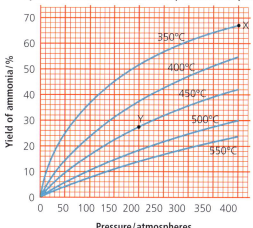

The yield of ammonia at different temperatures and pressures

Yield of ammonia / % (vertical axis, 0 to 70)
Pressure / atmospheres (horizontal axis, 0 to 400)

Curves labelled: 350°C, 400°C, 450°C, 500°C, 550°C. Point X at top right; point Y on the 450°C curve.

Making ammonia: a summary

$$N_2 + 3H_2\ \underset{\text{endothermic}}{\overset{\text{exothermic}}{\rightleftharpoons}}\ 2NH_3$$

4 molecules   2 molecules

To improve the yield:
- quite high pressure
- remove ammonia

To get a decent reaction rate:
- moderate temperature
- use a catalyst

1 Ammonia is made from nitrogen and hydrogen.
  a How are the nitrogen and hydrogen obtained?
  b What is the process for making ammonia called?
  c Write an equation for the reaction.

2 Look at the catalyst beds in the diagram on page 222.
  a What is in them?
  b Why are they arranged this way?

3 a Explain *why* high pressure and low temperature help the yield, in making ammonia. (Check Unit 9.6?)
  b 400 atmospheres and 250 °C would give a high yield. Why are these conditions *not* used in the Haber process?
  c What is the % yield of ammonia at 200 atmospheres and 450 °C? (Use the graph.)
  d What happens to the unreacted gases?

16.3 Fertilisers

What plants need

A plant needs carbon dioxide, light, and water. It also needs several different elements. The main ones are **nitrogen**, **potassium**, and **phosphorus**.

Plants need nitrogen for making chlorophyll, and proteins.

Potassium helps them to produce proteins, and to resist disease.

Phosphorus helps roots to grow, and crops to ripen.

Plants obtain these elements from compounds in the soil, which they take in through their roots as solutions. The most important one is nitrogen. Plants take it in as **nitrate** ions and **ammonium** ions.

Fertilisers

Every crop a farmer grows takes compounds from the soil. Some get replaced naturally. But in the end the soil gets worn out. New crops will not grow well. So the farmer has to add **fertilisers**.

A fertiliser is any substance added to the soil to make it more fertile.

Animal manure is a natural fertiliser. **Synthetic fertilisers** are made in factories, and sprinkled or sprayed on fields. Here are some examples.

ammonium nitrate, NH_4NO_3 ammonium sulfate, $(NH_4)_2SO_4$

potassium sulfate, K_2SO_4 ammonium phosphate, $(NH_4)_3PO_4$

▲ Nutrition for plants: these granules are made of animal manure, a natural fertiliser.

◀ Getting ready to apply fertiliser to fields. (Sometimes spelled *fertilizer*!)

◀ Synthetic fertilisers are made by reactions like those below. Then the solutions are evaporated to give solids. This shows fertiliser in storage.

Examples of reactions to make synthetic fertilisers

1 Ammonia reacts with nitric acid to give ammonium nitrate. This fertiliser is an excellent source of nitrogen:

$$NH_3 (aq) \quad + \quad HNO_3 (aq) \quad \longrightarrow \quad NH_4NO_3 (aq)$$
ammonia   nitric acid   ammonium nitrate

2 Ammonia reacts with sulfuric acid to give ammonium sulfate:

$$2NH_3 (aq) \quad + \quad H_2SO_4 (aq) \quad \longrightarrow \quad (NH_4)_2SO_4 (aq)$$
ammonia   sulfuric acid   ammonium sulfate

It's not all good news

Fertilisers help to feed the world. We could not grow enough crops without them. But there are drawbacks – as usual!

In the river   Fertilisers can seep into rivers from farmland. In the river, they help tiny water plants called **algae** to grow. These can cover the water like a carpet. When they die, bacteria feed on them, at the same time using up the oxygen dissolved in the water. So fish suffocate.

In the water supply   From rivers, the nitrate ions from fertilisers can end up in our water supply. They are converted to nitrite ions in our bodies. These combine with haemoglobin in blood, in place of oxygen, so the blood carries less oxygen around the body. This can cause illness, especially in infants. Their skin may take on a blue tinge.

So farmers should use fertilisers carefully. They should try to keep them away from river banks – and not spread them in wet weather.

What % of it is nitrogen?
Ammonium nitrate is rich in nitrogen. What % of it is nitrogen? Find out like this:

Formula: NH_4NO_3
A_r: N = 14, H = 1, O = 16
M_r: $(14 \times 2) + (4 \times 1) + (16 \times 3)$
$\quad = 80$
% of this that is nitrogen
$\quad = \dfrac{28}{80} \times 100\%$
$\quad = 35\%$

▲ Paddling through the algae. Fertiliser from farms helps these plants to grow.

Q

1 You can buy a mixture of fertilisers called *NPK fertiliser*. It contains elements plants need. Why do you think it is called *NPK*?

2 *Nitrogenous fertilisers* are fertilisers that contain nitrogen. Name three nitrogenous fertilisers.

3 Fertilisers can harm river life. Explain how.

4 The box in the margin above will remind you how to work out % composition.
  a Find the % of nitrogen in ammonium sulfate. (A_r: S = 32)
  b Which would provide more nitrogen: 1 kg of ammonium nitrate or 1 kg of ammonium sulfate?
  c Make sure *not* to add it in the rainy season. Why not?

16.4 Sulfur and sulfur dioxide

Where is sulfur found?

Sulfur is a non-metal. It is quite a common element in the Earth's crust.

- It is found, as the *element*, in large underground beds in several countries, including Mexico, Poland and the USA. It is also found around the rims of volcanoes.

- It occurs as a compound in many metal ores. For example in the lead ore **galena**, which is lead(II) sulfide, PbS.

- Sulfur compounds also occur naturally in the fossil fuels: coal, petroleum (crude oil), and natural gas.

Extracting the sulfur

From oil and gas Most sulfur is now obtained from the sulfur compounds found in petroleum and natural gas. These compounds are removed to help reduce air pollution.

For example natural gas is mainly methane. But it can have as much as 30% **hydrogen sulfide**. This is separated from the methane. Then it is reacted with oxygen, with the help of a catalyst, to give sulfur:

$$2H_2S\ (g)\ +\ O_2\ (g)\ \longrightarrow\ 2S\ (s)\ +\ 2H_2O\ (l)$$
hydrogen sulfide + oxygen $\longrightarrow$ sulfur + water

From sulfur beds About 5% of the sulfur we use comes from the underground sulfur beds. Superheated water is pumped down to melt the sulfur and carry it to the surface. (It melts at 115 °C.)

The properties of sulfur

1 It is a brittle yellow solid.
3 It has two different forms or **allotropes**, as shown on the right.
4 Because it is molecular, it has quite a low melting point.
5 Like other non-metals, it does not conduct electricity.
6 Like most non-metals, it is insoluble in water.
7 It reacts with metals to form sulfides. For example with iron it forms iron(II) sulfide:

$$Fe\ (s) + S\ (s) \longrightarrow FeS\ (s)$$

You can see photos of this reaction on page 42.

8 It burns in oxygen to form sulfur dioxide:
$$S\ (s) + O_2\ (g) \longrightarrow SO_2\ (g)$$

Uses of sulfur

- Most sulfur is used to make sulfuric acid.

- It is added to rubber, for example for car tyres, to toughen it. This is called **vulcanising** the rubber.

- It is used in making drugs, pesticides, dyes, matches, and paper.

- It is used in making cosmetics, shampoos, and body lotions.

- It is added to cement to make **sulfur concrete**. This is not attacked by acid. So it is used for walls and floors in factories that use acid.

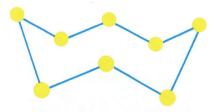

▲ A molecule of sulfur. It has 8 atoms – so the molecular formula of sulfur is S_8. But it is just called S in equations.

▲ This is a crystal of **rhombic** sulfur, the allotrope that is stable at room temperature.

▲ If you heat rhombic sulfur slowly to above 96 °C, the molecules rearrange themselves. The result is needle-shaped crystals of **monoclinic sulfur**.

Sulfur dioxide

Sulfur dioxide (SO_2) is a gas. It forms when sulfur burns in air.

1 It is a colourless gas, heavier than air, with a strong, choking smell.
2 Like most non-metal oxides, it is an acidic oxide. It dissolves in water, forming **sulfurous acid**, H_2SO_3:

$$H_2O\ (l)\ +\ SO_2\ (g)\ \longrightarrow\ H_2SO_3\ (aq)$$

This breaks down easily again to sulfur dioxide and water.
3 It acts as a bleach when it is damp or in solution. This is because it removes the colour from coloured compounds by **reducing** them.
4 It can kill bacteria.

Sulfur dioxide as a pollutant

Coal and petroleum contain sulfur compounds – even after petroleum is treated to remove them. Some coals contain a high % of sulfur.

When these fuels are burned in power stations, and factory furnaces, the sulfur compounds are oxidised to sulfur dioxide.

This escapes into the air, where it can cause a great deal of harm. It can attack your lungs, giving breathing problems. It also dissolves in rain to give **acid rain**. This attacks buildings and metal structures, and can kill fish and plants. (See page 210.)

Uses of sulfur dioxide

- Its main use is in the manufacture of sulfuric acid.
- It is used to bleach wool, silk, and wood pulp for making paper.
- It is used as a food preservative for things like dried fruit, and jam, to make them last longer. It stops the growth of bacteria and moulds.

▲ Sulfur dioxide is used to preserve dried fruit (like these apricots) and other foods. (*Sulfur* is often spelled *sulphur*.)

▲ Death by sulfur dioxide: this forest was killed by acid rain.

▲ The effect of acid rain on a limestone statue.

Q

1 Name three sources of sulfur in the Earth's crust.
2 Sulfur has quite a low melting point. Why is this?
3 Sulfur has two *allotropes*. What does that mean?
4 Sulfur reacts with iron to form iron(II) sulfide. Is this a redox reaction? Explain your answer.

5 a Sulfur dioxide is an *acidic* oxide. Explain.
  b What problems does this property cause, if sulfur dioxide escapes into the air from power stations?
6 Sulfur dioxide is a heavy gas. Do you think this contributes to air pollution? Explain your answer.

16.5 Sulfuric acid

Making sulfuric acid by the Contact process

More sulfuric acid is made than any other chemical! Most of it is made by the **Contact process**. The raw materials are:

- **sulfur, air, and water** … or
- **sulfur dioxide, air, and water.**
  The sulfur dioxide is obtained when sulfide ores, such as lead and zinc ores, are roasted in air to extract the metal from them.

Starting with sulfur, the steps in the Contact process are:

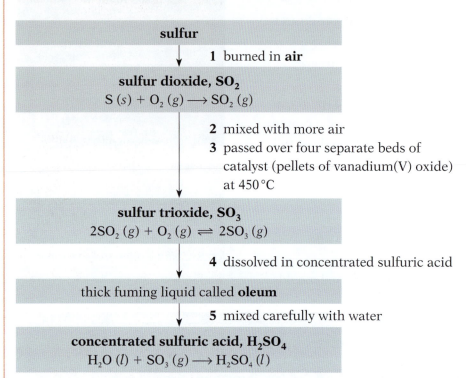

sulfur

↓ 1 burned in **air**

sulfur dioxide, SO$_2$
$$S\,(s) + O_2\,(g) \longrightarrow SO_2\,(g)$$

↓ 2 mixed with more air
3 passed over four separate beds of catalyst (pellets of vanadium(V) oxide) at 450 °C

sulfur trioxide, SO$_3$
$$2SO_2\,(g) + O_2\,(g) \rightleftharpoons 2SO_3\,(g)$$

↓ 4 dissolved in concentrated sulfuric acid

thick fuming liquid called **oleum**

↓ 5 mixed carefully with water

concentrated sulfuric acid, H$_2$SO$_4$
$$H_2O\,(l) + SO_3\,(g) \longrightarrow H_2SO_4\,(l)$$

▲ No swimming! A lake of natural sulfuric acid, in a volcano in Indonesia.

▲ You will see this sign on tanks of sulfuric acid. What is the message?

Things to note about the Contact process

- The reaction in step **3** is **reversible**. The sulfur trioxide continually breaks down again. So the mixture is passed over four separate beds of catalyst, to give the reactants further chances to react.

- Sulfur trioxide is removed between the last two beds of catalyst (using step **4**) in order to increase the yield.

- The reaction in step **3** is **exothermic**. So yield rises as temperature falls. But the catalyst will not work below 400 °C, and it works better at higher temperatures. So 450 °C is a compromise.

- To keep the temperature down to 450 °C, heat must be removed from the catalyst beds. So pipes of cold water are coiled around them to carry heat away. The heat makes the water boil. The steam is used to generate electricity for the plant, or for heating buildings.

- In step **4**, the sulfur trioxide is dissolved in concentrated acid instead of water, because with water, a thick, dangerous mist of acid forms.

Pressure in the Contact process
In step **3**:
$$2SO_2 + O_2 \rightleftharpoons 2SO_3$$
3 molecules   2 molecules

- So increasing the pressure will increase the yield of sulfur trioxide. (Page 123 explains why.)
- But in practice, the pressure is increased only a little (to less than two atmospheres).

Uses of sulfuric acid

Sulfuric acid is one of the world's most important chemicals. It has thousands of uses in industry. Its main uses are in making:

- fertilisers such as ammonium sulfate
- paints, pigments, and dyestuffs
- fibres and plastics
- soaps and detergents.

It is also the acid used in car batteries.

Dilute sulfuric acid

In the lab, dilute sulfuric acid is made by adding the concentrated acid to water. And never the other way round – because so much heat is produced that the acid could splash out and burn you.

Dilute sulfuric acid shows the usual reactions of acids:

1 acid + metal $\longrightarrow$ salt + hydrogen
2 acid + metal oxide or hydroxide $\longrightarrow$ salt + water
3 acid + carbonate $\longrightarrow$ salt + water + carbon dioxide

Its salts are called **sulfates**. And reactions **2** and **3** are neutralisations: water is produced as well as a salt.

For example dilute sulfuric acid reacts with iron like this:

$$H_2SO_4 \ (aq) \quad + \quad Fe \ (s) \quad \longrightarrow \quad FeSO_4 \ (aq) \quad + \quad H_2 \ (g)$$

sulfuric acid iron   iron(II) sulfate hydrogen

And with copper(II) oxide like this:

$$H_2SO_4 \ (aq) \quad + \quad CuO \ (s) \quad \longrightarrow \quad CuSO_4 \ (aq) \quad + \quad H_2O \ (l)$$

sulfuric acid copper(II) oxide copper(II) sulfate water

▲ Concentrated sulfuric acid was added to two teaspoons of sugar – and this is the result. It turned the sugar into carbon. Think what it could do to flesh!

> **Concentrated sulfuric acid – danger!** !
> - Concentrated sulfuric acid is a **dehydrating agent**. It removes water.
> - It 'likes' water so much that it removes hydrogen and oxygen atoms from other substances. For example from sugar (sucrose, $C_{12}H_{22}O_{11}$), leaving just carbon. Look at the photo above.
> - When it is mixed with water, the reaction gives out a great deal of heat.

▲ The teacher's sulfate. The white stick they call chalk is not really chalk – it is calcium sulfate.

Unit 9.6 will help you answer these questions.

Q

1 For making sulfuric acid, name:
 a the process
 b the raw materials
 c the catalyst
2 a The reaction between sulfur dioxide and oxygen is *reversible*. What does that mean?
 b Suggest a reason why a catalyst is needed.
 c At 500 °C, the catalyst makes sulfur trioxide form even faster. Why is this temperature *not* used?

3 Explain how these help to increase the yield of sulfur trioxide, in the Contact process:
 a Several beds of catalyst are used.
 b The sulfur trioxide is removed by dissolving it.
4 Identify two *oxidation* reactions in the manufacture of sulfuric acid.
5 a Write word equations for the reactions of zinc metal, zinc oxide (ZnO) and zinc carbonate ($ZnCO_3$) with dilute sulfuric acid.
 b Now write a balanced equation for each reaction in **a**.

16.6 Carbon and the carbon cycle

Carbon, the element

Some carbon is found in the Earth's crust as the free element, in two forms: **diamond** and **graphite**. Diamond is a hard, clear solid. Graphite is a dark, greasy solid. So diamond and graphite are **allotropes** (different forms of the same element).

Carbon compounds

There are *thousands* of carbon compounds in nature: in living things, in the soil, in the oceans, and in the atmosphere (carbon dioxide). You are around 75% water by mass – and around 20% carbon!

▲ **Charcoal**: a form of graphite made by heating coal or wood in a little air.

The carbon cycle

Carbon moves between compounds in the atmosphere, living things, the soil, and the ocean, in a non-stop journey called **the carbon cycle**:

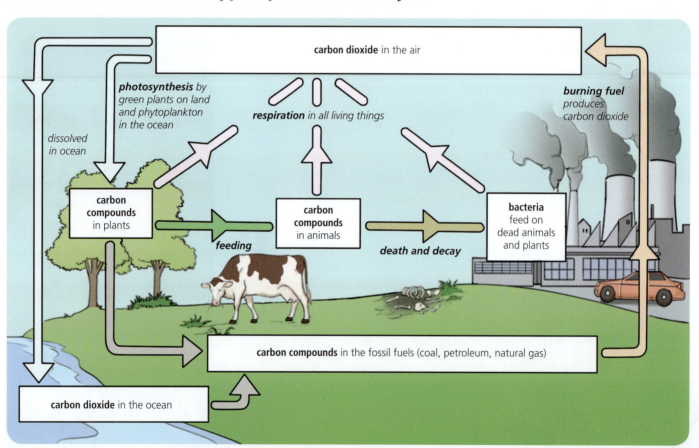

Note about the carbon cycle

- Carbon moves between the atmosphere, ocean, and living things, in the form of carbon dioxide.

- Carbon dioxide is …
  - removed from the atmosphere by **photosynthesis**, and **dissolving** in the ocean
  - added to it by **respiration**, and the **combustion** (burning) of fuels that contain carbon.

Two opposite reactions
Respiration, which goes on in your body, is the opposite of **photosynthesis** in plant leaves. You can compare their equations on the next page.

Removing carbon dioxide from the atmosphere

- **By photosynthesis** In this process, carbon dioxide and water react in plant leaves, to give glucose and oxygen. Chlorophyll, a green pigment in leaves, is a catalyst for the reaction. Sunlight provides the energy:

$$6CO_2\ (g) \quad + \quad 6H_2O\ (l) \quad \xrightarrow[\text{chlorophyll}]{\text{light}} \quad C_6H_{12}O_6\ (s) \quad + \quad 6O_2\ (g)$$
carbon dioxide water glucose oxygen

The plant uses the glucose to make the other carbon compounds it needs. Then animals eat the plants. So the carbon compounds get passed along the food chain. Many of them end up in your dinner!

Note that photosynthesis also goes on in **phytoplankton**, tiny plants that float in the ocean. These are eaten by fish and other organisms. So carbon is passed along food chains in the ocean too.

- **By dissolving** Some carbon dioxide from the air dissolves in the ocean. It provides carbonate ions, which shellfish use along with calcium ions from the water, to build their shells. (Shells are made of calcium carbonate.) Fish also use them in building their skeletons.

But only a certain % of carbon dioxide will dissolve. A balance is reached between its concentration in the air and the ocean.

▲ Rice plants: busy with photosynthesis. They will convert the glucose to starch.

Adding carbon dioxide to the atmosphere

- **By respiration** This is the process that takes place in our cells (and in the cells of plants and other animals) to provide energy:

$$C_6H_{12}O_6\ (aq) \quad + \quad 6O_2\ (g) \quad \longrightarrow \quad 6CO_2\ (g) \quad + \quad 6H_2O\ (l) \ + \text{energy}$$
glucose + oxygen $\longrightarrow$ carbon dioxide + water + energy

We get the glucose from food. The energy keeps us warm, and allows us to move, and enables other reactions to go on in our bodies.

- **By the combustion of fuels** Natural gas or **methane** burns like this:

$$CH_4\ (g) \quad + \quad 2O_2\ (g) \quad \longrightarrow \quad CO_2\ (g) \quad + \quad 2H_2O\ (l) \ + \text{energy}$$
methane + oxygen $\longrightarrow$ carbon dioxide + water + energy

The carbon cycle and fossil fuels

In the ocean, the remains of dead organisms fall to the ocean floor, and are buried. Over millions of years their soft parts turn into petroleum (oil) and natural gas. (Hard shells turn into limestone rock.)

Meanwhile, trees and other vegetation get buried in warm swamps. Over millions of years, they turn into coal.

In this way, carbon dioxide from the air ends up in the fossil fuels. And when we burn these, it is released again.

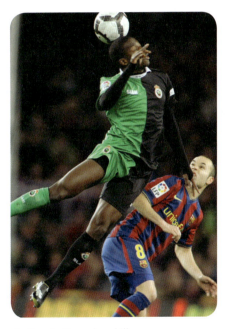

▲ Respiration plus skill …

1 What is the *carbon cycle*?
2 Compare respiration and the combustion of methane.
  a What is similar about the two reactions?
  b What do we use the energy from respiration for?
  c What do we use the energy from burning fuels for?
3 Now compare respiration and photosynthesis. What do you notice about these reactions?

4 See if you can draw a circular flowchart that shows:
  – how carbon dioxide from the air gets locked up as compounds in petroleum and natural gas
  – and how it is released again, millions of years later.
5 One part of the carbon cycle does *not* occur naturally (that is, without help from humans). Which part?
6 What part do *you* play in the carbon cycle?

16.7 Some carbon compounds

Carbon dioxide

The gas carbon dioxide (CO_2) occurs naturally in air. It is also a product in these three reactions:

1   The combustion of carbon compounds in plenty of air. For example, when natural gas (methane) burns in plenty of air, the reaction is:

$$CH_4\ (g)\ +\ 2O_2\ (g)\ \longrightarrow\ CO_2\ (g)\ +\ 2H_2O\ (l)$$

2   The reaction between glucose and oxygen, in your body cells:

$$C_6H_{12}O_6\ (aq)\ +\ 6O_2\ (g)\ \longrightarrow\ 6CO_2\ (g)\ +\ 6H_2O\ (l)$$

This is called **respiration**. You breathe out the carbon dioxide.

3   The reaction between dilute acids and carbonates. For example between hydrochloric acid and marble chips (calcium carbonate):

$$CaCO_3\ (s)\ +\ 2HCl\ (aq)\ \longrightarrow\ CaCl_2\ (aq)\ +\ CO_2\ (g)\ +\ H_2O\ (l)$$

Properties of carbon dioxide

1   It is a colourless gas, with no smell.
2   It is much heavier than air.
3   Things will not burn in it. We say it **does not support combustion**.
4   It is slightly soluble in water, forming carbonic acid, H_2CO_3.

Carbon monoxide

Carbon monoxide (CO) forms when carbon compounds burn in too little oxygen. For example, when methane burns in insufficient oxygen:

$$2CH_4\ (g)\ +\ 3O_2\ (g)\ \longrightarrow\ 2CO\ (g)\ +\ 4H_2O\ (l)$$

It is a deadly poisonous gas. It binds to the haemoglobin in red blood cells, and prevents it from carrying oxygen around the body. So victims die from oxygen starvation.

Carbon monoxide has no smell, which makes it hard to detect. So it is important to have gas heaters and boilers checked regularly, to make sure the air supply is not blocked by soot.

Carbonates

Carbonates are compounds that contain the carbonate ion, CO_3^{2-}. One example is calcium carbonate, $CaCO_3$, which occurs naturally as limestone, chalk and marble. These are the main properties of carbonates:

1   They are insoluble in water – *except for* sodium, potassium, and ammonium carbonates, which are soluble.
2   They react with acids to form a salt, water, and carbon dioxide.
3   Most of them break down on heating, to an oxide and carbon dioxide:

$$CaCO_3\ (s)\ \rightleftharpoons\ CaO\ (s)\ +\ CO_2\ (g)$$

calcium carbonate   calcium oxide   carbon dioxide
(limestone)   (lime)

But sodium and potassium carbonates do not break down, since the compounds of these reactive metals are more stable. (See page 185.)

▲ The 'fizz' in this soft drink is caused by carbon dioxide escaping.

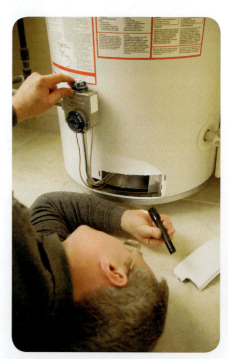

▲ Gas-fuelled water heaters and boilers should be checked regularly. Every year, hundreds of people are killed by carbon monoxide from faulty burners.

Methane

Methane is the compound CH_4.

- It is found in gas deposits in the ocean floor and on land, as **natural gas**. We use natural gas as a fuel.

- It also forms wherever bacteria break down plant material, in the absence of oxygen. For example in paddy fields, and swamps, and landfill sites (rubbish dumps).

- Some animals – including cattle, sheep, goats, and camels – give out methane as waste gas from digestion. Bacteria in their stomachs help to break down grass and other foods, and methane is one product.

Organic compounds

Methane is an **organic compound**. Organic compounds *all* contain carbon, and most contain hydrogen. Some contain elements such as sulfur and nitrogen too. Many are found in, or derived from, living things.

Methane is the simplest organic compound. There are millions more – far more than all the **inorganic** (non-organic) compounds. They include:

- the proteins, carbohydrates, and fats in your body

- the hundreds of different compounds in petroleum and coal

- the plastics and medical drugs made from the compounds in petroleum.

The study of these carbon compounds is called **organic chemistry**. The next two chapters in this book are about organic chemistry.

▲ This fire extinguisher sprays out carbon dioxide gas. (Things will not burn in an atmosphere of carbon dioxide.)

▲ The Amazon rainforest – packed full of organic compounds.

▲ Built up from organic compounds, and around 20% carbon!

Q

1 Give the word equation for the combustion of natural gas:
  a in a gas boiler, when the boiler is working well.
  b in a gas boiler, when the air inlet is partly blocked with soot.
2 Gas boilers should be checked regularly, to make sure air flows through the burner properly. Why?

3 a Write an equation to show what happens when lead(II) carbonate is heated.
  b What is this type of reaction called? (Page 185?)
4 a Name three sources of methane, CH_4.
  b Which do you think is the main source?
5 Is it organic, or inorganic? **a** sodium chloride **b** water

16.8 Greenhouse gases, and global warming

Carbon dioxide and methane are greenhouse gases

Carbon dioxide and methane are both **greenhouse gases**. That means they absorb heat in the atmosphere, and prevent it from escaping into space.

This is how greenhouse gases work:

1 The sun sends out energy as light and UV rays.

2 These warm the Earth, which reflects some of the energy away again, as heat.

3 Some of this heat escapes from the atmosphere.

4 But some is absorbed by greenhouse gases in the atmosphere. So the air, and Earth, are warmed.

atmosphere

Earth

There are several greenhouse gases. Carbon dioxide and methane are the two main ones we add to the atmosphere, through human activity. There is much more carbon dioxide than methane in the atmosphere. But the levels of both are rising:

- The level of carbon dioxide is rising because we burn more fossil fuel each year. The carbon dioxide from this goes into the atmosphere. It cannot escape into space, and the ocean can dissolve only some of it.

- The level of methane is rising because there is an increase in animal farming, and rice farming, around the world – and more and more landfill sites.

We need greenhouse gases. Without them, we would freeze to death at night, when the sun was not shining. But many scientists think the level of greenhouse gases is now so high that it is causing global warming.

▲ Two methane manufacturers.

Global warming

Measurements show that average temperatures around the world are rising. We call this **global warming**.

Why is it happening? Some scientists say it is a natural change, like similar changes in the past.

However, a panel of scientists from around the world examined all the data, and concluded that greenhouse gases are *almost certainly* the main cause.

They picked out carbon dioxide as the main culprit. The rise in average temperatures over time appears to match the rise in carbon dioxide levels over time. Compare the two graphs on the right.

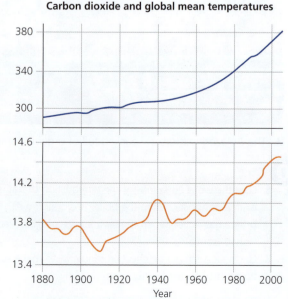

Carbon dioxide and global mean temperatures

Concentration of CO_2 in the atmosphere/ parts per million

Estimated global mean temperature/ °C

Year

Climate change

Air temperature affects rainfall, and cloud cover, and wind patterns.
So as average temperatures rise, climates around the world change too.

Scientists try to predict what will happen, using computer models.
They cannot make really good predictions yet, because they do not fully
understand the links between weather, and clouds, and the ocean.

But they do predict that:

- some places with quite a lot of rain will become very dry, and other
  places will get much wetter.
- melting land-ice in the Arctic and Antarctica will cause sea levels to
  rise, so low-lying countries will be at risk of flooding.
- storms, floods, and wildfires will be more frequent and severe.
- species that cannot adapt to the changing climate will die out.
- more drought is likely, which will led to famine – so more people will
  become refugees.

Most experts agree that climate change is already underway.

▲ Many countries are already having
more severe floods than usual.

What can we do?

If global warming is a natural change, we can do nothing to stop it.
We can only prepare for the consequences.

If we are causing global warming by burning fossil fuels, we still cannot
stop it, because the level of carbon dioxide already in the air is enough to
cause a further temperature rise. All we can do is cut back heavily on new
emissions of carbon dioxide, to stop warming getting out of control.

- Many people are trying to cut back on using fossil fuel, for example by
  using public transport or bikes, or walking, rather than going by car.
- Many countries have set targets for switching to clean ways to get
  electricity, such as windpower and solar power.
- Scientists are looking at ways to reduce the amount of carbon dioxide
  entering the atmosphere. For example by capturing it from power
  station chimneys, and burying it deep underground.

▲ Melting ice in the Arctic means polar
bears are under threat.

Some countries are starting to prepare for climate change. The poorest
countries are likely to suffer most, since they do not have enough money
to cope well with floods, drought, and other disasters.

Q

1 a We need greenhouse gases. Why?
  b So why are they becoming a problem?
2 The two main greenhouse gases we are adding to the
  atmosphere are …?
3 Global warming could lead to the extinction of some species
  of living things. Explain why.

4 The more carbon dioxide in the air, the more will dissolve in
  the ocean. (A balance is reached.)
  a Which type of oxide is carbon dioxide: acidic or basic?
  b i How might the pH of the ocean be affected by our
    burning of fossil fuels? Explain.
    ii Do you think this could cause problems? Explain.

16.9 Limestone

Limestone: from sea creatures

Most of the creatures that live in the sea have shells or skeletons made of calcium carbonate. When they die, their remains fall to the sea floor. Slowly, over millions of years, the layers of shells and bones become limestone rock. (The soft parts of sea organisms become oil and gas.)

Over millions of years, powerful forces raised some sea beds upwards, draining them to form land. That explains why plenty of limestone is found inland, miles from the sea!

Making use of limestone

Around 5 billion tonnes of limestone are quarried from the Earth's crust every year. This is what it is used for:

▲ Limestone – made from the shells and skeletons of sea creatures.

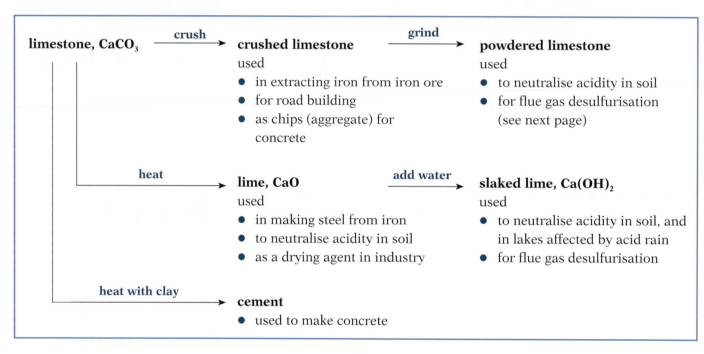

limestone, $CaCO_3$ —crush→ **crushed limestone** —grind→ **powdered limestone**

crushed limestone
used
- in extracting iron from iron ore
- for road building
- as chips (aggregate) for concrete

powdered limestone
used
- to neutralise acidity in soil
- for flue gas desulfurisation (see next page)

—heat→ **lime, CaO** —add water→ **slaked lime, $Ca(OH)_2$**

lime, CaO
used
- in making steel from iron
- to neutralise acidity in soil
- as a drying agent in industry

slaked lime, $Ca(OH)_2$
used
- to neutralise acidity in soil, and in lakes affected by acid rain
- for flue gas desulfurisation

—heat with clay→ **cement**
- used to make concrete

Lime

When limestone is heated, it breaks down to **lime** (or **quicklime**):

$$CaCO_3\,(s) \rightleftharpoons CaO\,(s) + CO_2\,(g)$$

calcium carbonate   calcium oxide   carbon dioxide
(limestone)   (lime)

This is **thermal decomposition**.

The drawing shows a lime kiln. The kiln is heated. Limestone is fed in at one end. Lime comes out the other.

The reaction is reversible. So the calcium oxide and carbon dioxide *could* combine again. But air is blown through the kiln to carry the carbon dioxide away before it has a chance to react.

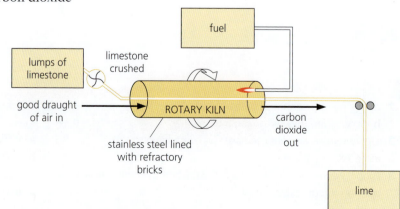

fuel

lumps of limestone

limestone crushed

good draught of air in

ROTARY KILN

carbon dioxide out

stainless steel lined with refractory bricks

lime

▲ A rotary kiln for making lime.

Slaked lime

Slaked lime forms when water is added to lime. The reaction is exothermic, so the mixture hisses and steams. Conditions are controlled so that the slaked lime forms as a fine powder:

$$CaO \ (s) \quad + \quad H_2O \ (l) \quad \longrightarrow \quad Ca(OH)_2 \ (s)$$

calcium oxide   calcium hydroxide

(lime)   (slaked lime)

Slaked lime is used to neutralise acidity in soil, and in lakes. In the lab, we use it to test for carbon dioxide. **Limewater** is a weak solution of calcium hydroxide, which is sparingly soluble in water. (See the test on page 279.)

Cement

Cement is made by mixing limestone with clay, heating the mixture strongly in a kiln, adding gypsum (hydrated calcium sulfate), and grinding up the final solid to give a powder.

▲ Cement: limestone heated with clay, and gypsum added to slow down the 'setting' process.

Flue gas desulfurisation

Flue gas desulfurisation means the removal of sulfur dioxide from the waste gases at power stations, before they go out the flue (chimney).

It is usually carried out using a runny mixture of powdered limestone, or slaked lime, and water. The mixture is sprayed through the waste gases, or the gases are bubbled through it.

When slaked lime is used, the reaction that removes the sulfur dioxide is:

$$Ca(OH)_2 \ (s) \quad + \quad SO_2 \ (g) \quad \longrightarrow \quad CaSO_3 \ (s) \quad + \quad H_2O \ (l)$$

calcium hydroxide   sulfur dioxide   calcium sulfite   water

Then the calcium sulfite can be turned into hydrated calcium sulfate:

$$2CaSO_3 \ (s) \quad + \quad O_2 \ (g) \quad + \quad 4H_2O \ (l) \quad \longrightarrow \quad 2 \ CaSO_4 . 2H_2O \ (s)$$

calcium sulfite   oxygen   water   hydrated

                                             calcium sulfate

Hydrated calcium sulfate is known as **gypsum**. It is used in making cement, plaster board, plaster for broken limbs, and other products. So the company that owns the power station can sell it, to earn some money.

▲ Flue gas desulfurisation removes sulfur dioxide from waste gases.

Q

1 How was limestone formed?

2 a How is lime made? Write the equation.

  b Why is it important to remove the carbon dioxide?

  c How is the carbon dioxide removed, in a lime kiln?

3 How is slaked lime made? Write the equation.

4 Give two uses each for lime and slaked lime.

5 Limewater is a solution of slaked lime. It is used for …?

6 Slaked lime is more soluble in water than limestone is. Which of the two might be a better choice, for controlling soil acidity in a rainy area? Explain your choice.

7 a Explain the term *flue gas desulfurisation*.

  b Name a material used for this process.

  c Calcium sulfite from the process is often turned into gypsum. What is gypsum, and why do they make it?

Checkup on Chapter 16

Revision checklist

Core syllabus content

Make sure you can …

☐ say how these can be prepared in the lab:
hydrogen   *ammonia*
and give two reactions for each of them

☐ give the equation for the reversible reaction between nitrogen and hydrogen

☐ name the three main elements plants need from the soil, and say why they need them

☐ explain what fertilisers are, and give examples of salts which act as fertilisers

☐ name three sources of sulfur

☐ say that sulfur is used in making sulfuric acid

☐ give three uses of sulfur dioxide

☐ give equations for four different types of reaction that produce carbon dioxide (including respiration, and the thermal decomposition of carbonates)

☐ give three sources of methane

☐ explain what a *greenhouse gas* is, and how it works

☐ explain these terms:
global warming   *climate change*

☐ name carbon dioxide and methane as greenhouse gases, which may contribute to climate change

☐ describe how limestone is converted to lime and slaked lime, and give equations

☐ give at least two uses each for limestone, lime and slaked lime

☐ explain what *flue gas desulfurisation* means, and describe how it is carried out

Extended syllabus content

Make sure you can also …

☐ explain why ammonia is an important chemical

☐ say how the raw materials (nitrogen and hydrogen) are obtained, for making ammonia

☐ state the conditions used in the manufacture of ammonia, and explain the choice of conditions

☐ describe the Contact process for making sulfuric acid, starting with sulfur, and state the conditions

☐ describe the properties and uses of dilute and concentrated sulfuric acid

☐ sketch the carbon cycle, and give equations for these three reactions linked to the carbon cycle:
respiration, combustion of a fuel such as methane, and photosynthesis

Questions

Core syllabus content

1 An NPK fertiliser contains the three main elements that plants need, for healthy growth.

   a Name the three elements.

   b Describe *how* each element helps plants.

   c Which of the three elements are provided by the following fertilisers?

     i ammonium phosphate

     ii potassium nitrate

     iii ammonium sulfate

   d Write a formula for each fertiliser in **c**.

2 **a** Copy the diagram below. Then fill in:

     i the common names of the substances

     ii their chemical formulae

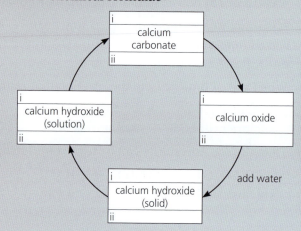

   b Beside each arrow say how the change is carried out. One example is shown.

   c Give three reasons why limestone is an important raw material.

3 Limestone is calcium carbonate, $CaCO_3$. It is quarried on a huge scale.

   a Which elements does it contain?

   b Much of the quarried limestone is turned into lime (CaO) for the steel industry.

     i What is the chemical name for lime?

     ii Describe how it is made from limestone.

   c Powdered limestone is used to improve the water quality in acidified lakes.

     i How might the lakes have become acidified?

     ii Why is limestone added?

     iii The limestone is used in powdered form, not lumps. Why? Try for more than one reason.

   d List other important uses of limestone.

4 Powdered limestone is used to treat the waste gases from power stations that burn coal and petroleum. The equation for the reaction that takes place is:

$$CaCO_3\,(s) + SO_2\,(g) \longrightarrow CaSO_3\,(s) + CO_2\,(g)$$

a **i** Name the gas that is removed by this reaction.
ii Why is it important to remove this gas?
b Why are large lumps of limestone *not* used?
c The process is called *flue gas desulfurisation*. Explain clearly what this means.
d The calcium sulfite is usually turned into gypsum, which has the formula $CaSO_4.2H_2O$.
i What is the *full* chemical name for gypsum?
ii Which type of chemical reaction occurs when $CaSO_3$ is converted into $CaSO_4$?
iii Give two uses for gypsum.
e Name two chemicals that could be used to make calcium sulfate by precipitation.

Extended syllabus content

5 This is about the manufacture of ammonia.
a Which two gases react to give ammonia?
b Why are the two gases scrubbed?
c Why is the mixture passed over iron?
d What happens to the *unreacted* nitrogen and hydrogen?
e In manufacturing ammonia, is the chosen pressure high, low, or moderate? Explain why.

6 Nitrogen and hydrogen are converted to ammonia in the Haber process:

$$N_2\,(g) + 3H_2\,(g) \rightleftharpoons 2NH_3\,(g)$$

Below is the energy level diagram for the reaction.

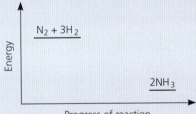

a What does this diagram tell you?
b Explain why high temperatures are *not* used in the manufacture of ammonia.
c The reaction is *reversible*, and *reaches equilibrium*. Explain very clearly what the two terms in italics mean.
d **i** What effect does a catalyst have on an equilibrium reaction?
ii Which catalyst is used in the Haber process?
iii What effect does this catalyst have on the % yield of ammonia?

7 Sulfuric acid is made by the Contact process. The first stage is to make sulfur trioxide, like this:

$$2SO_2\,(g) + O_2\,(g) \rightleftharpoons 2SO_3\,(g)$$

The energy change in the reaction is – 97 kJ/mol.

a Name the catalyst used in this reaction.
b Is the reaction exothermic, or endothermic?
c What are the reaction conditions for making sulfur trioxide?
d Will the yield of sulfur trioxide increase, decrease, or stay the same, if the temperature is raised? Explain your answer.
e Describe how sulfur trioxide is changed into concentrated sulfuric acid.

8 Below is a flow chart for the Contact process:

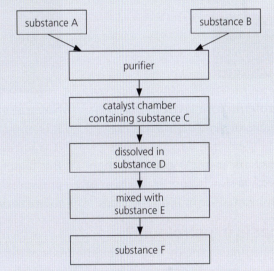

a Name the substances **A**, **B**, **C**, **D**, **E**, and **F**.
b Why is a catalyst used?
c Write a chemical equation for the reaction that takes place on the catalyst.
d The production of substance **F** is very important. Why? Give three reasons.
e Copy out the flow chart, and write in the full names of the different substances.

9 Dilute sulfuric acid has typical acid properties. An excess of it is added to test-tubes **W, X, Y** and **Z**, which contain these powdered substances:
W copper(II) oxide **X** magnesium
Y calcium hydroxide **Z** sodium carbonate
a In which test-tubes will you observe fizzing?
b In which test-tube will a coloured solution form?
c In which of the test-tubes does neutralisation take place?
d Name the four salts obtained, after reaction.
e Write balanced equations for the four reactions.

17.1 Petroleum: a fossil fuel

The fossil fuels

The **fossil fuels** are **petroleum** (or crude oil), **coal**, and **natural gas**. They are called fossil fuels because they are the remains of plants and animals that lived millions of years ago.

Petroleum formed from the remains of dead organisms that fell to the ocean floor, and were buried under thick sediment. High pressures slowly converted them to petroleum, over millions of years.

Natural gas is mainly **methane**. It is often found with petroleum. It is formed in the same way. But high temperatures and high pressures caused the compounds to break down to gas.

Coal is the remains of lush vegetation that grew in ancient swamps. The dead vegetation was buried under thick sediment. Pressure and heat slowly converted it to coal, over millions of years.

What is in petroleum?

Petroleum is a smelly mixture of hundreds of different compounds. They are **organic compounds**, which means they contain carbon, and usually hydrogen.

In fact most are **hydrocarbons** – they contain *only* carbon and hydrogen. These drawings show molecules of three different hydrocarbons:

This is a molecule of pentane, C_5H_{12}. It has a straight chain of 5 carbon atoms.

This is a molecule of cyclohexane, C_6H_{12}. Here a chain of 6 carbon atoms form a ring.

This is a molecule of 3-methyl pentane, C_6H_{14}. Here 6 carbon atoms form a **branched** chain.

A formula drawn out in this way is called a **structural formula**.

Notice how the carbon atoms are bonded to each other, to make the spine of each molecule. The hydrogen atoms are bonded to the carbon atoms. In petroleum you will find hydrocarbon molecules of different shapes and sizes, with different numbers of carbon atoms, from 1 to over 70.

How we use petroleum

Over 13 billion litres of petroleum are used around the world every day.

Around half the petroleum pumped from oil wells is used for transport. It provides the fuel for cars, trucks, planes, and ships. You won't get far without it!

Most of the rest is burned for heat, in factories, homes, and power stations, as above. In a power station, the heat is used to turn water to steam, to drive turbines.

A small % is used as the starting chemicals to make many other things: plastics, shampoo, paint, thread, fabric, detergents, makeup, medical drugs, and more.

Many of the things you use every day were probably made from petroleum. Toothbrush, comb, and shampoo just for a start!

A non-renewable resource

Petroleum is still forming, very slowly, under the oceans. But we are using it up much faster than it can form, which means it will run out one day.

So petroleum is called a **non-renewable resource**.

It is hard to tell when it will run out. At the present rate of use, some experts say the world's reserves will last about 40 more years. What will we do then?

▲ A platform for pumping petroleum from under the ocean.

Q

1 The other name for petroleum is … ?
2 Why is petroleum called a *fossil fuel*?
3 What is a *hydrocarbon*?
4 What is petroleum made of?

5 Explain why petroleum is such a valuable resource.
6 Petroleum is called a *non-renewable* resource. Why?
7 What do you think we will use for fuel, when petroleum runs out?

17.2 Refining petroleum

What does *refining* mean?

Petroleum contains *hundreds* of different hydrocarbons. But a big mixture like this is not very useful.

So the first step is to separate the compounds into groups with molecules of a similar size. This is called **refining** the petroleum. It is carried out by **fractional distillation**.

Refining petroleum in the lab

The apparatus on the right can be used to refine petroleum in the lab.

1 As you heat the petroleum, the compounds start to evaporate. The ones with smaller lighter molecules go first, since it takes less energy to free these from the liquid.

2 As the hot vapours rise, so does the thermometer reading. The vapours condense in the cool test-tube.

3 When the thermometer reading reaches 100 °C, replace the first test-tube with an empty one. The liquid in the first test-tube is your first **fraction** from the distillation.

4 Collect three further fractions in the same way, replacing the test-tube at 150 °C, 200 °C, and 300 °C.

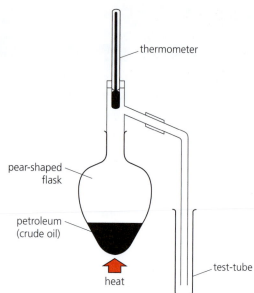

Comparing the fractions

Now compare the fractions – how runny they are, how easily they burn, and so on. You can burn samples on a watch glass, like this:

fraction 1	fraction 2	fraction 3	fraction 4
It catches fire easily. The flame burns high, which shows that the liquid is **volatile** – it evaporates easily.	This catches fire quite easily. The flame burns less high – so this fraction is less volatile than fraction 1.	This seems less volatile than fraction 2. It does not catch fire so readily or burn so easily – it is not so **flammable**.	This one does not ignite easily. You need to use a wick to keep it burning. It is the least flammable of the four.

This table summarises the results:

Fraction	Boiling point range	How easily does it flow?	How volatile is it?	How easily does it burn?	Size of molecules
1	up to 100 °C	very runny	volatile	very easily	small
2	100 – 150 °C	runny	less volatile	easily	
3	150 – 200 °C	not very runny	even less volatile	not easily	
4	200 – 300 °C	viscous (thick and sticky)	least volatile	only with a wick	large

The trends the fractions show

Those results show that, the larger the molecules in a hydrocarbon:

– the higher its boiling point will be
– the less volatile it will be
– the less easily it will flow (or the more viscous it will be)
– the less easily it will burn.

These trends help to dictate what the different fractions will be used for, as you will see below.

In the petroleum refinery

In a refinery, the fractional distillation is carried out in a tower that is kept very hot at the base, and cooler towards the top. Look at the drawing.

Petroleum is pumped in at the base. The compounds start to boil off. Those with the smallest molecules boil off first, and rise to the top of the tower. Others rise only part of the way, depending on their boiling points, and then condense.

▲ The new road surface is bitumen mixed with fine gravel.

The table shows the fractions that are collected.

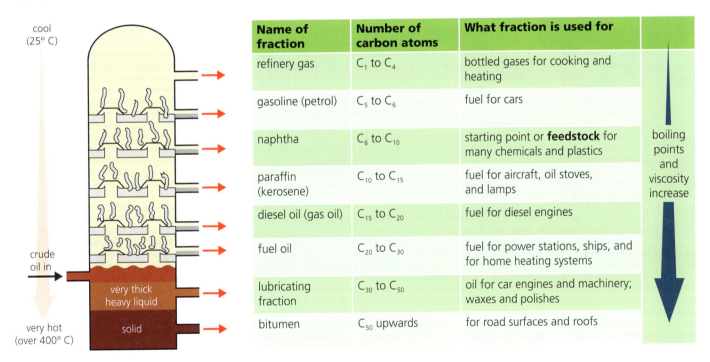

cool (25° C)

crude oil in

very hot (over 400° C)

very thick heavy liquid

solid

Name of fraction	Number of carbon atoms	What fraction is used for	
refinery gas	C_1 to C_4	bottled gases for cooking and heating	boiling points and viscosity increase
gasoline (petrol)	C_5 to C_6	fuel for cars	
naphtha	C_6 to C_{10}	starting point or **feedstock** for many chemicals and plastics	
paraffin (kerosene)	C_{10} to C_{15}	fuel for aircraft, oil stoves, and lamps	
diesel oil (gas oil)	C_{15} to C_{20}	fuel for diesel engines	
fuel oil	C_{20} to C_{30}	fuel for power stations, ships, and for home heating systems	
lubricating fraction	C_{30} to C_{50}	oil for car engines and machinery; waxes and polishes	
bitumen	C_{50} upwards	for road surfaces and roofs	

As the molecules get larger, the fractions get less runny, or more viscous: from gas at the top of the tower to solid at the bottom. They also get less flammable. So the last two fractions in the table are not used as fuels.

Q

1 Which two opposite processes take place, during fractional distillation?

2 A group of compounds collected during fractional distillation is called a .... ?

3 What does it mean? **a** volatile **b** viscous

4 List four ways in which the properties of different fractions differ.

5 Name the petroleum fraction that: **a** is used for petrol
   b has the smallest molecules **c** is the most viscous
   d has molecules with 20 to 30 carbon atoms

243

17.3 Cracking hydrocarbons

After fractional distillation …

Petroleum is separated into fractions by fractional distillation. But that is not the end of the story. The fractions all need further treatment before they can be used.

1 They contain impurities – mainly sulfur compounds. If left in the fuels, these will burn to form harmful sulfur dioxide gas.
2 Some fractions are separated further into single compounds, or smaller groups of compounds. For example the gas fraction is separated into methane, ethane, propane, and butane. (We buy butane in canisters.)
3 Part of a fraction may be **cracked**.
   Cracking breaks molecules down into smaller ones.

Cracking a hydrocarbon in the lab

This experiment is carried out using a hydrocarbon oil from petroleum. The product is a gas, collected over water in the inverted test-tube:

▲ This sulfur was obtained from sulfur compounds removed from natural gas.

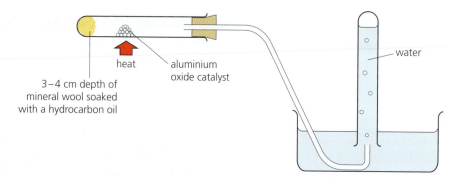

heat — aluminium oxide catalyst — water

3–4 cm depth of mineral wool soaked with a hydrocarbon oil

The moment heating is stopped, the delivery tube must be lifted out of the water. Otherwise water will get sucked up into the hot test-tube. Now compare the reactant and product:

	The reactant	The product
Appearance	thick colourless liquid	colourless gas
Smell	no smell	pungent smell
Flammability	difficult to burn	burns readily
Reactions	few chemical reactions	many chemical reactions

So the product is quite different from the reactant. Heating has caused the hydrocarbon to break down. A **thermal decomposition** has taken place. Note that:

● the reactant had a high boiling point and was not flammable – which means it had large molecules, with long chains of carbon atoms.

● the product has a low boiling point and is very volatile – so it must have small molecules, with short carbon chains.

● the product must also be a hydrocarbon, since nothing new was added.

So the molecules of the starting hydrocarbon have been cracked. And since the product is reactive, it could be a useful chemical.

▲ Some of the naphtha fraction from refining will be piped to the cracking plant.

Cracking in the refinery

In the refinery, cracking is carried out in a similar way.

- The long-chain hydrocarbon is heated to vaporise it.
- The vapour is usually passed over a hot catalyst.
- Thermal decomposition takes place.

Why cracking is important

- Cracking helps you make the best use of petroleum. Suppose you have too much of the naphtha fraction, and too little of the gasoline fraction. You can crack some naphtha to get molecules the right size for petrol.
- Cracking *always* produces short-chain compounds with a carbon–carbon double bond. This bond makes the compounds **reactive**. So they can be used to make plastics and other substances.

Examples of cracking

1 **Cracking the naphtha fraction** Compounds in the naphtha fraction are often cracked, since this fraction is used as the feedstock for making many useful chemicals. This is the kind of reaction that occurs:

decane, $C_{10}H_{22}$ from naphtha fraction

540°C, catalyst

pentane, C_5H_{12} suitable for petrol

propene, C_3H_6

ethene, C_2H_4

So decane has been broken down into three smaller molecules. The propene and ethene molecules have carbon–carbon double bonds. These two compounds belong to the **alkene** family, and they are very reactive.

2 **Cracking ethane** Ethane has very short molecules – but even it can be cracked, to give ethene and hydrogen:

ethane

steam
>800 °C

ethene   + H_2

hydrogen

The hydrogen from cracking is used to make ammonia – see page 222.

▲ Plastic furniture: it all began with cracking.

▲ Decane is one of the hydrocarbons in **white spirit**, a solvent used to thin oil-based paint, and clean paintbrushes.

Q

1 What happens during cracking?
2 Cracking is a *thermal decomposition*. Explain why.
3 Describe the usual conditions needed for cracking a hydrocarbon in the petroleum refinery.
4 What is *always* produced in a cracking reaction?
5 Explain why cracking is so important.
6 a A straight-chain hydrocarbon has the formula C_5H_{12}. Draw the structural formula for its molecules.
  b Now show what might happen when the compound is cracked.

17.4 Families of organic compounds

What their names tell you

There are *millions* of organic compounds. That could make organic chemistry confusing – but to avoid this, the compounds are named in a very logical way.

The rest of this chapter is about four families of organic compounds. **For these families, the name of the organic compound tells you:**

- **which family it belongs to**
- **how many carbon atoms are in it.**

Look at these two tables:

If the name ends in …	… the compound belongs to this family …	Example
-ane	the alkanes	ethane, C_2H_6
-ene	the alkenes	ethene, C_2H_4
-ol	the alcohols	ethanol, C_2H_5OH
-oic acid	the carboxylic acids	ethanoic acid, CH_3COOH

This in the name …	… means this many carbon atoms …	Example from the alkane family
meth-	1	methane, CH_4
eth-	2	ethane, C_2H_6
prop-	3	propane, C_3H_8
but-	4	butane, C_4H_{10}
pent-	5	pentane, C_5H_{12}
hex-	6	hexane, C_6H_{14}

▲ Natural gas burning at a cooker hob: it is mainly methane, the simplest alkane.

The alkanes: the simplest family

Here again are the first four members of the alkane family. Note that methane is the simplest member. What patterns do you notice?

Compound	methane	ethane	propane	butane
Formula	CH_4	C_2H_6	C_3H_8	C_4H_{10}
Structural formula	H H—C—H H	H H H—C—C—H H H	H H H H—C—C—C—H H H H	H H H H H—C—C—C—C—H H H H H
Number of carbon atoms in the chain	1	2	3	4
Boiling point/°C	−164	−187	−42	−0.5

boiling point increases with chain length

Comparing families

This table shows one member from each of the four families. Compare them.

Family	A member	Structural formula	Comments
alkanes	ethane, C_2H_6	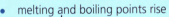	• The alkanes contain only carbon and hydrogen, so they are **hydrocarbons**. • The bonds between their carbon atoms are all **single bonds**.
alkenes	ethene, C_2H_4		• The alkenes are **hydrocarbons**. • All alkenes contain **carbon – carbon double bonds**. • The C=C bond is called their **functional group**.
alcohols	ethanol, C_2H_5OH		• The alcohols are not hydrocarbons. • They are like the alkanes, but with an **OH group**. • The OH group is their functional group.
carboxylic acids	ethanoic acid, CH_3COOH		• The carboxylic acids are not hydrocarbons. • All carboxylic acids contain the **COOH group**. • The COOH group is their functional group.

Functional groups

A functional group is the part of a molecule that largely dictates how the molecule will react.

For example, all the alkenes have similar reactions because they all have the same functional group, the C=C bond.

Homologous series

Look back at the alkanes in the table at the bottom of page 246. They form a **homologous series**. In a homologous series:

- All the compounds fit the same general formula.
  For the alkanes the general formula is C_nH_{2n+2}, where n is a number.
  For methane n is 1, giving the formula CH_4.
  For ethane n is 2, giving C_2H_6.
  For propane n is 3, giving C_3H_8.

- The chain length increases by 1 each time.

- As the chain gets longer, the compounds show a gradual change in properties. For example, their boiling points rise, and they burn less easily.

As you will see, all four families in the table form homologous series.

In a homologous series …
As the chain gets longer:
- melting and boiling points rise
- viscosity increases – the compounds flow less easily
- flammability decreases – the compounds burn less easily.

1 *Propanol* is an organic compound.
  a How many carbon atoms does it contain?
  b Which family does it belong to?
  c See if you can draw a structural formula for it.

2 Draw a structural formula for the alkane called *hexane*.
3 An alkane has 32 carbon atoms in each molecule. Give its formula.
4 Try to draw the structural formula for *propanoic acid*.

17.5 The alkanes

Alkanes: a reminder

This is what you have learned about the alkanes so far:

- The alkanes are the simplest family of organic compounds.
- They are hydrocarbons: they contain only carbon and hydrogen.
- Their carbon – carbon bonds are all single bonds.
- They form a homologous series, with the general formula C_nH_{2n+2}.

This table shows the first four members of the alkane family. What patterns do you notice?

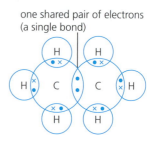

one shared pair of electrons (a single bond)

▲ The bonding in ethane.

Compound	methane	ethane	propane	butane
Formula	CH_4	C_2H_6	C_3H_8	C_4H_{10}
Structural formula	H–C–H with H above and H below	H–C–C–H with H H above and H H below	H–C–C–C–H with H H H above and H H H below	H–C–C–C–C–H with H H H H above and H H H H below
Number of carbon atoms in the chain	1	2	3	4
Boiling point/°C	−164	−87	−42	−0.5

boiling point increases with chain length →

Key points about the alkanes

1 They are found in petroleum and natural gas. Petroleum contains alkanes with up to 70 carbon atoms. Natural gas is mainly methane, with small amounts of ethane, propane, butane, and other compounds.

2 The first four alkanes are gases at room temperature. The next twelve are liquids. The rest are solids. Boiling points increase with chain length because attraction between the molecules increases – so it takes more energy to separate them.

3 Since all their carbon – carbon bonds are single bonds, the alkanes are called **saturated**. Look at the bonding in ethane on the right above.

4 Generally, the alkanes are quite unreactive.

5 But alkanes do burn well in a good supply of oxygen, forming carbon dioxide and water vapour, and giving out plenty of heat. So they are used as **fuels**. Methane burns the most easily. Like this:

$$CH_4\,(g) + 2O_2\,(g) \longrightarrow CO_2\,(g) + 2H_2O\,(l) + \text{heat energy}$$

6 If there is not enough oxygen, the alkanes undergo **incomplete combustion**, giving poisonous carbon monoxide. For example:

$$2CH_4\,(g) + 3O_2\,(g) \longrightarrow 2CO\,(g) + 4H_2O\,(l) + \text{less heat energy}$$

▲ Butane is used as fuel for cooking and heating, in many homes.

7 Alkanes also react with chlorine in sunlight. For example:

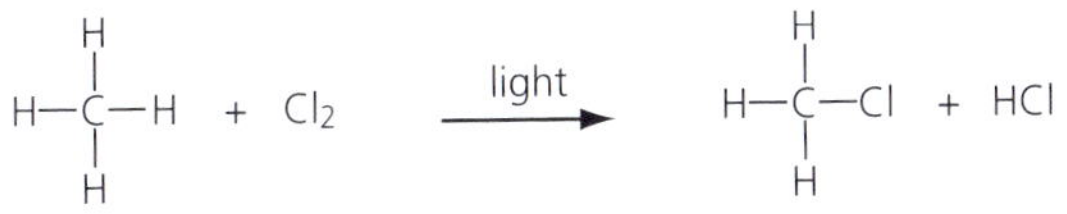

H—C—H (methane) + Cl₂ →[light] H—C—Cl (chloromethane) + HCl

methane   chloromethane

This is called a **substitution** reaction, because a chlorine atom takes the place of a hydrogen atom. If there is enough chlorine, all four hydrogen atoms will be replaced, one by one. Look at the panel on the right.

The reaction can be explosive in sunlight. But it will not take place in the dark, because it is also a **photochemical reaction**: light energy is needed to break the bonds in the chlorine molecules, to start the reaction off.

Chlorine and methane

The hydrogen atoms can be replaced one by one:

chloromethane CH_3Cl
dichloromethane CH_2Cl_2
trichloromethane $CHCl_3$
tetrachloromethane CCl_4

All four are used as solvents. But they can cause health problems, so are being used less and less.

Isomers

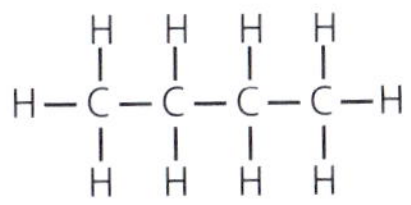

H—C—H
H—C—C—C—H

H—C—C—C—C—H

butane
boiling point 0 °C

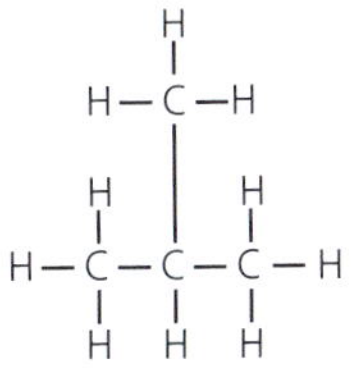

2-methylpropane
boiling point −10 °C

Compare these alkane molecules. Both have the same formula, C_4H_{10}. But they have different structures. The first has a **straight** or unbranched chain. In the second, the chain is **branched**.

The two compounds are **isomers**.
Isomers are compounds with the same formula, but different structures.

The more carbon atoms in a compound, the more isomers it has. There are 75 isomers with the formula $C_{10}H_{22}$, for example.

Since isomers have different structures, they also have slightly different properties. For example branched isomers have lower boiling points, because the branches make it harder for the molecules to get close. So the attraction between them is less strong, and less heat is needed to overcome it.

▲ Branched-chain hydrocarbons burn less easily than their straight-chain isomers. So they are used in petrol to control combustion and stop engine 'knock' – and especially for racing cars.

1 Describe the bonding in ethane. A drawing will help!

2 Why are alkanes such as methane and butane used as fuels? See if you can give *at least* two reasons.

3 Butane burns in a similar way to methane. See if you write a balanced equation for its combustion.

4 a The reaction of chlorine with methane is called a *substitution* reaction. Why?
  b What special condition is needed, for this reaction?

5 Ethane reacts with chlorine, in a substitution reaction.
  a Draw the structural formula for each compound that can form, as the reaction proceeds. (Isomers too!)
  b Write the formula for each compound in **a**.

6 The compound C_5H_{12} has *three* isomers.
  a Draw the structures of these three isomers.
  b Their boiling points are 9.5, 28, and 36 °C. Match these to your drawings, and explain your choice.

17.6 The alkenes

The alkene family

- The alkenes are hydrocarbons.
- They form a homologous series, with the general formula C_nH_{2n}.
- They all contain the C=C double bond. This is their functional group, and largely dictates their reactions. Look at the bonding in ethene.
- Because they contain C=C double bonds, they are called **unsaturated**. (Alkanes have only single carbon – carbon bonds, so are **saturated**.)

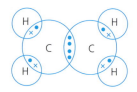

▲ The bonding in ethene.

Here are the first three members of the family. Note how ethene is drawn:

Compound	ethene	propene	but-1-ene
Formula	C_2H_4	C_3H_6	C_4H_8
Structural formula			
Number of carbon atoms	2	3	4
Boiling point/°C	−102	−47	−6.5

Key points about the alkenes

1 The alkenes are made from alkanes by cracking. For example ethene is formed by cracking ethane. Hydrogen is also produced:

ethane $\xrightarrow[>800\ °C]{steam}$ ethene + H_2 hydrogen

2 Alkenes are much more reactive than alkanes, because the double bond can break, to add on other atoms.

For example, ethene can add on hydrogen again, to form ethane:

ethene (g) + H_2(g) $\xrightarrow[a\ catalyst]{heat,\ pressure}$ ethane (g)

It also adds on water (as steam) to form ethanol, an alcohol:

ethene (g) + H_2O(g) $\xrightleftharpoons[a\ catalyst]{heat,\ pressure}$ ethanol (l)

These reactions are called **addition reactions**. Can you see why?
An addition reaction turns an unsaturated alkene into a saturated compound.

▲ Ethanol is a good solvent. It also kills many germs, so it is used in disinfectant gels for wiping your hands, in hospital.

Polymerisation

Alkene molecules undergo a very useful **addition reaction,** where they add on *to each other* to form compounds with very long carbon chains.

The alkene molecules are called **monomers**. The long-chain compounds that form are called **polymers**. The reaction is called **polymerisation**. For example ethene polymerises like this:

ethene molecules (monomers)

polymerisation

part of a polythene molecule (a polymer)

The product is **poly(ethene)** or **polythene**. The chain can be many thousands of carbon atoms long!

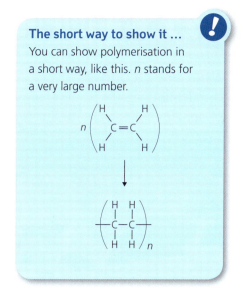

The short way to show it ...
You can show polymerisation in a short way, like this. *n* stands for a very large number.

A test for unsaturation

You can use **bromine water** to test whether a hydrocarbon is unsaturated. It is an orange solution of bromine in water. If a C=C bond is present, a reaction takes place, and the colour disappears. For example:

ethene (colourless)   (g)   + Br_2 (aq) (orange)   →   1,2–dibromoethane (colourless)

This is another example of an addition reaction. The bromine adds on.

Isomers in the alkene family

In alkenes, the chains can branch in different ways, *and* the double bonds can be in different positions.

Compare the three compounds below. All three have the formula C_4H_8, but they have different structures. So they are isomers.

but-1-ene   but-2-ene   2-methylpropene

Now look at the numbers in their three names. What do these tell you?

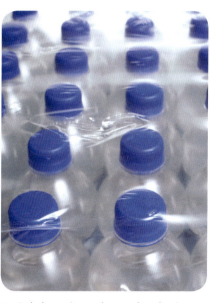

▲ Polythene is used to make plastic bottles, plastic bags, plastic sheeting ...

1 a Name the two simplest alkenes.
  b Now draw their structural formulae.
2 What makes alkenes react so differently from alkanes?
3 Ethene can *polymerise*. What does that mean?

4 a Propene is *unsaturated*. What does that mean?
  b Write an equation for its reaction with bromine.
5 How would you turn propene into:
  a propane?   **b** propanol?

17.7 The alcohols

What are alcohols?

The **alcohols** are the family of organic compounds that contain the **OH** group. This table shows the first four members:

Alcohol	methanol	ethanol	propan-1-ol	butan-1-ol
Formula	CH_3OH	C_2H_5OH	C_3H_7OH	C_4H_9OH
Structural formula	H H—C—O—H H	H H H—C—C—O—H H H	H H H H—C—C—C—O—H H H H	H H H H H—C—C—C—C—O—H H H H H
Number of carbon atoms	1	2	3	4
Boiling point/°C	65	78	87	117

Note that:

- they form a homologous series, with the general formula $C_nH_{2n+1}OH$.
- their OH functional group means they will all react in a similar way.
- two of the names above have -1- in. This tells you that the OH group is attached to a carbon atom at one end of the chain.

Ethanol, an important alcohol

- Ethanol is a good solvent. It dissolves many substances that do not dissolve in water.
- It evaporates easily – it is **volatile**. That makes it a suitable solvent to use in glues, printing inks, perfumes, and aftershaves.
- It burns readily in oxygen, and the reaction gives out plenty of energy, as heat. So it is used as a fuel. See the next page for more.

Two ways to make ethanol

Ethanol is made in two ways, one biological and one chemical.

1 By fermentation – the biological way

It is made from glucose (a sugar) using **yeast**, in the absence of air. Yeast is a mass of living cells. The enymes in it catalyse this reaction:

$$C_6H_{12}O_6 \ (aq) \ \xrightarrow[\text{in yeast}]{\text{enzymes}} \ 2C_2H_5OH \ (aq) \ + \ 2CO_2 \ (g) \ + \ \text{energy}$$

glucose   ethanol   carbon dioxide

- The process is called **fermentation**, and it is exothermic.
- You can start with anything that contains sugars, starch, or cellulose, because they all break down to glucose. For example you can use maize, sugarcane, potatoes, and wood, to make ethanol.
- The yeast stops working when the % of ethanol reaches a certain level, or if the mixture gets too warm.
- The ethanol is separated from the final mixture by fractional distillation.

▲ Ethanol is used as a solvent for perfume. Why?

> **Alcoholic drinks**
> Ethanol is the ingredient that makes people drunk, in alcoholic drinks.

▲ Corn (maize) is widely grown in the USA to make ethanol, for car fuel. Fuel made from plant material, using yeast or bacteria, is called **biofuel**.

2 By the hydration of ethene – the chemical way

Hydration means water is added on. This is an **addition reaction**.

ethene + H_2O (g) → ethanol

570 °C, 60–70 atm
a catalyst
(phosphoric acid)

- The reaction is reversible, and exothermic.
- High pressure and a low temperature would give the best yield. But in practice the reaction is carried out at 570 °C, to give a decent rate.
- A catalyst is also used, to speed up the reaction.

▲ Traffic in Rio de Janeiro in Brazil: running mainly on ethanol made from sugar cane.

Comparing the two methods

Each method of making ethanol has advantages – and disadvantages! Look at this table:

Ethanol by fermentation	Ethanol by hydration
Advantages	Advantages
• It uses renewable resources. You can grow more maize, and sugar cane, for example. • You can use waste plant material too.	• The reaction is fast. • You can run it continuously. Just keep removing the ethanol. • It gives pure ethanol. No need for fractional distillation.
Disadvantages	Disadvantages
• You need a lot of plant material to make a litre of ethanol. So a lot of land is needed to grow enough crops. • Fermentation is slow – and you start all over again for each new batch of material. • The yeast stops working when the % of ethanol reaches a certain level – even if there's still glucose left. • You need heat, for fractional distillation. Heat costs money!	• Ethene is made from oil – and that is not a renewable resource. (We may run out of it one day.) • A great deal of heat is needed: to make the steam, and keep the reaction running at 570 °C. Expensive! • It is a reversible reaction, and at 570 °C the yield is not high. So you must keep recycling the unreacted ethene and steam.

Ethanol as a fuel

Ethanol burns well in oxygen. The reaction gives out plenty of heat:

$$C_2H_5OH \ (l) + 3O_2 \ (g) \longrightarrow 2CO_2 \ (g) + 3H_2O \ (l) + heat$$

It is increasingly used as a fuel for car engines because:

- it can be made quite cheaply, and even from waste plant material.
- many countries have no petroleum of their own, and have to buy it from other countries; it costs a lot, so ethanol is an attractive option.
- it can help in the fight against global warming. Ethanol gives out carbon dioxide (a greenhouse gas) when it burns – but at the same time, plants being grown to make ethanol take carbon dioxide in. So overall, ethanol has less impact on the environment than petrol does.

Cars … or people?
- If lots of crops are grown for ethanol, it means less land for food crops.
- A shortage of food crops means a rise in food prices.
- When food prices rise, it affects poor people the most.

Many people are against growing crops to make ethanol. They say: *Feed people, not cars!*

1 All alcohols react in a similar way. Why?
2 Draw the structural formula for ethanol.
3 Give two uses of ethanol.
4 In Brazil, sugarcane is used to make ethanol. Name the process used, and say what the catalyst is.

5 a Write a word equation for the combustion (burning) of:
   i ethanol   ii methane
  b Compare the equations. What do you notice?
6 There is another isomer with the same formula as propan-1-ol. Draw its structure, and suggest a name.

253

17.8 The carboxylic acids

The carboxylic acid family

Now we look at the family of organic acids: the carboxylic acids.
Here are the first four members of the family:

Name of acid	methanoic	ethanoic	propanoic	butanoic
Formula	HCOOH	CH_3COOH	C_2H_5COOH	C_3H_7COOH
Structural formula				
Number of carbon atoms	1	2	3	4
Boiling point/°C	101 °C	118 °C	141 °C	164 °C

- The family forms a homologous series with the general formula
  $C_nH_{2n}O_2$. Check that this fits with the formulae in the table above.
- The functional group **COOH** is also called the **carboxyl group**.

We focus on ethanoic acid in the rest of this unit. But remember that
other carboxylic acids behave in a similar way, because they all contain
the carboxyl group.

Two ways to make ethanoic acid

Ethanoic acid is made by oxidising ethanol:

The oxidation can be carried out in two ways.

1 By fermentation – the biological way

When ethanol is left standing in air, bacteria bring about its oxidation
to ethanoic acid. This method is called **acid fermentation**.

Acid fermentation is used to make vinegar (a dilute solution of
ethanoic acid). The vinegar starts as foods such as apples, rice, and
honey, which are first fermented to give ethanol.

2 Using an oxidising agent – the chemical way

Ethanol is oxidised much faster by warming it with the powerful
oxidising agent potassium manganate(VII), in the presence of acid.
The manganate(VII) ions are themselves reduced to Mn^{2+} ions, with a
colour change. The acid provides the H^+ ions for the reaction:

$$MnO_4^- + 8H^+ + 5e^- \longrightarrow Mn^{2+} + 4H_2O$$
purple   colourless

▲ Organic chemistry at the dinner
table. Vinegar (left) is mainly ethanoic
acid and water. Olive oil (right) is made
of **esters**.

Ethanoic acid: typical acid reactions

1 A solution of ethanoic acid in water turns litmus red.

2 It also reacts with metals, bases, and carbonates, to form **salts**.
For example, it reacts with sodium hydroxide like this:

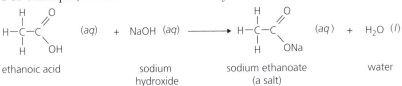

ethanoic acid   sodium hydroxide   sodium ethanoate (a salt)   water

or CH_3COOH (aq) + NaOH (aq) $\longrightarrow$ CH_3COONa (aq) + H_2O (l)

Like all salts, sodium ethanoate is an ionic compound.

The ethanoate ion
This is its structural formula:
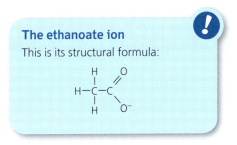

A typical weak acid

A solution of ethanoic acid is acidic because it contains H^+ ions. They are present because some ethanoic acid molecules dissociate in water:

$$CH_3COOH \ (aq) \xrightarrow{\text{some molecules}} CH_3COO^- \ (aq) + H^+ \ (aq)$$

ethanoic acid   ethanoate ions   hydrogen ions

But only *some* molecules dissociate. So ethanoic acid is a **weak acid**.

Esters

Ethanoic acid also reacts with alcohols, to give compounds called **esters**.
The alcohol molecule is reversed below, to help you see what is happening:

this group is called an **ester** linkage

ethanoic acid   propanol   propyl ethanoate (an ester)   water

or CH_3COOH (l) + C_3H_7OH (l) $\overset{\text{conc. } H_2SO_4}{\rightleftharpoons}$ $CH_3COOC_3H_7$ (l) + H_2O (l)

Note these points:

- Two molecules have joined to make a larger molecule, with the loss of a small molecule, water. So this is called a **condensation reaction**.

- The reaction is reversible, and sulfuric acid acts a catalyst.

- The alcohol part comes *first* in the name – but second in the formula. (Compare with sodium ethanoate above.)

- Propyl ethanoate smells of pears. In fact many esters have attractive smells and tastes. So they are added to shampoos and soaps for their smells, and to ice cream and other foods as flavourings.

▲ The smell and taste of the apple come from natural esters. Synthetic esters are used in the shampoo.

1 What is the functional group of the carboxylic acids?

2 Copy and complete. (Page 148 may help!)
carboxylic acid + metal → _____ + _____
carboxylic acid + alkali → _____ + _____
carboxylic acid + alcohol ⇌ _____ + _____

3 Carboxylic acids are *weak* acids. Explain why.

4 Draw structural formulae to show the reaction between ethanol and ethanoic acid, and name the products.

5 What is a *condensation reaction*?

6 Esters are important compounds in industry. Why?

Checkup on Chapter 17

Revision checklist

Core syllabus content

Make sure you can …

- ☐ name the fossil fuels and say how they were formed
- ☐ explain what a *hydrocarbon* is
- ☐ explain why petroleum has to be refined, and:
  - – describe the refining process
  - – name the different fractions
  - – say what these fractions are used for
- ☐ explain what *cracking* is, and why it is so useful, and give the equation for the cracking of ethane
- ☐ say what a *structural formula* is, and draw structural formulae for
  methane ethane ethene ethanol ethanoic acid
- ☐ say what family a compound belongs to, from its structural formula, and its name
- ☐ give the functional groups for the alkenes, alcohols, and carboxylic acids
- ☐ describe alkanes as unreactive except for burning, and give an equation for the combustion of methane
- ☐ explain what *unsaturated* means and describe a test to identify an unsaturated hydrocarbon
- ☐ describe how ethene monomers add on to each other to form the polymer poly(ethene)
- ☐ describe the two ways of making ethanol
- ☐ give two uses for ethanol (as solvent, and fuel)
- ☐ give the reaction for the combustion of ethanol
- ☐ give at least two advantages of ethanol as a fuel
- ☐ give examples of reactions to show that ethanoic acid is a typical acid

Extended syllabus content

Make sure you can also …

- ☐ name, and draw the structural formulae for, the four simplest members of each family: alkanes, alkenes, alcohols, and carboxylic acids
- ☐ give the general properties of a homologous series
- ☐ explain what *isomers* are, and draw examples
- ☐ describe the substitution reactions of alkanes with chlorine
- ☐ describe the addition reactions of alkenes with hydrogen, steam, and bromine
- ☐ describe the two ways to make ethanoic acid
- ☐ explain why ethanoic acid is a weak acid
- ☐ describe the reaction of ethanoic acid with an alcohol, and name the products that form

Questions

Core syllabus content

1 Petroleum is separated into fractions, like this:

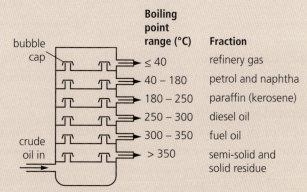

Boiling point range (°C)	Fraction
≤ 40	refinery gas
40 – 180	petrol and naphtha
180 – 250	paraffin (kerosene)
250 – 300	diesel oil
300 – 350	fuel oil
> 350	semi-solid and solid residue

a i What is this process called?

ii It uses the fact that different compounds have different ……… ………. What is missing?

b i Is naphtha just one compound, or a group of compounds? Explain.

ii Using the terms *evaporation* and *condensation*, explain how naphtha is produced.

c Give one use for each fraction obtained.

d A hydrocarbon has a boiling point of 200 °C.

i Are its carbon chains shorter, or longer, than those found in naphtha?

ii Is it more viscous, or less viscous, than the compounds found in naphtha?

2

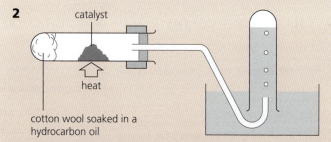

A hydrocarbon can be cracked in the lab using the apparatus above.

a What is cracking?

b Which two things are needed, to crack the hydrocarbon?

c The first tube of collected gas is discarded. Why? (What else is in the heated tube?)

d At the end of the experiment, the delivery tube must be removed from the water immediately. Why is this?

e Ethane, C_2H_6, can be cracked to give ethene, C_2H_4, and hydrogen. Write an equation for this.

3 Answer these questions about the alkanes.

a Which two elements do alkanes contain?

b Which alkane is the main compound in natural gas?

c After butane, the next two alkanes in the series are *pentane* and *hexane*. How many carbon atoms are there in a molecule of:
  i pentane?   **ii** hexane?

d Will pentane react with bromine water? Explain.

e Alkanes burn in a good supply of oxygen. Name the gases formed when they burn.

f Write the word equation for the complete combustion of pentane in oxygen.

g Name a harmful substance formed during *incomplete* combustion of pentane in air.

4 When ethanol vapour is passed over heated aluminium oxide, a dehydration reaction occurs, and the gas ethene is produced.

a Draw a diagram of suitable apparatus for carrying out this reaction in the lab.

b What is meant by a *dehydration reaction*?

c Write an equation for this reaction, using the structural formulae.

d **i** What will you see if the gas that forms is bubbled through bromine water?

  ii You will not see this if ethanol vapour is passed through bromine water. Why not?

5 a Which of these could be used as monomers for addition polymers? Explain your choice.
  i ethene, $CH_2=CH_2$
  ii ethanol, C_2H_5OH
  iii propane, C_3H_8
  iv styrene, $C_6H_5CH=CH_2$
  v chloropropene, $CH_3CH=CHCl$

b Suggest a name for each polymer obtained.

Extended syllabus content

6 The saturated hydrocarbons form a homologous series with the general formula C_nH_{2n+2}.

a What is a *homologous series*?

b Explain what the term *saturated* means.

c Name the series described above.

d **i** Give the formula and name for a member of this series with two carbon atoms.

  ii Draw its structural formula.

e **i** Name a homologous series of *unsaturated* hydrocarbons, and give its general formula.

  ii Give the formula and name for the member of this series with two carbon atoms.

  iii Draw the structural formula for the compound.

7 Ethanol is a member of a homologous series.

a Give two general characteristics of a homologous series.

b **i** Which homologous series is ethanol part of?

  ii What is the general formula for the series?

  iii What does *functional group* mean?

  iv What is the functional group in ethanol's homologous series?

c Write down the formula of ethanol.

d **i** Draw the structural formula for the fifth member of the series, pentan-1-ol.

  ii Draw the structural formula for an isomer of pentan-1-ol.

  iii Describe how pent-1-ene could be made from pentan-1-ol.

  iv Name the organic product formed when pentan-1-ol is oxidised using acidified potassium manganate(VII).

8 Ethanoic acid is a member of the homologous series with the general formula $C_nH_{2n}O_2$.

a Name this series.

b What is the functional group of the series?

c Ethanoic acid is a *weak* acid. Explain what this means, using an equation to help you.

d Ethanoic acid reacts with carbonates.

  i What would you *see* during this reaction?

  ii Write a balanced equation for the reaction with sodium carbonate.

e **i** Name the member of the series for which $n = 3$, and draw its structural formula.

  ii Give the equation for the reaction between this compound and sodium hydroxide.

9 Ethanoic acid reacts with ethanol in the presence of concentrated sulfuric acid.

a Name the organic product formed.

b Which type of compound is it?

c How could you tell quickly that it had formed?

d What is the function of the sulfuric acid?

e The reaction is *reversible*. What does this mean?

f Write an equation for the reaction.

10 Hex-1-ene is an unsaturated hydrocarbon. It melts at $-140°C$ and boils at $63°C$. Its empirical formula is CH_2. Its relative molecular mass is 84.

a **i** To which family does hex-1-ene belong?

  ii What is its molecular formula?

b **i** Hex-1-ene reacts with bromine water. Write an equation to show this reaction.

  ii What is this type of reaction called?

  iii What would you *see* during the reaction?

18.1 Introducing polymers

What is a polymer?

A polymer is any substance containing very large molecules, formed when lots of small molecules join together.

For example, look what happens when ethene molecules join:

This test tube contains ethene gas. When ethene is heated to 50 °C, at a few atmospheres pressure, and over a special catalyst …

… it turns into a liquid that cools to a waxy white solid. This is found to contain very long molecules, made by the ethene molecules joining.

And it is really useful. It can be used to make toys, dustbins, tables and chairs, water pipes, buckets, crates, washing-up bowls, and so on.

The reaction that took place is:

ethene molecules (monomers)

polymerisation

part of a polythene molecule (a polymer)

The drawing above shows just six ethene molecules joining together. In fact many thousands of ethene molecules join together, giving very large molecules.

The product is called a **polymer**. The polymer made from ethene is called **poly(ethene)** or **polythene**. Poly- means *many*.

The reaction is called a **polymerisation**.

In a polymerisation reaction, thousands of small molecules join to give one large molecule. The small molecules are called monomers. The product is called a polymer.

Synthetic polymers

Polythene is a **synthetic polymer**. *Synthetic* means it is made in a factory. Other synthetic polymers include nylon, Terylene, lycra, chewing gum, and plastics such as polystyrene and perspex. Hair gels and shower gels contain water-soluble polymers.

▲ Hair gel: a water-soluble polymer. When you put it on, the water in it evaporates so the gel gets stiff.

Natural polymers

Polythene was first made in 1935. But for billions of years, nature has been busy making natural polymers. Look at these examples:

Starch is a polymer made by plants. The starch molecules are built from molecules of **glucose**, a sugar. We eat plenty of starch in rice, bread, and potatoes.

Plants also use glucose to make another polymer called **cellulose**. Cotton T-shirts and denim jeans are almost pure cellulose, made by the cotton plant.

Your skin, hair, nails, bones and muscles are mostly polymers, made of large molecules called **proteins**. Your body builds these up from **amino acids**.

The wood in trees is about 50% cellulose. Paper is made from wood pulp, so this book is mainly cellulose. The polymer in your hair and nails, and in wool and silk, and animal horns and claws, is called **keratin**. The polymer in your skin and bones is called **collagen**.

So – you contain polymers, you eat polymers, you wear polymers, and you use polymers. Polymers play a big part in your life!

The reactions that produce polymers

All polymers, natural and synthetic, consist of large molecules, formed by small molecules joined together.

But polymers are not all formed in the same way. There are two quite different types of polymerisation reaction: **addition polymerisation**, and **condensation polymerisation**. You can find out more about these in the next two units. Watch out for the differences!

▲ Wood: over 50% cellulose. This wood may end up as paper.

1 What is:
  a a polymer?
  b a natural polymer?
  c a synthetic polymer?
  d polymerisation?
  e a monomer?

2 Name the natural polymer found in:
  a your hair
  b this book

3 Name at least three items you own, that are made of polymers.

18.2 Addition polymerisation

Another look at the polymerisation of ethene

Here again is the reaction that produces polythene:

ethene molecules (monomers)

$$H_2C=CH_2 \quad H_2C=CH_2 \quad H_2C=CH_2 \quad H_2C=CH_2 \quad H_2C=CH_2 \quad H_2C=CH_2$$

polymerisation

part of a polythene molecule (a polymer)

$$-C-C-C-C-C-C-C-C-C-C-C-C-$$

The reaction can be shown in a short form like this:

$$n \left(\begin{array}{c} H \\ C=C \\ H \end{array} \begin{array}{c} H \\ \\ H \end{array} \right) \xrightarrow[\text{a catalyst}]{\text{heat, pressure,}} \left(\begin{array}{c} H \ H \\ -C-C- \\ H \ H \end{array} \right)_n$$

where n stands for a large number. It could be many thousands. The catalyst for the reaction is usually a mixture of titanium and aluminium compounds.

It's an addition reaction

The reaction above takes place because the C=C bond in the ethene molecules breaks, allowing the molecules to *add on* to each other. So it is called **addition polymerisation**.

In addition polymerisation, double bonds in molecules break, and the molecules add on to each other.

The monomer

The small starting molecules in a polymerisation are called **monomers**. In the reaction above, ethene is the monomer.

For addition polymerisation to take place, the monomers must have C=C double bonds.

The chain lengths in polythene

In polythene, all the molecules have long chains of carbon atoms, with hydrogen atoms attached. So they are all similar. But they are not all identical. The chains are not all the same length. That is why we can't write an exact formula for polythene.

By changing the reaction conditions, the *average* chain length can be changed. But the chains will never be all the same length.

The relative atomic mass (M_r) of an ethene molecule is 28. The *average* M_r of the polymer molecules in a sample of polyethene can be 500 000 or more. In other words, when making polythene, at least 17 000 ethene molecules join, on average!

▲ Polythene for packaging is made and sold as pellets like these. Later they will be heated to soften or melt them, and turned into plastic bottles and bags.

▲ To make a bottle, polythene pellets are melted. A little molten polymer is fed into a mould. A jet of air forces it into the shape of the mould. Then the mould is opened – and out comes a bottle!

Making other polymers by addition

Look at the polymers in this table. You have probably heard of them all.
They are all made by addition polymerisation. Compare them:

The monomer	Part of the polymer molecule	The equation for the reaction
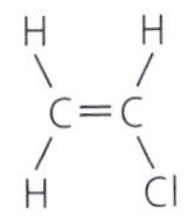 chloroethene (vinyl chloride)	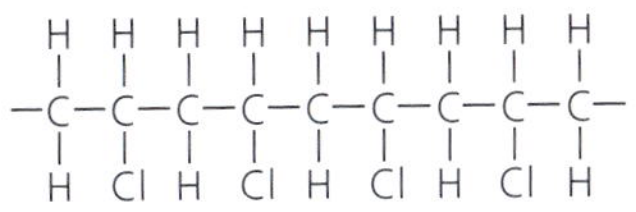 poly(chloroethene) or polyvinyl chloride (PVC)	 n stands for a large number!
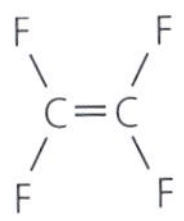 tetrafluoroethene	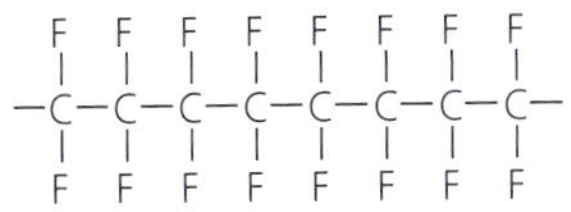 poly(tetrafluoroethene) or Teflon	
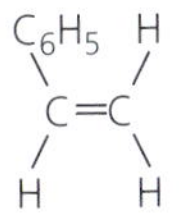 phenylethene (styrene)	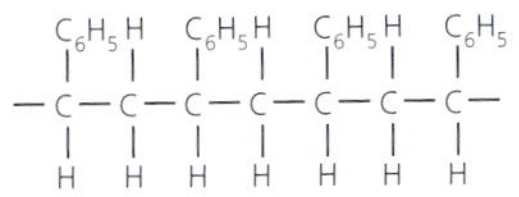 poly(phenylethene) or poly(styrene)	

Identifying the monomer

If you know the structure of the addition polymer, you can work out what
the monomer was. Like this:

● Identify the repeating unit. (It has two carbon atoms side by side, in
the main chain.) You could draw brackets around it.

● Then draw the unit, but put a double bond between the two carbon
atoms. That is the monomer.

For example:

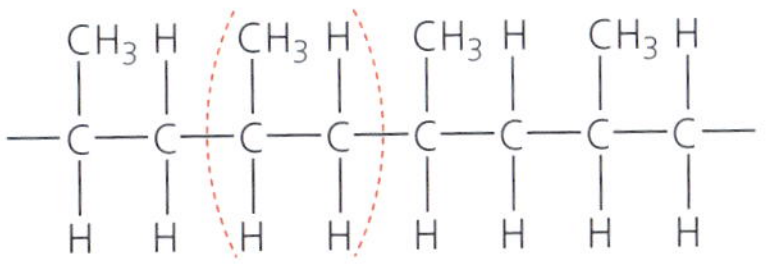

This shows part of a molecule of
poly(propene). The unit within
brackets is the repeating unit.

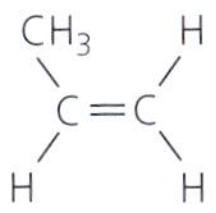

So this is the monomer that was
used. It is the alkene **propene**.
Note the C=C double bond.

▲ PVC is light and flexible so is widely
used for hoses and water pipes, and as
an insulating cover for electrical wiring.

Q

1 Why is it called *addition* polymerisation?
2 Could methane (CH_4) be used as a monomer for addition
polymerisation? Explain your answer.
3 Draw a diagram to show the polymeristion of ethene.
4 It is not possible to give an exact formula for the polymer
molecules in polythene. Why not?

5 Draw a diagram to show the polymerisation of:
 a chloroethene   b phenylethene
6 A polymer has the general formula
shown on the right.
Draw the monomer that was used
to make it.

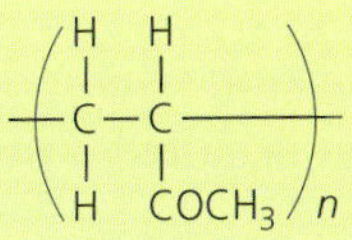

18.3 Condensation polymerisation

Condensation polymerisation is different!

Addition polymerisation requires one type of monomer, with C=C bonds. The molecules add to each other to form one product: the polymer. But condensation polymerisation is different.

- It does not depend on C=C bonds.
- *Two* types of monomer join. Each has *two functional groups*.
- They join at their functional groups, by getting rid of or **eliminating** a small molecule. So there are *two* products: the polymer, and another.

Let's look at two examples.

1 Making nylon

Below are the two monomers used in making nylon. We will call them **A** and **B**, for convenience:

A 1,6-diaminohexane   **B** hexan-1,6-dioyl chloride

A has an NH_2 group *at each end*. **B** has a COCl group *at each end*. Only these functional groups take part in the reaction. So we will show the rest of the molecules as blocks, for simplicity.

The reaction

This shows the reaction between the two monomer molecules:

HCl is lost

Then another **B** reacts here …   … and another **A** reacts here … and so on.

So the nitrogen atom at one end of **A** has joined to the carbon atom at one end of **B**, by eliminating a molecule of hydrogen chloride.

The reaction continues at the other ends of A and B. In this way, thousands of molecules join, giving the polymer molecule **nylon**. Here is part of it:

this linking group of atoms is called an **amide linkage**

The group where the monomers joined is called the **amide** linkage. So nylon is called a **polyamide**. (Proteins have this link too, as you will see.)

Nylon can be drawn into tough strong fibres that do not rot away. So it is used for thread, ropes, fishing nets, car seat belts, and carpets.

contains **A**

contains **B**

▲ Making nylon in the school lab. **A** is dissolved in water. **B** is dissolved in an organic solvent that does not mix with water. Nylon forms where the solutions meet.

▲ Nylon thread: tough, strong, great for flying kites.

2 Making Terylene

Like nylon, Terylene is made by condensation polymerisation, using two different monomers. This time we call them **C** and **D**:

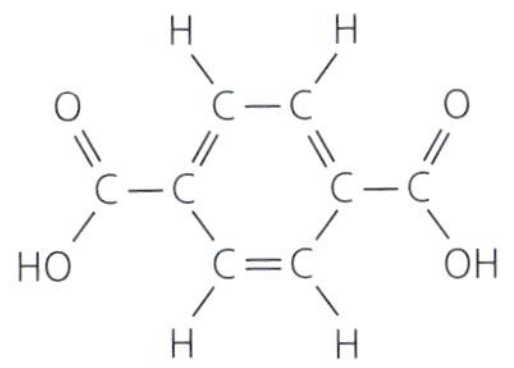

C benzene-1,4-dicarboxylic acid

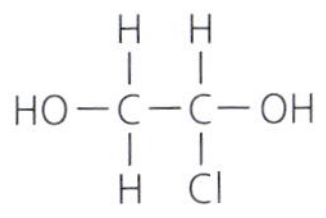

D ethane-1,2-diol

C has two COOH (carboxyl) groups, and **D** has two OH (alcohol) groups. Only these functional groups take part in the reaction.
So once again we can show the rest of the molecules as blocks.

The reaction

This shows the reaction between the two monomer molecules:

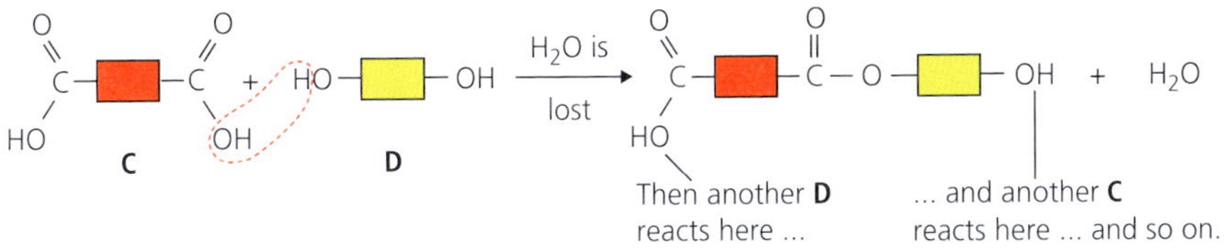

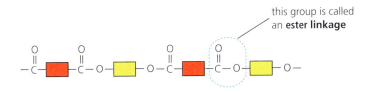

Then another **D**   ... and another **C**
reacts here ...   reacts here ... and so on.

So a carbon atom at one end of **C** has joined to an oxygen atom at one end of **D**, by eliminating a water molecule.

The reaction continues at the other ends of **C** and **D**. In this way thousands of molecules join, giving the polymer molecule **Terylene**. Here is part of it:

this group is called
an **ester linkage**

In fact the reaction is the same as the reaction between the acid and alcohol on page 255, giving an ester. (See the last section on that page.)

So the group where the monomers have joined is called an **ester linkage**. Terylene is called a **polyester**.

Terylene is used for shirts and other clothing, and for bedlinen. It is usually woven with cotton. The resulting fabric is more hard-wearing than cotton, and does not crease so easily. Terylene is also sold as polyester thread.

▲ Fibres of nylon and Terylene are made by pumping the melted polymer through a spinneret (like a shower head). As it comes out through the holes, the polymer hardens into long threads.

▲ Shirts made from Terylene woven with cotton.

Q

1 How many products are there, in condensation polymerisation?

2 In condensation polymerisation, each monomer molecule must have *two* functional groups. Explain why.

3 List the differences between condensation and addition polymerisation.

4 a Draw a diagram to show the reaction that produces nylon. (You can show the carbon chains as blocks.)
   b Circle the *amide linkage* in your drawing.
   c Nylon is called a *polyamide*. Why?

5 Draw part of a Terylene molecule in a simple way, using blocks as above. Circle the *ester linkage*.

18.4 Making use of synthetic polymers

Plastics are synthetic polymers

Synthetic polymers are usually called **plastics**. (*Plastic* means *can be moulded into shape without breaking*, and this is true of all synthetic polymers while they are being made.) But when they are used in fabrics, and for thread, we still call them synthetic polymers.

Most plastics are made from chemicals found in the naphtha fraction of petroleum (pages 243 and 245). They are usually quite cheap to make.

The properties of plastics

Most plastics have these properties:

1 They do not usually conduct electricity or heat.

2 They are unreactive. Most are not affected by air, water, acids, or other chemicals. This means they are usually safe for storing things in, including food.

3 They are usually light to carry – much lighter than wood, or stone, or glass, or most metals.

4 They don't break when you drop them. You have to hammer most rigid plastics quite hard, to get them to break.

5 They are strong. This is because their long molecules are attracted to each other. Most plastics are hard to tear or pull apart.

6 They do not catch fire easily. But when you heat them, some soften and melt, and some char (go black as if burned).

Changing the properties

By choosing monomers and reaction conditions carefully, you can make plastics with exactly the properties you want. For example, look at how you can change the properties of polythene:

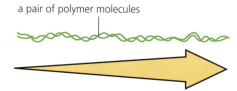

▲ A synthetic polymer for sewing. (It is polyester.)

a pair of polymer molecules

as molecules get longer, the force of attraction between them increases

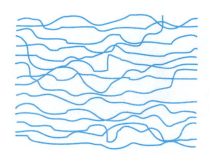

At about 50 °C, 3 or 4 atmospheres pressure, and using a catalyst, you get long chains like these. They are packed close together so the polythene is quite **dense**.

At about 200 °C, 2000 atmospheres pressure, and with a little oxygen present, the chains will branch. Now they can't pack closely, so the polythene is far less dense.

So by choosing the right conditions, you can change the density of the polythene, and make it 'heavy' or 'light' to suit your needs.

The **high-density** polythene is hard and strong, which is why it is used for things like bowls and dustbins. The **low-density** polythene is ideal for things like plastic bags, and 'cling film' for wrapping food.

Uses for synthetic polymers

Given all those great properties, it is not surprising that plastics have thousands of uses. Here are some examples.

Polymer	Examples of uses
polythene	plastic bags and gloves, clingfilm (*low density*) mugs, bowls, chairs, dustbins (*high density*)
polychloroethene (PVC)	water pipes, wellingtons, hoses, covering for electricity cables
polypropene	crates, ropes
polystyrene	used as expanded polystyrene in fast-food cartons, packaging, and insulation for roofs and walls (to keep homes warm)
Teflon	coated on frying pans to make them non-stick, fabric protector, windscreen wipers, flooring
nylon	ropes, fishing nets and lines, tents, curtains
Terylene	clothing (especially mixed with cotton), thread

▲ Another use of nylon: for parasails like this one, and parachutes.

▲ Polystyrene is an insulator: it helps to prevent heat loss. So it is used under floors, and in fast-food cartons.

▲ Teflon – a slippery polymer. It is used to coat irons to help them glide, and on frying pans to stop food sticking.

Q

1 Look at the properties of plastics, on page 264. Which *three* properties do you think are the most important for:
   a plastic bags?
   b kitchen bowls and utensils?
   c water pipes?
   d fishing nets?
   e hair dryers?
   f polystyrene fast-food containers?

2 What is *low-density* polythene, and how is it made?

3 Teflon is used to coat frying pans, to make them non-stick. So what properties do you think Teflon has? List them.

4 a What is *expanded polystyrene*?
   b Give three uses of this material.

5 a Now make a table with these headings:

Item	Properties of the plastic in it	Disadvantages of this plastic	Name of this plastic

   b i Fill in the first column of your table, giving three or four plastic items you own or use.
   ii In the second column, give the properties you observe, for that plastic. (You are a scientist!) For example is the plastic rigid? Or flexible?
   iii In the third column give any disadvantages you notice, for this plastic.
   iv Then see if you can name it. If you can, well done!

18.5 Plastics: here to stay?

Plastics: the problem

It is hard to imagine life without plastics. They are used everywhere.

One big reason for their success is their unreactivity.

But this is also a problem. They do not break down or rot away in the soil. We say they are **non-biodegradeable**. Most of the plastics thrown out in the last 50 years are still around – and may still be here 50 years from now. A mountain of waste plastic is growing.

Polythene: the biggest problem

Polythene is the biggest problem.

It is the most-used plastic in the world, thanks to its use in plastic bags. Around 5 trillion polythene bags are made every year. (That's 5 million million.) Most are used only once or twice, then thrown away.

In many places, rubbish is collected and brought to **landfill sites**. The plastic bags fill up these sites. In other places, rubbish is not collected. So the plastic bags lie around and cause many problems. For example:

- they choke birds, fish and other animals that try to eat them. Or they fill up the animals' stomachs so that they cannot eat proper food, and starve to death. (Animals cannot digest plastics.)
- they clog up drains, and sewers, and cause flooding.
- they collect in rivers, and get in the way of fish. Some river beds now contain a thick layer of plastic.
- they blow into trees and onto beaches. So the place looks a mess. Tourists are put off – and many places depend on tourists.

Because of these problems, plastic bags have been banned in many places. For example in Bangladesh, Rwanda, and several states in India.

▲ Plastic bags – here today, still here tomorrow …

▲ A stomach full of plastic means the bird will starve to death.

▲ Nice for visitors …

Recycling plastics

Some waste plastics do get reused. For example:

- some are melted down and made into new plastic bags, and things like soles for shoes, and fleeces.

- some are melted and their long chains cracked, to make small molecules that can be polymerised into new plastics.

- some are burned, and the heat is used to produce electricity.

But only a small % of waste plastic is reused in these ways. One reason is that there are many different types of plastic. These must be separated before reusing them, but that is not easy to do. Burning also poses problems, since some plastics give off poisonous gases.

▲ A degradeable plastic bag: it will break down along with the vegetable peelings and scrap paper inside.

Degradeable plastics

Degradeable polythene is already here. Some is **biodegradeable**: it contains additives such as starch that bacteria can feed on. Some is **photodegradeable**: it contains additives that break down in sunlight. In both cases, the polythene breaks down into tiny flakes.

The amount of additive can be varied for different purposes – for example to make rubbish sacks that will break down within weeks.

Bio-polymers: the future?

In future, the plastics you use could be **bio-polymers** – grown inside plants, or made in tanks by bacteria.

For example, one strain of bacteria can feed on sugar from crops such as maize, to produce polyesters.

Plants that can make plastics in their cells have already been developed. When the plants are harvested, the plastic is extracted using a solvent. Then the solvent is evaporated.

Work on bio-polymers is still at an early stage. But when oil runs out, we will be glad of bio-polymers. And they have two advantages for the environment: they are a renewable resource, and biodegradeable.

▲ This little cress plant has been genetically modified to produce a plastic in its cells.

 Q

1 Describe some negative effects of plastics on the environment.

2 Polythene is responsible for most of the environmental problems caused by plastics. Explain why.

3 See if you can come up with some ideas, to help prevent pollution by plastic bags.

4 Explain what these are, in your own words:
   a photodegradable polythene   b bio-polymers

18.6 Natural polymers in food (part I)

The two natural polymers in your food

Your food contains two natural polymers: **carbohydrates** and **proteins**. Their story begins with plants. Plants are polymer factories!

- Plants make glucose, a sugar, from carbon dioxide and water, by **photosynthesis**. This needs sunshine. Chlorophyll in their leaves acts as a catalyst:

$$6CO_2 \, (g) \, + \, 6H_2O \, (l) \, \longrightarrow \, C_6H_{12}O_6 \, (s) \, + \, 6O_2 \, (g)$$

carbon   water   glucose   oxygen
dioxide   (a sugar)

- Then they use glucose molecules as monomers, to make **starch** and **cellulose**. These are **carbohydrates**.
- They use glucose, plus nitrates and other compounds from the soil, to make **amino acids**.
- Then they use amino acid molecules as monomers, to make **proteins**.

Then animals eat the plants. And you eat food from plants and animals.

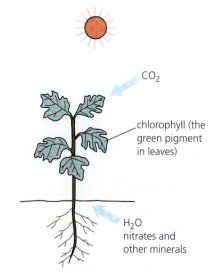

▲ Busy making polymers. A plant uses cellulose for stems and other structures, and starch as an energy store. It uses proteins for many different purposes.

From glucose to carbohydrates

We call glucose a **simple sugar** because it is just a single sugar unit.

Look at the drawing of the glucose molecule on the right. As you can see, it has five OH groups. But only two of them take part in polymerisation. So we can draw the molecule like this, to focus on those groups:

HO ⬜ OH

Let's see how glucose molecules join.

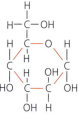

▲ A molecule of glucose, $C_6H_{12}O_6$.

---

1 Small numbers of glucose molecules can join to make larger molecules. Here two join to make the sugar **maltose**.

HO ⬜ OH   HO ⬜

↓ water molecule eliminated

HO ⬜ O ⬜ OH

A small molecule is eliminated (water). So this is a condensation reaction.

2 Hundreds or thousands of glucose molecules can join in the same way, to give **starch**. Because it is built from a large number of sugar units, starch is called a **complex carbohydrate**.

HO ⬜ OH   HO ⬜ OH   HO ⬜ OH   HO ⬜ OH

↓ water molecules eliminated

—O ⬜ O ⬜ O ⬜ O ⬜

Water molecules are eliminated during the polymerisation. So this is a condensation polymerisation.

As you can tell from its formula, glucose contains only carbon, hydrogen, and oxygen atoms. Since starch is built up from glucose molecules, it contains only those atoms too.

▲ Oranges and other fruit contain a simple sugar called **fructose**, which is an isomer of glucose.

Cellulose

Cellulose is also made by the condensation polymerisation of glucose. At least 1000 glucose units join to make it. But they join up in a different way than for starch, so cellulose has quite different properties.

The cell walls in plants are made of cellulose. So we eat cellulose every time we eat cereals, vegetables, and fruit. We can't digest it, but it helps to clean out our digestive systems. We call it **fibre**.

▲ The chemical name for white sugar is **sucrose**. A sucrose molecule is a unit of glucose joined to a unit of fructose.

When you eat starchy foods

When you eat starchy foods (such as rice, potatoes, bread) the starch breaks down to glucose again, in your digestive system. An enzyme called **amylase** act as a catalyst.

The glucose reacts with oxygen in your cells, giving out energy. The process is called **respiration**. (See page 230.) It is the opposite of photosynthesis.

You also build glucose into a polymer called **glycogen**, and store it in your liver and muscles, to use for energy later.

▲ It can do something we can't do: digest cellulose. (Grass is mainly cellulose.) It breaks it down with the help of catalysts called **cellulases**.

◀ Before races, marathon runners eat lots of pasta and other starchy foods. Their bodies build up stores of glycogen, which will provide energy during the run.

18.7 Natural polymers in food (part II)

Proteins

As you saw in the last unit, proteins are natural **polymers**, built up from molecules of **amino acids**.

All amino acids contain carbon, hydrogen, oxygen, and nitrogen. (Some contain other elements too.) Twenty-two different amino acids are used in building proteins. We can represent them like this:

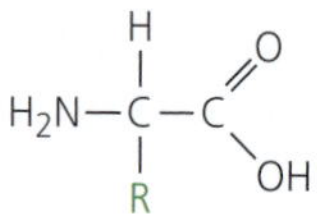

Only the COOH and NH$_2$ groups take part in the polymerisation. As you see, both are attached to the same carbon atom. Now look at R. It stands for 'the rest of the molecule'. Some examples of R are shown on the right.

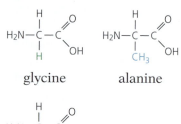

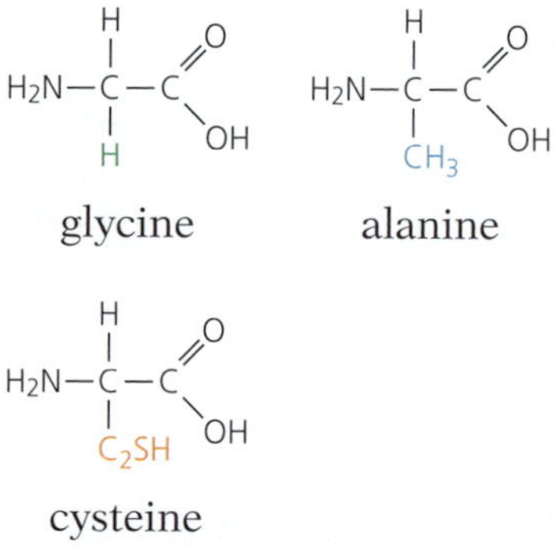

glycine   alanine

cysteine

▲ Three different amino acids – to show you some groups that R may represent.

How amino acids join up to make proteins

This drawing shows four *different* amino acid monomers joining, to form part of a protein molecule. We show the C=O bond straight up, to make it easier for you to see what's happening:

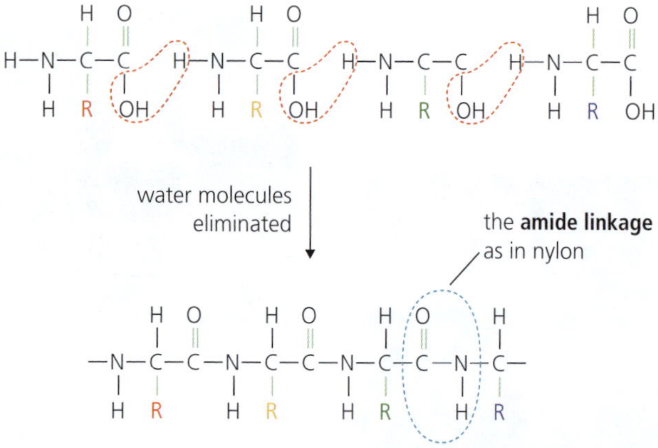

water molecules eliminated

the **amide linkage** as in nylon

From 60 to 6000 amino acid units can join to make a protein. They can be different amino acids, as in the drawing – and they can join in a different orders. So there are an enormous number of proteins!

Just as for carbohydrates, the reaction is a **condensation polymerisation**, with the loss of water molecules. Note the **amide linkage**, as in nylon. (Look back at page 262.)

When you eat proteins

When you eat foods containing proteins, the proteins break down to amino acids again, in your digestive system. Enzymes called **proteases** act as catalysts for the reaction.

Your body uses the amino acids to build up the new proteins it needs. For example, keratin, which is the main substance in your hair, nails, and outer skin; and collagen, which is needed for skin, bones, blood vessels, tendons, and muscles. Enzymes are also proteins.

▲ Tidying up my keratin.

Digesting polymers: a summary

As you saw, when monomers join to form carbohydrates and proteins, water molecules are eliminated.

When you digest food, *the opposite happens*. These polymers are broken down again to their monomers, by reaction with water.

A reaction in which a molecule breaks down through reaction with water is called **hydrolysis**. During hydrolysis in your body:

- **starch** is broken down to glucose, with amylase as catalyst.
- **proteins** are broken down to amino acids, with proteases as catalysts.

Hydrolysis in the lab

You can also carry out the hydrolysis of carbohydrates and proteins in the lab. There are two ways: using acid only, and using enzymes.

Using enzymes

You can *buy* enzymes: amylase, proteases, and cellulases (which catalyse the hydrolysis of cellulose). They will break down the polymers in the lab, just as in the digestive systems of humans and animals.

The temperature and pH must match those at which the enzymes work in digestive systems. For example, amylase works at temperatures of 32 °C – 40 °C, and a pH of 6.5 – 7.5. Too high a temperature, or too high or low a pH, will destroy it.

Using acid

Without enzymes, the conditions have to be tougher. Look at this table:

Polymer	Conditions for the hydrolysis	Complete hydrolysis gives ...
carbohydrates	heat with dilute hydrochloric acid	simple sugars
proteins	boil with 6M hydrochloric acid for 24 hours	amino acids

Identifying the products

You can use paper chromatography to identify the products of hydrolysis, as shown in Unit 2.5. For example, you can find out which amino acids were in your starting protein. Amino acids and sugars are colourless, so you will need to use locating agents to show them up.

▲ The hydrolysis of starch to glucose starts in your mouth, because your saliva contains amylase. (So bread begins to taste sweet, if you keep chewing it!)

▲ In industry, vegetable proteins are hydrolysed by boiling with hydrochloric acid. Then sodium hydroxide is added to neutralise the solution. The products are added to sauces and other processed foods, to improve the flavour. Yummy?

Q

1 What is: **a** an amino acid?   **b** a protein?
2 **a** Draw the structure of part of a protein molecule, and label one amide linkage in it.
   b If five amino acids join, how many water molecules will be eliminated?
3 Compare the reactions that produce carbohydrates and proteins. What do they have in common?
4 What does *hydrolysis* mean?

5 There are two ways to carry out the hydrolysis of a protein, or a carbohydrate, in the lab. What are they?
6 **a** See if you can draw a diagram to show the complete hydrolysis of starch to glucose, in the lab.
   b See if you can explain why conditions must be tougher, when acid is used instead of amylase, for the hydrolysis.
7 Hydrolysis of a protein in the lab gives a mixture of products.
   a Why?   **b** How will you identify them?

Checkup on Chapter 18

Revision checklist

Core syllabus content

Make sure you can …

☐ explain these terms:

monomer polymer   *polymerisation*
natural polymer   *synthetic polymer*

☐ draw a labelled diagram to show the addition polymerisation of ethene to make polyethene

☐ explain that plastics are synthetic polymers

☐ give at least five general properties of plastics

☐ give examples of uses for at least three plastics

☐ explain what *non-biodegradeable* means

☐ describe the problems that non-biodegradeable plastics can cause, in the environment

☐ name two natural polymers in food, and say what monomers they are built from

Extended syllabus content

☐ describe *addition polymerisation*, and
  – say what the key feature of the monomer is
  – draw part of a polymer molecule, formed from a given monomer
  – identify the monomer, for a given polymer

☐ name at least three polymers formed by addition polymerisation, and give uses for them

☐ describe *condensation polymerisation*, and
  – say what the key features of the monomers are
  – state the differences between condensation and addition polymerisation

☐ draw simple diagrams to show the monomers, and part of the polymer molecule, for:

nylon   *Terylene*

using blocks to represent carbon chains

☐ explain what the *amide* and *ester* linkages are, and identify them on a drawing

☐ give uses for nylon and Terylene

☐ draw simple diagrams to show how these are formed by condensation polymerisation:

starch   *proteins*

☐ explain what *hydrolysis* is

☐ say how the hydrolysis of carbohydrates and proteins can be carried out in the lab, and name the products of complete hydrolysis

☐ describe how to carry out paper chromatography to identify products of hydrolysis (Unit 2.5)

Questions

Core syllabus content

1 In organic chemistry, many small units may join to form a large molecule. This represents a small unit:

a What are the small units called?

b What general name is given to a large molecule obtained when the small units join?

c This type of reaction is called p.......... ?

d Draw a diagram to represent the large molecule.

2 Nylon and Terylene are both synthetic polymers. At sea, nylon is used for ropes and fishing nets. A lot of Terylene is used for plastic drinks bottles. These plastics do not decay, if left in the soil.

a Explain this term:
  i synthetic   ii polymer

b State two properties that make nylon suitable for ropes and fishing nets.

c Give two advantages of using Terylene instead of glass, for drinks bottles.

d i What name is given to a plastic which does not decay in the soil?
  ii Give two examples of problems it can cause.

Extended syllabus content

3 This diagram represents two units of an addition polymer called polyacrylamide:

$$
\begin{array}{ccccccc}
& H & & H & & H & & H \\
& | & & | & & | & & | \\
-\!\!-C & -\!\!- & C & -\!\!- & C & -\!\!- & C\!\!-\!\!- \\
& | & & | & & | & & | \\
& H & & CONH_2 & & H & & CONH_2
\end{array}
$$

a Draw the structure of the monomer.

b Suggest a name for the monomer.

c Is the monomer saturated, or unsaturated?

4 The polymer 'Teflon' is made from this monomer, tetrafluoroethene:

$$
\begin{array}{ccc}
F & & F \\
\diagdown & & \diagup \\
C & = & C \\
\diagup & & \diagdown \\
F & & F
\end{array}
$$

a Which feature of the monomer makes polymerisation possible?

b Which type of polymerisation occurs?

c Draw three units in the structure of the polymer molecule that forms.

d Give the chemical name for this polymer.

5 The polymer poly(dichloroethene) was once used to make 'cling film', for covering food to keep it fresh. (But not now, because of health fears.) Its structure is:

$$\left(\begin{array}{c} H \quad Cl \\ | \quad | \\ -C-C- \\ | \quad | \\ H \quad Cl \end{array}\right)_n$$

a What does *n* represent?

b Name the monomer, and draw its structural formula.

c Which type of polymerisation takes place?

d One property of poly(dichlorothene) is its *low permeability* to moisture and gases.

   i See if you can explain what the term in italics means.

   ii That property is important in keeping food fresh. Why?

   iii Give three other *physical* properties a polymer would need, to be suitable for use as 'cling film'.

e Poly(dichloroethene) is *non-biodegradable*.

   i Explain the term in italics.

   ii When used as cling film, this plastic was not suitable for recycling. Suggest a reason.

6 **Polyamides** are polymers made by condensation polymerisation. One polyamide was developed for use in puncture-resistant bicycle tyres. The two monomers for it are:

$$H_2N-\bigcirc-NH_2 \qquad \begin{array}{c} O \quad\quad O \\ \| \quad\quad \| \\ C-\bigcirc-C \\ / \quad\quad \backslash \\ Cl \quad\quad Cl \end{array}$$

The hexagon with the circle in the middle stands for a ring of 6 carbon atoms, with 3 double bonds.

a What is *condensation polymerisation*?

b Show in detail how the monomers join.

c Name the other product of the reaction.

d i In what way is this polymer similar to nylon? (See page 262.)

   ii But its properties are different from those of nylon. Why?

A similar polymer has been developed as a fabric for fireproof clothing. Its structure is:

$$\left(\begin{array}{c} O \quad\quad O \\ \| \quad\quad \| \\ -C-\bigcirc-C-N-\bigcirc-N- \\ \quad\quad\quad | \quad\quad\quad | \\ \quad\quad\quad H \quad\quad\quad H \end{array}\right)_n$$

e Draw the structures of the two monomers that could be used to make this polymer.

7 Many synthetic polymers contain the amide linkage.

a Draw the structure of the amide linkage.

b Which important natural polymers also contain the amide linkage?

The substances in **b** will undergo hydrolysis in the laboratory, in the presence of acid.

c i What does *hydrolysis* mean?

   ii What are the products of the hydrolysis?

   iii How can the products be separated?

8 One very strong polymer has this structure:

a Which type of polymerisation produced it?

b Which type of linkage joins the monomers?

c Draw the structures of the two monomers from which this polymer could be made.

d Compare the structure above with that for Terylene (page 263). What may be responsible for the greater strength of this polymer?

9 Starch is a carbohydrate. It is a natural polymer. This shows part of a starch molecule:

$$-\boxed{}-O-\boxed{}-O-\boxed{}-O-$$

a What is a *natural polymer*?

b What is a *carbohydrate*?

c Which type of polymerisation gives starch?

d What do the blocks represent, above?

e i Draw a diagram showing the structure of the monomer for starch. (Use a block.)

   ii Name this monomer.

f Starch is also called a *polysaccharide*. Why?

g Starch can be broken down by hydrolysis.

   i Describe two ways in which the hydrolysis is carried out. (One occurs in your body.)

   ii One takes place at a far lower temperature than the other. What makes this possible?

10 In the lab, *partial* hydrolysis of starch gives a mixture of colourless products. They can be identified using chromatography. A locating agent is needed.

a Draw diagrams showing at least two of the products. (Use blocks like those in question **7**.)

b What is a *locating agent* and why is it needed?

c Outline the steps in carrying out the chromatography. (Page 20 may help.)

19.1 Chemistry: a practical subject

The lab: the home of chemistry

All the information in this book has one thing in common. It is all based on real experiments, carried out in labs around the world, over the years – and even over the centuries. The lab is the home of chemistry!

How do chemists work?

Like all scientists, chemists follow the **scientific method**. This flowchart shows the steps. The handwritten notes are from a student.

1 You observe something that makes you ask yourself a question.
   Kitchen cleaner X is better at removing grease than kitchen cleaner Y. Why?

2 Come up with a **hypothesis** – a reasonable statement that you can test. You might need to do some research in books or on the internet, to help you.
   Sodium hydroxide is used in kitchen cleaners to help remove grease.
   The labels on X and Y say they contain sodium hydroxide.
   My hypothesis: X may contain more sodium hydroxide than Y does.

3 Plan an experiment to test the hypothesis.
   I plan to do a titration to test my hypothesis. See the details in the next unit.

4 Carry out the experiment, and record the results.
   See the results in the next unit.

5 Analyse the results.
   You can help me do this, in the next unit.

6 Did the results support your hypothesis?
   See the next unit.

7 Share your conclusions with other people.
   The teacher wants to see them!

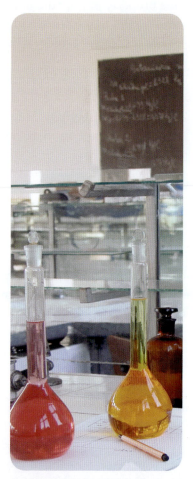

▲ Step into the lab, and try the scientific method ….

Planning an experiment: the variables

Suppose you want to investigate how the rate of a reaction changes with temperature.

- The temperature is under your control. So it is called the **independent variable**. It is the *only* thing *you* change as you do the experiment.

- If the rate changes as you change the temperature, the rate is a **dependent variable**. It **depends on** the temperature.

In many experiments you do, there will be an independent variable. You control it – and keep everything else unchanged.

That golden rule …
When you investigate something in the lab, **change only one thing at a time**, and see what effect it has.

The skills you use

When you plan and carry out an experiment, you use many different skills:

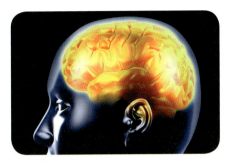

Thinking Use your brain before, during, and after the experiment. That is what brains are for. (They really like being used.)

Observing This is a very important skill. Chemists have made some amazing discoveries just by watching very carefully.

Using apparatus and techniques Weigh things, measure out volumes of liquids, measure temperature, do titrations, prepare crystals ....

Working accurately Sloppy work will ruin an experiment. Follow the instructions. Measure things carefully. Think about safety too.

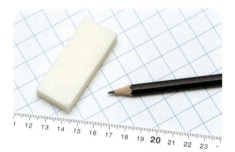

Doing some maths You often have to do some calculations using your results. And drawing a graph can help you see what is going on.

Writing up You may have to write a report on your experiment, and give conclusions. And say how the experiment could be improved?

The experiments you do

Often, you will not get a chance to plan an experiment for yourself. Instead, the teacher will tell you what to do. So you might miss out steps **1–3** in the flowchart on page 274.

But even if you pick up at step **4**, you are still using the scientific method, and gaining practice in it. And you are following in the footsteps of many famous scientists, who have changed our lives by their careful work in the lab.

One day, you may become a scientist yourself. Even a famous one!

Q

1 Do you think this counts as a *hypothesis*?
  a I am late for class again.
  b If I add more yeast, the fermentation may go faster.
  c December follows November.
  d The rate of photosynthesis may change with temperature.

2 Explain in your own words what an *independent variable* is.
3 Which would be the independent variable, in an experiment to test the statement in **1b**?
4 Do you think the scientific method would be useful to:
  a a doctor?   **b** a detective?
  Explain your answer.

19.2 Example of an experiment

Comparing those kitchen cleaners

In step **2** of the scientific method in the last unit, a student put forward a hypothesis. Here you can read how the student tested the hypothesis. But the report is not quite finished. That is *your* task.

An experiment to compare the amount of sodium hydroxide in two kitchen cleaners

Introduction

I noticed that kitchen cleaner X is better at removing grease than kitchen cleaner Y is. The labels show that both kitchen cleaners contain sodium hydroxide. This chemical is used in many cleaners because it reacts with grease to form soluble sodium salts, which go into solution in the washing-up water.

My hypothesis

Kitchen cleaner X may contain more sodium hydroxide than kitchen cleaner Y does.

Planning my experiment

I plan to titrate a sample of each cleaner against dilute hydrochloric acid, using methyl orange as indicator. This is a suitable method because the sodium hydroxide in the cleaner will neutralise the acid. The indicator will change colour when neutralisation is complete.

To make sure it is a fair test, I will use exactly the same volume of cleaner, and the same concentration of acid, and the same number of drops of indicator each time, and swirl the flask in the same way. The only thing I will change is the type of cleaner.

I will wear safety goggles, since sodium hydroxide and hydrochloric acid are corrosive.

The experiment

25 cm³ of cleaner X were measured into a conical flask, using a pipette. 5 drops of methyl orange were added, and the solution turned yellow.

A burette was filled to the upper mark with hydrochloric acid of concentration 1 mol/dm³. The initial level of the acid was noted.

The acid was allowed to run into the conical flask. The flask was continually and carefully swirled. As the acid dripped in, the solution showed flashes of pink. When the end point was near the acid was added drop by drop. When the solution changed from yellow to pink, the titration was stopped. The final level of the acid was recorded.

The experiment was repeated with cleaner Y.

▲ The colour of the solution has changed: the titration is complete.

The results

For X:

Initial level of acid in the burette	0.0 cm³
Final level	22.2 cm³
Volume of acid used	22.2 cm³

For Y:

Initial level of acid in the burette	22.2 cm³
Final level	37.5 cm³
Volume of acid used	15.3 cm³

Analysis of the results

The same volume of each cleaner was used. The sodium hydroxide in X neutralised 22.2 cm³ of acid. The sodium hydroxide in Y neutralised 15.3 cm³ of acid. This means that solution …

My conclusion

These results …

To improve the reliability of the results

I would …

▲ One is better at removing grease. Might it have a higher concentration of sodium hydroxide?

In the question section below, you will have the chance to complete the student's analysis and conclusions, and come up with suggestions for ensuring that the results were reliable.

Q

1 In this experiment, was there:
  a an independent variable? If so, what was it?
  b a dependent variable? If so, what was it?

2 a Look at the apparatus below.
   Which pieces did the student use in the experiment?
   Give their letters and names.
  b When measuring out solutions for titration, a pipette is used instead of a measuring cylinder. Why is this?
  c Why is a conical flask used rather than a beaker, for the titration?
  d Why are burettes used for titrations?
  e Which is *more* accurate for measuring liquids?
     i a burette   **ii** a pipette
   Explain clearly why you think so.

3 Why is an indicator needed, for titrations?
4 a Suggest another indicator the student could have used, in place of methyl orange. (Hint: page 145.)
  b What colour change would be observed at the end-point, for the indicator you suggested?
5 Now complete the student's *Analysis of the results*.
6 Complete the *Conclusion*, by saying whether or not the results supported the hypothesis.
7 How would you improve the reliability of the results?
8 How would you modify the experiment, to compare liquid scale-removers for kettles? (They contain acid.)
9 Next week the student will do an experiment to see whether neutralisation is exothermic or endothermic. Which item below will the student *definitely* use?

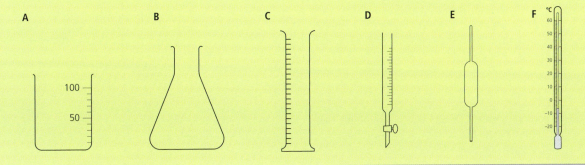

19.3 Working with gases in the lab

Preparing gases in the lab

You might have to prepare a gas in the lab, one day. The usual way to make a gas is to displace it from a solid or solution, using apparatus like this. The table below gives some examples.

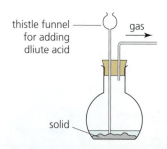

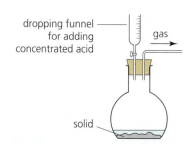

To make ...	Place in flask	Add	Reaction
carbon dioxide	calcium carbonate (marble chips)	dilute hydrochloric acid	$CaCO_3 (s) + 2HCl (aq) \longrightarrow CaCl_2 (aq) + H_2O (l) + CO_2 (g)$
hydrogen	pieces of zinc	dilute hydrochloric acid	$Zn (s) + 2HCl (aq) \longrightarrow ZnCl_2 (aq) + H_2 (g)$
oxygen	manganese(IV) oxide (as a catalyst)	hydrogen peroxide	$2H_2O_2 (aq) \longrightarrow 2H_2O (l) + O_2 (g)$

But to make ammonia, you can heat any ammonium compound with a base such as sodium hydroxide or calcium hydroxide – using both reactants in solid form.

Collecting the gases you have prepared

The table below shows four ways of collecting a gas you have prepared. The method depends on whether the gas is heavier or lighter than air, whether you need it dry, and what you want to do with it.

Using a measuring cylinder
- You can use a gas jar to collect a gas over water.
- But if you want to measure the volume of the gas, roughly, use a measuring cylinder instead.
- If you want to measure its volume accurately, use a gas syringe.

Method	upward displacement of air	downward displacement of air	over water	gas syringe
Use when ...	the gas is heavier than air	the gas is lighter than air	the gas is sparingly soluble in water	you want to measure the volume accurately
Apparatus	gas jar	gas jar	gas jar / water	gas syringe
Examples	carbon dioxide, CO_2 sulfur dioxide, SO_2 hydrogen chloride, HCl	ammonia, NH_3 hydrogen, H_2	carbon dioxide, CO_2 hydrogen, H_2 oxygen, O_2	any gas

Tests for gases

You have a sample of gas. You think you know what it is, but you're not sure. So you need to do a test. Below are some tests for common gases. Each is based on particular properties of the gas, including its appearance, and sometimes its smell.

Remember

Aqueous solution just means a solution in water.

Gas	Description and test details
Ammonia, NH₃ Properties Test Result	Ammonia is a colourless alkaline gas with a strong sharp smell. Hold damp red litmus paper in it. The indicator paper turns blue. (You may also notice the sharp smell.)
Carbon dioxide, CO₂ Properties Test Result	Carbon dioxide is a colourless, weakly acidic gas. It reacts with **limewater** (an aqueous solution of calcium hydroxide) to give a white precipitate of calcium carbonate: $CO_2\ (g) + Ca(OH)_2\ (aq) \longrightarrow CaCO_3\ (s) + H_2O\ (l)$ Bubble the gas through limewater. Limewater turns cloudy or milky.
Chlorine, Cl₂ Properties Test Result	Chlorine is a green poisonous gas which bleaches dyes. Hold damp indicator paper in the gas, *in a fume cupboard*. Indicator paper turns white.
Hydrogen, H₂ Properties Test Result	Hydrogen is a colourless gas which combines violently with oxygen when lit. Collect the gas in a tube and hold a lighted splint to it. The gas burns with a squeaky pop.
Oxygen, O₂ Properties Test Result	Oxygen is a colourless gas. Fuels burn much more readily in it than in air. Collect the gas in a test-tube and hold a glowing splint to it. The splint immediately bursts into flame.
Sulfur dioxide, SO₂ Properties Test Result	Sulfur dioxide is a colourless, poisonous, acidic gas, with a choking smell. It will reduce the purple potassium manganate(VII) ion to the colourless potassium manganese(II) ion. Soak a piece of filter paper in acidified aqueous potassium manganate(VII). Place it in the gas. (*Acidified* means that a little dilute acid – usually hydrochloric acid – has been added.) The colour on the paper changes from purple to colourless.

Q

1 a Sketch the *complete* apparatus you will use to prepare and collect carbon dioxide. Label all the parts.
  b How will you then test the gas to confirm that it is carbon dioxide?
  c Write the equation for a positive test reaction.
2 a Hydrogen cannot be collected by upward displacement of air. Why not?
  b Hydrogen burns with a squeaky pop. Write a balanced equation for the reaction that takes place.

3 a Name two substances you could use to make ammonia.
  b Ammonia cannot be collected over water. Why not?
  c The test for ammonia is ...... ?
4 It is not a good idea to rely on smell, to identify a gas. Suggest at least two reasons why.
5 To measure the rate of the reaction between magnesium and hydrochloric acid, you will collect the hydrogen that forms. Which is better to use for this: a measuring cylinder over water, or a gas syringe? Give more than one reason.

19.4 Testing for ions in the lab: cations

Testing a salt

You are given a tiny amount of a salt, and asked to find out what it is. Don't panic! Just turn into a detective, and do some tests!

Remember, a salt is made up of positive ions, or **cations**, and negative ions, or **anions**. There are different tests for each.

Tests for cations

In this unit we look at tests for cations: metal ions, and the ammonium ion. Let's start with flame tests.

Flame tests

A **flame test** works well for some metal cations – and especially Group I. You start with the solid salt. These are the steps:

- First, clean a platinum or nichrome wire. To do this, dip it into concentrated hydrochloric acid, and hold it in a hot bunsen flame.

- Next, moisten the clean wire by dipping it into the acid again. Then dip it into the salt, so that some sticks to it.

- Now hold it in the clear part of a blue bunsen flame, and observe the colour.

The colours for lithium, sodium, potassium, and copper cations are shown in the photo below.

> **Remember CAP!**
> **C**ations **A**re **P**ositive.
> They'd go to the **cat**hode (−).

> **More about flame tests**
> In the flame, electrons in the cations take in heat energy, and jump to higher energy levels. Then they fall back again, giving out energy as light of a specific colour.

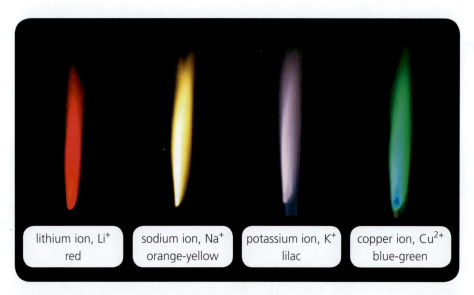

| lithium ion, Li⁺ red | sodium ion, Na⁺ orange-yellow | potassium ion, K⁺ lilac | copper ion, Cu²⁺ blue-green |

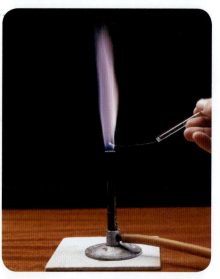

▲ Carrying out a flame test. The lilac colour shows that the salt contains the potassium cation (K^+).

Chemical tests

The table on the next page shows chemical tests for identifying cations.

- To test for the ammonium ion, you add aqueous sodium hydroxide to the solid salt, or its solution, to drive off ammonia gas. Then test the gas.

- To test for metal cations, you add dilute aqueous sodium hydroxide, or aqueous ammonia, to a solution of the salt – to precipitate a metal hydroxide. (Most hydroxides are insoluble, or only slightly soluble.) Then observe the colour and behaviour of the precipitate.

▲ To test, you need only small amounts!

Cation	Test	If the cation is present	Ionic equation for the reaction
ammonium NH_4^+	Add dilute aqueous sodium hydroxide. Heat gently.	Ammonia gas is given off. (It turns litmus blue.)	$NH_4^+ (aq) + OH^- (aq) \longrightarrow NH_3 (g) + H_2O (l)$
copper(II) Cu^{2+}	Add dilute aqueous sodium hydroxide.	A pale blue precipitate forms. Adding *more* (excess) sodium hydroxide has no effect.	$Cu^{2+} (aq) + 2OH^- (aq) \longrightarrow Cu(OH)_2 (s)$
	Add aqueous ammonia.	A pale blue precipitate forms. It dissolves again if you add *more* (excess) ammonia, giving a deep blue solution.	The precipitate dissolves again in excess ammonia because a soluble **complex ion** forms. (That's an ion with several negative ions, or molecules, around the positive metal ion.)
iron(II) Fe^{2+}	Add dilute aqueous sodium hydroxide, *or* aqueous ammonia.	A pale green precipitate forms.	$Fe^{2+} (aq) + 2OH^- (aq) \longrightarrow Fe(OH)_2 (s)$
iron(III) Fe^{3+}	Add dilute aqueous sodium hydroxide, *or* aqueous ammonia.	A red-brown precipitate forms.	$Fe^{3+} (aq) + 3OH^- (aq) \longrightarrow Fe(OH)_3 (s)$
aluminium Al^{3+}	Add dilute aqueous sodium hydroxide.	A white precipitate forms. It dissolves again If you add *more* (excess) sodium hydroxide, giving a colourless solution.	$Al^{3+} (aq) + 3OH^- (aq) \longrightarrow Al(OH)_3 (s)$ The precipitate dissolves in excess sodium hydroxide because aluminium hydroxide is *amphoteric*. (It reacts with acids *and* alkalis.)
	Add aqueous ammonia.	A white precipitate forms. Adding *more* (excess) ammonia has no effect.	
zinc Zn^{2+}	Add dilute aqueous sodium hydroxide, *or* aqueous ammonia.	A white precipitate forms in each case. And in each case it dissolves again if you add *more* (excess) of the reagent, giving a colourless solution.	$Zn^{2+} (aq) + 2OH^- (aq) \longrightarrow Zn(OH)_2 (s)$ The precipitate dissolves again in excess sodium hydroxide because zinc hydroxide is amphoteric. It dissolves again in excess ammonia because a soluble complex ion forms.
calcium Ca^{2+}	Add dilute aqueous sodium hydroxide.	A white precipitate forms. Adding *more* (excess) sodium hydroxide has no effect.	$Ca^{2+} (aq) + 2OH^- (aq) \longrightarrow Ca(OH)_2 (s)$
	Add aqueous ammonia.	No precipitate, or a very slight white precipitate.	
chromium Cr^{3+}	Add dilute aqueous sodium hydroxide.	A grey-green precipitate forms. It dissolves again If you add *more* sodium hydroxide, giving a green solution.	$Cr^{3+} (aq) + 3OH^- (aq) \longrightarrow Cr(OH)_3 (s)$ The precipitate dissolves in excess sodium hydroxide because chromium hydroxide is amphoteric.
	Add aqueous ammonia.	A grey-green precipitate forms. It dissolves again if you add *more* ammonia, giving a purple solution.	The precipitate dissolves again in excess ammonia because a soluble complex ion forms.

Q

1 The other name for a positive ion is … ?
2 How would you carry out a *flame test*?
3 Which metal cation will give this flame colour?
  a lilac   b blue-green   c red
4 In the table above, the precipitates that form are all *h*……?

5 a Two cations in the table above can't be identified using sodium hydroxide alone. Which two? And why not?
  b Which further test would you do?
6 Sodium hydroxide, and ammonia, cannot be used to identify potassium ions. Why not? See if you can explain.

19.5 Testing for ions in the lab: anions

Now for the anions

In the last unit you saw how to test an unknown salt, to see if it contains certain positive ions, or **cations** – for example NH_4^+, K^+, Al^{3+}.

In this unit we look at tests to identify the negative ion, or **anion**, that the unknown salt contains.

- Some of the tests work by producing a gas. So you can start with the solid salt, or its aqueous solution. Then you test the gas.

- The others work by producing a precipitate. For those, you start with the aqueous solution.

The tests are listed below, starting with those very common anions, the halide ions.

Tests for anions

Halide ions (Cl^-, Br^-, I^-)

- To a small amount of the solution, add an equal volume of dilute nitric acid. Then add aqueous silver nitrate.

- Silver halides are insoluble. So if halide ions are present, a precipitate will form. The colour tells you which one. Look at this table:

Precipitate	Indicates presence of ...	Ionic equation for the reaction
white	chloride ions, Cl^-	$Ag^+ (aq) + Cl^- (aq) \longrightarrow AgCl (s)$
cream	bromide ions, Br^-	$Ag^+ (aq) + Br^- (aq) \longrightarrow AgBr (s)$
yellow	iodide ions, I^-	$Ag^+ (aq) + I^- (aq) \longrightarrow AgI (s)$

The colours are shown in the photo below.

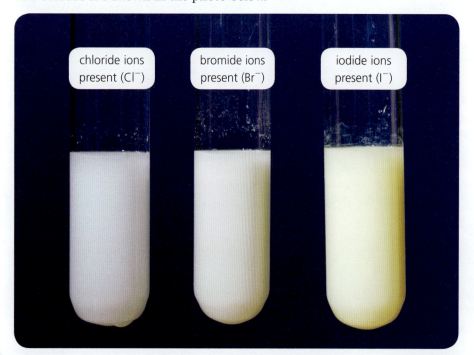

◄ The precipitates in the test for halides.

Remember AN!

Anions are **N**egative.
They'd go to the **an**ode (+).

Precipitates

A *precipitate* is an insoluble product of a reaction. It's easy to see – which is useful in tests.

- You mix a known solution with the unknown one.
- If an insoluble product appears, that gives you useful clues about the unknown substance.

Look back at the table about the solubility of salts on page 156.

Sulfate ions (SO_4^{2-})

- To a small amount of the solution add an equal volume of dilute hydrochloric acid. Then add barium nitrate solution.
- Barium sulfate is insoluble. So if sulfate ions are present a **white** precipitate will form. The ionic equation for the reaction is:

$$Ba^{2+}(aq) + SO_4^{2-}(aq) \longrightarrow BaSO_4(s)$$

> **Sulfates and sulfites**
> - Sulfates are salts of sulfuric acid, H_2SO_4.
> - Sulfites are salts of sulfurous acid, H_2SO_3.

Sulfite ions (SO_3^{2-})

Sulfites are salts that contain the sulfite ion, SO_3^{2-}. It is similar to the sulfate ion, but with just three oxgyen atoms.

- To a small amount of the solution add an equal volume of dilute hydrochloric acid.
- Heat the mixture gently. A gas, sulfur dioxide, is given off. The ionic equation for the reaction is:

$$SO_3^{2-}(aq) + 2H^+(aq) \longrightarrow H_2O(l) + SO_2(g)$$

- To confirm that the gas is sulfur dioxide, soak a small piece of filter paper in acidified aqueous potassium manganate(VII), which is purple. Place it in the mouth of the test-tube. The colour will turn from purple to colourless. (See page 279.)

▲ Sulfites such as sodium sulfite and calcium sulfite are added to dried fruit and other foodstuffs, to preserve them. (They prevent the growth of bacteria.)

Nitrate ions (NO_3^-)

- To a small amount of the solid or solution, add a little dilute aqueous sodium hydroxide, and then some small pieces of aluminium foil. Heat gently.
- If ammonia gas is given off, the unknown substance contained nitrate ions. The ionic equation for the reaction is:

$$8Al(s) + 3NO_3^-(aq) + 5OH^-(aq) + 2H_2O(l) \longrightarrow 3NH_3(g) + 8AlO_2^-(aq)$$

Carbonate ions (CO_3^{2-})

- To a small amount of the solid or solution, add a little dilute hydrochloric acid.
- If the mixture bubbles and gives off a gas that turns limewater milky, the substance contained carbonate ions. The gas is carbon dioxide. The ionic equation for the reaction is:

$$2H^+(aq) + CO_3^{2-}(aq) \longrightarrow CO_2(g) + H_2O(l)$$

▲ Testing for carbonates. The test-tube on the right contains limewater. The carbon dioxide is turning it milky.

Q

1. What are anions? Give two examples from this unit.
2. A test will produce a *pctpiteeiar* if the product is *ilbsoenul*. Unjumble the words in italics!
3. Silver nitrate is used in the test for halides. Why?
4. How would you test for the anion in calcium sulfite?
5. Nitrates are not tested by forming a precipitate. Why not?
6. Where do the OH^- ions come from, in the test for nitrate ions?
7. a Why is acid used, in testing for carbonates?
   b Limewater is also used in the test.
      i What is limewater?   ii Why is it a good choice?

Checkup on Chapter 19

Revision checklist

For all students

Make sure you can …

☐ identify these common pieces of laboratory
apparatus, and say what they are used for:

beaker	*test-tube*	*conical flask*
pipette	*burette*	*measuring cylinder*
gas jar	*gas syringe*	*condenser*
thermometer	*filter funnel*	*water trough*

☐ arrange these pieces of apparatus in order of accuracy
(as here) for measuring out a volume of liquid:
beaker   measuring cylinder   burette   pipette

☐ describe how to carry out these procedures:

filtration	*crystallisation*
simple distillation	*fractional distillation*
titration	*paper chromatography*

☐ explain what these are, in experiments:
an independent variable   a dependent variable

☐ explain why measurements are often repeated, in
experimental work

☐ describe how to prepare these gases in the lab:
hydrogen   oxygen   carbon dioxide   ammonia
   – and name suitable reactants to use
   – give the equations for the reactions
   – draw the apparatus

☐ give the test for these gases:

hydrogen	*oxygen*	*carbon dioxide*
ammonia	*chlorine*	*sulfur dioxide*

☐ give another term for: *cation   anion*

☐ say how to carry out a flame test for metal cations,
and state the flame colour for the cations of:
lithium   sodium   potassium   copper

☐ explain that in the chemical tests for anions and
cations, either a precipitate is formed, or a gas is
given off

☐ describe chemical tests to identify these cations:
Cu^{2+} Fe^{2+} Fe^{3+} Al^{3+} Zn^{2+} Ca^{2+} Cr^{3+} NH_4^+

☐ describe chemical tests to identify these anions:
halide ions (Cl^-, Br^-, I^-)   *sulfate ion*, SO_4^{2-}
sulfite ion, SO_3^{2-} *nitrate ion*, NO_3^-
carbonate ion, CO_3^{2-}

Make sure you can also …

☐ describe a test for water (page 120)

☐ explain that melting and boiling points can be
used to test whether a substance is pure (page 15)

Questions

For all students

1 A sample of soil from a vegetable garden was
thoroughly crushed, and water added as shown:

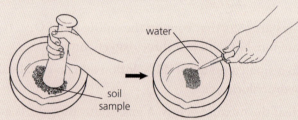

  a Using a conical flask, filter funnel, filter paper,
universal indicator, and dropping pipette, show
how you would measure the pH of the soil.

  b How would you check that the results for this
sample were valid for the whole garden?

2 This apparatus is used to collect gases in the lab.

  a Make a drawing of the apparatus, labelling the
water, trough, measuring cylinder, delivery tube,
flask, and dropping funnel.

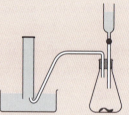

  b This apparatus can be used for preparing the
gases hydrogen and carbon dioxide, but *not*
sulfur dioxide. Explain why.

3 This apparatus is used to measure rate of a reaction.

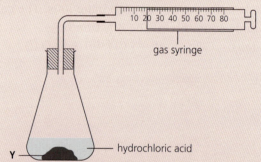

  a Suggest a suitable reagent to use as Y.

  b Which other piece of apparatus is needed?

  c Outline the procedure for this experiment.

  d You must be careful not to use too much of the
reagents. Why?

4 Two solutions, W and Y, are tested with universal indicator paper.

Solution W: the indicator paper turns red.
Solution Y: the indicator paper turns orange.

a i Which solution could have a pH of 1, and which could have a pH of 5?
ii Which type of solution is Y?

Further tests are carried out in test-tubes.

TEST A

A piece of magnesium is added to solution W.

b i What will you observe in the test-tube?
ii What is formed as a result of the reaction.
iii How will solution Y compare, in this reaction?

TEST B

A solid, which is a sodium compound, is added to solution W. A gas is given off. It turns limewater milky.

c i What colour will the solid be?
ii Name the gas released.
iii Suggest a name for the solid.

TEST C

A few drops of barium nitrate solution are added to a solution of W. A white precipitate forms.

d i Name the white precipitate.
ii Identify solution W.

5 In some countries, potassium sulfite is used as a food preservative. It is not permitted in the UK. An unknown white solid food preservative is to be tested, to see if it is potassium sulfite.

a First, a flame test is carried out on the sample.
i Describe how you would do the flame test.
ii What colour will be seen in the flame, if the sample is potassium sulfite?
b The sample is heated gently with dilute hydrochloric acid.
i Name the colourless gas that will be released, if the sample is potassium sulfite.
ii Write the formula for this gas.
iii What test would you carry out, to confirm this gas?

6 Ammonium nitrate (NH_4NO_3) is an important fertiliser. The ions in it can be identified by tests.

a Name the cation present, and give its formula.
b Describe a test that will confirm the presence of this cation.
c Name the anion present and give its formula.
d Describe a test that will confirm the presence of this anion.

7 This table shows the result of adding aqueous sodium hydroxide and aqueous ammonia to an aqueous solution of a metal chloride.

Test	Observation
aqueous sodium hydroxide added	white precipitate forms; it dissolves in excess of the sodium hydroxide
ammonia solution added	white precipitate forms; it dissolves in excess of the ammonia

a The above describes a test for a particular cation
i What is a cation?
ii Which cation has been identified in the test?
iii Name a cation that would give a similar result for sodium hydroxide, but not for the ammonia solution.
iv Name a cation that forms a white precipitate with aqueous sodium hydroxide, but does not dissolve in excess sodium hydroxide.
b i Give the formula of the anion present in the compound tested.
ii Describe and give the result of the test you would carry out to confirm the presence of this anion.

8 A sample of mineral water contained these ions:

Name of ion	Concentration (milligrams / dm^3)
calcium	55
chloride	37
hydrogen carbonate	248
magnesium	19
nitrate	0.05
potassium	1
sodium	24
sulfate	13

a Make two lists, one for the anions and the other for the cations present in this mineral water.
b i Which metal ion is present in the highest concentration?
ii What mass of that metal would be present in a small bottle of water, volume 50 cm³?
c Which of the ions will react with barium nitrate solution to give a white precipitate?
d Of the metal ions, only calcium can be identified by a precipitation test. Why is this?
e A sample of the water is heated with sodium hydroxide and aluminium foil. Ammonia gas could not be identified, even though the nitrate ion is present. Suggest a reason.

Answers for Chapters 1 – 19

Unless otherwise indicated, the questions, example answers, marks awarded and/or comments that appear in this book were written by the author(s). In examination, the way marks would be awarded to answers like these may be different.

Chapter 1

page 3   **1** They follow a random zig-zag path. This happens because they collide with each other and bounce off in new directions. **2** It is named after Robert Brown.   **3** The purple colour spreading through the water shows that particles left the solid. That means the solid must be made of particles.   **4** It spreads upwards because the bromine particles are struck by the particles in air, and bounce away in all directions, including upwards.   **5 a** Diffusion means the way particles mix and spread, through colliding with each other.   **b** The smell is carried by particles from the perfume. These collide with gas particles in the air and bounce off in all directions.

page 5   **1** any two from the properties for each, at the top of page 4 **2 a** condensing   **b** freezing   **3** oxygen   **4** ethanol **5 a** under a minute (about four-fifths of a minute) **b** around five minutes   **c** It must take a lot more heat energy to separate the particles from each other to form a gas, than to melt the solid.   **6** Sublimation means the change from the solid state to the gas state on heating, without melting first; carbon dioxide (dry ice) sublimes when it is left to warm up at room temperature

page 7   **1** You could copy drawings from page 6 of your book. (For the solid and liquid, use the simpler drawings in the bottom row.) Add notes like those in the table on page 6.   **2 a** You cannot pour solids because the particles are held in a lattice, and cannot slide past each other.   **b** You can pour liquids because the particles can slide past each other.   **3** The drawing could be like this one:

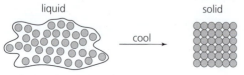

liquid   solid

cool →

As the liquid cools the particles lose energy and move more and more slowly …

… and eventually they settle into a regular pattern. A solid has formed.

4 a As the solid is heated, the particles vibrate more and more. At a certain point they gain enough energy to leave the lattice and form a gas.   **b** The particles gain enough heat energy from the sun to evaporate from the clothing. They go off into the air as a gas. **5 a i** gas   **ii** liquid   **b** keep cooling it

page 9   **1** particles hitting the walls of the container   **2** The more gas particles you blow in, the more often the balloon walls are struck. The balloon bursts when the pressure on it gets too much. **3** As the container cools, the particles will have less energy and move more slowly. They will hit the walls less often, with less force. So the pressure will fall.   **4** The pressure will fall because the particles now have more space to travel in. So they will hit the walls less often. Your drawing could be like this:

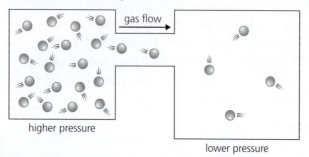

gas flow →

higher pressure

lower pressure

5 a because when perfume particles collide with gas particles in the air, they bounce off in all directions   **b** because the perfume particles evaporate faster in warm weather, and travel away faster   **6** Hydrogen is the lightest gas.   **7** The smoke will form sooner because the particles travel faster. (But it will still form closer to B.)

page 10   **Core**   **1 a** melting   **b** solidifying (freezing) **c** condensing   **d** subliming   **2 a i** At the bottom of the beaker, the liquid is purple.   **ii** All the liquid is now purple.   **b** Purple potassium manganate(VII) particles leave the crystal. Water particles collide with them, causing them to spread.   **c** dissolving and diffusion   **3 a** The particles are held in a fixed position. **b** The particles can slide around but are attracted to each other so stay close.   **c** The particles collide and bounce away to fill all the available space.   **d** The particles cannot be pushed closer together.   **e** As you blow more gas particles into the balloon, the sides of the balloon are struck more often. The pressure makes the balloon expand. (Eventually it will burst.)   **4 a** 17 °C **b** remains steady   **c** More energy is needed to separate the particles completely from each other (in boiling) than to break down the lattice structure (in melting).   **d** melting point is not 0 °C, boiling point is not 100 °C   **e** like the one shown, but with the horizontal parts at exactly 0 °C and 100 °C

page 11   **Core**   **5 a** liquid   **b** ethanol   **c** like the one shown, but with the horizontal parts at exactly 0 °C and 100 °C   **6 a** The particles are held in fixed positions.   **b** The particles will take in the heat energy, and travel faster. They will strike the walls more often, and with more energy, so pressure rises.   **c** The poisonous particles diffuse through the air in all directions, as a result of collisions.   **d** The dust particles are being struck by the gas particles in air.   **7 a** v   **b** See the drawing for potassium manganate(VII) on page 3   **c** Blue particles leave the crystal. Water particles collide with them, causing them to spread.

Extended 8 a You'll see the charcoal particles jiggling around.   **b** The charcoal particles are being struck by water particles.   **c** The individual particles would be too small to see, under the microscope.   **9 a** Hydrogen is leaving the tube faster than air particles are entering.   **b** It is faster.   **c** The mass of a hydrogen particle is less than the masses of the particles in air.   **d** The water level will fall, because air particles enter the tube faster than carbon dioxide particles leave it.   **10 a** helium and methane: they have the lowest relative masses   **b** helium   **c** Yes. Its relative molecular mass is greater than those of nitrogen and oxygen, the main gases in air.   **d** greater that 16, but less than 28

Chapter 2

page 13   **1 a** dissolves in a solvent (the particles separate and spread out among the solvent particles)   **b** does not dissolve in a solvent (the particles do not separate and spread)   **c** a solution with water as the solvent   **2 a** silver nitrate   **b** about 20 times more soluble   **c** a colourless solution with some undissolved (white) solid lying at the bottom   **d** The undissolved solid will gradually dissolve.   **3 a** sodium and potassium   **b** ammonium salts   **4** Examples are: **a** sugar, salt, instant coffee, soap powder   **b** flour, pepper, plastic, wood, stainless steel.   **5** Examples are: propanone, in nail polish remover; white spirit or turpentine to thin oil-based paint; ethanol in perfumes and aftershaves.   **6** Water contains dissolved oxygen, which fish then take in through their gills; carbon dioxide is dissolved under pressure to make soft drinks fizzy.

page 15 **1** It contains particles of only one substance. **2** No. An impurity is an *unwanted* substance. **3** Because the presence of an impurity changes the melting and boiling points of a substance. **4** Yes. It can contain particles of other gases. For example air can contain harmful gases from car exhausts.

page 17 **1 a** the liquid obtained from filtering **b** the solid trapped in the filter paper; when you filter a mixture of chalk and water, water is the filtrate and chalk the residue. **2 a** The tiny sugar particles are all separated and can pass through the filter paper. **b** heat the solution to evaporate the water **3** Heat the solution to evaporate water, until it is saturated; let it cool; filter off the crystals that form. **4** Add ethanol to the salt / sugar mixture, to dissolve the sugar. Filter to separate the salt. Heat the filtrate over a water bath to evaporate the ethanol and obtain solid sugar. Take care: ethanol catches fire easily. **5** Use a magnet to remove the iron.

page 19 **1** by simple distillation; see the apparatus at the top of page 18 **2** Their purpose is to allow the solvent to condense. The cold water keeps the temperature down, to allow condensation to take place. **3** because this is a mixture of two liquids, not a solution containing dissolved salts **4** Taking the fractional distillation of a water / ethanol mixture as example: the glass beads in the fractionating column provide a large surface area on which condensation can take place. Once the beads warm up to 78 °C, the ethanol passes over into the condenser. But the water vapour condenses on the beads and drips back into the flask. **5** because X has no green dot to match the dot for C **6** You could cut out each coloured circle separately, and soak it in a little water. The dye will move into the water. You could decant the liquid (or use a centrifuge) to separate it from the paper, and evaporate the water to obtain the solid dye. (You will be working with very tiny quantities!)

page 21 **1** It depends on how the substances interact with both the paper and solvent. For example substance X may be more soluble in the solvent than substance Y, and less attracted to the paper. So it will travel further. Eventually X and Y will separate. **2 a** It shows the location of a colourless substance. **b** because amino acids are colourless **3** A substance has a fixed R_f value for a given solvent. So we can identify a substance by looking up a table of R_f values. **4 a** Only C. A has a spot at the same height as the C spot. **b and c** These are the R_f values and amino acid names: **A** 0.14, lysine; 0.26, glycine; 0.73, leucine **B** 0.38, alanine **C** 0.73, leucine **D** 0.43, proline **E** 0.6, valine.

page 24 **Core** **1 a** A iii B v C vi D i E iv F ii **2 a** to condense water vapour **b** to prevent steam escaping before it can condense **c i** in the test-tube that is being heated **ii** in the lower part of the glass arm (the delivery tube) **d** distillation **e** salt

page 25 **Core** **3 a i** 100 °C **ii** It is lower than that of seawater. (Impurities raise the boiling point.) **b** A flask **c i** C **ii** at the lower end of the condenser **d** Filtration would remove only insoluble impurities. **4 a** Bunsen burner, filter funnel, tripod, conical flask, gauze, stirring rod, filter paper, beaker **b** place sample in beaker and add water / place beaker on gauze on tripod / heat to boiling, stirring with glass rod / put filter paper in filter funnel, in conical flask / carefully filter the heated mixture / remove the solid residue and dry in an oven **5 a** mixture **b** Liquids have to be separated from each other, so simple distillation would not work well. **c** Nitrogen has the lowest boiling point of the three, and oxygen the highest. **6 a** You will separate it by filtering; see the diagram on page 16. **b** Use simple distillation. **c** See page 18 (top diagram). **7 a** B **b** A **c** red

Extended **8 a** proline and valine **b** proline **c** They will not change. (The spots will travel less far too.) **9** See the instructions on pages 20 and 21.

Chapter 3

page 27 **1 a** An atom is a particle of matter that cannot be broken down further by chemical means; everything is made up of atoms. **b** An element contains only one kind of atom. **2** mostly empty space, with a tiny nucleus at the centre, and a cloud of electrons whizzing around the nucleus, at a distance from it **3 a** sodium **b** iron **c** lead **d** silver **4 a** calcium **b** magnesium **c** nitrogen **5** Einsteinium, Es **6** You could choose: **a** any three metals from Groups I or II **b** any three non-metals from Group 0 or Group VII. (In the other groups there is a change from non-metallic to metallic character down the group.)

page 29 **1** protons, electrons, neutrons **2 a** proton **b** neutron **c** electron **3** fluorine **4** They have equal numbers of protons and electrons so the charges cancel. **5 a** the number of protons in the nucleus of an atom **b** the total number of protons and neutrons in the nucleus of an atom **6** In order: carbon atom with 6p, 6e, 6n; oxygen atom with 8p, 8e, 8n; magnesium atom with 12p, 12e, 12n; aluminium atom with 13p, 13e, 14n; copper atom with 29p, 29e, 35n

page 31 **1 a** atoms of the same element, with different numbers of neutrons **b** carbon-12, $_6^{12}$C; carbon-13, $_6^{13}$C; carbon-14, $_6^{14}$C **2** Its atoms are unstable, and break down, giving out radiation. **3** short for 'a radioactive isotope'; for example carbon-14, cobalt-60, uranium-235 **4 a** It can kill cells, so our organs get damaged. **b** Radiation kills cancer cells more readily than it kills healthy cells. **5 a** Radioactive material is mixed with the liquid or gas in the pipe. The radiation from it cannot be detected through the pipe, but can be detected where it leaks out, using a Geiger counter. **b** No radiation will be detected outside the pipe. **6 a** to kill off bacteria or other micro-organisms that cause decay **b** cobalt-60 and caesium-137

page 33 **1 a** The diagrams should show 3 electron shells, with 2 electrons in the first (inner) shell, 8 in the second, and 3 in the third. **b** 2+8+3 **c** aluminium **2 a** 2+1 **b** 2+8+2 **c** 1 **3** Group V **4** three **5 a** 2+8+18+8 **b** It is unreactive, because the outer shells of its atoms have a stable arrangement of electrons.

page 39 **1** Note that you must know the names and symbols for the first 20 elements. You should also know names and symbols for the more common transition elements, and other noble gases. **2** You can self-check your answers using the glossary. **3** good conductor, light, the most abundant metal, quite strong, will not rust away (it forms a protective coating – see page 187) **4** Any three of these: metals conduct electricity, form positive ions when they react (but H^+ is a non-metal exception), tend to be hard and strong, tend to have high melting points. **5** For a physical property: any of the first seven in the second column in the table on page 38. For a chemical property: any of the final two in the table.

page 40 **Core** **1 a** A, B and E **b** D, 2− **c** C, 2+ **d** A and B **e** A and B = Mg, C = Mg^{2+}, D = O^{2-}, E = F **2 A** p **B** n **C** n **D** e **E** e **F** e **G** n **H** n **I** p **J** p **3 a i** the symbol of the element **ii** proton number **iii** nucleon number **b i** 60 **ii** 34 **iii** 0 **iv** 10 **v** 146 **c** $_{35}^{81}$Br

page 41 **Core** **4 a** This table gives the answers:

	Al	B	N	O	P	S
i	3	2	2	2	3	3
ii	3	3	5	6	5	6
iii	13	5	7	8	15	16
iv	13	5	7	8	15	16
v	2.8.3	2.3	2.5	2.6	2.8.5	2.8.6
vi	3	3	5	6	5	6

b valency **c** Elements with the same number of valency electrons will have similar properties: Al and B; N and P; O and S.
5 a B has an extra neutron **b** isotopes **c** A, $_5^{10}$B; B, $_5^{11}$ **d** 10
e heavier **f i** 2 + 3 **ii** Isotopes have the same number of electrons, and the same electronic configuration.
6 a i same **ii** different **b** sodium, since its relative atomic mass is a whole number (and note that the value for magnesium is usually rounded off to 24, for school chemistry) **7 a** 38 **b** 5 **c** 2
8 a i 2 + 8 **ii** It is a stable arrangement (8 outer electrons) so the element will be unreactive. **b** Group 0 **c** helium, argon, krypton, xenon or radon (the question shows an atom of neon)
9 a i 38 **ii** 40 **b** It has an unstable nucleus. **c** See uses on page 31. **10** *electrical conductors* − allow electricity to pass through them; *density* − how 'heavy' a substance is; *ductile* − can be drawn into a wire; *malleable* − can be beaten into shape; *sonorous* − makes a ringing noise when hit **b** in making electrical wiring **c** malleability **d** Possible examples are: sodium, lithium, potassium, aluminium. **e** They must be sonorous. **f** heat
g For example you could choose *strength* (needed for structures such as bridges and aeroplanes) or *hardness* (needed for things like railway lines, coins, hammers). **h** non conductor of heat / electricity, brittle, non-shiny, low density **i** Metals generally form basic oxides, and non-metals form acidic oxides.

Chapter 4

page 43 **1** In a mixture of iron and sulfur, the iron and sulfur are not chemically combined. – but in the compound iron sulfide they are. **2** Energy is given out − in the form of dazzling light and a fizzing sound, as well as heat. The white ash that forms looks very different from magnesium. Also note that you need heat to start the reaction off. This is the case for many chemical changes (but not all).
3 a physical (the glass could be melted down and reformed into a bottle) **b** chemical **c** physical (you could unravel the fabric)
d chemical

page 45 **1** They have a stable arrangement of outer electrons. They do not need to gain or lose electrons, so there is no need to react.
2 They react with atoms of other elements in order to gain, lose, or share outer electrons, and reach a stable arrangement.
3 The drawings should be as on page 45.
4 a It has 11 protons but only 10 electrons, so it has a charge of 1+.
b It has 17 protons and 18 electrons so it has a charge of 1−.
5 an atom or group of atoms with a positive or negative charge
6 They already have a stable arrangement of valency electrons, so there is no need to gain or lose any.

page 47 **1** The drawing should be as on page 46.
2 the attraction between ions of opposite charge
3 a giant structure containing millions of ions, in which sodium ions alternate with chloride ions in a regular lattice; its formula is NaCl because there is one Na^+ ion for each $Cl^−$ ion
4 a The atom loses two electrons to form an ion; so the ion has two more protons than electrons, which gives a charge of 2+.
b Opposite charges attract. **c** The charge on each Mg^{2+} ion is balanced by the charge on each pair of $Cl^−$ ions, so the charges cancel out. **d** There are two $Cl^−$ ions for each Mg^{2+} ion.

page 49 **1** The atom loses two electrons to form an ion. So the ion has two more protons than electrons, giving a charge of 2+.
2 The atom loses three electrons to form an ion.
3 a K^+, $Cl^−$ **b** Ca^{2+}, $S^{2−}$ **c** Li^+, $S^{2−}$ **d** Mg^{2+}, $F^−$
4 a KCl **b** CaS **c** Li_2S **d** MgF_2 **5 a** $CuCl_2$ **b** Fe_2O_3
6 copper(I) chloride, iron(II) sulfide, magnesium nitrate, ammonium nitrate, calcium hydrogen carbonate
7 a Na_2SO_4 **b** KOH **c** $AgNO_3$

page 51 **1 a** covalent bond **b** the force of attraction between their nuclei and the shared electrons

2 a unit made of two or more atoms joined by covalent bonds
3 five from: hydrogen, H_2; oxygen, O_2; nitrogen, N_2; fluorine, F_2; chlorine, Cl_2; bromine, Br_2; iodine, I_2; sulfur, S_8 (usually referred to as S in equations)
4 The drawings should be like those on pages 50 and 51.
5 The two nitrogen atoms in the molecule share three pairs of electrons.

page 53 **1 a** a compound where atoms of different elements bond by sharing electrons **b** any five of the examples on pages 52 and 53; others include ethanol, C_2H_5OH, and hydrogen bromide, HBr **2** The drawings should be like those on page 52.
3 Three hydrogen atoms share electrons with one nitrogen atom, giving 2 electrons for each hydrogen and 8 for nitrogen.
4 The drawing should be like that on page 53. **5** Each carbon atom shares two electrons with the other carbon atom. The result is a double bond.

page 55 **1** The particles are held in a fixed pattern extending in all directions. **2 a** ions **b** molecules **3** Solid sodium chloride will not conduct electricity because the charged particles are not free to move. But in water, the lattice breaks up and the ions are released. The solution can conduct because the ions are now free to move.
4 a Molecular: the low melting point suggests that the forces holding the particles in the lattice are weak. **b** No, because the liquid contains molecules, not charged particles. **5** The water molecules are held in a regular lattice, with weak forces between them. As the ice is heated the molecules vibrate more vigorously about their positions. Evenutally they gain enough energy to break away from the lattice, to form a liquid (water).

page 57 **1** Molecular; the melting point of a giant structure would be much higher, since the atoms are held in the lattice by strong covalent bonds. **2** They are two different forms of the same element. **3** Each carbon atom in it is held in the lattice by four strong covalent bonds. **4** Because of their different bonding and structure. Diamond is hard with a high melting point for the reason given in **3**. But in graphite each carbon atom is bonded to only *three* others, forming flat sheets that can slide over each other. So graphite is soft and slippery. The non-bonded electrons are free to move, so graphite can conduct. **5 a** an oxide of silicon, with the formula SiO_2 **b** Both are hard, with high melting points; they do not conduct electricity. **c** All the atoms are held in a giant lattice by strong covalent bonds. **6 a** It takes a great deal of heat energy to break the covalent bonds that hold the atoms in the lattice. **b** It may be because silica contains two different types of atoms, of different sizes. That must make the lattice weaker in some way.

page 59 **1** a regular lattice of metal ions in a sea of electrons
2 the force of attraction between the metal ions and free electrons
3 can be hammered into shape **4** The layers of ions can slide over each other; the electrons move too so the bonds do not break.
5 a The electrons can move through the lattice, carrying the current. **b** Yes; the electrons are still free to move. **6** Some examples are: **a** car and aircraft bodies; metal signs; metal roofs **b** wiring in computers; metal cables; metal guitar strings **c** terminals in torch batteries; wiring in computers and phones; outdoor electricity cables **7** The diagram should be like the one for copper on page 58.

page 60 **Core 1 a** 3; the drawing should show 2 shells, with 2 electrons in the inner shell and 1 in the outer shell **b** by losing its partly-filled outer shell **c** The drawing should show just one shell with 2 electrons; the symbol is Li^+. **d** 9; the drawing should show 2 shells, with 2 electrons in the inner shell and 7 in the outer shell **e** by gaining electrons, to obtain a stable outer shell **f** the drawing should be as for **d** but with 8 electrons In the outer shell; the symbol is $F^−$ **g** The diagram should show the lithium atom's outer electron moving to the fluorine atom. **h** lithium + fluorine → lithium fluoride **2 a** ammonia, NH_3 **b** electrons **c** covalent **d** hydrogen chloride, water, methane, or any other compound formed between non-metals **3 a** a gas **b** made of molecules; the evidence is its low melting and boiling points **c i** covalent **ii**

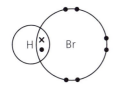

d HBr **e i** hydrogen fluoride, hydrogen chloride or hydrogen iodide **ii** HF, HCl, HI

page 61 **Core** **4 a** D and G − good conductors in both solid and liquid states **b** B and E − soluble, conduct electricity when liquid but not when solid **c** Ionic compounds have strong bonds between ions, leading to high melting points. **d** A and C **e i** F **ii** macromolecule **f i** ionic **ii** covalent **iii** ionic **iv** covalent

Extended 5 a i loses **ii** 3 **iii** positive **iv** 3+ **b i** gains **ii** 3 **iii** negative **iv** 3− **c i** 2+8 for both **ii** They are the same. By forming ions both gain a stable outer shell of 8 electrons. **d** phosphorus (or arsenic) **6 a** both solid **b** giant covalent **c** giant covalent, given its high melting and boiling points **d** Carbon dioxide is a gas, silicon dioxide a solid. **e** molecular **f** No. Its very high melting point indicates a giant structure. **7 a** giant structure **b** ionic **c i** 4 **ii** 4 **d i** ZnS **ii** Yes. The charges on the ions, 2+ and 2−, balance out. **e** any combination of a Group II metal (or transition element with ions with a charge of 2+) with a group VI non-metal, **OR** any combination of a Group I metal (or transition element with ions with a charge of 1+) with a Group VII non-metal, **OR** any combination of a Group III metal (or transition element with ions with a charge of 3+) with a group V non-metal **8 a** The metal consists of a lattice of metal ions in a sea of free electrons. The ions are held together by their attraction to the free electrons between them. **b i** The free electrons can move through the lattice of metal ions, carrying a charge. **ii** Layers of metal ions can slide over each other without breaking the metallic bond, because the free electrons move too.

Chapter 5

page 63 **1** hydrogen oxide **2 a** sodium fluoride **b** hydrogen fluoride **c** hydrogen sulfide **d** beryllium bromide **3** It contains two oxygen atoms for every silicon atom. **4 a** HBr **b** NaCl **c** $CaCl_2$ **d** BaO **5** BaI_2 **6** phosphorus chloride, PCl_3 (also called phosphorus trichloride)

page 65 **1** + means *reacts with*, → means *to give* **2 a** $2Na\,(s) + Cl_2\,(g) \rightarrow 2NaCl\,(s)$ **b** $H_2\,(g) + I_2\,(g) \rightarrow 2HI\,(g)$ **c** $2Na\,(s) + 2H_2O\,(l) \rightarrow 2NaOH\,(aq) + H_2\,(g)$ **d** $2NH_3\,(g) \rightarrow N_2\,(g) + 3H_2\,(g)$ **e** $C\,(s) + CO_2\,(g) \rightarrow 2CO\,(g)$ **f** $4Al\,(s) + 3O_2\,(g) \rightarrow 2Al_2O_3\,(s)$ **3** $2Al\,(s) + 3Cl_2\,(g) \rightarrow 2AlCl_3\,(s)$

page 67 **1 a** The relative atomic mass A_r for an element is the average mass of its naturally-occurring isotopes, relative to the mass of a carbon-12 atom. **b** because *relative* means *compared with* something else **2** 127 **3** The term *relative molecular mass* is used for molecules while *formula mass* is used for ionic compounds. **4 a** 32 **b** 254 **c** 16 **d** 71 **e** 58 **f** 46 **g** 132 ; all are relative molecular masses except for **g**.

page 69 **1 a** 95 g **b** 35.5 g **c** 47.5 g **2** 75% carbon, 25% hydrogen **3 a** 90% pure **b** 1.5 g

page 70 **Core** **1 a** H_2O **b** CO **c** CO_2 **d** SO_2 **e** SO_3 **f** NaCl **g** $MgCl_2$ **h** HCl **i** CH_4 **j** NH_3 **2 a** PbO_2 **b** Pb_3O_4 **c** KNO_3 **d** N_2O **e** N_2O_4 **f** $NaHCO_3$

g Na_2SO_4 **h** $Na_2S_2O_3$ **3 a** 1 copper, 1 oxygen **b** 2 copper, 1 oxygen **c** 1 aluminium, 3 chlorine **d** 1 hydrogen, 1 nitrogen, 3 oxygen **e** 1 calcium, 2 oxygen, 2 hydrogen **f** 2 carbon, 2 oxygen, 4 hydrogen **g** 2 nitrogen, 4 hydrogen, 3 oxygen **h** 2 nitrogen, 8 hydrogen, 1 sulfur, 4 oxygen **i** 3 sodium, 2 phosphorus, 8 oxygen **j** 1 iron, 1 sulfur, 11 oxygen, 14 hydrogen **k** 1 cobalt, 2 chlorine, 12 hydrogen, 6 oxygen

page 71 **Core** **4 a** HBr **b** PCl_3 **c** H_2O_2 **d** C_2H_2 **e** N_2H_4 **f** XeF_6 **g** H_2SO_4 **h** H_2SO_3 **5 a** calcium chloride **b** $CaCl_2$ **6 a** phosphorus oxide **b** P_2O_3 **7 a** zinc + hydrogen chloride (hydrochloric acid) → zinc chloride + hydrogen **b** sodium carbonate + sulfuric acid → sodium sulfate + carbon dioxide + water **c** magnesium + carbon dioxide → magnesium oxide + carbon **d** zinc oxide + carbon → zinc + carbon monoxide **e** chlorine + sodium bromide → sodium chloride + bromine **f** copper(II) oxide + nitric acid → copper(II) nitrate + water **8 a** $N_2 + 2O_2 \rightarrow 2NO_2$ **b** $K_2CO_3 + 2HCl \rightarrow 2KCl + CO_2 + H_2O$ **c** $C_3H_8 + 5O_2 \rightarrow 3CO_2 + 4H_2O$ **d** $Fe_2O_3 + 3CO \rightarrow 2Fe + 3CO_2$ **e** $Ca(OH)_2 + 2HCl \rightarrow CaCl_2 + 2H_2O$ **f** $2Al + 6HCl \rightarrow 2AlCl_3 + 3H_2$ **9 a** $MgSO_4 + 6H_2O \rightarrow MgSO_4.6H_2O$ **b** $2C + O_2 \rightarrow 2CO$ **c** $2CuO + C \rightarrow 2Cu + CO_2$ **d** $C_2H_6 \rightarrow C_2H_4 + H_2$ **e** $ZnO + C \rightarrow Zn + CO$ **f** $NiCO_3 \rightarrow NiO + CO_2$ **g** $CO_2 + 2H_2 \rightarrow CH_4 + O_2$ **h** $NaOH + HNO_3 \rightarrow NaNO_3 + H_2O$ **i** $C_2H_6 \rightarrow C_2H_4 + H_2$ **10 a** 18 **b** 17 **c** 46 **d** 80 **e** 98 **f** 36.5 **g** 142 **11 a** 40 **b** 239 **c** 78 **d** 58.5 **e** 170 **f** 132 **g** 138 **h** 278 **12 a i** 27.2 g **ii** 2.72 g **b** 50 g **c** 80% **13 a** 17.5% **b i** 2185 kg (or 2.185 tonnes) **ii** 375 kg **c** 91.7%

Chapter 6

page 73 **1** 6.02×10^{23} atoms **2** 6.02×10^{23} molecules **3** Avogadro constant **4 a** 1 g **b** 127 g **c** 35.5 g **d** 71 g **5 a** 32 g **b** 64 g **6** 138 g **7 a** 9 moles **b** 3 moles **8 a** 6.02×10^{23} sodium ions **b** 35.5 g

page 75 **1 a** magnesium + oxygen → magnesium oxide **b** 2 moles **c i** 32 g **ii** 8 g **2 a** copper(II) carbonate → copper(II) oxide + carbon dioxide **b** $CuCO_3$ 124 g, CuO 80 g, CO_2 44 g **c i** 11 g **ii** 20 g

page 77 **1** room temperature and pressure (20 °C and 1 atmosphere) **2** the volume occupied by 1 mole of the gas molecules: 24 dm³ **3** 24 dm³ **4 a** 168 dm³ **b** 12 dm³ **c** 0.024 dm³ (24 cm³) **5 a** 12 dm³ **b** 2.4 dm³ **6 a** 12 dm³ **b** 12 dm³ **7 a** 12 dm³ **b** 6 dm³

page 79 **1 a** 1 mole **b** 1 mole **2 a** 2 mol/dm³ **b** 1.5 mol/dm³ **3 a** 0.5 dm³ (500 cm³) **b** 0.005 dm³ (5 cm³) **4 a** 20 g **b** 0.5 g **5 a** 0.5 moles per litre **b** 0.25 moles per litre

page 81 **1 a** 4 **b** 4 **2** found by experiment **3** FeS **4** SO_3

page 83 **1** $MgCl_2$ **2** AlF_3 **3** The empirical formula gives the *simplest ratio* in which atoms combine; the molecular formula gives the *exact numbers of atoms* in the molecule. They may be the same, for example for CH_4. **4** CH **5** C_2H_4 **6 a** C_7H_{16} **b** C_7H_{16} **7** P_4O_6

page 85 **1 a** the actual amount obtained, given as a % of the theoretical amount (which is calculated from the equation) **b** the mass of pure substance in the product, given as a % of the total mass of the product **2** 76.7% **3** 63% **4** 172 g **5** 88%

page 86 **Extended** **1 a** iron(III) oxide + carbon monoxide → iron + carbon dioxide **b** 160 **c** 2000 moles **d** 2 moles **e** 4000 moles **f** 224 kg **2 a** calcium carbonate → calcium oxide + carbon dioxide **b** 0.5 moles **c i** 11.2 g **ii** 8.8 g **iii** 4.8 dm³ or 4800 cm³

3 a i 4 moles **ii** 19 moles **b** 4.75 moles **c i** 114 dm^3
d 227 g **e** 502.2 dm^3 **f** A small amount of liquid produces a very large volume of gas. This creates a massive pressure wave, which causes damage.

page 87 **Extended** **4 a** 0.5 moles **b** 25 cm^3 **c i** 75 cm^3
ii 50 cm^3 **5 a** 1.4 g **b** 0.025 moles **c** 0.025 moles
d Fe + H$_2$SO$_4$ → FeSO$_4$ + H$_2$; Fe^{2+} **e** 0.6 dm^3
6 a 106.5 g **b** 3 moles **c** 1 mole **d** AlCl$_3$ **e** 0.1 mol / dm^3
7 a 45.5 cm^3 **b** 41.7 cm^3 **c** 62.5 cm^3
8 a P$_2$O$_3$ **b** 41.3 g **c** P$_4$O$_{10}$ **d** It could be P$_4$O$_6$ to match the other oxide (or P$_2$O$_3$). **9 a** Zn$_3$P$_2$ **b** 24.1%
10 a 64 g **b** 4 moles **c** 2 moles **d** MnO$_2$ **e** 632.2 g
11 The missing molecular formulae are: **a** N$_2$H$_4$ **b** C$_2$N$_2$ **c** N$_2$O$_4$
d C$_6$H$_{12}$O$_6$ **12 a** carbon and hydrogen
b CH$_2$ **c** A is C$_3$H$_6$ and B is C$_6$H$_{12}$.
13 a 217 g **b** 20.1 g of mercury, 1.6 g of oxygen **c** 94.5 %
14 a 0.0521 moles **b** 4.375 g **c** 87.5%

Chapter 7

page 89 **1 a** ... oxygen is gained **b** ... oxygen is lost **c** ... take place together **2** Your arrows should show that magnesium is oxidised to magnesium oxide, and sulfur dioxide is reduced to sulfur.
3 from **red**uction + **ox**idation
4 Methane gains oxygen – it is oxidised. Since oxidation and reduction always take place together, this means that oxygen is reduced.
5 Your arrows should show that magnesium is oxidised, and oxygen is reduced.

page 91 **1** gain of oxygen, or loss of electrons **b** loss of oxygen, or gain of electrons **2** It shows how a substance gains or loses electrons, during a reaction. **3 a** Potassium loses electrons to chlorine. **b** 2K → 2K$^+$ + 2e$^-$; Cl$_2$ + 2e$^-$ → 2Cl$^-$
4 a Br$_2$ + 2e$^-$ → 2Br$^-$; 2I$^-$ → I$_2$ + 2e$^-$
b The ionic equation is: Br$_2$ + 2I$^-$ → 2Br$^-$ + I$_2$

page 93 **1 a** hydrogen + oxygen → water **b** Oxidation states: in H$_2$ and O$_2$, both 0; in H$_2$O, +I for H, −II for O. **c** Yes, oxidation states have changed. **d** Hydrogen has been oxidised and oxygen reduced.
2 i a potassium bromide → potassium + bromine **b** Oxidation states: in KBr, +I for K and −I for Br; in the products K and Br$_2$, both 0. **c** Yes, oxidation states have changed. **d** Bromine has been oxidised and potassium has been reduced.
ii a potassium iodide + chlorine → potassium chloride + iodine.
b Oxidation states: in KI, +I for K and −I for I; in Cl$_2$, 0; in KCl, +I for K and −I for Cl; in I$_2$, 0. **c** Yes, oxidation states have changed.
d Iodine has been oxidised and chlorine reduced.
3 b C +IV (and O −II) **c** C + O$_2$ → CO$_2$
d The oxidation states for the reactants are both 0. The oxidation states for the products are as in **b**. There has been a change in oxidation states, so this is a redox reaction; the oxygen is reduced.

4 Yes. Oxidation states are *always* zero in elements, but *never* zero in compounds. So there is always a change in oxidation states during the reaction, which means it is a redox reaction.

page 95 **1 a** An oxidising agent oxidises another substance during a redox reaction – and is itself reduced. **b** A reducing agent reduces another substance during a redox reaction – and is itself oxidised.
2 a oxidising agent, oxygen; reducing agent, magnesium
b oxidising agent, iron(III) oxide; reducing agent, carbon monoxide
3 a oxidising agent, chlorine; reducing agent, iron
b oxidising agent, copper(II) sulfate; reducing agent, iron
4 a In KMnO$_4$, manganese is in oxidation state +VII. It has a strong drive to become the more stable ion with oxidation state +II. To achieve this reduction, it acts as a strong oxidising agent.
b When aqueous potassium iodide is oxidised by an oxidising

agent, the solution goes from colourless to red-brown. So this is a good visual test for the presence of an oxidising agent.

page 96 **Core** **1 a** A calcium + oxygen → calcium oxide
B carbon monoxide + oxygen → carbon dioxide
C methane + oxygen → carbon dioxide + water
D copper(II) oxide + carbon → copper + carbon dioxide
E iron + oxygen → iron(III) oxide
F iron(III) oxide + carbon monoxide → iron + carbon dioxide
b In each equation above, the underlined reactant is oxidised, and the other reactant is reduced.
2 a A yes (oxygen gained and lost) **B** yes (oxygen gained and lost) **C** no (it is a neutralisation) **D** yes (oxygen gained and lost) **E** yes (oxygen gained and lost)
b i Oxidised: **A** Mg, **B** C, **D** Fe, **E** C.
ii Reduced: **A** CO$_2$, **B** SiO$_2$, **D** CuO, **E** PbO.
Extended 3 a magnesium chloride **b** Mg (s) + Cl$_2$ (g) → MgCl$_2$ (s)
c i magnesium **ii** Mg → Mg^{2+} + 2e$^-$
d i chlorine **ii** Cl$_2$ + 2e$^-$ → 2Cl$^-$

page 97 **Extended** **4 a** lithium fluoride
b 2Li (s) + F$_2$ (g) → 2LiF (s)
c

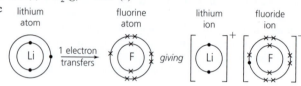

d i lithium **ii** Li → Li$^+$ + e$^-$ **e** F$_2$ + 2e$^-$ → 2F$^-$
5 a displacement **b** solution turns orange **c i** Cl$_2$ + 2e$^-$ → 2Cl$^-$ **ii** reduced (it gains electrons) **d** 2Br$^-$ → Br$_2$ + 2e$^-$
e chlorine; it wins the competition to exist as ions **f** the iodide ion **ii** Br$_2$ (aq) + 2I$^-$ (aq) → 2Br$^-$ (aq) + I$_2$ (aq)
6 a −I **b** The reactants are colourless but iodine is coloured.
c i oxidised **ii** 2I$^-$ → I$_2$ + 2e$^-$ **d i** −I **ii** It goes down to −II in the compound water. **iii** H$_2$O$_2$ (aq) + 2H$^+$ (aq) + 2e$^-$ → 2H$_2$O (l)
7 a i +III **ii** −III **iii** +IV **iv** +III **v** +I **vi** +II
b Copper is a transition element, and transition elements can have more than one oxidation state, in their compounds.
8 a An acid must be present. **b** from purple to colourless
c the Fe^{2+} ion **d** when the purple colour remains **e** Fe^{2+} → Fe^{3+} + e$^-$
9 a i +VI **ii** +VI **b** There is no change in oxidation state.
10 a i silver chloride **ii** insoluble **b** It is not a redox reaction: there is no change in oxidation state.
c 2AgCl (s) → 2Ag (s) + Cl$_2$ (g). There are changes in oxidation state: Ag +I to 0, Cl −I to 0. So this is a redox reaction.

Chapter 8

page 99 **1 a** substance that allows electricity to pass through it
2 The drawing could be like the first one on page 98, but with mercury in the beaker.
3 The free electrons can move through the metal, as a current.
4 Electrolysis is the breakdown of an ionic compound, when molten or in aqueous solution, by the passage of electricity.
5 No; molecular substances do not conduct electricity.
6 a An electrolyte is a molten compound or solution that contains ions, so can conduct electricity. Examples are molten lead bromide, aqueous sodium chloride, aqueous copper(II) sulfate, or any other molten or dissolved ionic compound. **b** A non-electrolyte will not conduct electricity whether molten or in solution. Examples are ethanol, an aqueous solution of sugar, petrol, paraffin, molten naphthalene.

page 101 **1 a** ionic compounds **b** molten or in solution

2 a sodium and chlorine **b** aluminium and oxygen
c calcium and fluorine **d** lead and sulfur
3 a hydrogen and chlorine **b** hydrogen and chlorine
c copper and oxygen

page 103 **1** at the cathode
2 a $2Cl^- \rightarrow Cl_2 + 2e^-$ **b** $2O^{2-} \rightarrow O_2 + 4e^-$
3 In the answers for **b** and **c** below, the two half-equations have been balanced to give the same number of electrons in each. But you can also give answers without balancing the two half-equations.
a $2H^+ + 2e^- \rightarrow H_2$; $2Cl^- \rightarrow Cl_2 + 2e^-$ **b** $4H^+ + 4e^- \rightarrow 2H_2$; $4OH^- \rightarrow O_2 + 2H_2O + 4e^-$ **c** $2Cu^{2+} + 4e^- \rightarrow 2Cu$; $4OH^- \rightarrow O_2 + 2H_2O + 4e^-$

page 105 **1 a** concentrated solution of sodium chloride, obtained by pumping water into salt mines, or evaporating seawater
2 brine → hydrogen + chlorine + sodium hydroxide
3 a Your sketch should be based on the drawing on page 104.
b Oxidation takes place at the anode, and reduction at the cathode. See the equations on page 104.
4 The membrane lets ions through, but keeps the gases apart.
5 a because the products have many important uses
b You can choose uses from the panels on page 105.
6 For example, make sure that: poisonous chlorine cannot escape within the factory or out to the atmosphere; hydrogen does not catch fire; corrosive sodium hydroxide does not leak out; the membrane is working properly so that the gases cannot mix.

page 107 **1 a** The solution loses its colour because the copper ions (which give the colour) are reduced to copper atoms at the cathode.
b The solution keeps its colour because while copper ions are discharged at the cathode, the anode dissolves to form more copper ions. So their concentration in the solution does not change.
2 the anode – because the anode dissolves.
3 Anode, impure copper; cathode, pure copper; electrolyte, aqueous copper(II) sulfate. The anode dissolves to form copper ions in solution. At the cathode, copper ions are reduced to copper atoms, and a layer of pure copper builds up.
4 using electricity to coat one metal with another
5 to make it look shiny and attractive – and if the cutlery is not stainless steel, the nickel coat will also prevent rust
6 a nickel **b** the cutlery **c** a solution of a soluble nickel salt such as nickel nitrate (all nitrates are soluble)

page 108 **Core 1** **a** The ions In the solid are not free to move. **b** bubbles of gas **c** bromine **d** lead
2 a E **b** C **c i** F **ii** A **iii** C **iv** E
d i B and D **ii** hydrochloric acid; sodium chloride (or the chloride of any other metal above hydrogen in the order of reactivity – except lead chloride, which is insoluble)

page 109 **Extended** **3 a** It contains ions which are free to move.
b A is chlorine, which bleaches damp litmus paper. B is hydrogen, which burns in air with a squeaky pop. **c i** and **ii** See page 105.
d i $2Cl^- \rightarrow Cl_2 + 2e^-$, $2H^+ + 2e^- \rightarrow H_2$ **ii** Electrons are transferred.
e i sodium hydroxide **ii** It will neutralise an acid, giving a salt and water. **4 a i** Na^+, Cl^-, H^+, OH^- **ii** Cu^{2+}, Cl^-, H^+, OH^- **b** The ions are attracted to the electrode of opposite charge. So negative ions move towards the anode, and positive ions towards the cathode.
c For the sodium chloride solution: **i** $2Cl^- \rightarrow Cl_2 + 2e^-$
ii $2H^+ + 2e^- \rightarrow H_2$. For the copper(II) chloride solution: **i** $2Cl^- \rightarrow Cl_2 + 2e^-$ **ii** $Cu^{2+} + 2e^- \rightarrow Cu$. **d** Both contain chloride ions in a concentrated solution. **e i** Oxygen will then be given off in preference to chlorine. **ii** $4OH^- \rightarrow O_2 + 2H_2O + 4e^-$ **f** Each time, the ions of the *less reactive* substance are discharged. So hydrogen ions are discharged in preference to sodium ions, and copper ions in preference to hydrogen ions. **g** Concentrated hydrochloric acid, or a concentrated solution of any reactive metal chloride (such as potassium chloride) will give same result.
h copper(II) chloride **5 a i** Positive ions move through the

solution to the cathode (−), where they accept electrons, and negative ions to the anode (+), where they give up electrons. **ii** In the wires, the electrons flow from the anode, through the battery, to the cathode. **b i/ii** $Li^+ + e^- \rightarrow Li$, reduction; $2Cl^- \rightarrow Cl_2 + 2e^-$, oxidation; $2LiCl\,(l) \rightarrow 2Li\,(l) + Cl_2\,(g)$, redox **6 a** Li^+, Cl^-, H^+, OH^- **b i** Oxygen is formed at the anode: $4OH^- \rightarrow O_2 + 2H_2O + 4e^-$ **ii** Hydrogen is formed at the cathode: $2H^+ + 2e^- \rightarrow H_2$
c dilute hydrochloric acid, dilute sulfuric acid, sodium hydroxide solution, or any compound of a reactive metal, in a dilute solution **d** Chlorine will form at the anode, instead of oxygen.
7 a chromium sulfate, nitrate or chloride **b i** like the diagram on page 107, but with chromium in place of silver, and iron as cathode, in a solution of the chromium compound **ii** They will travel through the wires, from the chromium anode to the iron cathode. **c** The anode dissolves: $Cr \rightarrow Cr^{3+} + 3e^-$. At the cathode: $Cr^{3+} + 3e^- \rightarrow Cr$ **d** at the anode **e** Because for each chromium ion deposited at the cathode, another forms at the anode.
8 a The anode dissolves: $Ni \rightarrow Ni^{2+} + 2e^-$. At the cathode: $Ni^{2+} + 2e^- \rightarrow Ni$. **b** at the cathode, where the nickel ions gain electrons **c** The anode gets smaller as it dissolves. **d** Because for each nickel ion deposited at the cathode, another forms at the anode.
e electroplating metal objects, for example steel cutlery

Chapter 9

page 111 **1 a** exothermic (but needs energy to start it off) **b** exothermic **c** endothermic
2 kilojoule (kJ) **3** It is exothermic.
4 See the diagrams on pages 110 and 111.

page 113 **1** Bonds must be broken, then new bonds must form.
2 The energy that has to be put in to break bonds is greater than the energy given out when new bonds form.
3 $2H-H + O=O \rightarrow 2H-O-H$ **4** -486 kJ/mole

page 115 **1 a i** A good fuel gives out plenty of energy. So on the energy level diagram the products should be at a much lower energy level than the reactants. **ii** A very poor fuel will give out very little energy, so the energy gap should be very small. **b** Think about: pollution, reliability of supply, ease and safety of storage and use, cost. **2** Hydrogen gives out most heat per gram, and produces only water. This makes it an attractive fuel. (But it is highly flammable, so safety is an issue.)
3 $2C_4H_{10}\,(g) + 13O_2\,(g) \rightarrow 8CO_2\,(g) + 10H_2O\,(l)$

page 117 **1** No. Electrons must be given up during the reaction, to supply a current. So it must be a redox reaction. **2** the more reactive one **3** H^+ ions (from water) accept electrons at the copper strip, to form hydrogen gas. **4** No, the two strips must be of different metals. The more reactive metal will give up electrons, to form a current. **5 a** hydrogen **b** At the negative pole, hydrogen and OH− ions (from the electrolyte) react together, releasing electrons. These flow through the wires to the positive pole where they take part in a reaction between oxygen and water, to form more OH− ions. **c** The ions in it carry the current, to complete the circuit.

page 121 **1** one that can go backwards as well as forwards
2 anhydrous copper(II) sulfate + water → hydrated copper(II) sulfate
3 heat it (to expel the water of crystallisation)
4 It will turn blue (as water is driven off).
5 *Dynamic* means the forward and back reactions are both going on. But they are going on at exactly the same rate, so the quantities of reactants and products do not change. A balance or *equilibrium* has been reached. **6 a** $N_2 + 3H_2 \rightarrow 2NH_3$, and $2NH_3 \rightarrow N_2 + 3H_2$.
b No. Equilibrium is not established until the forward and back reactions are taking place at the same rate.

page 123 **1** Because it is not possible to convert *all* the nitrogen and hydrogen to ammonia. So special steps must be taken to increase the yield. **2** If a reversible reaction has reached a state of equilibrium and you change a condition, it will act to oppose the change, and establish a new equilibrium. **3 a** The system acts to oppose the change by reducing the number of molecules present – so more ammonia forms. **b** When ammonia is removed, the system acts to restore equilibrium by forming more ammonia. **4 a** $2SO_2 + O_2 \rightleftharpoons 2SO_3$. **b i** The yield increases (because an increase in pressure favours the side of the equation with fewer molecules). **ii** The yield decreases (because a rise in temperature favours the back reaction, which is endothermic).

page 124 **Core** **1 a** neutralisation **b** It gives out energy. **c** It will rise. **d** The diagram should be like the one on page 110, but with NaOH (*aq*) and HCl (*aq*) as reactants, and NaCl (*aq*) and H_2O (*l*) as products. **2 a** beaker, spatula, thermometer **b** a fall of 4 °C for ammonium nitrate, and a rise of 20 °C for calcium chloride **c i** calcium chloride **ii** because of the temperature rise **iii** The energy level of the ions in solution is lower than their energy level in the solid.
d The answers are:

	Estimated temperature of solution / °C	
solution	ammonium nitrate	calcium chloride
i	17	65
ii	23	35
iii	21	45

3 a to condense water vapour **b** The blue solid turns white. **c** It can go backwards as well as forwards. **d** Add water to the white solid left in the test-tube after heating. **e** $CuSO_4.5H_2O$ (*s*) $\rightleftharpoons$ $CuSO_4$ (*s*) + $5H_2O$ (l)

page 125 **Extended** **4 a** The energy given out when new bonds form is *greater than* the energy needed to break bonds.
b i

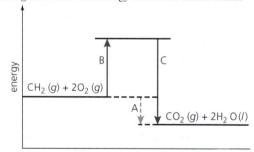

ii Some energy must be put in, to start the bonds breaking. A spark or flame can provide this energy. **c i** Energy is given out, overall. **ii** exothermic **d** 55.6 kJ
5 a to remove impurities (for example the magnesium is likely to be coated with a fine layer of magnesium oxide) **b** sulfuric acid **c** because electrons were flowing through it **d** magnesium **e** Chemical energy is being changed into electrical energy. **f** a cell **g** The acid could spill out – and it is corrosive. The set-up would be difficult to carry around.
6 a Bonds broken: 1 N−N, 4 N−H, 1 O=O **b** New bonds formed: 1 N≡N, 4 O−H. **c i** 2220 kJ/mol **ii** 2801 kJ/mol **d** − 581 kJ/mol **e** exothermic **f** from the chemicals to the surroundings. **g** Since energy is released during combustion, it could be used a fuel. (It is used in some rocket fuels.) But it is corrosive, unstable, and highly toxic, which means it is dangerous.
7 a i and **iii** **b** The number of molecules is the same on both sides of the equation, so equilibrium will not shift in response to a change in pressure. **c** Increasing the pressure will speed up both the forward and back reactions, so equilibrium is reached faster. (The molecules are pushed closer together so more collisions occur.)
8 a N_2 (*g*) + $3H_2$ (*g*) $\rightleftharpoons$ $2NH_3$ (*g*) **b** exothermic; the energy change has a negative sign **c i** In this reaction, 4 molecules of reactants give 2 molecules of product. When pressure is increased, molecules are pushed into a smaller space. So the system acts to oppose this change by forming fewer molecules (that is, more product). **ii** Raising the temperature favours the endothermic reaction – in this case the back reaction. So the yield falls. **d i** The rate at which ammonia is made increases. (Increasing the pressure increase the rate of both reactions. This is because molecules are pushed closer together, so more collisions occur.) **ii** The rate at which ammonia is made increases. (Increasing the temperature increases the rate of both reactions, because collisions have more energy, so more are successful.) **e** to give an acceptable rate
9 a The yellow solution will turn orange. **b** Add alkali, to remove the H^+ ions in a neutralisation reaction. The colour change will reverse.

Chapter 10

page 127 **1** c a d b e **2** 60 km/hour **3** Of these, only **b** is suitable (but it is much easier to measure the volume of hydrogen formed). **a** is unsuitable because zinc is a solid. **c** is unsuitable because the zinc is in irregular pieces, with reaction occurring all over the surface. **4 a** iron + sulfuric acid → iron(II) sulfate + hydrogen **b** Measure the rate at which iron or sulfuric acid is used up, or iron(II) sulfate or hydrogen is formed.

page 129 **1 a** to remove the layer of magnesium oxide or any other impurity on its surface (which would lead to false results) **b** because the reaction starts the moment the reactants are mixed **c** to prevent hydrogen from escaping into the air **2** It is over at the point where the curve goes flat. **3 a i** 29 cm³ **ii** 39 cm³ **b** 1.5 minutes **c i** 5 cm³ of hydrogen per minute **ii** 0 cm³ of hydrogen per minute (reaction over)

page 131 **1 a i** 60 cm³ **ii** 60 cm³ **b** Faster reaction (B) gives steeper curve. **2** because the same amount of magnesium was used each time (with the acid in excess) **3** A reaction goes *faster* when the concentration of a *reactant* is increased. It also goes *faster* when the *temperature* is raised. **4** For example turning up the heat when something is cooking too slowly; using gloss paint in warm weather, when the hardening reactions take place faster; keeping unripe fruit in a warm place to help it ripen. **5** The rate slows. Examples of how we make use of this: storing food and medical vaccines in fridges, to stop them going off; deep-freezing some foods (such as pizza); transporting food in refrigerated trucks; storing tissue in liquid nitrogen (very cold)

page 133 **1** For experiment 1 (large chips): **a** 0.55 g **b** 0.33 g / minute. For experiment 2 (small chips): **a** 0.95 g **b** 0.5 g /minute. **2 a** 1 g of small marble chips **b** the small chips, since the reaction will be faster **3** It is very fine, so has a very large surface area. It is also flammable. So it will react explosively fast with the oxygen in the air, once energy is provided to start the combustion reaction. (For example, provided by a spark from machinery, or a lit match.)

page 135 **1** Two particles can react together only if they *collide* and the *collision* has enough *energy* to be *successful*. **2 a** one which results in reaction **b** one where no reaction takes place **3 a** There are twice as many acid particles, so there will be many more collisions between acid particles and magnesium atoms. So the number of *successful* collisions will increase too. **b** The particles take in heat energy and move faster – so they collide more often, *and* the collisions have more energy. So a greater number of collisions are successful. **c** Stirring helps to remove bubbles of hydrogen from the metal surface, *and* brings more acid particles to it – so the number of collisions increases. **d** Many more magnesium atoms are exposed when the metal is powdered, so the number of collisions greatly increases – which means the number of successful collisions increases too.

page 137 **1** a substance that speeds up a reaction while remaining chemically unchanged itself **2** It does not change **b** or **c**. **3 a** a protein made in living cells, which acts as a catalyst for biological reactions; for example catalase, or amylase **4** Without enzymes, most reactions taking place in our bodies would be far too slow at body temperature. **5** The product forms faster, so more product can be produced in a given time. Also a catalyst may allow a reaction to run at an acceptable rate at a lower temperature, and this cuts fuel costs. **6** iron in making ammonia, vanadium(V) oxide in making sulfuric acid **7** The enzymes in it will probably become denatured above 60 °C: they lose their shape and stop working.

page 141 **1** one that depends on energy from light; examples are photosynthesis, and the decomposition of silver halides in photographic film
2 a $6CO_2$ (g) + $6H_2O$ (l) $\xrightarrow[\text{chlorophyll}]{\text{light}}$ $C_6H_{12}O_6$ (s) + $6O_2$ (g)
b It acts as a catalyst. **c** It provides more energy.
3 a because light brings about its decomposition to bromine and fine particles of silver **b** Electrons are lost by bromide ions (oxidation) and gained by silver ions (reduction). **4** When the camera shutter opens, the light that strikes the film varies in intensity, depending on the scene. So the amount of silver bromide that decomposes varies too. The brighter the light, the more silver particles form. The varying density of the silver particles is what creates the image.

page 142 **Core** **1 a** It holds the magnesium ribbon until you are ready to start the reaction. **ii** to measure gas volume **b** Shake the flask to get the magnesium into the acid.
2 a more than is needed to react with that amount of magnesium **c i** 14 cm³ of hdrogen in the first minute **ii** 9 cm³ of hdrogen in the second minute **iii** 8 cm³ of hdrogen in the third minute **d** As the reaction proceeds, magnesium is used up, so there is less for the acid to react with. **e** 40 cm³ **f** 5 minutes **g** 8 cm³ of hydrogen per minute **h** lower the temperature
3 a Catalase acts as a catalyst. **b** The concentration of the acid is increased, so reaction rate increases too. **c** The powder has a much larger total surface area than the ribbon, so burns much faster.

page 143 **Core** **4 a i** a gas syringe **ii** a stopclock **b** A – steeper curve on the graph **c** Here curves X and Y have been added to the graph:

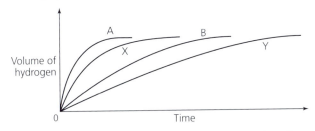

5 a substance that speeds up a chemical reaction, without being used up in the process **b** See the diagram on page 128. **c** oxygen **d** $2H_2O_2$ (aq) → $2H_2O$ (l) + O_2 (g) **e** Your graph should follow the usual shape, steepest at the start and flattening when the reaction is over. **f** The rate decreases as the reaction proceeds. **g** It decreases. **h** water and copper(II) oxide **i** There will still be 0.5 g. **j** The sketch should show a steeper curve, but still ending at 58 cm³ of oxygen. **k** manganese(IV) oxide, or the enzyme catalase
Extended **6 a** carbon dioxide **b** Measure the loss of mass of the flask (i.e. the mass of gas lost) over time, or collect the gas and measure its volume over time **c i** It will increase the rate. **ii** It will reduce the rate. **d i** As the temperature rises the particles gain energy, move faster, and collide more often with more energy.

So more collisions are successful. So the rate rises. **ii** Adding water reduces the concentration of the acid particles. So there is less chance of acid particles colliding with calcium carbonate particles. So there will be fewer successful collisions, so the rate decreases. **e** The rate will increase: more particles are exposed so there will be more collisions. **7 a** It will reduce. **b** zinc: all the iodine has been used up **c** As the reaction proceeds the number of iodine particles falls, so there is less chance of successful collisions between iodine and zinc.
d

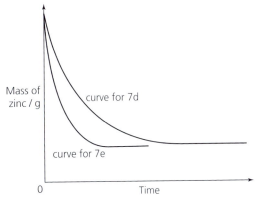

e The curve showing mass loss is steeper, as shown above – but note that the final mass is the same for both temperatures.
f At the higher temperature the particles have more energy and move faster. They collide more often and more of the collisions have enough energy for reaction to occur.
8 a i oxygen **ii** glucose **b** It is a photochemical reaction: it needs the energy from light. **c** Use a measuring cylinder in place of the test-tube, and record the volume of oxygen collected over time. (Or use a test-tube with a side arm, leading to a gas syringe.) **d** The intensity of the light striking the pondweed would increase. Since the light provides energy for it, the rate of the photosynthesis reaction would also increase.

Chapter 11

page 145 **1** reacts with substances it comes into contact with (including flesh) **2** use blue litmus paper (or pH paper/a pH meter) **3** H_2SO_4, HNO_3, $Ca(OH)_2$, $NH_3(aq)$
4 It indicates, by its colour, whether a solution is acidic or alkaline.
5 a weakly alkaline **b** weakly acidic **c** neutral **d** strongly acidic **e** strongly alkaline **f** more acidic than **b** but not strongly acidic
6 green – sugar is a neutral substance

page 147 **1** HCl (aq) → $H^+(aq)$ + Cl^- (aq) **2** Their solutions contain hydrogen ions. **3 a** The solution contains fewer ions. **b** The solution contains fewer hydrogen ions. (It is a weaker acid.) **4** They contain hydroxide ions. **5** NH_3 (aq) → NH_4^+ (aq) + OH^- (aq) **6 a** The solution contains fewer ions. **b** The solution contains fewer hydroxide ions.

page 149 **1 a** zinc + sulfuric acid → zinc sulfate + hydrogen **b** sodium carbonate + sulfuric acid → sodium sulfate + water + carbon dioxide **2 a** **3** calcium chloride, Ca^{2+}, Cl^- **4** Use this reaction: zinc oxide + nitric acid → zinc nitrate + water **5 a** Both reactions produce calcium chloride and water. **b** Only the reaction with the carbonate produces carbon dioxide. **6 a** It is a base, so it will neutralise the acid. **b** water (since neutralisation takes place) **7** Zinc loses electrons: $Zn(s)$ → Zn^{2+} (aq) + $2e^-$ (oxidation). Hydrogen ions gain electrons: $2H^+$ (aq) + $2e^-$ → H_2 (g) (reduction).

page 151 **1 a** It is an equation showing only the ions involved in the reaction. **b** H^+ (aq) + OH^- (aq) → H_2O (l) **2** ions that are present in the solution but not involved in the reaction **3** The hydrogen atom contains 1 proton and 1 electron (no neutron). When it becomes an ion it loses the electron, leaving just a proton.

4 a They transfer a hydrogen ion to a base.
b They accept a hydrogen ion from an acid.
5 During neutralisation a proton is transferred, but no electrons.
6 a i H^+ (aq), Cl^- (aq), Mg^{2+} (aq), O^{2-} (aq) → Mg^{2+} (aq), Cl^- (aq)
ii The spectator ions are Mg^{2+} (aq) and Cl^- (aq).
iii $2H^+$ (aq) + O^{2-} (s) → H_2O (l)
7 $2H^+$ (aq) + CO_3^{2-} (aq) → H_2O (l) + CO_2 (g)

page 153 **1** Show that it reacts with an acid to form a salt. **2** basic, acidic **3** phosphorus, sulfur, carbon (most reactive first) **4** It will turn red. **5** It will react with both acids and bases; zinc and aluminium oxides **6** Test the solution using Universal indicator paper. The paper will go green. Or use a pH meter, which will show a pH of 7.

page 155 **1** zinc / zinc oxide / zinc carbonate and hydrochloric acid **2 a** reaction too slow, since lead is not very reactive **b** Use lead oxide / carbonate and nitric acid. **3** There was more zinc than was needed to react with the acid (so some will remain unreacted). **4** to find the exact volumes of acid and alkali that react together in a neutralisation reaction **5** These give more accurate volume measurements. **6** ammonia solution and nitric acid

page 157 **1** The reaction produces an insoluble chemical. **2** any four of the soluble salts shown in the table on page 156 **3** Choose suitable soluble salts using the table on page 156. For example:
a calcium nitrate + sodium sulfate **b** magnesium sulfate + potassium carbonate **c** zinc chloride + sodium carbonate
d lead nitrate + potassium chloride
4 For the examples given above, for question **3**:
a $Ca(NO_3)_2$ (aq) + Na_2SO_4 (aq) → $CaSO_4$ (s) + $2NaNO_3$ (aq)
b $MgSO_4$ (aq) + K_2CO_3 (aq) → $MgCO_3$ (s) + K_2SO_4 (aq)
c $ZnCl_2$ (aq) + Na_2CO_3 (aq) → $ZnCO_3$ (s) + $2NaCl$ (aq)
d $Pb(NO_3)_2$ (aq) + $2KCl$ (aq) → $PbCl_2$ (s) + $2KNO_3$ (aq)
5 a It is an ion present in the solution, but not involved in the reaction.
b For the examples given above, for question **3**: **a**, NO_3^- (aq), Na^+ (aq)
b, SO_4^{2-} (aq), K^+ (aq) **c**, Cl^- (aq), Na^+ (aq) **d**, NO_3^- (aq), K^+ (aq)
6 For the examples given above, for question **3**:
a Ca^{2+} (aq) + SO_4^{2-} (aq) → $CaSO_4$ (s)
b Mg^{2+} (aq) + CO_3^{2-} (aq) → $MgCO_3$ (s)
c Zn^{2+} (aq) + CO_3^{2-} (aq) → $ZnCO_3$ (s)
d Pb^{2+} (aq) + $2Cl^-$ (aq) → $PbCl_2$ (s)

7 The film has to be coated with very fine particles of silver bromide, which is insoluble. So it is deposited by precipitation.

page 159 **1** one of known concentration **2** 50 cm³ **3** 1.6 mol/dm³

page 160 **Core** **1** The correct choices are, in order: bases, sulfates, metals, hydrogen, carbonates, carbon dioxide, good, red, colourless, pH number, lower. **2 a** B **b** carbon dioxide **c** A and B react together to form a soluble salt. **3 a i** copper(II) oxide **ii** calcium oxide **iii** phosphorus pentoxide **b** The paper turns blue **c** ammonia **d i** copper(II) oxide + sulfuric acid → copper(II) sulfate + water **ii** neutralisation **e** sulfur dioxide, sulfur trioxide, nitrogen dioxide

page 161 **Core** **4 i** The missing chemicals are, in order: **b** zinc chloride, hydrogen **c** sulfuric acid **d** hydrochloric acid, sodium carbonate, carbon dioxide **e** sulfuric acid, iron, hydrogen **f** alkali, sodium nitrate, water **g** sulfuric acid, water **h** carbonate, sulfuric acid, copper(II) carbonate, water
ii a $Ca(OH)_2$ (aq) + $2HNO_3$ (aq) → $Ca(NO_3)_2$ (aq) + $2H_2O$ (l)
b Zn (s) + $2HCl$ (aq) → $ZnCl_2$ (aq) + H_2 (g)
c $2KOH$ (aq) + H_2SO_4 (aq) → K_2SO_4 (aq) + $2H_2O$ (l)
d Na_2CO_3 (s) + $2HCl$ (aq) → $2NaCl$ (aq) + H_2O (l) + CO_2 (g)
e Fe (s) + H_2SO_4 (aq) → $FeSO_4$ (aq) + H_2 (g)

f $NaOH$ (aq) + HNO_3 (aq) → $NaNO_3$ (aq) + H_2O (l)
g CuO (s) + H_2SO_4 (aq) → $CuSO_4$ (aq) + H_2O (l)
h $CuCO_3$ (s) + H_2SO_4 (aq) → $CuSO_4$ (aq) + H_2O (l) + CO_2 (g)
5 a carbon dioxide **b i** copper(II) carbonate + ethanoic acid → copper(II) ethanoate + water **ii** No more gas bubbles off, and solid remains in the beaker. **c i** copper(II) carbonate; after reaction, some remains in the beaker **ii** ethanoic acid **d** to make sure the reaction is quick and complete **e** copper(II) carbonate
f • put some ethanoic acid into a beaker
• add a spatula measure of copper(II) carbonate, and stir
• continue to add copper(II) carbonate until no further reaction
• filter the mixture into an evaporating basin, to remove unreacted copper(II) carbonate
• put the evaporating basin on a tripod and gauze, and evaporate some of the water (to obtain a saturated solution)
• leave to cool and crystallise
• remove the crystals by filtering or decanting
• rinse and dry the crystals
g copper(II) oxide or copper(II) hydroxide
Extended **6 a** sulfuric acid **b** MgO (s) + H_2SO_4 (aq) → $MgSO_4$ (aq) + H_2O (l) **c i** strong **ii** the hydrogen ion, H^+
d i a chemical that neutralises an acid to form a salt and water
ii O^{2-} + $2H^+$ → H_2O **7 i** e.g. sodium chloride (or any other soluble chloride) and silver nitrate **ii** precipitation **b i** Ag^+ (aq) + Cl^- (aq) → $AgCl$ (s) **ii** sodium ion (or ion of other chosen metal), nitrate ion **8 a** contains water of crystallisation
b Na_2CO_3 (aq) + $2HCl$ (aq) → $2NaCl$ (aq) + H_2O (l) + CO_2 (g)
c 0.014 moles **d** 0.007 moles **e** 0.742 g **f** 1.258 g
g 0.07 moles **h** 10 moles **i** $Na_2CO_3.10H_2O$

Chapter 12

page 163 **1 a** e.g. iron, aluminium, copper **b** oxygen, nitrogen (or a noble gas other than helium) **2 a** non-metal, atom contains 7 protons, 7 electrons, 7 neutrons, two electron shells so electron distribution is 2.5 **b** metal, atom contains 12 protons, 12 electrons, 12 neutrons, three electron shells so electron distribution is 2.8.2
3 VII and 0 **4 a** lithium, potassium, rubidium, caesium
b chlorine, bromine, iodine **5** oxygen; krypton has a stable outer shell of electrons already so is unreactive **6 a** europium
b mendelevium **c** americium **7** It gives order to the chemistry of the elements, and allows predictions to be made about properties.

page 165 **1 a** alkali metals **b** Their oxides and hydroxides form alkaline solutions. **2 a** soft **b** reactive **3 a** Melting points change gradually down or up the group. **b** There are trends in boiling point, density, softness. **c** density of potassium
4 a potassium chloride **b** white **c** A colourless solution forms. **d** Yes; it contains potassium and chloride ions, which are free to move. **5** lithium; it is the less reactive of the two metals
6 a a very vigorous reaction; the products are hydrogen, which bursts into flame, and an alkaline solution of rubidium hydroxide **b** burns very brightly; the product is an ionic white solid, rubidium chloride

page 167 **1** Fluorine is a yellow gas, chlorine a green gas, bromine a red liquid, and iodine a black solid. **2 a** Reactivity decreases down the group. **b** No, it is the opposite. **3 a** All are coloured solids that contain the Fe^{3+} ion. **b** ionic **4** Their atoms need just 1 more electron for a stable outer shell. They have a strong drive to obtain it.
5 a bromine + potassium iodide → potassium bromide + iodine; the solution will turn red-brown **b** Bromine has a stronger drive to obtain electrons than iodine does, so it displaces iodide ions from the solution. **6 a** a solid, to fit with the trend down the group
b coloured **c** harmful

page 169 **1** because their atoms all have the same (stable)

arrangement of outer-shell electrons   **2** They already have a stable
outer shell of electrons, so don't need to react to obtain one.
3 a Both increase down the group.   **b i** yes   **ii** yes   **4** Their
unreactivity makes them safe to use. See uses on page 169.
5 a a gas   **b** heavier   **c** No.

page 171   **1** iron, copper, nickel etc.   **2 a** hard   **b** high
density   **c** high melting point   **d** unreactive with water   **3** It is a
liquid at room temperature.   **4** They are used for their colour.
5 the transition elements (which have variable oxidation states)
6 Iron rusts, but copper does not corrode – and copper is a better
conductor of heat.

page 173   **1 a** It increases by 1 each time.   **b** The elements
change from metal to non-metal, with a metalloid in the middle;
and their oxides change from basic to acidic, with an amphoteric
one in between.   **2** It decreases. As you go along the period, the
metal atoms need to lose more electrons to form stable outer shells,
and this gets more difficult (takes more energy).
3 the number of electrons an atom loses, gains, or shares, to form a
compound; see the table on page 172 for examples
4 an element with both metal and non-metal properties
5 a metalloid that can conduct electricity (its electrical conductivity
lies between that of a metal and a non-metal); silicon or germanium
6 a Here is the table:

I	II	III	IV	V	VI	VII	0
lithium	beryllium	boron	carbon	nitrogen	oxygen	fluorine	neon
1	2	3	4	5	6	7	8
LiCl	$BeCl_2$	BCl_3	CCl_4	NCl_3	OCl_2	ClF	–
1	2	3	4	3	2	1	0

(The last three compounds in the table are called nitrogen
trichloride, oxygen dichloride, and chlorine fluoride.)
b As guidance, the actual boiling points in °C are: Li 181, Be 1287,
B 2076, C sublimes at 3642 °C, N –210, O –219, F –220, Ne –248

page 176   **Core**   **1 a** Period 2   **b** Neon (Ne); its atoms already have
a stable outer electron shell so do not need to react to obtain one.
c Group I (the alkali metals)   **d** K
2 a Melting point and boiling points decrease, reactivity increases.
b The actual values for rubidium are: melting point 39 °C, boiling
point 688 °C. It is extremely reactive.   **c i** 5   **ii** 37   **iii** 1
3 a helium   **b** chlorine   **c** neon   **d** bromine   **e** fluorine   **f** iodine

page 177   **Core**   **4 a** the halogens   **b i** even more
explosively   **ii** hydrogen + fluorine → hydrogen fluoride
c i less explosively than with chlorine
ii hydrogen + bromine → hydrogen bromide
5 a the second row of the Periodic Table   **b i** The modern Group I
does not contain H, Cu, or Ag.   **ii** All the elements in it are reactive,
with similar properties – but copper and silver are unreactive.
iii alkali metals   **c i** halogens   **ii** It is a metal.   **iii** It is one of the
transition elements.   **d** to allow for elements not yet discovered at
that time   **e** The noble gases were not discovered until later –
largely because they are unreactive.
Extended 6 a A, alkali metal; B, halogen; C, noble gas; D, transition
element; E, Group II (alkaline earth metal); F, halogen; G, transition
element   **b i** A: towards the bottom of the group; B: top of the
group; C: towards the bottom of the group   **ii** A: contains 1 electron;
B: contains 7 electrons; C: contains 8 electrons   **c** Atoms react in
order to gain a stable outer shell of electrons. The easier it is to gain
or lose electrons to achieve this, the more reactive an element will
be. But if its atoms already have a stable outer shell, an element will
be unreactive.   **d** A: rubidium or caesium; B: fluorine; C: argon,
krypton, xenon, or radon; D: any transition element, e.g. iron; E:
magnesium; F: bromine; G, iron   **e** D and G

7 a four of helium, neon, argon, krypton, xenon, radon
b i consists of single atoms   **ii** Their atoms already have a stable
outer shell, so do not need to share electrons.   **c i** Their atoms
gain, lose, or share electrons, and in this way obtain stable outer
electron shells, like the noble gases.   **ii** unreactive

Chapter 13

page 179   **1 a** mercury, or any alkali metal (e.g.
sodium)   **b** mercury   **2 a** tin   **b** 2.64 times more dense
3 a unreactive, malleable , ductile, shiny   **b** ductile and a very
good conductor of electricity   **4** bells, musical instruments (e.g.
steel drums)   **5 a** It remains liquid across a suitable temperature
range (from well below to well above room temperature), expands a
lot when heated, and flows easily without sticking to the
glass.   **b** does not corrode, light, malleable

page 181   **1** $2K\,(s) + 2H_2O\,(l) → 2KOH\,(aq) + H_2\,(g)$
2 a potassium – it shows a more vigorous reaction with water
b zinc – it reacts with hydrochloric acid, but copper does
not   **3** hydrogen   **4** Iron loses electrons (oxidation), and hydrogen
ions gain electrons (reduction).

page 183   **1** carbon   **2 a i** no reaction   **ii** The carbon will reduce
the zinc oxide.   **b** zinc oxide + carbon → zinc + carbon dioxide
3 aluminium   **4** Iron takes copper's place, in the compound.
5 Copper is the more reactive metal, so it displaces silver. It forms
blue copper ions, while the silver is deposited on the wire.
6 a $Cu\,(s) → Cu^{2+}\,(aq) + 2e^-$, $Ag^+\,(aq) + e^- → Ag\,(s)$
b $Cu\,(s) + 2Ag^+\,(aq) → Cu^{2+}\,(aq) + 2\,Ag\,(s)$

page 185   **1 a** See the list on page 184.   **b** K, Na: Group I; Ca,
Mg: Group II; Al, Group III; all the rest are transition elements
c Group I   **d** in the block of transition elements   **2** It is too
reactive to remain uncombined.   **3** Gold is found native, while
aluminium had to wait for the invention of electrolysis, for
extraction and widespread use.
4 silver nitrate, since it is a compound of a less reactive metal
5 $2Pb(NO_3)_2\,(s) → 2PbO\,(s) + 4NO_2\,(g) + O_2\,(g)$

page 187   **1 a** Iron is more reactive than copper, so when the two
metals are connected it gives up electrons, which flow to the copper
as a current.   **b** copper   **c** Use copper, or a less reactive metal than
copper (such as silver) with one that is more reactive than iron
(such as zinc or magnesium).   **2** 1.6 V   **3 a** coated with
zinc   **b** The coating keeps air and moisture away. And if it is
damaged, the zinc continues to protect by sacrificial
protection.   **4** The aluminium quickly forms a fine coat of
aluminium oxide, which prevents further corrosion.
5 a aluminium + iron(III) oxide → aluminium oxide + iron
b to allow it to mix well with the iron(III) oxide to ensure complete
reaction, and also to give a large surface area, to speed up the
reaction

page 188   **Core**   **1 a** sodium, calcium, magnesium, aluminium,
iron, copper, lead, gold   **b i** how heavy a substance is (or mass
per cm^3)   **ii** iron   **c** sodium, calcium, magnesium, aluminium,
iron, lead, copper, gold   **d i** 0.97 g/cm^3 **ii** 19.3 g/cm^3
iii Yes, in general the more reactive the metal, the lower the density.
e less mass to move along, and lower force on road surfaces   **f**
aluminium: unreactive, malleable, abundant so not too expensive
2 a sodium   **b** sodium, calcium and magnesium
c

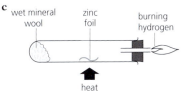

wet mineral wool   zinc foil   burning hydrogen

heat

d i hydrogen **ii** dilute hydrochloric acid, or sulfuric acid
page 189 Core 3 a zinc (or iron, magnesium);
zinc + copper(II) sulfate → zinc sulfate + copper
b iron (or zinc); iron + hydrochloric acid →
iron(II) chloride + hydrogen **c** sodium (or potassium, lithium);
sodium + water → sodium hydroxide + hydrogen
d magnesium; magnesium + steam → magnesium oxide + hydrogen
4 a ii, iii **b** carbon + <u>copper(II) oxide</u> → copper + carbon dioxide;
magnesium + <u>carbon dioxide</u> → magnesium oxide + carbon
Extended 5 a magnesium + copper(II) sulfate → magnesium
sulfate + copper **b** because magnesium is more reactive that copper
c i Mg → Mg^{2+} + 2e$^-$ **ii** oxidation **d i** Cu^{2+} + 2e$^-$ → Cu
ii reduction **iii** magnesium **e i** Mg + Cu^{2+} → Mg^{2+} + Cu
ii redox **f i** yes **ii** no **iii** yes **g i** For iron + copper(II) sulfate:
brown copper appears, solution goes from blue to pale green.
For zinc + lead(II) nitrate: zinc disappears and silvery lead forms,
but the solution remains colourless.
ii Fe + Cu^{2+} → Fe^{2+} +Cu; Zn + Pb^{2+} → Zn^{2+} + Pb
6 a <u>reduction</u> and <u>oxidation</u> **b i** magnesium **ii** copper(II) oxide
c 2 electrons are transferred from each magnesium atom to each
copper ion **d** Magnesium is more reactive than copper so has a
stronger drive to give up electrons. **e i** calcium, sodium,
potassium **ii** loses them **7 a i** the breaking down of a compound
on heating **ii** Cu(OH)$_2$ (s) → CuO (s) + H$_2$O (l) **iii** blue to black
b sodium or potassium hydroxide **c i** oxygen **ii** copper(II) nitrate
iii 2Cu(NO$_3$)$_2$ (s) → 2CuO (s) + 4NO$_2$ (g) + O$_2$ (g) **d** The nitrate of
sodium, the more reactive metal, breaks down only to its nitrite, but
the nitrate of the less reactive metal breaks down to its oxide.
8 a The three unknown metals all form the negative pole.
b A, C, B **c i** 2.45 volts **ii** 0.65 volts **d i** A **ii** C
9 a electrolysis **b** In cells, chemical energy produces a current;
in electrolysis, a current brings about chemical change.

Chapter 14

page 191 1 oxygen **2** aluminium, iron **3** It is found
uncombined with other elements, because it is unreactive.
4 No. **5** They are scarce, expensive, and often kept as a form of
wealth; gold, silver, platinum, palladium **6** Iron; it is abundant and
not too expensive, and it can be turned into steel which is hard and
strong. **7** the rock from which the metal is extracted **8 a** rock
salt; sodium chloride **b** bauxite; aluminium oxide

page 193 1 It occurs naturally as the element. **2 a** because
carbon is more reactive than lead **b** lead oxide **c** carbon
3 because one substance is reduced, and the other is oxidised **4**
a Sodium is very reactive, so its compounds are very stable – so a
powerful method of extraction is needed. **b** 2NaCl (l) → 2Na (l) +
Cl$_2$ (g) **5** zinc sulfide **6** Zinc sulfide is roasted to give zinc oxide,
which is dissolved in dilute sulfuric acid to give zinc sulfate solution.
Electrolysis of this gives zinc at the cathode.

page 195 1 Fe$_2$O$_3$ (s) + 3CO (g) → 2Fe (l) + 3CO$_2$ (g)
2 hot air **3** nitrogen and carbon dioxide
4 It breaks down to the base calcium oxide, which removes acidic
impurities: CaO (s) +SiO$_2$ (s) → CaSiO$_3$ (s)
5 Slag is used for building materials; the gases heat the blast of air.

page 197 1 aluminium oxide, which decomposes to give
aluminium; cryolite, to dissolve the aluminium oxide
2 a See the diagram on page 197.
b because they are positive ions, and the cathode is negative
c Aluminium forms: Al^{3+}(l) + 3e$^-$ → Al (l) (the equation may be
multiplied × 4) **d** The carbon anode reacts with the released
oxygen, giving carbon dioxide. **3 a** can be beaten into
shape **b** can be drawn into wires **c** not harmful to
health **d** light for its volume **e** not affected by substances in the

environment (e.g. oxygen, water) **4** See page 196.

page 199 1 Answers will vary. **2 a** an alloy **b** They are likely to
be different from the properties of copper, or tin. **c** church bells,
statues

page 201 1 iron **2 a** to remove carbon and other non-metal
impurities as oxides **b** carbon + oxygen → carbon dioxide, carbon
+ oxygen → carbon monoxide **3** It is a base, so will remove the
acidic oxide impurities.
4 a

carbon atom
iron atoms

pure iron contains
only iron atoms in
the metal lattice

mild steel has
carbon atoms within
the metal lattice

b In pure iron, the layers of atoms can slide over each other. But in
the alloy, the new atoms in the lattice prevent the layers from
sliding, and this makes the steel stronger. **5** Atoms of other
elements are mixed all through the iron atoms, like a solute
dissolved in water. **6** iron, nickel, chromium

page 204 Core 1 a i iron(III) oxide, aluminium oxide, sodium
chloride **ii** hematite, bauxite, rock salt **b** sodium, aluminium, iron
c i by electrolysis **ii** The metal ions gain electrons to become
metal atoms. **d i** by heating with carbon (carbon monoxide) **ii** The
metal oxide loses oxygen. **iii** Those metals are more reactive than
carbon. **e i** electrolysis **ii** heating with carbon **iii** electrolysis
f not combined with any other element **g** last **h** copper,
platinum **2 a** See page 194. **b i** hematite, coke (coal), limestone
ii Hematite is the iron ore; coke is used to produce carbon
monoxide, the gas which brings about the reduction; limestone is
used to remove acidic impurities such as sand. **c i** slag **ii** used
for road building **d i** nitrogen or carbon dioxide **ii** used to heat up
the air blast **e i** Fe$_2$O$_3$ (s) + 3CO (g) → 2Fe (l) + 3CO$_2$ (g) **ii** The
iron(III) oxide loses oxygen. **iii** carbon monoxide

page 205 Extended 3 a rock from which a metal is
extracted **b** contains a low percentage of the metal **c i** 990
kg **ii** about 25% **d** Copper has many important uses, but it is
not abundant in the Earth's crust. If demand for it grows, the price
will rise, making a low-grade ore economical to mine.
e i reduction **ii** carbon or carbon monoxide
f i electrolysis **ii** electricity **iii** The impure copper may contain
small amounts of scarce / precious metals. They will fall to the
bottom of the electrolysis tank, below the anode. (See page 106.)
g landscape dug up, affecting local people and wildlife; a huge
amount of waste rock to be disposed of; harmful substances released
at different stages; pollution from traffic related to mining, and from
burning fossil fuels for electricity **4 a** zinc sulfide, lead
sulfide **b** ZnS, PbS
c i 2ZnS (s) + 3O$_2$ (g) → 2ZnO (s) + 2SO$_2$ (g)
2PbS (s) + 3O$_2$ (g) → 2PbO (s) + 2SO$_2$ (g)
ii oxidation **iii** It is a harmful acidic oxide that causes breathing
difficulties, and dissolves in rain to give acid rain.
d i ZnO (s) + C (s) → Zn (s) + CO (g); PbO (s) + C (s) → Pb (s) + CO (g)
ii zinc oxide, lead oxide **5 a** reduced **b i** to remove acidic
impurities **ii** to form the electrodes **iii** to dissolve the aluminium
oxide (alumina) **c** Both involve the reduction of the oxide. In each
case molten metal forms, and is drawn off. **d** road building
e Impurities must be removed from aluminium ore (bauxite) before
electrolysis, and that costs money. The electrolysis requires
electricity, which is expensive. By contrast impurities can be
removed from hematite in the blast furnace. The fuel for the
extraction is coal, which is cheap. **f i** Steels have more useful

properties than the pig iron from the blast furnace.   **ii** First, carbon and other impurities are removed. Then controlled amounts of other metals are added, to form alloys with the required properties.   **g** Metals are a finite resource. Recycling helps to ensure they last longer. It is cheaper to recycle them than to mine the ores and extract the metals. It also has less impact on the environment in terms of scarred landscapes, and waste material.

Chapter 15

page 207   **1 a** 78%   **b** 21%   **c** 99%   **2** about 3.7 times   **3** 1%   **4** The atmosphere thins out as altitude increases; at high altitudes there is not enough oxygen, so breathing becomes difficult.   **5** oxygen; a great many things burn in it   **6 a** copper(II) oxide, CuO   **b** Pass hydrogen over the hot oxide, or heat it with carbon or a more reactive metal such as zinc.

page 209   **1 a** Expansion causes its temperature to fall, so eventually it cools to a liquid.   **b** Argon has a lower boiling point.   **c** Compressing the gas needs equipment that can withstand high pressures, and this is expensive. (Electricity is needed to power the pumps for compressing, and it is also expensive.)   **2** See page 209.   **3** Air is only 21% oxygen, so acetylene burns much faster in pure oxygen than in air, giving a hotter flame.   **4** Liquid nitrogen is very cold (colder than –196 °C). Nitrogen is inert, so it is safe to transport, and it does not react with food.   **5** See page 209.

page 211   **1 a** carbon monoxide, lead compounds   **b** nitrogen oxides, formed when nitrogen and oxygen react together in hot engines and furnaces
2 $CH_4 (g) + 2O_2 (g) \rightarrow CO_2 (g) + 2H_2O (l)$
3 $2CH_4 (g) + 3O_2 (g) \rightarrow 2CO (g) + 4H_2O (l)$ **4 a** carbon monoxide + oxygen → carbon dioxide   **b** They act as catalysts for the reactions.

page 213   **1** the corrosion of iron in damp air   **2** oxygen and water   **3** Place iron nails / turnings in two test-tubes, one containing damp oxygen, and the other damp nitrogen. Observe the extent of rusting.   **4** The coating of tin prevents oxygen from reaching the iron.   **5** You could cover the moving parts in grease, and paint the frame. Both prevent rusting by keeping out oxygen and water.   **6 a** Another metal corrodes instead of iron, to protect the iron.   **b** They are more reactive than iron, so they give up electrons more readily.   **c** It is less reactive than iron, so the iron will corrode.

page 215   **1 a** water trapped in rocks below ground   **b** underground rocks that contain a large volume of water   **c** a very tiny organism, for example a bacterium or virus   **2** to make small particles stick together, so that they can be filtered off   **3** It kills bacteria.   **4** It helps to fight tooth decay.   **5** It may contain dissolved harmful chemicals.   **6** See the blue panel at the bottom of page 215.   **7 a** Dip the blue cobalt(II) chloride paper into it; the paper will turn pink.   **b** Measure the temperature at which it boils and freezes; only pure water boils at 100 °C and freezes at 0°C.   **c** Carry out a simple distillation.

page 218   **Core**   **1** The missing terms are: mixture, nitrogen, oxygen, carbon dioxide, noble gases, 0, oxygen, respiration, acetylene   **2 a** nitrogen   **b** oxygen   **c** argon   **d i** sulfur dioxide   **ii** It causes breathing problems, and forms acid rain.   **e** See page 210.
3 a $4Fe (s) + 3O_2 (g) + 4H_2O (l) \rightarrow 2Fe_2O_3.2H_2O (s)$ (The equation without water molecules is also acceptable.)   **b** to speed up the rusting process (and not rely only on water vapour)   **c** Rusting uses up oxygen from the air in the test-tube.   **d** 20.8%   **e** the same result, since the amount of oxygen in the test-tube won't have changed

page 219   **Core**   **4 a i** 36%   **ii** It is greater.
b 78%   **c** oxygen   **5 a** filtering   **b** makes small particles stick together   **c i** insoluble impurities   **ii** soluble impurities   **d i** adding

chlorine   **ii** to kill bacteria   **e** limestone, or slaked lime   **f** fluorine
Extended   **6 a** O_2, N_2 **b** liquid   **c** by expanding cold compressed air   **d** The liquid air is warmed up. When it reaches –196 °C nitrogen boils off and is collected. When it reaches –183 °C oxygen boils off and is collected.   **f** argon, krypton, or xenon
7 a i in the exhaust pipe   **ii** to convert harmful gases to harmless gases   **iii** transition elements   **b i** by reaction between nitrogen and oxygen, in the hot air in the car engine   **ii** from the petrol
c i nitrogen   **ii** carbon dioxide   **iii and iv** carbon dioxide, water vapour   **8 a i** It is a poisonous gas.   **ii** It causes breathing problems / forms acid rain.   **b** a reaction with a catalyst present, to increase its rate   **c** $2NO (g) + 2CO (g) \rightarrow N_2 (g) + 2CO_2 (g)$
9 a i It must be more reactive than iron.   **ii** zinc or magnesium
iii $Zn \rightarrow Zn^{2+} + 2e^-$, $Mg \rightarrow Mg^{2+} + 2e^-$ **b** sacrificial protection

Chapter 16

page 221   **1** magnesium, hydrochloric acid   **2 a** Hydrogen is more reactive than copper.   **b** redox   **3** oxygen   **4** one in which a reactive element takes the place of a less reactive one, in a compound   **5** sodium hydroxide + ammonium sulfate → sodium sulfate + water + ammonia

page 223   **1 a** nitrogen by removing oxygen from the air; hydrogen by reacting methane with steam, or cracking ethane   **b** the Haber process   **c** $N_2 (g) + 3H_2 (g) \rightarrow 2NH_3 (g)$
2 a iron   **b** to give several chances for the gases to react
3 a High pressure favours the side with fewer molecules (the product); the forward reaction is exothermic, so is favoured by low temperature.   **b** It would be dangerous and expensive to work at such a high pressure, and at 250 °C the reaction would be too slow.
c approximately 28%   **d** They are recycled, to give the gases another chance to react.

page 225   **1** N stands for nitrogen, P for phosphorus, and K for potassium.   **2** Examples: ammonium nitrate, potassium nitrate, ammonium sulfate.   **3** They promote the growth of algae. When these die, bacteria feed on them, while using up the oxygen dissolved in the water. This starves fish and other river life of oxygen.
4 21.2%   **b** 1 kg of ammonium nitrate   **c** It is very soluble. Rain water will carry it off into rivers.

page 227   **1** It occurs as the element in underground beds and around volcanoes; and as compounds in metal ores and petroleum.   **2** The molecules are held in the lattice by weak forces.   **3** two different solid forms with different structures   **4** Yes: iron loses electrons to form Fe^{2+} and sulfur gains electrons to form S^{2-}.   **5 a** It dissolves in water to form an acid.   **b** It causes acid rain.   **6** Yes: it remains in the lower atmosphere, causing problems.

page 229   **1 a** the Contact process   **b** sulfur and air   **c** vanadium(V) oxide
2 a It can go both ways.   **b** To speed up the reaction, so as to make it more economic.   **c** A higher temperature favours the back reaction, so more sulfur trioxide will also break down.   **3 a** to give further chances for the gases to react   **b** Removing the product means more product will form, to restore equilibrium.   **4** the reactions in steps 1 and 3 in the diagram on page 228   **5 a** zinc + sulfuric acid → zinc sulfate + hydrogen, zinc oxide + sulfuric acid → zinc sulfate + water, zinc carbonate + sulfuric acid → zinc sulfate + water + carbon dioxide
b $Zn (s) + H_2SO_4 (aq) \rightarrow ZnSO_4 (aq) + H_2 (g)$
$ZnO (s) + H_2SO_4 (aq) \rightarrow ZnSO_4 (aq) + H_2O (l)$
$ZnCO_3 (s) + H_2SO_4 (aq) \rightarrow ZnSO_4 (aq) + H_2O (l) + CO_2 (g)$

page 231   **1** It is the continual circulation of carbon between the atmosphere, living things, the soil, and the ocean.   **2 a** Both produce carbon dioxide and water, and give out energy.   **b** for warmth, movement, and energy for other reactions going on inside us   **c** heating, cooking, generating electricity   **3** They are opposite reactions.   **4** Here is one example:

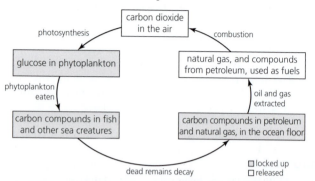

5 burning fossil fuels   **6** You convert carbon compounds from food into carbon dioxide during respiration. You may burn fossil fuels too.

page 233   1 a methane + oxygen → carbon dioxide + water
b methane + oxygen → carbon monoxide + water
2 to ensure compete combustion of the gas, and prevent carbon monoxide from forming   **3 a** $PbCO_3(s) → PbO (s) + CO_2 (g)$
b thermal decomposition   **4 a** wells of natural gas, breakdown of plant material in the absence of oxygen, waste gas from animals
b natural gas   **5 a** inorganic   **b** inorganic

page 235   1 a Without them the temperature on Earth would be too low to support life.   **b** An increase in the level of greenhouse gases means an increase in global air temperatures, causing climate change.   **2** carbon dioxide and methane   **3** Plants and animals that cannot adapt to a changing climate will face extinction.
4 a acidic   **b i** Its pH could decrease as more carbon dioxide dissolves in it.   **ii** Increasing acidity will affect ocean species, and especially where shells or outer skeletons are made of calcium carbonate (e.g. coral).

page 237   1 from the shells and skeletons of dead ocean creatures falling to the ocean floor; they are buried under pressure, and changed to rock   **2 a** by heating limestone: $CaCO_3(s) → CaO (s) + CO_2 (g)$
b to prevent the back reaction   **c** by blowing air through the kiln
3 by adding water to lime: $CaO (s) + H_2O (l) → Ca(OH)_2(aq)$
4 See page 236.   **5** testing for carbon dioxide   **6** Limestone is better: slaked lime could get washed away before the acidity of the soil is neutralised.   **7 a** the removal of sulfur dioxide from the waste gases produced in a power station, before they are released through a chimney or flue   **b** limestone, or slaked lime   **c** hydrated calcium sulfate; it is used in making cement, and plaster products

page 238   Core   1 a nitrogen, phosphorus, potassium
b nitrogen: for making chlorophyll and the proteins plants need for growth; phosphorus: root production and helping fruit to ripen; potassium: for disease resistance
c i N and P   **ii** K and N   **iii** N   **d i** $(NH_4)_3PO_4$ **ii** KNO_3
iii $(NH_4)_2SO_4$ **2 a** calcium carbonate: **i** limestone **ii** $CaCO_3$; calcium oxide: **i** lime (quicklime) **ii** CaO; calcium hydroxide (solid): **i** slaked lime **ii** $Ca(OH)_2(s)$; calcium hydroxide (solution):
i limewater **ii** $Ca(OH)_2(aq)$ **b** heat $CaCO_3 → CaO →$ add water →
$Ca(OH)_2(s) →$ add water → $Ca(OH)_2(aq) →$ react with $CO_2 → CaCO_3$.
c It has many uses – see page 236.   **3 a** calcium, carbon and oxygen
b i calcium oxide   **ii** by heating in a kiln; thermal decomposition occurs   **c i** by pollutants dissolving in the water, and acid rain falling into the lakes   **ii** It neutralises the acidity.   **iii** The powder has a greater surface area, so reaction is faster. It can be spread more easily, so less limestone is needed.   **d** See page 236.

page 239   Core   4 a i sulfur dioxide   **ii** It causes acid rain.
b The reaction would be incomplete, due to the small surface area exposed to the gases over a short period of time.   **c** Sulfur dioxide is removed from the waste gases that go out of the flue.
d i hydrated calcium sulfate   **ii** oxidation   **iii** cement-making, plaster board, plaster of Paris   **e** calcium nitrate or chloride + sodium sulfate (or any soluble sulfate, or sulfuric acid)
Extended 5 a nitrogen and hydrogen   **b** to remove impurities
c Iron is a catalyst for the reaction.   **d** They are recycled for another chance to react.   **e** high: this favours the side with fewer molecules
6 a that the reaction is exothermic   **b** High temperatures favour the reverse reaction, which is endothermic.   **c** *reversible*: reaction can go both ways; *equilibrium*: forward and back reactions are taking place at the same rate, so there is no overall change   **d i** The reaction reaches equilibrium faster.   **ii** iron   **iii** no effect
7 a vanadium(V) oxide   **b** exothermic   **c** 450 °C, pressure close to atmospheric, catalyst   **d** It will increase, because a lower temperature favours the exothermic reaction.   **e** It is dissolved in concentrated sulfuric acid to form oleum, which is then mixed carefully with water.
8 a A/B − sulfur dioxide/oxygen, C vanadium(V) oxide,
D − concentrated sulfuric acid, E − water, F − concentrated sulphuric acid   **b** to speed up the reaction   **c** $2SO_2 (g) + O_2 (g) ⇌ 2SO_3 (g)$
d It has many uses. See page 229.   **9 a** X, Z   **b** W   **c** W, Y, Z
d W copper(II) sulfate, X magnesium sulfate, Y calcium sulfate, Z sodium sulfate   **e** W: $CuO (s) + H_2SO_4 (aq) → CuSO_4 (aq) + H_2O (l)$
X: $Mg (s) + H_2SO_4 (aq) → MgSO_4 (aq) + H_2 (g)$
Y: $Ca(OH)_2 (s) + H_2SO_4 (aq) → CaSO_4 (aq) + 2H_2O (l)$
Z: $Na_2CO_3 (s) + H_2SO_4 (aq) → Na_2SO_4 (aq) + H_2O (l) + CO_2(g)$

Chapter 17

page 241   1 crude oil   **2** It was formed from the remains of sea organisms that lived millions of years ago.   **3** a compound containing only carbon and hydrogen   **4** It is a mixture of organic compounds, mainly hydrocarbons.   **5** It provides fuels for heating and transport and feedstock for the chemicals used to make plastics and many other useful products.   **6** It is being used up very fast, and forms very slowly, so in effect it could run out one day.   **7** We will use energy sources such as: wind and solar power; nuclear fuel; fuels made from plants.

page 243   1 evaporating and condensing   **2** fraction
3 a evaporates easily to form a vapour   **b** thick and sticky
4 boiling point, volatility, viscosity, flammability (how easy it will burn)   **5 a** gasoline   **b** refinery gas   **c** bitumen   **d** fuel oil

page 245   1 Long chain hydrocarbons are broken down into smaller molecules.   **2** Heat is needed to make the molecules break down.
3 heat and a catalyst   **4** a molecule with a double bond (an alkene)
5 It provides the hydrocarbons that industry requires, including reactive alkenes that can be used to make plastics.
6 a See the drawing of pentane on page 245.   **b** Cracking will produce two molecules, with a total of 5 C atoms and 12 H atoms. One of the molecules will have a double bond.

page 247   1 a 3   **b** alcohols   **c** See propan-1-ol on page 252.
2   H H H H H H
     | | | | | |
  H—C—C—C—C—C—C—H
     | | | | | |
     H H H H H H
3 $C_{32}H_{66}$ **4**   H H O
                          | | ⟋⟍
                       H—C—C—C
                          | | ⟍
                          H H   OH

page 249   1 The diagram is shown at the top of page 248. Each C atom shares electrons with 3 H atoms and one other C atom.   **2** They burn readily and cleanly, providing plenty of heat energy, and they are easy to transport by pipeline, or in cylinders under pressure.   **3** $2C_4H_{10} + 13O_2 → 8CO_2 + 10H_2O$
4 a Chlorine takes the place of a hydrogen atom.   **b** sunlight
5 a and **b** See the table below. Note that a molecule of HCl is also produced in each subsitution reaction.

one H substituted: C_2H_5Cl 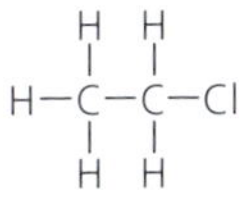	two H substituted: $C_2H_4Cl_2$ 
H H | | H—C—C—Cl | | H H	H H H H | | | | Cl—C—C—Cl and H—C—C—Cl | | | | H H H Cl
three H substituted: $C_2H_3Cl_3$ 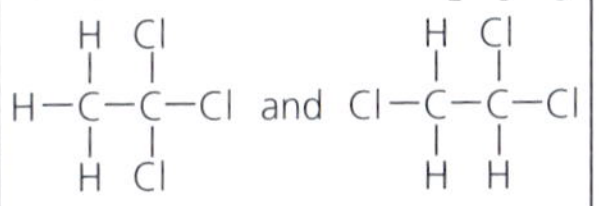	four H substituted: $C_2H_2Cl_4$ 
H Cl H Cl | | | | H—C—C—Cl and Cl—C—C—Cl | | | | H Cl H H	H Cl Cl Cl | | | | Cl—C—C—Cl and Cl—C—C—Cl | | | | H Cl H H
five H substituted: C_2HCl_5 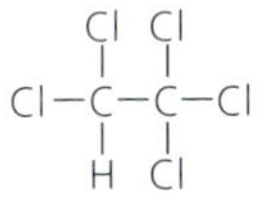	all six H substituted: C_2Cl_6 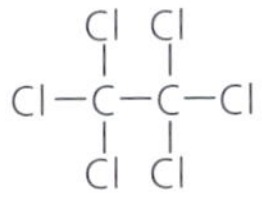
Cl Cl | | Cl—C—C—Cl | | H Cl	Cl Cl | | Cl—C—C—Cl | | Cl Cl

6 a and **b** The structures are shown below. Boiling points decrease as branching increases. This is because branching prevents the molecules from getting close, so there is less attraction between them. Therefore less heating is needed to overcome the attraction.

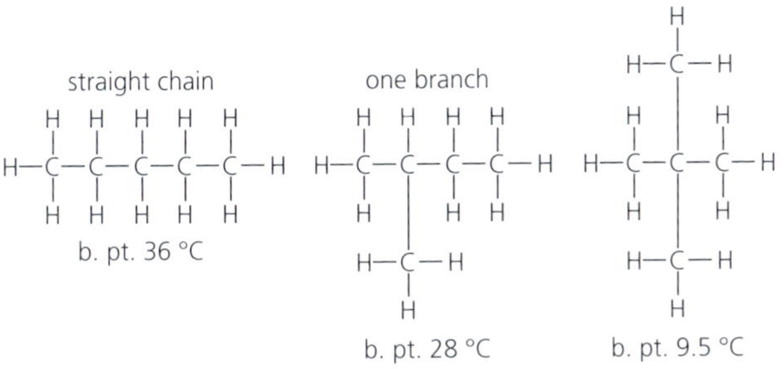

straight chain (b. pt. 36 °C) one branch (b. pt. 28 °C) two branches (b. pt. 9.5 °C)

page 251 **1 a, b** ethene, propene; for their structures see page 250
2 Their C=C bonds break easily, allowing other atoms to add on.
3 Ethene molecules can join up to form very long carbon chains.
4 a It contains a C=C bond. **b** $C_2H_4 + Br_2 \rightarrow C_2H_4Br_2$
5 a add hydrogen **b** add water (as steam)

page 253 **1** because they all contain the OH functional group **2** See the drawing in the table on page 252. **3** as a solvent, and a fuel **4** fermentation, catalysed by enzymes in yeast
5 a i ethanol + oxygen → carbon dioxide + water
ii methane + oxygen → carbon dioxide + water
b same products **6** propan-2-ol, with the structure shown here:

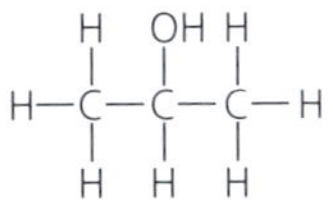

H OH H
\| \| \|
H—C—C—C—H
\| \| \|
H H H

page 255 **1** COOH **2** salt + hydrogen, salt + water, ester + water
3 Only some of the molecules dissociate in water, giving H+ ions.
4

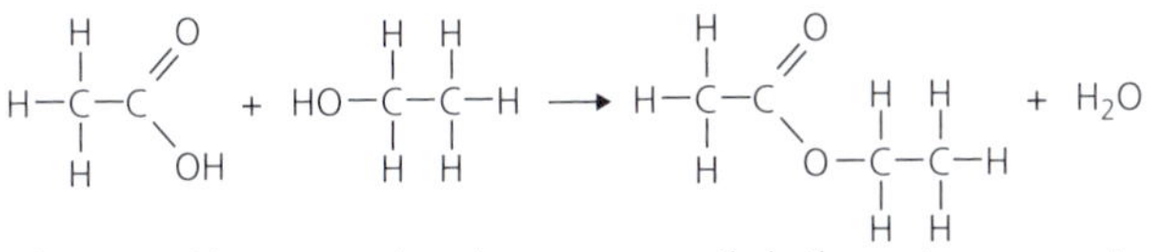

H O H H   H O H H
\| // \| \| \| // \| \|
H—C—C + HO—C—C—H → H—C—C + H₂O
\| \OH \| \| \| \O—C—C—H
H   H H   H \| \|
             H H

ethanoic acid   ethanol   ethyl ethanoate   water

5 Two molecules join together with the loss of a water molecule.
6 They have attractive smells and tastes, so are used in making soaps, cosmetics, soft drinks, ice-cream, and many other foodstuffs.

page 256 **Core** **1 a i** fractional distillation **ii** boiling points
b i The range of boiling points shows that it is a group of compounds. **ii** As the crude oil is heated, the compounds in it evaporate. As they rise through the tower they condense to form a liquid. A mixture of these liquids is drawn off together, as naphtha. **c** See page 243. **d i** longer
ii more viscous **2 a** breaking down long-chain molecules into smaller ones **b** heat and catalyst **c** It will contain some air.
d to prevent water being sucked back into hot test-tube
e C_2H_6 (g) → C_2H_4 (g) + H_2 (g)

page 257 **Core** **3 a** carbon and hydrogen **b** methane
c i 5 **ii** 6 **d** No, because it has no C=C bond. It is saturated.
e carbon dioxide and water vapour **f** pentane + oxygen → carbon dioxide + water **g** carbon monoxide
4 a The apparatus should be like that on page 244, with ethanol instead of hydrocarbon oil **b** Water is removed from the compound. **c** The reaction is the reverse of that shown at the top of page 253. **d i** The orange bromine water turns colourless. **ii** because ethanol does not contain a C=C bond
5 a ethene, styrene and chloropropene, because all three molecules contain a C=C bond **b** polyethene, polystyrene, polychloropropene
Extended **6 a** a family of organic compounds that share the same general formula **b** All the C-C bonds in the compound are single bonds. **c** the alkanes **d i** C_2H_6, ethane **ii** See page 248.
e i the alkenes, C_nH_{2n} **ii** C_2H_4, ethene **iii** See page 250.
7 a Members have the same general formula, the same functional group, and similar chemical properties. Chain length increases by 1 C at a time, leading to an increase in boiling points. **b i** the alcohols
ii $C_nH_{2n-1}OH$ **iii** the part of the molecule that determines its reactions **iv** the OH group **c** C_2H_5OH
d i You should draw out all the C-H bonds in this structure:
$CH_3-CH_2-CH_2-CH_2-CH_2-OH$

ii Here are the structures of three isomers. You should draw out all the C–H bonds, in your chosen isomer: $CH_3-CH(CH_3)-CH_2-CH_2-OH$
$CH_3-CH_2-CH(CH_3)-CH_2-OH$ $CH_3-C(CH_3)_2-CH_2-OH$

iii by dehydration: pass its vapour over a hot catalyst
iv pentanoic acid **8 a** the carboxylic acids **b** COOH **c** Only some of its molecules dissociate to form ions:

CH_3COOH (aq) $\xrightarrow{\text{some molecules}}$ CH_3COO^- (aq) + H⁺ (aq)

d i fizzing, as carbon dioxide bubbles off
ii $2CH_3COOH$ (aq) + Na_2CO_3 (s) → $2CH_3COONa$ (aq) + H_2O (l) + CO_2 (g)
e i propanoic acid, C_2H_5COOH (or $C_3H_6O_2$). Its structure is on page 254.
ii C_2H_5COOH (aq) + NaOH (s) → C_2H_5COONa (aq) + H_2O (l)
9 a ethyl ethanoate **b** an ester **c** by its sweet smell
d It acts as a catalyst. **e** It can go both ways.
f CH_3COOH (l) + C_2H_5OH (l) $\rightleftharpoons$ $CH_3COO\,C_2H_5$ (l) + H_2O (l)
10 a i alkenes **ii** C_6H_{12} **b i** C_6H_{12} (l) + Br_2 (aq) → $C_6H_{12}Br_2$ (l)
ii addition **iii** a colour change from orange to colourless

Chapter 18

page 259 **1 a** a substance made of very large molecules, which were formed by polymerisation **b** a polymer found in living things **c** a polymer made in a factory **d** the joining of thousands of small molecules, to form very large molecules **e** the small molecule from which a polymer is made **2 a** keratin **b** cellulose

page 261 **1** because small molecules add on to each other, with no other product **2** No, because it does not contain a C=C (double) bond. **3** See the diagram on page 260. **4** because they all have different chain lengths

5 Look at this table:

	Monomer	Part of the polymer molecule
a	H H C=C with H and Cl	polymer chain —C—C—C—C—C—C—C— with H, Cl, H, Cl, H, Cl, H
b	C_6H_5 H C=C with H and Cl	polymer chain —C—C—C—C—C—C—C— with C_6H_5, H and H across

6 H H C=C with H and $COCH_3$

page 263 **1** two – the very large molecules, and the eliminated small molecules **2** so that each end of it can join to another molecule, to form a chain **3** Condensation: two different monomers / join by eliminating a small molecule / two products. Addition: one monomer / join by addition at a C=C bond / one product.
4 a/b See the lower diagram on page 262.
c The monomer units are linked by an amide group (—CONH—).
5 See the lower diagram on page 263.

page 265 **1** Choose the most suitable properties for each group of items. ('Unreactive' is important for all those uses!) **2** It is a low-density plastic formed when the polymer chains branch, and cannot pack closely together. **3** high melting point, unreactive, non-toxic, slippery **4 a** Polystyrene, which is 'expanded' by blowing a gas through it during manufacture, to make a light rigid foam. **b** packaging, insulation, fast-food cartons

page 267 **1** See the list on page 266. **2** It is used all over the world for packaging and plastic bags, which get thrown away.
4 a breaks down in sunlight **b** plastics produced by living organisms

page 269 **1** carbohydrates and proteins **2 a** It is a single sugar unit. **b** fructose **3** Your drawing should look like the one on page 268, but with five unit of glucose instead of four. **4** four
5 a Both are formed by the polymerisation of glucose. **b** The glucose units in them are joined in a different way, giving different properties. **6 a** A single sugar unit; for example, a molecule of glucose or fructose. **b** A substance formed from many sugar units; for example starch or cellulose.

page 271 **1 a** a carboxylic acid with an NH_2 group; the general formula for an amino acid is $NH_2CH(R)COOH$ **b** a very large molecule built up from amino acid units **2 a** Your drawing could be like the lower part of the last drawing on page 270. **b** four
3 Both are condensation polymerisations, and water molecules are eliminated. **4** the breaking down of a compound by reaction with water **5** using heat and hydrochloric acid, or using enzymes
6 a Your diagram could look like this:

part of a starch molecule

—☐—O—☐—O—☐—O—☐—O—

hydrolysis: water molecules added | H_2O

HO—☐—OH HO—☐—OH HO—☐—OH HO—☐—OH

molecules of glucose

b Amylase is a catalyst, allowing the reaction to go faster at low temperatures. Without the catalyst, more heating is needed.
7 a It breaks down to the different amino acids that formed it.
b by paper chromatography, using a locating agent

page 272 **Core** **1 a** monomers **b** polymer **c** polymerisation
d You could show it like this:
Or like this: —(☐)$_n$— ☐—☐—☐—☐—☐
2 a i made in a factory **ii** consists of very large molecules
b any two of: tough, strong, flexible, light, unreactive, does not rot away **c** light to carry, does not break if you drop it
d i non-biodegradeable **ii** any two from the list on page 266

Extended 3 a H H C=C with H and $CONH_2$ **b** acrylamide **c** unsaturated

4 a the C=C bond **b** addition
c —C—C—C—C—C—C— with F F F F F F top and F F F F F F bottom **d** polytetrafluroethene

page 273 **5 a** a very large number **b** 1,1-dichloroethene H Cl C=C with H and Cl

c addition **d i** Moisture and gases cannot pass through it easily.
ii Water and oxygen can react with substances in food, and spoil it.
iii strong, transparent, flexible, able to be made into thin sheets, able to 'cling' to itself or other surfaces **e i** cannot be broken down by bacteria or other organisms in the soil **ii** very thin, and clung to itself and other things, so difficult to collect; often contaminated with food **6 a** Monomers join to form long chains by eliminating small molecules. **b** The reaction is the same as for nylon on page 262, but show the rings instead of blocks. **c** hydrogen chloride
d i Both contain the amide link. **ii** The chains of atoms in the middle of each monomer are different. That will affect the properties.
e Cl—C(=O)—⬡—C(=O)—Cl H_2N—⬡—NH_2

7 a the amide linkage: —C(=O)—N(H)— **b** proteins

c i the breaking down of a compound by reaction with water
ii amino acids **iii** chromatography
8 a condensation polymerisation **b** ester linkage
c Cl—C(=O)—⬡⬡—C(=O)—Cl and HO—CH_2—CH_2—OH

d Monomer C on page 263 has only one ring of atoms. The first monomer above has two. This must account for the difference. (The two rings are flat so chains can get closer to each other.)
9 a one that is produced in / by living things **b** It consists of very large molecules formed from sugar units. **c** condensation
d the part of the monomer not directly involved in the polymerisation
e i HO—☐—OH **ii** glucose

f It is made from many sugar units. (*Poly-* means many.)
g i In the lab: heating with dilute hydrochloric acid, or using amylase (an enzyme). In the digestive system: at body temperature, catalysed by amylase.   **ii** Amylase increases the rate of the reaction at low (body) temperatures.
10 a HO—[ ]—OH HO—[ ]—O—[ ]—OH   and so on.
b a substance used to show up colourless substances; it is needed because the products are colourless   **c** See pages 20 and 21.

Chapter 19

page 275   **1 a** no   **b** yes   **c** no   **d** yes   **2** a variable that is under your control, which you can change when you do an experiment
3 the amount of yeast added   **4 a** Yes, e.g. carrying out tests to confirm a diagnosis.   **b** Yes, e.g. collecting / checking DNA samples or fingerprints to confirm that the suspect carried out the crime.

page 277   **1 a** Yes, the brand of kitchen cleaner.   **b** Yes, the amount of acid needed to neutralise the alkali.
2 a B, conical flask; D, burette; E, pipette   **b** It delivers the required volume more accurately.   **c** to allow the liquid to be swirled (to mix it) without spilling   **d** They allow you to deliver liquids drop by drop, and to measure the delivered volume accurately   **e** a pipette; it is easy to fill it to the exact volume needed, because the mark is on its very slim stem
3 to tell you when neutralisation is complete
4 a phenolphthalein   **b** pink to colourless
5 … X contains a higher concentration of sodium hydroxide than solution Y does.
6 … support my hypothesis that X is a better at removing grease than Y, because it has a higher concentration of sodium hydroxide.
7 Make sure to shake the cleaners well. Repeat each titration to check the results.
8 Titrate the scale remover with an alkaline solution (e.g. sodium hydroxide) of known concentration, using a suitable indicator (e.g. methyl orange)   **9** F, the thermometer

page 279   **1 a** The apparatus for preparing the gas should be as on top left of page 278, and it can be collected by upward displacement of air, or over water, as on the bottom of page 278.   **b** It will turn limewater milky.   **c** CO_2 (g) + $Ca(OH)_2$ (aq) → $CaCO_3$ (s) + H_2O (l)
2 a it is lighter than air.   **b** $2H_2$ (g) + O_2 (g) → $2H_2O$ (l)
3 a ammonium chloride or sulfate or nitrate, and calcium or sodium hydroxide   **b** It is very soluble in water.   **c** It turns damp red litmus turn blue.   **4** may be too little present to detect; may be harmful   **5** gas syringe: more accurate, less messy to use

page 281   **1** cation   **2** See the instructions on page 280.
3 a the potassium ion   **b** the copper(II) ion   **b** the lithium ion
4 hydroxides   **5 a** the zinc and aluminium cations; the precipitates that form both dissolve again on adding excess sodium hydroxide   **b** Repeat the test, using aqueous ammonia. If the

precipitate redissolves on adding excess ammonia, the zinc cation is present.   **6** because potassium hydroxide is soluble

page 283   **1** negative ions; for example Cl^-, Br^-, I^-, CO_3^{2-}, NO_3^-
2 precipitate, insoluble   **3** because silver halides are insoluble, with different colours for Cl^-, Br^- and I^- **4** See the instructions on page 283.
5 All nitrates are soluble in water.
6 from the sodium hydroxide solution
7 a because a gas (carbon dioxide) is produced in the reaction
b i a solution of calcium hydroxide   **ii** A positive result is easy to see: the limewater turns milky.

page 284   **1 a** Place the filter funnel, with filter paper, in the conical flask. Pour the mixture into the filter funnel. Add a few drops of universal indicator to the filtrate, using the dropping pipette. The final colour of the indicator gives the pH of the soil sample.
b by sampling other areas of the garden
2 a If you do not recognise these pieces of apparatus, look through the textbook to find examples. For example, page 278.
b because sulfur dioxide is soluble in water
3 a a metal such as magnesium, zinc, or iron, or any carbonate
b a stopclock
c • Weigh out the required mass of solid Y.
  • Using a measuring cylinder, place a known volume of
    acid in the flask.
  • Add Y to the acid, and immediately stopper the flask and
    start the stopclock.
  • Record the volume of gas collected, at regular time intervals.
d More gas may be produced than the syringe can hold.

page 285   **4 a i** pH 1 – W, pH 5 – Y   **ii** It is a weak acid.
b i fizzing, and the metal disappears   **ii** a magnesium salt, and hydrogen   **iii** The rate of the reaction will be much slower.
c i white   **ii** carbon dioxide   **iii** sodium carbonate (or sodium hydrogen carbonate)
d i barium sulfate   **ii** sulfuric acid
5 a i See instructions on page 280.   **ii** lilac   **b i** sulfur dioxide
ii SO_2 **iii** See the test for sulfur dioxide on page 279.
6 a ammonium ion, NH_4^+ **b** See the test for the ammonium ion on page 281.   **c** nitrate ion, NO_3^- **d** See the test for the nitrate ion on page 283.
7 a i a positive ion   **ii** the zinc cation, Zn^{2+} **iii** the aluminium cation, Al^{3+} **iv** the calcium cation, Ca^{2+} **b i** the chloride ion, Cl^-
ii See the test for the chloride ion on page 282.
8 a anions: chloride, hydrogen carbonate, nitrate, sulfate; cations: calcium, magnesium, sodium potassium   **b i** calcium
ii 2.75 milligrams   **c** sulfate   **d** All the compounds of the other metals are soluble.   **e** The nitrate concentration is probably too low to give a positive identification.

1 This question is about hydrogen and some compounds containing hydrogen.

a Hydrogen is a gas at room temperature. Describe the arrangement and motion of the molecules in hydrogen gas. [2]

b Draw the electronic structure of a hydrogen molecule. [1]

c The symbols for two isotopes of hydrogen are:

$$^1_1\text{H} \qquad ^3_1\text{H}$$

i What do you understand by the term *isotope*? [1]

ii Copy and complete the table to show the number of subatomic particles in these two isotopes of hydrogen.

isotope	^1_1H	^3_1H
number of electrons		
number of neutrons		
number of protons		

[4]

d When hydrogen burns, energy is given out. State the name given to a reaction which gives out energy. [1]

e Hydrochloric acid reacts both with metals and with metal carbonates.

i A student observed the reaction of hydrochloric acid with four different metals. The student used the same concentration of hydrochloric acid and the same mass of metal in each experiment.

metal	observations
cobalt	dissolves very slowly and very few bubbles produced
iron	dissolves slowly and a few bubbles produced slowly
magnesium	dissolves very quickly and many bubbles produced rapidly
zinc	dissolves very quickly and many bubbles produced rapidly

Use the information in the table to suggest the order of reactivity of these metals.

most reactive ⟶ least reactive [2]

ii State the names of the three products formed when hydrochloric acid reacts with calcium carbonate. [3]

Cambridge IGCSE Chemistry 0620 Paper 21 Q2 June 2010

2 Lavandulol is found in lavender plants. The formula of lavandulol is shown on the right.

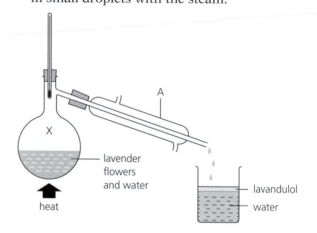

a Which is the alcohol functional group in this formula? [1]

b Is lavandulol a saturated or unsaturated compound? Give a reason for your answer. [1]

c State the names of the two products formed when lavandulol is burnt in excess oxygen. [2]

d Lavandulol can be extracted from lavender flowers by distillation using the apparatus shown below. The lavandulol is carried off in small droplets with the steam.

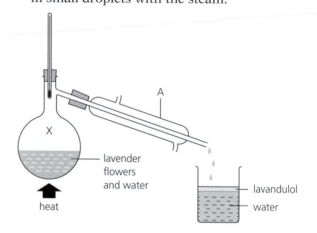

i State the name of the piece of apparatus labelled A. [1]

ii What is the temperature of the water at point X in the diagram? [1]

iii The lavandulol and water are collected in the beaker. What information in the diagram shows that lavandulol is less dense than water? [1]

e Lavender flowers contain a variety of different pigments (colourings). A student separated these pigments using paper chromatography. The results are shown in the diagram below.

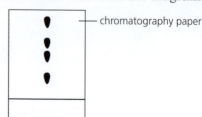

chromatography paper

i Copy the diagram and put an X to show where the mixture of pigments was placed at the start of the experiment. [1]

ii How many pigments have been separated? [1]

iii Draw a diagram to show how the chromatography apparatus was set up. Label
   - the solvent
   - the origin line [1]

iv During chromatography, the solvent evaporates and then diffuses through the chromatography jar. What do you understand by the term *diffusion*? [1]

v Ethanol can be used as a solvent in chromatography. Draw the formula for ethanol showing all atoms and bonds. [1]

vi Which two of the following statements about ethanol are true?
   It is a carboxylic acid.
   It is a product of the fermentation of glucose.
   It is an unsaturated compound.
   It is formed by the catalytic addition of steam to ethene. [1]

Cambridge IGCSE Chemistry 0620 Paper 2 Q3 November 2006

3 Hydrogen chloride, HCl, is an acidic gas.

a Draw a dot and cross diagram of a molecule of hydrogen chloride. Show only the outer electrons. [2]

b Hydrogen chloride dissolves in water to form a solution of hydrochloric acid. A student titrated aqueous ammonia with hydrochloric acid using the apparatus shown below.

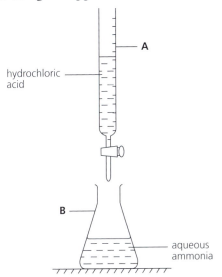

i State the names of the pieces of apparatus labelled A and B. [2]

ii Describe how the pH value of the solution in B changes as hydrochloric acid is added until the acid is in excess. [3]

iii Complete the word and symbol equations for this reaction.

ammonia + hydrochloric acid ⟶ ....

.... + HCl ⟶ NH_4Cl [2]

c Aqueous ammonia is used to test for copper(II) ions. Describe what happens when you add aqueous ammonia to a solution of copper(II) sulfate until the aqueous ammonia is in excess. [4]

Cambridge IGCSE Chemistry 0620 Paper 21 Q2 November 2012

4 The table shows some properties of four substances, A, B, C and D.

	melting point / °C	does the solid conduct electricity?	does a solution of the solid conduct electricity?
A	962	yes	does not dissolve
B	747	no	dissolves and conducts
C	113	no	does not dissolve
D	3550	no	does not dissolve

a Which one of these substances has
   i a giant covalent structure,
   ii a simple molecular structure,
   iii a metallic structure? [3]

b A student carried out an experiment to determine the rate of reaction of calcium carbonate with excess hydrochloric acid.

$$CaCO_3 (s) + 2HCl (aq) \longrightarrow$$
$$CaCl_2 (aq) + CO_2 (g) + H_2O (l)$$

He recorded the loss of mass of the reaction mixture over a period of time.

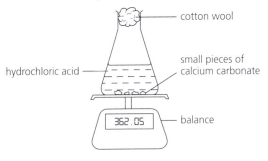

i Explain why the reaction mixture decreases in mass. [1]

He carried out the reaction at constant temperature using 2 g of calcium carbonate in small pieces. The hydrochloric acid was in excess. He plotted his results on a grid. This is shown below.

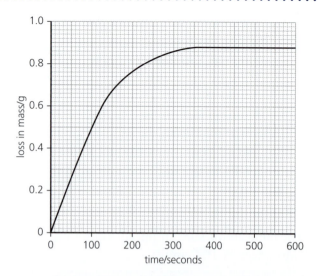

ii At what time has the reaction just finished? [1]

iii From the graph, deduce the loss in mass in the first 100 seconds. [1]

The student repeated the experiment keeping everything the same except for the size of the pieces of calcium carbonate. He used smaller pieces of calcium carbonate but the mass used was the same.

iv Sketch the grid and graph, and draw in a line to show how the loss of mass changes with time when smaller pieces of calcium carbonate are used. [2]

v State the effect of increasing the concentration of hydrochloric acid on the rate (speed) of this reaction when all other factors remain constant. [1]

Cambridge IGCSE Chemistry 0620 Paper 21 Q5 June 2013

5 Bromine is a red-brown liquid. When warmed, it forms an orange vapour.

a Describe what happens to the arrangement and motion of the particles when bromine changes state from a liquid to a vapour. [3]

b Bromine can be obtained from bromide ions in seawater.

i The symbol equation for this reaction is:

$$Cl_2 + 2Br^- \longrightarrow 2Cl^- + Br_2$$

Complete the word equation for this reaction

.... + bromide ions $\longrightarrow$ .... + ... [1]

ii Bromine is very volatile, so it can be removed from solution by bubbling air through the solution. What do you understand by the term *volatile*? [1]

c Hydrogen reacts with bromine in the presence of a hot platinum catalyst to form hydrogen bromide.

i Define the term *catalyst*. [1]

ii Hydrogen bromide reduces hydrogen peroxide, H_2O_2.

$$2HBr + H_2O_2 \longrightarrow Br_2 + 2H_2O$$

Explain how this equation shows that hydrogen peroxide is reduced. [1]

iii A solution of hydrogen bromide in water is called hydrobromic acid.

Hydrobromic acid has similar reactions to hydrochloric acid.

State the names of three products formed when hydrobromic acid reacts with sodium carbonate. [2]

Cambridge IGCSE Chemistry 0620 Paper 21 Q8 June 2011

6 The table shows observations about the reactivity of various metals with dilute hydrochloric acid.

metal	observations
calcium	many bubbles produced rapidly with much spitting
copper	no bubbles formed
iron	a few bubbles produced very slowly
magnesium	many bubbles produced rapidly with no spitting

a Put the metals in order of increasing reactivity. [1]

b Zinc is between iron and magnesium in reactivity. Suggest what observations will be made when zinc reacts with dilute hydrochloric acid. [1]

c Magnesium is extracted by the electrolysis of molten magnesium chloride.

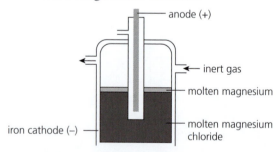

i What information in the diagram suggests that magnesium is less dense than molten magnesium chloride? [1]

ii Magnesium is extracted by electrolysis rather than by heating its oxide with carbon. Why? [1]

iii Suggest why a stream of inert gas is blown over the surface of the molten magnesium. [1]

iv Name a gaseous element which is inert. [1]

d In some old magnesium manufacturing plants, coal gas is blown over the surface of the magnesium. The list shows the main substances in coal gas.

carbon monoxide   ethene   hydrogen

hydrogen sulfide   methane

  i Draw the structure of ethene showing all atoms and bonds. [1]

  ii Suggest two hazards of using coal gas by referring to two specific substances in the list. [2]

e Carbon monoxide can be removed from coal gas by mixing it with steam and passing the mixture over a catalyst of iron(III) oxide at 400 °C.

$$CO + H_2O \rightleftharpoons CO_2 + H_2$$

  i Write a word equation for this reaction. [1]

  ii What does the symbol $\rightleftharpoons$ mean? [1]

  iii Iron(III) oxide reacts with acids to form a solution containing iron(III) ions. Describe a test for aqueous iron(III) ions. Give the result. [2]

Cambridge IGCSE Chemistry 0620 Paper 2 Q2 June 2009

7 Iron is extracted from its ore in a blast furnace.

a State the name of the ore from which iron is extracted. [1]

b The diagram shows a blast furnace.

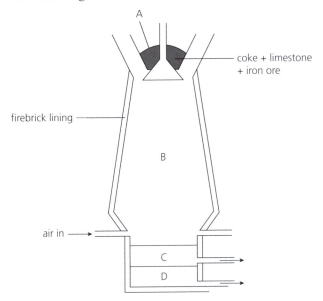

  i Which one of the raw materials is added to the blast furnace to help remove the impurities from the iron ore? [1]

  ii The impurities are removed as a slag. Which letter on the diagram shows the slag? [1]

c Carbon monoxide is formed in the blast furnace by reaction of coke with oxygen.

  i Complete the equation for this reaction.

...... C + ... $\longrightarrow$ ...... CO   [2]

  ii State the adverse affect of carbon monoxide on human health. [1]

d In the hottest regions of the blast furnace the following reaction takes place.

$$Fe_2O_3 + 3C \longrightarrow 2Fe + 3CO$$

Which two of these five sentences correctly describe this reaction?

The iron oxide gets reduced.

The reaction is a thermal decomposition.

The carbon gets oxidised.

The carbon gets reduced.

Carbon neutralises the iron oxide.   [1]

e Aluminium cannot be extracted from aluminium oxide in a blast furnace. Explain why aluminium cannot be extracted in this way. [2]

f **i** State the name of the method used to extract aluminium from its oxide ore. [1]

  ii State one use of aluminium. [1]

Cambridge IGCSE Chemistry 0620 Paper 2 Q4 November 2008

8 The table shows the mass of various compounds obtained when 500 cm³ of seawater is evaporated.

compound	ions present	mass of compound / g
sodium chloride	Na^+ and Cl^-	14.0
magnesium chloride	Mg^{2+} and Cl^-	3.0
magnesium sulfate	Mg^{2+} and SO_4^{2-}	2.0
calcium sulfate	Ca^{2+} and SO_4^{2-}	0.5
potassium chloride	K^+ and Cl^-	
potassium bromide		0.5
calcium carbonate	Ca^{2+} and CO_3^{2-}	
sodium iodide	Na^+ and I^-	
		total mass = 20.0 g

a Which negative ion is present in seawater in the highest concentration? [1]

b Write the symbols for the two ions present in potassium bromide. [1]

c Calculate the mass of sodium chloride present in 5 g of the solid left by evaporating the seawater. [1]

d Describe a test for the iodide ion, and the result. [2]

e Aqueous chlorine reacts with aqueous sodium iodide.

   i Complete the equation for this reaction.

     $Cl_2 + 2NaI \longrightarrow$ .... $+ 2NaCl$ [1]

   ii What colour is the solution when the reaction is complete? [1]

   iii An aqueous solution of iodine does not react with aqueous potassium bromide. Explain why there is no reaction. [1]

f Calculate the relative formula mass of magnesium chloride, $MgCl_2$. [1]

Cambridge IGCSE Chemistry 0620 Paper 21 Q4 June 2010

9 Choose from the following list of oxides to answer the questions below. You can use each oxide once, more than once or not at all.

    carbon dioxide   carbon monoxide
    magnesium oxide   nitrogen dioxide
    sulfur dioxide   water

a Which one of these oxides is a basic oxide? [1]

b Which two oxides cause acid rain? [2]

c Which two oxides are formed when a hydrocarbon undergoes complete combustion? [2]

d Which one of these oxides turns white copper(II) sulfate blue? [1]

e Which oxide is formed when calcium carbonate undergoes thermal decomposition? [1]

Cambridge IGCSE Chemistry 0620 Paper 22 Q1 November 2010

10 The diagram shows a basic oxygen converter.

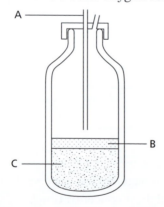

This is used to convert impure iron from the blast furnace into steel.

During this process, some of the impurities in the iron are converted into a slag.

a Which letter, A, B, or C, shows:

- where the oxygen enters
- the slag
- the molten steel? [3]

b In the converter, the oxygen oxidises sulfur, carbon and phosphorus to their oxides.

   i Explain why sulfur dioxide and carbon dioxide are easily removed from the converter. [1]

   ii Explain how calcium oxide is used to remove phosphorus(V) oxide from the converter. [3]

c Stainless steel is an alloy.

   i Which **one** of the diagrams, A, B, C or D, best represents an alloy? [1]

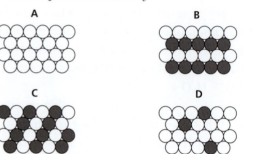

   ii State one use of stainless steel. [1]

Cambridge IGCSE Chemistry 0620 Paper 21 Q7 June 2011

11 a Match the fuel on the left with the information on the right. The first one has been done for you.

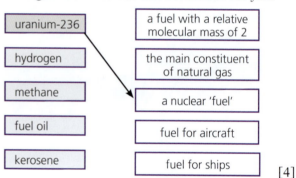

[4]

b Two students investigated some fuels to find which gave off the most energy. They tested four liquid fuels using the apparatus shown below.

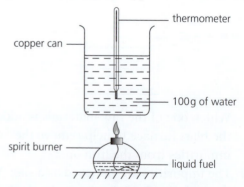

   i In each experiment, the amount of fuel burnt was the same. Suggest one other factor that should be kept the same in each experiment. [1]

ii The students used the thermometer to stir the water. Suggest why it is important to keep the water stirred. [1]

iii The results are shown in the table below.

fuel	temperature of the water / °C	
	initial	final
ethanol	24	40
propanol	24	42
paraffin	22	33
petroleum spirit	20	40

Which fuel transfers the most energy to the water? Explain your answer. [2]

c Air is needed for fuels to burn. The pie chart below shows the composition of the air.

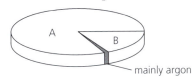

mainly argon

State the name of: gas A gas B [2]

d Argon is a noble gas.

i State one use for argon. [1]

ii To which period in the Periodic Table does argon belong? [1]

iii Describe the chemical properties of argon. [1]

Cambridge IGCSE Chemistry 0620 Paper 21 Q5 November 2012

12 The formulae of four organic compounds are:

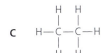

a i State the name of the type of bonding between the atoms in these four compounds. [1]

ii Which one of these compounds, A, B, C or D, is a saturated hydrocarbon? [1]

iii Which one of these compounds is acidic? [1]

iv State the name of compound D. [1]

v Compound A contains a C=C double bond. Describe a test for a C=C double bond, and the result. [2]

b Compound C is a member of the alkane homologous series.

i State two features of an homologous series. [2]

ii State the formula and name of another alkane in the same homologous series as compound C. [2]

c The alkanes present in petroleum can be separated by fractional distillation.

This diagram shows a fractional distillation column.

On a copy of the diagram:

i label where the temperature in the column is the lowest. Mark this with the letter X. [1]

ii Label where the bitumen fraction is collected. Mark this with the letter Y. [1]

Cambridge IGCSE Chemistry 0620 Paper 22 Q4 November 2010

1 The following techniques are used to separate mixtures.

A simple distillation   B fractional distillation

C evaporation   D chromatography

E filtration   F diffusion

From this list, choose the most suitable technique to separate the following.

a methane from a mixture of the gases, methane and ethane [1]

b water from aqueous magnesium sulfate [1]

c glycine from a mixture of the amino acids, glycine and lysine [1]

d iron filings from a mixture of iron filings and water [1]

e zinc sulfate crystals from aqueous zinc sulfate [1]

f hexane from a mixture of the liquids, hexane and octane [1]

Cambridge IGCSE Chemistry 0620 Paper 31 Q1 June 2011

2 For each of the following unfamiliar elements predict one physical and one chemical property.

a caesium (Cs) [2]

b vanadium (V) [2]

c fluorine (F) [2]

Cambridge IGCSE Chemistry 0620 Paper 32 Q1 June 2010

3 Across the world, food safety agencies are investigating the presence of minute traces of the toxic hydrocarbon, benzene, in soft drinks. It is formed by the reduction of sodium benzoate by vitamin C.

a Sodium benzoate is a salt. It has the formula C_6H_5COONa. It can be made by neutralising benzoic acid using sodium hydroxide.

i Deduce the formula of benzoic acid. [1]

ii Write a word equation for the reaction between benzoic acid and sodium hydroxide. [1]

iii Name **two** other compounds that would react with benzoic acid to form sodium benzoate. [2]

b Benzene contains 92.3% of carbon and its relative molecular mass is 78.

i What is the percentage of hydrogen in benzene? [1]

ii Calculate the ratio of moles of C atoms: moles of H atoms in benzene. [2]

iii Calculate its empirical formula and then its molecular formula. [2]

c This shows the structural formula of Vitamin C.

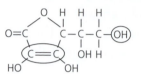

i What is its molecular formula? [1]

ii Name the two functional groups that are circled. [2]

Cambridge IGCSE Chemistry 0620 Paper 31 Q4 November 2008

4 An ore of copper is the mineral, chalcopyrite. This is a mixed sulfide of iron and copper.

a Analysis of a sample of this ore shows that 13.80 g of the ore contained 4.80 g of copper, 4.20 g of iron and the rest sulfur.

i Copy and complete the table. [3]

	copper	iron	sulfur
composition by mass /g	4.80	4.20	
number of moles of atoms			
simplest mole ratio of atoms			

ii Find the empirical formula of chalcopyrite. [1]

b Impure copper is extracted from the ore. This copper is refined by electrolysis.

i Name

A: the material used for the positive electrode (anode),

B: the material used for the negative electrode (cathode),

C: a suitable electrolyte. [3]

ii Write an ionic equation for the reaction at the negative electrode. [1]

iii One use of this pure copper is electrical conductors, another is to make alloys. Name the metal that is alloyed with copper to make brass. [1]

c Two of the elements in chalcopyrite are the metal, copper, and the non-metal, sulfur.
These have different properties.
Copper is an excellent conductor of electricity and is malleable.
Sulfur is a poor conductor and is not malleable, it is brittle.
Explain, in terms of their structures, why this is so.

i difference in electrical conductivity [2]

ii difference in malleability [2]

Cambridge IGCSE Chemistry 0620 Paper 3 Q6 November 2006

5 There are three types of giant structure – ionic, metallic and macromolecular.

a Sodium nitride is an ionic compound.
Draw a diagram that shows the formula of the compound, the charges on the ions, and the arrangement of the valency electrons around the negative ion. Use × to represent an electron from a sodium atom. Use ○ to represent an electron from a nitrogen atom. [3]

b i Describe metallic bonding. [3]

ii Use the above ideas to explain why metals are
A: good conductors of electricity [1]
B: malleable. [2]

c Silicon(IV) oxide has a macromolecular structure.

i Describe the structure of silicon(IV) oxide (a diagram is not acceptable). [3]

ii Diamond has a similar structure and consequently similar properties. Give two physical properties common to both diamond and silicon(IV) oxide. [2]

Cambridge IGCSE Chemistry 0620 Paper 31 Q2 November 2008

6 Until recently, arsenic poisoning, either deliberate or accidental, has been a frequent cause of death. The symptoms of arsenic poisoning are identical with those of a common illness, cholera.
A reliable test was needed to prove the presence of arsenic in a body.

a In 1840, Marsh devised a reliable test for arsenic.

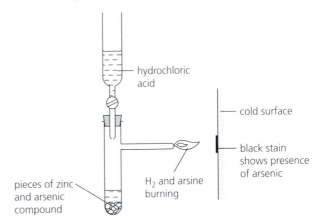

Hydrogen is formed in this reaction.
Any arsenic compound reacts with this hydrogen to form arsine which is arsenic hydride, AsH_3. The mixture of hydrogen and arsine is burnt at the jet and arsenic forms as a black stain on the glass.

i Write an equation for the reaction which forms hydrogen. [2]

ii Draw a diagram which shows the arrangement of the outer (valency) electrons in one molecule of the covalent compound arsine. The electron distribution of arsenic is $2 + 8 + 18 + 5$. Use × to represent an electron from an arsenic atom. Use ○ to represent an electron from a hydrogen atom. [2]

b Another hydride of arsenic has this composition:
arsenic 97.4 % hydrogen 2.6 %

i Calculate the empirical formula of this hydride from the above data. Show your working. [2]

ii The mass of one mole of this hydride is 154 g. What is its molecular formula? [1]

iii Deduce the structural formula of this hydride. [1]

c Hair is a natural protein. Hair absorbs arsenic from the body. Analysis of the hair provides a measurement of a person's exposure to arsenic. To release the absorbed arsenic for analysis, the protein has to be hydrolysed.

i What is the name of the linkage in proteins? [1]

ii Name a reagent which can be used to hydrolyse proteins. [1]

iii What type of compound is formed by the hydrolysis of proteins? [1]

d In the 19th Century, a bright green pigment, copper(II) arsenate(V) was used to kill rats and insects. In damp conditions, micro-organisms can act on this compound to produce the very poisonous gas, arsine.

i Suggest a reason why it is necessary to include the oxidation states in the name of the compound. [1]

ii The formula for the arsenate(V) ion is AsO_4^{3-}. Complete the ionic equation for the formation of copper(II) arsenate(V).

......Cu^{2+} + ...AsO_4^{3-} $\longrightarrow$ .... [2]

Cambridge IGCSE Chemistry 0620 Paper 33 Q6 November 2012

7 The diagram shows a cell. This is a device which produces electrical energy. The reaction in a cell is a redox reaction and involves electron transfer.

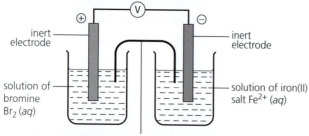

filter paper soaked in sodium chloride solution to complete the circuit

i Complete the sentence.

A cell will change .... energy into electrical energy. [1]

ii Draw an arrow on (*a copy of*) the diagram to show the direction of the electron flow. [1]

iii In the left hand beaker, the colour changes from brown to colourless. Complete the equation for the reaction.

Br_2 + ... $\longrightarrow$ .... [2]

iv Is the change in iii oxidation or reduction? Give a reason for your choice. [1]

v Complete the following description of the reaction in the right hand beaker.

Fe^{2+} changes into .... [1]

vi When a solution of bromine is replaced by a solution of chlorine, the voltage increases. When a solution of bromine is replaced by a solution of iodine, the voltage decreases. Suggest an explanation for this difference. [1]

Cambridge IGCSE Chemistry 0620 Paper 31 Q3 November 2010

8 The reactivity series shows the metals in order of reactivity.

a The reactivity series can be established using displacement reactions. A piece of zinc is added to aqueous lead nitrate. The zinc becomes coated with a black deposit of lead.

$Zn + Pb^{2+} \longrightarrow Zn^{2+} + Pb$

Zinc is more reactive than lead.

The reactivity series can be written as a list of ionic equations. The most reactive metal is the best reductant (reducing agent).

$Zn \longrightarrow Zn^{2+} + 2e^-$

$Fe \longrightarrow Fe^{2+} + 2e^-$

$Pb \longrightarrow Pb^{2+} + 2e^-$

$Cu \longrightarrow Cu^{2+} + 2e^-$

$Ag \longrightarrow Ag^+ + e^-$

i Write the ionic equation for a metal which is more reactive than zinc. [1]

ii Write an ionic equation for the reaction between aqueous silver(I) nitrate and zinc. [2]

iii Explain why the positive ions are likely to be oxidants (oxidising agents). [1]

iv Deduce which ion is the best oxidant (oxidising agent). [1]

v Which ion(s) in the list can oxidise lead? [1]

b A reactivity series can also be established by measuring the voltage of simple cells.
The diagram shows a simple cell.

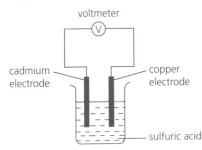

Results from cells using the metals tin, cadmium, zinc and copper are given in the table below.

cell	positive electrode	negative electrode	voltage / volts
1	copper	cadmium	0.74
2	copper	tin	0.48
3	copper	zinc	1.10

Write the four metals in order of increasing reactivity and explain how you used the data in the table to determine this order. [3]

Cambridge IGCSE Chemistry 0620 Paper 31 Q5 June 2013

9 A major ore of zinc is zinc blende, ZnS. A by-product of the extraction of zinc from this ore is sulfur dioxide which is used to make sulfuric acid.

a i Zinc blende is heated in air. Zinc oxide and sulfur dioxide are formed. Write the balanced equation for this reaction. [2]

ii Zinc oxide is reduced to zinc by heating with carbon. Name *two* other reagents which could reduce zinc oxide. [2]

iii The zinc obtained is impure. It is a mixture of metals. Explain *how* fractional distillation could separate this mixture.
zinc bp = 908 °C, cadmium bp = 765 °C, lead bp = 1751 °C [2]

b Sulfur dioxide is used to make sulfur trioxide in the Contact Process.
$$2SO_2\,(g) + O_2\,(g) \rightleftharpoons 2SO_3\,(g)$$

The forward reaction is exothermic.
The conditions used are: temperature 450 °C, pressure 2 atmospheres, catalyst vanadium(V) oxide. Explain, mentioning both position of equilibrium and rate, why these conditions give the most economic yield. [4]

Cambridge IGCSE Chemistry 0620 Paper 31 Q4 June 2011

10 Hydrogen reacts with the halogens to form hydrogen halides.

a Bond energy is the amount of energy, in kJ, that must be supplied (endothermic) to break one mole of a bond.

bond	bond energy in kJ / mol
H−H	+436
F−F	+158
H−F	+562

Use the above data to show that the following reaction is exothermic.
$$H—H + F—F \longrightarrow 2H—F \qquad [3]$$

b The hydrogen halides react with water to form acidic solutions.
$$HCl + H_2O \rightleftharpoons H_3O^+ + Cl^-$$
$$HF + H_2O \rightleftharpoons H_3O^+ + F^-$$

i Explain why water behaves as a base in both of these reactions. [2]

ii At equilibrium, only 1% of the hydrogen chloride exists as molecules, the rest has formed ions. In the other equilibrium, 97% of the hydrogen fluoride exists as molecules, only 3% has formed ions. What does this tell you about the strength of each acid? [2]

iii How would the pH of these two solutions differ? [1]

Cambridge IGCSE Chemistry 0620 Paper 31 Q7 June 2009

11 The structural formula of cyclohexane is drawn here.

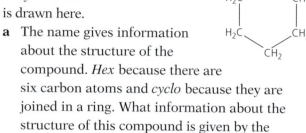

a The name gives information about the structure of the compound. *Hex* because there are six carbon atoms and *cyclo* because they are joined in a ring. What information about the structure of this compound is given by the ending *ane*? [2]

b What are the molecular and empirical formulae of cyclohexane? [2]

c Draw the structural formula of cyclobutane. [1]

d i Deduce the molecular formula of hexene. [1]

   ii Explain why cyclohexane and the alkene, hexene, are isomers. [2]

e Describe a test which would distinguish between cyclohexane and the unsaturated hydrocarbon hexene. Give the result of the test for each.

Cambridge IGCSE Chemistry 0620 Paper 31 Q4 June 2013

12 The structural formula of a butanol is:

$CH_3 - CH_2 - CH_2 - CH_2 - OH$

a Butanol can be made from petroleum and also by fermentation.

   i Describe the chemistry of making butanol from petroleum by the following route.

     petroleum $\longrightarrow$ butene $\longrightarrow$ butanol [3]

   ii Explain, in general terms, what is meant by *fermentation*. [3]

b Butanol can be oxidised to a carboxylic acid by heating with acidified potassium manganate(VII). Give the name and structural formula of the carboxylic acid. [1]

c Butanol reacts with ethanoic acid to form a liquid, X, which has the sweet smell of bananas. Its empirical formula is C_3H_6O and its M_r is 116.

   i What type of compound is liquid X? [1]

   ii Give the molecular formula of liquid X. [1]

   iii Draw the structural formula of X. Show all the individual bonds. [2]

Cambridge IGCSE Chemistry 0620 Paper 31 Q6 June 2011

13 Monomers polymerise to form polymers.

a i Explain the term *polymerise*. [1]

   ii There are two types of polymerisation – addition and condensation. What is the difference between them? [2]

b An important monomer is chloroethene which has the structural formula shown here.

It is made by the following method.

$C_2H_4 + Cl_2 \longrightarrow C_2H_4Cl_2$ dichloroethane

This is heated to make chloroethene.

$C_2H_4Cl_2 \longrightarrow C_2H_3Cl + HCl$

   i Ethene is made by cracking alkanes. Complete the equation for cracking dodecane.

     $C_{12}H_{26} \longrightarrow$ .... $+ 2C_2H_4$ [1]

Another method of making dichloroethane is from ethane.

$C_2H_6 + 2Cl_2 \longrightarrow C_2H_4Cl_2 + 2HCl$

   ii Suggest a reason why the method using ethene is preferred. [1]

   iii Describe an industrial method of making chlorine. [2]

   iv Draw the structural formula of poly(chloroethene). Include three monomer units. [2]

Cambridge IGCSE Chemistry 0620 Paper 31 Q5 November 2010

14 The hydrolysis of complex carbohydrates to simple sugars is catalysed by enzymes called carbohydrases and also by dilute acids.

a i They are both catalysts. How do enzymes differ from catalysts such as dilute acids? [1]

   ii Explain why ethanol, C_2H_6O, is not a carbohydrate but glucose, $C_6H_{12}O_6$, is a carbohydrate. [2]

b Draw the structure of a complex carbohydrate, such as starch. The formula of a simple sugar can be represented by OH —☐— OH [3] c Iodine reacts with starch to form a deep blue colour.

  i In the experiment illustrated below, samples are removed at intervals and tested with iodine in potassium iodide solution.

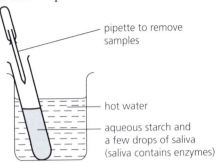

pipette to remove samples

hot water

aqueous starch and a few drops of saliva (saliva contains enzymes)

Typical results of this experiment are shown in the table.

time / min	colour of sample tested with iodine in potassium iodide solution
0	deep blue
10	pale blue
30	colourless

Explain these results. [3]

  ii If the experiment was repeated at a higher temperature, 60 °C, all the samples stayed blue. Suggest an explanation. [1]

Cambridge IGCSE Chemistry 0620 Paper 32 Q2 June 2010

1 Zinc blende is an ore of zinc containing zinc sulfide, ZnS. A student attempted to obtain a sample of zinc metal from this ore. The diagram shows the procedure followed in four stages.

Stage 1
A lump of zinc blende was heated to form zinc oxide.

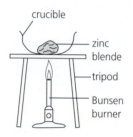

Stage 2
The zinc oxide was crushed.

Stage 3
Dilute acid was added.

Stage 4
The mixture was separated to give a solution of zinc sulfate.

a State the name of the apparatus labelled X. [1]
b Explain why the student should have carried out stage 2 before stage 1. [2]
c Identify the dilute acid used in stage 3. [1]
d Name the process used in stage 4. [1]
e Suggest how the student could have obtained a sample of zinc from the zinc sulfate solution. [1]

Cambridge IGCSE Chemistry 0620 Paper 6 Q1 June 2012

2 Describe a chemical test to distinguish between each of the following pairs of substances.

Example: hydrogen and carbon dioxide
test: lighted splint
result: with hydrogen gives a pop
result: with carbon dioxide splint is extinguished

a zinc carbonate and zinc chloride [2]
b ammonia and chlorine [3]
c aqueous iron(II) sulfate and aqueous iron(III) sulfate [3]

Cambridge IGCSE Chemistry 0620 Paper 6 Q3 June 2009

3 A student investigated the addition of four different solids, A, B, C and D, to water. Five experiments were carried out.

Experiment 1
By using a measuring cylinder, 30 cm³ of distilled water was poured into a polystyrene cup and the initial temperature of the water was measured. 4 g of solid A was added to the cup and the mixture stirred with a thermometer. The temperature of the solution was measured after 2 minutes.

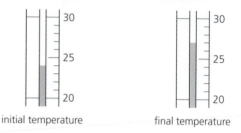

Experiment 2
Experiment 1 was repeated using 4 g of solid B.

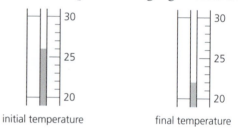

Experiment 3
Experiment 1 was repeated using 4 g of solid C.

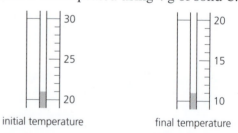

Experiment 4
Experiment 1 was repeated using 4 g of solid D.

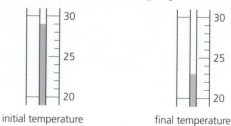

Experiment 5

A little of the solution from Experiment 4 was added to a little of the solution from Experiment 2 in a test-tube. The observations were recorded.

observations

a fast reaction
vigorous effervescence and bubbles produced

a Copy out the table and use the thermometer diagrams for Experiments 1–4 to record the initial and final temperatures. Calculate and record the temperature difference in the table.

expt	initial temperature/°C	final temperature/°C	difference /°C
1			
2			
3			
4			

[4]

b Draw a labelled bar chart of the results to Experiments 1, 2, 3 and 4 on graph paper. Label your bar chart as shown below. [4]

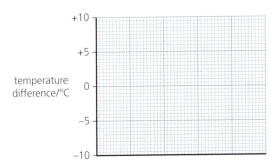

Use the results and observations from Experiments 1 – 5 to answer the following questions.

c i Which solid dissolves in water to produce an exothermic reaction? [1]

ii Give a reason why you chose this solid. [1]

d Which Experiment produced the largest temperature change? [1]

e Predict the temperature change that would happen if

i 8 g of solid B was used in Experiment 2, [1]

ii 60 cm³ of water was used in Experiment 4. [1]

iii Explain your answer to **e ii**. [2]

f Suggest an explanation for the observations in Experiment 5. [2]

Cambridge IGCSE Chemistry 0620 Paper 6 Q4 November 2008

4 Hydrogen peroxide breaks down to form oxygen. The volume of oxygen given off can be measured using this apparatus.

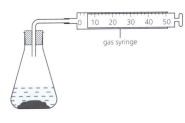

gas syringe

Solids W and X both catalyse the breakdown of hydrogen peroxide. The syringe diagrams show the volume of oxygen formed every 20 seconds using these catalysts at 25 °C.

time/s	using catalyst W	using catalyst X
0	0 10 20 30 40	0 10 20 30 40
20	0 10 20 30 40	0 10 20 30 40
40	0 10 20 30 40	0 10 20 30 40
60	0 10 20 30 40	0 10 20 30 40
80	0 10 20 30 40	0 10 20 30 40
100	0 10 20 30 40	0 10 20 30 40

a Copy the table. Use the gas syringe diagrams to complete it.

time/s	volume of oxygen/cm³	
	catalyst W	catalyst X
0		
20		
40		
60		
80		
100		

[3]

b Plot a graph to show each set of results. Clearly label the curves. [6]

c Which solid is the better catalyst in this reaction? Give a reason for your choice. [2]

d Why is the final volume of oxygen the same in each experiment? [1]

e Sketch a line on the grid to show the shape of the graph you would expect if the reaction with catalyst X was repeated at 40 °C. [2]

Cambridge IGCSE Chemistry 0620 Paper 6 Q6 June 2007

5 An experiment was carried out to determine the solubility of potassium chlorate at different temperatures. The solubility is the mass of potassium chlorate that dissolves in 100 g of water. The results obtained are shown in the table below.

temperature / °C	0	10	20	30	40	50	60
solubility in g / 100 g of water	14	17	20	24	29	34	40

a Draw a smooth line graph to show the solubility of potassium chlorate at different temperatures. Label your graph as shown below. [4]

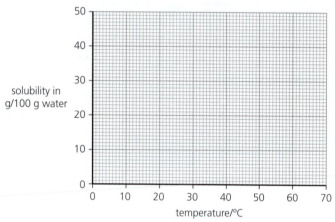

b Use your graph to determine the solubility of potassium chlorate at 70 °C. Show clearly on the graph how you obtained your answer. [2]

c What would be the effect of cooling a saturated solution of potassium chlorate from 60 °C to 20 °C? [2]

Cambridge IGCSE Chemistry 0620 Paper 6 Q6 November 2008

6 A student investigated the reaction of methane, CH_4, and copper(II) oxide. She passed methane gas over hot copper(II) oxide using the apparatus shown. The solid changed colour to red-brown and drops of liquid condensed in the cold part of the tube.

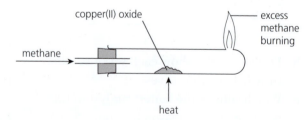

a What was the original colour of the solid? [1]

b Suggest the identity of
  i the red-brown solid **ii** the drops of liquid [2]

c Suggest a physical test to identify the liquid, and give the result. [2]

Cambridge IGCSE Chemistry 0620 Paper 62 Q2 June 2013

7 Electricity was passed through molten lead iodide as shown below.

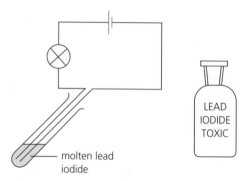

A purple gas was observed coming from the positive electrode (anode).

a What piece of apparatus is missing from the diagram? [1]

b Copy the diagram and clearly label the electrodes. [1]

c Give one other expected observation
  i during the electrolysis
  ii when the molten lead iodide cools and solidifies. [2]

d Suggest why a stopper is not used in the top of the boiling tube. [1]

e Explain the observation at the positive electrode. [2]

f Give one safety precaution necessary when carrying out this experiment. [1]

Cambridge IGCSE Chemistry 0620 Paper 63 Q6 June 2012

8 a A student investigated the reaction between dilute hydrochloric acid and two different alkaline solutions, F and G.
Two experiments were carried out.

Experiment 1

A burette was filled up to the 0.0 cm³ mark with dilute hydrochloric acid.

Using a measuring cylinder, 25 cm³ of solution F was placed into a conical flask with a few drops of phenolphthalein indicator.

The hydrochloric acid was added to the flask until the colour of the phenolphthalein changed. Copy the table and use the burette diagram to record the final volume.

Experiment 2

Experiment 1 was repeated using solution G. Use the burette diagrams to record the volumes and complete the table of results.

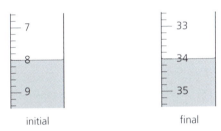

Table of results

	burette readings / cm³	
	Experiment 1	Experiment 2
final reading		
initial reading		
difference			[4]

b Which ion is present in all alkaline solutions? [1]

c i In which Experiment was the greatest volume of hydrochloric acid used? [1]

   ii Compare the volumes of hydrochloric acid used in Experiments 1 and 2. [1]

   iii Suggest explanations for the difference in volumes. [2]

d Predict the volume of hydrochloric acid which would be needed to react completely with 12.5 cm³ of solution G. Explain your answer. [3]

e i State two sources of error in the experimental procedure. [2]

   ii Suggest two improvements to reduce the sources of error in the experimental procedure. [2]

Cambridge IGCSE Chemistry 0620 Paper 63 Q4 November 2010

9 The notes below show the steps taken by a student to prepare crystals of hydrated nickel nitrate, $Ni(NO_3)_2 \cdot 6H_2O$.

Step 1 Place 25 cm³ of dilute nitric acid in a beaker.

Step 2 Add nickel carbonate powder to the beaker until it is in excess.

Step 3 Separate the solution of nickel nitrate from the mixture.

Step 4 Heat the solution to obtain crystals of hydrated nickel nitrate.

a i Name the piece of apparatus used to measure the nitric acid in Step 1. [1]

   ii Why is it **not** necessary to heat the dilute nitric acid before adding the nickel carbonate? [1]

b How would the student know when excess nickel carbonate was present in Step 2? [1]

c Draw a diagram to show the separation method used in Step 3.

d How could the student make sure a good sample of crystals was obtained when carrying out Step 4? Explain your answer... [2]

Cambridge IGCSE Chemistry 0620 Paper 63 Q2 November 2010

10 The label shows some information on a bottle of liquid sink and drain cleaner.

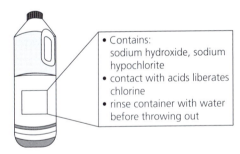

a Give a chemical test, and the result, for the presence of sodium hydroxide. [2]

b Suggest why it could be dangerous to pour fizzy drinks into a sink containing this liquid cleaner. [2]

c Why should the container be rinsed with water before throwing out? [1]

d Give a chemical test, and the result, for chlorine. [2]

Cambridge IGCSE Chemistry 0620 Paper 61 Q7 June 2011

11 A mixture of two solids, **R** and **S**, was analysed.
Solid **R** was the water-soluble salt aluminium
sulfate, $Al_2(SO_4)_3$, and solid **S** was an insoluble salt.

The tests on the mixture and some of the
observations are in the following table. Write down
the observations that are missing from the table.

tests	observations
Distilled water was added to the mixture in a boiling tube. The boiling tube was shaken and the contents of the boiling tube filtered, keeping the filtrate and residue for the following tests. The filtrate was divided into five test-tubes.	

tests on the filtrate	observations
a Appearance of the first portion of the filtrate.	.... [1]
b Drops of aqueous sodium hydroxide were added to the second portion of the solution and the test-tube shaken. Excess aqueous sodium hydroxide was then added to the test-tube.	.... [3]
c Aqueous ammonia was added to the third portion, dropwise and then in excess.	.... [2]
d Dilute nitric acid was added to the fourth portion of the solution followed by aqueous silver nitrate.	.... [1]
e Dilute nitric acid was added to the fifth portion of the solution and then aqueous barium nitrate.	.... [2]

tests on the residue	observations
f Dilute hydrochloric acid was added to the residue. The gas given off was tested.	rapid effervescence limewater turned milky
Excess aqueous sodium hydroxide was added to the mixture in the test-tube.	white precipitate, insoluble in excess

g Name the gas given off in test **f**. [1]

h What conclusions can you draw about solid **S**? [2]

Cambridge IGCSE Chemistry 0620 Paper 62 Q4 June 2013

12 Two solids, **S** and **V**, were analysed.

S was copper(II) oxide.

The tests on the solids, and some of the observations, are in the following table. Write down the observations that are missing from the table.

test	observation
tests on solid S	
a Appearance of solid **S**	black solid
b Hydrogen peroxide was added to solid S in a test-tube. A glowing splint was inserted into the tube	slow effervescence splint relit
c Dilute sulfuric acid was added to solid S in a test-tube. The mixture was heated to boiling point. The solution was divided into three equal portions into test-tubes.	blue solution formed
i To the first portion of the solution excess sodium hydroxide was added.	.... [1]
ii To the second portion of the solution, about 1 cm^3 of aqueous ammonia solution was added.	.... [2]
Excess ammonia solution was then added.	.... [2]
iii To the third portion of the solution, dilute hydrochloric acid was added followed by barium chloride solution.	.... [2]
tests on solid V	
d Appearance of solid **V**	black solid
e Hydrogen peroxide was added to solid **V** in a test-tube. A glowing splint was inserted into the tube.	rapid effervescence splint relit

   f i Compare the reactivity of solid **S** and solid **V** with hydrogen peroxide. [1]

     ii Identify the gas given off in test **e**. [1]

   g What conclusions can you draw about solid **V**? [2]

Cambridge IGCSE Chemistry 0620 Paper 6 Q5 June 2009

13 Two metals, A and B, each react with dilute sulfuric acid to produce hydrogen. Plan an investigation to show which metal, A or B, is the more reactive metal. You may include a diagram in your answer. You are provided with:
- standard laboratory equipment
- powdered metals A and B
- dilute sulfuric acid.

Cambridge IGCSE Chemistry 0620 Paper 62 Q6 June 2013

Glossary

A

A_r short for *relative atomic mass of an element* – the average mass of its isotopes, relative to the mass of a carbon-12 atom)

acetylene a gas (formula C_2H_2) used as a fuel, for example in the oxy-acetylene torch

acid rain rain that is acidic because gases such as sulfur dioxide are dissolved in it (from burning fossil fuels)

acidic solution has a pH less than 7; an acidic solution contains H^+ ions

acid fermentation the process in which bacteria convert ethanol to ethanoic acid

addition reaction where a molecule adds onto an alkene, and the C = C double bond of the alkene changes to a single bond

addition polymerisation where small molecules join to form a very large molecule, by adding on at double bonds

alcohols a family of organic compounds, similar to the alkanes but with the OH functional group; ethanol is an example

alkali a soluble base; for example sodium hydroxide

alkali metals the Group I elements of the Periodic Table

alkaline earth metals the Group II elements of the Periodic Table

alkaline solution has a pH above 7; alkaline solutions contain OH^- ions

alkanes a family of saturated hydrocarbons with the general formula C_nH_{2n+2}; 'saturated' means they have only single C–C bonds

alkenes a family of unsaturated hydro-carbons with the general formula C_nH_{2n}; their molecules contain a carbon = carbon double bond

allotropes different forms of an element; diamond and graphite are allotropes of carbon

alloy a mixture where at least one other substance is added to a metal, to improve its properties; the other substance is often a metal too (but not always)

amphoteric can be both acidic and basic in its reactions; for example aluminium oxide is an amphoteric oxide

anion another name for a negative ion

anode the positive electrode of a cell

aquifer underground rocks holding a large volume of water; it can be pumped out to give a water supply

atmosphere the layer of gases around the Earth; here at the Earth's surface, we call it air

atomic number the number of protons in the nucleus of an atom; it is also called the *proton number*

atoms elements are made up of atoms, which contain protons, neutrons, and electrons

Avogadro constant the number of particles in one mole of an element or compound; it is 6.02×10^{23}

B

back reaction the reaction in which the product breaks down again, in a reversible reaction

bacteria tiny organisms, some of which can cause disease; others break down dead plant and animal material

balanced equation a chemical equation in which the number of each type of atom is the same on both sides of the arrow

base a metal oxide or hydroxide; a base will neutralise an acid, to form a salt and water

battery a portable electrical cell; for example a torch battery

biodegradable will decay naturally in the soil, with the help of bacteria

biopolymer a polymer made by bacteria

blast furnace the chemical plant in which iron is extracted from its ore, iron(III) oxide

boiling the change from a liquid to a gas, which takes place at the boiling point

boiling point the temperature at which a substance boils

bond energy the energy needed to break a bond, or released when the bond is formed; it is given in kilojoules (kJ) per mole

bonding how the atoms are held together in an element or compound; there are three types of bonds: ionic, covalent, and metallic

brittle breaks up easily when struck

brine the industrial name for a concentrated solution of sodium chloride in water; it can be made by dissolving rock salt

Brownian motion the random motion of particles suspended in a liquid or a gas; it occurs because the particles are continually bombarded by molecules

burette a piece of laboratory equipment for delivering a measured volume of liquid

burning an exothermic chemical reaction in which the reactant combines with oxygen to form an oxide; also called combustion

C

carbon cycle the way carbon moves non-stop between the atmosphere, living things, the land, and the ocean; it moves in the form of carbon dioxide

carboxylic acids a family of organic acids, which have the COOH functional group

cast iron iron from the blast furnace that is run into molds to harden; it contains a high % of carbon, which makes it brittle

catalyst a substance that speeds up a chemical reaction, without itself being used up in the process

catalytic converter a device in a car exhaust, in which catalysts are used to convert harmful gases to harmless ones

catalytic cracking where large molecules of hydrocarbons are split up into smaller ones, with the help of a catalyst

cathode the negative electrode of an electrolysis cell

cation another name for a positive ion

cell (biological) the building blocks for animals and plants

cell (electrical) a device that converts chemical energy to electrical energy

cement a substance used in building, made from limestone and clay

ceramic a hard, ureactive material that can withstand high temperatures, made by baking clay in a kiln; ceramics are non-conductors

chalk a rock made of calcium carbonate

change of state a change in the physical state of a substance – for example from solid to liquid, or liquid to gas

chemical change a change in which a new chemical substance forms

chemical equation uses chemical symbols to describe a chemical reaction

chemical reaction a process in which chemical change takes place

chromatogram the paper showing the separated coloured substances, after paper chromatography has been carried out

climate change how climates around the Earth are changing, because of the rise in average air temperatures

coagulant a substance that makes small particles stick together; used in cleaning up water, ready for piping to homes

coke a form of carbon made by heating coal

combination where two or more substances react to form a single substance

combustible catches fire and burns easily

combustion another name for burning

compound fertiliser it provides nitrogen, potassium, and phosphorus for plants

compound ion an ion containing more than one element; for example the nitrate ion NO_3^-

compound a substance in which two or more elements are chemically combined

concentration tells you how much of one substance is dissolved in another; usually given as grams or moles per dm^3

condensation the physical change in which a gas turns into a liquid on cooling

condensation polymerisation where molecules join to make very large molecules, by eliminating small molecules (such as water molecules)

condenser a piece of lab equipment used to cool a gas rapidly, and turn it into a liquid

conductor a substance that allows heat or electricity to pass through it easily

Contact process the industrial process for making sulfuric acid

corrosion where a substance is attacked by air or water, from the surface inwards; the corrosion of iron is called rusting

covalent bond the chemical bond formed when two atoms share electrons

covalent compound a compound made of atoms joined by covalent bonds

cracking reactions in which long-chain hydrocarbon molecules are broken down to shorter, more useful molecules

crude oil the fossil fuel formed over millions of years from the remains of tiny sea plants and animals; it is also called petroleum

crystallisation the process in which crystals form, as a saturated solution cools

D

decomposition reaction where a substance breaks down to give two or more products

denature to destroy the structure of an enzyme by heat, or a change in pH

density tells you how 'heavy' something is; the density of a substance is its mass per unit volume; for water it is 1 g/cm^3

diatomic its molecules contain two atoms joined by a covalent bond

diffusion the process in which particles mix by colliding randomly with each other, and bouncing off in all directions

displacement reaction a reaction in which a more reactive element takes the place of a less reactive one, in a compound

dissolving the process in which a soluble substance forms a solution

distillation separating a liquid from a mixture by boiling it off, then condensing it

double bond a covalent bond in which two atoms share two pairs of electrons

ductile can easily be drawn out into a wire

dynamic equilibrium where forward and back reactions take place at the same rate, so there is no *overall* change

E

electrodes the conductors used to carry current into and out of an electrolyte; they could be graphite rods, for example

electrolysis the process of breaking down a compound by passing a current through it

electrolyte the liquid through which the current is passed, in electrolysis; the current is carried by ions in the electrolyte

electron distribution how the electrons in an atom are arranged in shells (2 + 8 + ...)

electron shells the different energy levels which electrons occupy, around the nucleus

electronic configuration another term for electron distribution

electrons the particles with a charge of 1– and almost no mass, in an atom

electroplating coating one metal with another, using electrolysis

element a substance that cannot be split into anything simpler, in a chemical reaction

empirical found by experiment

empirical formula shows the simplest ratio in which the atoms in a compound are combined

endothermic takes in energy from the surroundings

enzymes proteins made by living cells, that act as biological catalysts

equation it uses symbols to describe a chemical reaction (but a *word equation* uses just words)

equilibrium the state where the forward and back reactions are taking place at the same rate, in a reversible reaction; so there is no *overall* change

ester a compound formed when an alcohol reacts with a carboxylic acid; esters often smell of fruit or flowers

evaporation the physical change where a liquid turns to a gas at a temperature below its boiling point

exothermic gives out energy

extract to remove a metal from its ore

F

fermentation the process in which the enzymes in yeast break down sugars, to form ethanol and carbon dioxide

fertilisers substances added to soil to help crops grow well

filtering separating solids from liquids by pouring the mixture through filter paper

filtrate the liquid obtained from filtration (after the solid has been removed)

flammable burns easily

flue gas desulfurisation the removal of sulfur dioxide from the waste gases at power stations, to stop it getting into the atmosphere

formula uses symbols and numbers to tell you what elements are in a compound, and the ratio in which they are combined

forward reaction the reaction in which the product is made, in a reversible reaction

fossil fuels petroleum (crude oil), natural gas, and coal; they are called the fossil fuels because they were formed from the remains of living things, millions of years ago

fractional distillation a method used to separate two or more liquids that have different boiling points

fractions the different groups of compounds that a mixture is separated into, by fractional distillation; fractions are collected one by one

freezing the change from liquid to solid, that occurs at the freezing point (= melting point)

fuel a substance we use to provide energy; most fuels are burned to release their energy (but nuclear fuels are not)

fuel cell a cell in which a chemical reaction provides electricity

functional group the part of the molecule of an organic compound, that largely dictates how it reacts; for example the OH group in molecules of the alcohol family

G

galvanising coating iron with zinc, to prevent the iron from rusting

giant structure where a very large number of atoms or ions are held in a lattice by strong bonds; metals, diamond and ionic solids such as sodium chloride are all giant structures

global warming the rise in average temperatures taking place around the world; many scientists believe that carbon dioxide (from burning fossil fuels) is the main cause

greenhouse gas a gas in the atmosphere that traps heat, preventing its escape into space; carbon dioxide and methane are examples

group a column of the Periodic Table; elements in a group have similar properties

H

Haber process the process for making ammonia from nitrogen and hydrogen, in industry

half-equation an equation that shows the reaction taking place at an electrode

halogens the Group VII elements of the Periodic Table

heating curve a graph showing how the temperature of a substance changes on heating, while it goes from solid to liquid to gas

homologous series a family of organic compounds, that share the same general formula and have similar properties

hydrated has water molecules built into its crystal structure; for example copper(II) sulfate: $CuSO_4.5H_2O$

hydrocarbon a compound containing *only* carbon and hydrogen

hydrogenation adding hydrogen

hydrogen fuel cell it uses the reaction between hydrogen and oxygen to give an electric current

hydrolysis the breaking down of a compound by reaction with water

hypothesis a statement you can test by doing an experiment and taking measurements

321

I

incomplete combustion the burning of fuels in a limited supply of oxygen; it gives carbon monoxide instead of carbon dioxide

indicator a chemical that shows by its colour whether a substance is acidic or alkaline

inert does not react (except under extreme conditions)

inert electrode is not changed during electrolysis; all it does is conduct the current

in excess more than is needed for a reaction; some will be left at the end

insoluble does not dissolve in a solvent

insulator a poor conductor of heat or electricity

intermolecular forces forces between molecules

ion a charged atom or group of atoms formed by the gain or loss of electrons

ionic bond the bond formed between ions of opposite charge

ionic compound a compound made up of ions, joined by ionic bonds

ionic equation shows only the ions that actually take part in a reaction, and ignores any other ions present; the other ions are called *spectator ions*

isomers compounds that have the same formula, but a different arrangement of atoms

isotopes atoms of the same element, that have a different numbers of neutrons

L

lattice a regular arrangement of particles

lime the common name for calcium oxide

limewater a solution of the slightly soluble compound calcium hydroxide

locating agent used to show up colourless substances, in chromatography; it reacts with them to give coloured substances

M

M_r short for *relative molecular mass* or *relative formula mass* for a compound; see the definitions for these

malleable can be bent or hammered into shape

mass number the total number of protons and neutrons in the nucleus of an atom; it is also called the *nucleon number*

mass spectrometer an instrument used to find the masses of atoms and molecules

melting point the temperature at which a solid substance melts

melting the physical change from a solid to a liquid

metal an element that shows metallic properties (for example conducts electricity, and forms positive ions)

metallic bond the bond that holds the atoms together in a metal

metalloid an element that has properties of both a metal and a non-metal

microbe a microscopic (very tiny) living organism, such as a bacterium or virus

minerals compounds that occur naturally in the Earth; rocks contain different minerals

mixture contains two or more substances that are not chemically combined

molar solution contains one mole of a substance in 1 dm^3 (1 litre) of water

mole the amount of a substance that contains the same number of elementary units as the number of carbon atoms in 12g of carbon-12; you obtain it by weighing out the A_r or M_r of the substance, in grams

molecular made up of molecules

molecule a unit of two or more atoms held together by covalent bonds

monatomic made up of single atoms; for example neon is a monatomic element

monomers small molecules that join together to form polymers

N

native describes a metal that is found in the Earth as the element

negative electrode another name for the cathode, in an electrolysis cell

negative ion an ion with a negative charge

neutral (electrical) has no charge

neutral (oxide) is neither acidic nor basic; carbon monoxide is a neutral oxide

neutral (solutions) neither acidic nor alkaline; neutral solutions have a pH of 7

neutralisation the chemical reaction between an acid and a base or a carbonate, giving a salt and water

neutron a particle with no charge and a mass of 1 unit, found in the nucleus of an atom

nitrogenous fertiliser it provides nitrogen for plants, in the form of nitrate ions or ammonium ions

noble gases the Group VIII elements of the Periodic Table; they are called 'noble' because they are so unreactive

non-metal an element that does not show metallic properties: the non-metals lie to the right of the zig-zag line in the Periodic Table, (except for hydrogen, which sits alone)

non-renewable resource a resource such as petroleum that we are using up, and which will run out one day

non-toxic not harmful health

nucleon number the total number of protons and neutrons in the nucleus of an atom; it is also called the *mass number*

nuclear fuel contains radioisotopes such as uranium-235; these are forced to break down, giving out energy

nucleus the centre part of the atom, made up of protons and neutrons

O

ore rock containing a metal, or metal compounds, from which the metal is extracted

organic chemistry the study of organic compounds (there are millions of them!)

organic compound a compound containing carbon, and usually hydrogen; petroleum is a mixture of many organic compounds

oxidation a chemical reaction in which a substance gains oxygen, or loses electrons

oxidation state every atom in a formula can be given a number that describes its oxidation state; for example in NaCl, the oxidation states are +I for sodium, and – I for chlorine

oxide a compound formed between oxygen and an other element

oxidising agent a substance that brings about the oxidation of another substance

ozone a gas with the formula O_3

ozone layer the layer of ozone up in the atmosphere, which protects us from harmful UV radiation from the sun

P

paper chromatography a way to separate the substances in a mixture, using a solvent and special paper; the substances separate because they travel over the paper at different speeds

percentage composition it tells you which elements are in a compound, and what % of each is present by mass

period a horizontal row of the Periodic Table; its number tells you how many electron shells there are

periodicity the pattern of repeating properties that shows up when elements are arranged in order of proton number; you can see it in the groups in the Periodic Table

Periodic Table the table showing the elements in order of increasing proton number; similar elements are arranged in columns called groups

petroleum a fossil fuel formed over millions of years from the remains of tiny sea plants and animals; it is also called *crude oil*

pH scale a scale that tells you how acidic or alkaline a solution is; it is numbered 0 to 14

photochemical reaction a reaction that depends on light energy; photosynthesis is an example

photodegradeable can be broken down by light

photosynthesis the process in which plants convert carbon dioxide and water to glucose and oxygen

physical change a change in which no new chemical substance forms; melting and boiling are physical changes

physical properties properties such as density and melting point (that are not about chemical behaviour)

pipette a piece of laboratory equipment used to deliver a known volume of liquid, accurately

plastics a term used for synthetic polymers (made in factories, rather than in nature)

pollutant a substance that causes harm if it gets into the air or water

pollution when harmful substances are released into the environment

polymer a compound containing very large molecules, formed by polymerisation

polymerisation a chemical reaction in which many small molecules join to form very large molecules; the product is called a polymer

positive ion an ion with a positive charge

precipitate an insoluble chemical produced during a chemical reaction

precipitation reaction a reaction in which a precipitate forms

product a chemical made in a chemical reaction

protein a polymer made up of many different amino acid units joined together

proton number the number of protons in the nucleus of an atom; it is also called the *atomic number*

proton a particle with a charge of 1+ and a mass of 1 unit, found in the nucleus of an atom

pure there is only one substance in it

Q

quicklime another name for calcium oxide

R

radioactive isotopes (radioisotopes) unstable atoms that break down, giving out radiation

random motion the zig-zag path a particle follows as it collides with other particles and bounces away again

rate of reaction how fast a reaction is

reactant a starting chemical for a chemical reaction

reactive tends to react easily

reactivity how readily a substance reacts

reactivity series the metals listed in order of their reactivity

recycling reusing resources such as scrap metal, glass, paper and plastics

redox reaction any reaction in which electrons are transferred; one substance is oxidised (it loses electrons) and another is reduced (it gains electrons)

reducing agent a substance which brings about the reduction of another substance

reduction when a substance loses oxygen, or gains electrons

refining (petroleum) the process of separating petroleum (crude oil) into groups of compounds with molecules fairly close in size; it is carried out by fractional distillation

refining (metals) the process of purifying a metal; copper is refined using electrolysis

relative atomic mass (A_r) the average mass of the isotopes of an element, relative to the mass of an atom of carbon-12

relative formula mass (M_r) the mass of one formula unit of an ionic compound; you find it by adding together the relative atomic masses of the atoms in the formula

relative molecular mass the mass of a molecule; you find it by adding the relative atomic masses of the atoms in it

renewable resource a resource that will not run out; for example water, air, sunlight

residue the solid you obtain when you separate a solid from a liquid by filtering

respiration the reaction between glucose and oxygen that takes place in the cells of all living things (including you) to provide energy

reversible reaction a reaction that can go both ways: a product can form, then break down again; the symbol $\rightleftharpoons$ is used to show a reversible reaction

rusting the special term used for the corrosion of iron; oxygen and water attack the iron, and red-brown rust forms

S

sacrificial protection allowing one metal to corrode, in order to protect another metal

salt an ionic compound formed when an acid reacts with a metal, a base, or a carbonate

saturated compound an organic compound in which all the bonds between carbon atoms are single covalent bonds

saturated solution no more of the solute will dissolve in it, at that temperature

single bond the bond formed when two atoms share just one pair of electrons

slaked lime another name for calcium hydroxide

solubility the amount of solute that will dissolve in 100 grams of a solvent, at a given temperature

soluble will dissolve in a solvent

solute the substance you dissolve in the solvent, to make a solution

solution a mixture obtained when a solute is dissolved in a solvent

solvent the liquid in which a solute is dissolved, to make a solution

sonorous makes a ringing noise when struck

spectator ions ions that are present in a reaction mixture, but do not actually take part in the reaction

stable unreactive

state symbols these are added to an equation to show the physical states of the reactants and products (g = gas, l = liquid, s = solid, aq = aqueous)

structural formula the formula of a compound displayed to show the bonds between the atoms as lines; we often show organic compounds this way

T

thermal decomposition the breaking down of a compound by heating it

thermite process the redox reaction between iron oxide and aluminium, which produces molten iron

titration a laboratory technique for finding the exact volume of an acid solution that will react with a given volume of alkaline solution, or vice versa

toxic poisonous

transition elements the elements in the wide middle block of the Periodic Table; they are all metals and include iron, tin, copper, and gold

trend a gradual change; the groups within the Periodic Table show trends in their properties; for example as you go down Group I, reactivity increases

triple bond the bond formed when two atoms share three pairs of electrons; a nitrogen molecule has a triple bond

U

universal indicator a paper or liquid you can use to find the pH of a solution; it changes colour across the whole range of pH

unreactive does not react easily

unsaturated compound an organic compound with at least one double bond between carbon atoms

V

valency a number that tells you how many electrons an atom gains, loses or shares, in forming a compound

valency electrons the electrons in the outer shell of an atom

variable oxidation state where atoms of an element can lose different numbers of electrons, in forming compounds; for example copper forms Cu^+ and Cu^{2+} ions

viscosity a measure of how runny a liquid is; the more runny it is, the lower its viscosity

viscous thick and sticky

volatile evaporates easily, to form a vapour

W

water of crystallisation water molecules built into the crystal structure of a compound; for example in copper(II) sulfate, $CuSO_4.5H_2O$

weak acids acids in which only some of the molecules are dissociated, to form H^+ ions; ethanoic acid is a weak acid

Y

yield the actual amount of a product obtained in a reaction; it is often given as a % of the theoretical yield (which you can work out from the equation)

The periodic table

I	II											III	IV	V	VI	VII	VIII
1 H hydrogen 1																	**2** He helium 4
3 Li lithium 7	**4** Be beryllium 9											**5** B boron 11	**6** C carbon 12	**7** N nitrogen 14	**8** O oxygen 16	**9** F fluorine 19	**10** Ne neon 20
11 Na sodium 23	**12** Mg magnesium 24											**13** Al aluminium 27	**14** Si silicon 28	**15** P phosphorus 31	**16** S sulfur 32	**17** Cl chlorine 35.5	**18** Ar argon 40
19 K potassium 39	**20** Ca calcium 40	**21** Sc scandium 45	**22** Ti titanium 48	**23** V vanadium 51	**24** Cr chromium 52	**25** Mn manganese 55	**26** Fe iron 56	**27** Co cobalt 59	**28** Ni nickel 59	**29** Cu copper 64	**30** Zn zinc 65	**31** Ga gallium 70	**32** Ge germanium 73	**33** As arsenic 75	**34** Se selenium 79	**35** Br bromine 80	**36** Kr krypton 84
37 Rb rubidium 85	**38** Sr strontium 88	**39** Y yttrium 89	**40** Zr zirconium 91	**41** Nb niobium 93	**42** Mo molybdenum 96	**43** Tc technetium –	**44** Ru ruthenium 101	**45** Rh rhodium 103	**46** Pd palladium 106	**47** Ag silver 108	**48** Cd cadmium 112	**49** In indium 115	**50** Sn tin 117	**51** Sb antimony 122	**52** Te tellurium 128	**53** I iodine 127	**54** Xe xenon 131
55 Cs caesium 133	**56** Ba barium 137	**57–71** lanthanoids	**72** Hf hafnium 178	**73** Ta tantalum 181	**74** W tungsten 184	**75** Re rhenium 186	**76** Os osmium 190	**77** Ir iridium 192	**78** Pt platinum 195	**79** Au gold 197	**80** Hg mercury 201	**81** Tl thallium 204	**82** Pb lead 207	**83** Bi bismuth 209	**84** Po polonium –	**85** At astatine –	**86** Rn radon –
87 Fr francium –	**88** Ra radium –	**89–103** actinoids	**104** Rf rutherfordium –	**105** Db dubnium –	**106** Sg seaborgium –	**107** Bh bohrium –	**108** Hs hassium –	**109** Mt meitnerium –	**110** Ds darmstadtium –	**111** Rg roentgenium –	**112** Cr copernicium –		**114** Fl flerovium –		**116** Lv livermorium –		

lanthanoids

57 La lanthanum 139	**58** Ce cerium 140	**59** Pr praseodymium 141	**60** Nd neodymium 144	**61** Pm promethium –	**62** Sm samarium 150	**63** Eu europium 152	**64** Gd gadolinium 157	**65** Tb terbium 159	**66** Dy dysprosium 163	**67** Ho holmium 165	**68** Er erbium 167	**69** Tm thulium 169	**70** Yb ytterbium 173	**71** Lu lutetium 175

actinoids

89 Ac actinium –	**90** Th thorium 232	**91** Pa protactinium 231	**92** U uranium 238	**93** Np neptunium –	**94** Pu plutonium –	**95** Am americium –	**96** Cm curium –	**97** Bk berkelium –	**98** Cf californium –	**99** Es einsteinium –	**100** Fm fermium –	**101** Md mendelevium –	**102** No nobelium –	**103** Lr lawrencium –

The symbols and proton numbers of the elements

Element	Symbol	Proton number	Element	Symbol	Proton number	Element	Symbol	Proton number	Element	Symbol	Proton number
actinium	Ac	89	calcium	Ca	20	francium	Fr	87	lawrencium	Lw	103
aluminium	Al	13	californium	Cf	98	gadolinium	Gd	64	lead	Pb	82
americium	Am	95	carbon	C	6	gallium	Ga	31	lithium	Li	3
antimony	Sb	51	cerium	Ce	58	germanium	Ge	32	lutetium	Lu	71
argon	Ar	18	chlorine	Cl	17	gold	Au	79	magnesium	Mg	12
arsenic	As	33	chromium	Cr	24	hafnium	Hf	72	manganese	Mn	25
astatine	At	85	cobalt	Co	27	helium	He	2	mendelevium	Md	101
barium	Ba	56	copper	Cu	29	holmium	Ho	67	mercury	Hg	80
berkelium	Bk	97	curium	Cm	96	hydrogen	H	1	molybdenum	Mo	42
beryllium	Be	4	dysprosium	Dy	66	indium	In	49	neodymium	Nd	60
bismuth	Bi	83	einsteinium	Es	99	iodine	I	53	neon	Ne	10
boron	B	5	erbium	Er	68	iridium	Ir	77	neptunium	Np	93
bromine	Br	35	europium	Eu	63	iron	Fe	26	nickel	Ni	28
cadmium	Cd	48	fermium	Fm	100	krypton	Kr	36	niobium	Nb	41
caesium	Cs	55	fluorine	F	9	lanthanum	La	57	nitrogen	N	7

The symbols and proton numbers of the elements (continued)

Element	Symbol	Proton number	Element	Symbol	Proton number	Element	Symbol	Proton number
nobelium	No	102	rhodium	Rh	45	thallium	Tl	81
osmium	Os	76	rubidium	Rb	37	thorium	Th	90
oxygen	O	8	ruthenium	Ru	44	thulium	Tm	69
palladium	Pd	46	samarium	Sm	62	tin	Sn	50
phosphorus	P	15	scandium	Sc	21	titanium	Ti	22
platinum	Pt	78	selenium	Se	34	tungsten	W	74
plutonium	Pu	94	silicon	Si	14	uranium	U	92
polonium	Po	84	silver	Ag	47	vanadium	V	23
potassium	K	19	sodium	Na	11	xenon	Xe	54
praseodymium	Pr	59	strontium	Sr	38	ytterbium	Yb	70
promethium	Pm	61	sulfur	S	16	yttrium	Y	39
protactinium	Pa	91	tantalum	Ta	73	zinc	Zn	30
radium	Ra	88	technetium	Tc	43	zirconium	Zr	40
radon	Rn	86	tellurium	Te	52			
rhenium	Re	75	terbium	Tb	65			

Relative atomic masses (A_r) for calculations

Element	Symbol	A_r
aluminium	Al	27
bromine	Br	80
calcium	Ca	40
carbon	C	12
chlorine	Cl	35.5
copper	Cu	64
fluorine	F	19
helium	He	4
hydrogen	H	1
iodine	I	127
iron	Fe	56
lead	Pb	207
lithium	Li	7
magnesium	Mg	24
manganese	Mn	55
neon	Ne	20
nitrogen	N	14
oxygen	O	16
phosphorus	P	31
potassium	K	39
silver	Ag	108
sodium	Na	23
sulfur	S	32
zinc	Zn	65

Index

acetylene 209
acid rain 210, 227
acids 144
    acid reactions 148, 149
    acids produce hydrogen ions 146
    comparing acids 146
    corrosive acids 144
    hydrolysis 271
    indicators 145
    pH 145
    proton donors 151
    redox reactions 149
    carboxylic acids 254–255
    strong acids 146
    weak acids 144, 146, 255
addition polymerisation 259
    chain lengths in polythene 260
    identifying the monomer 261
    making other polymers by addition 261
    monomers 260
    polymerisation of ethene 260
addition reactions 250, 251, 253, 260
air 18, 206, 228
    air pollutants 210
    fractional distillation of liquid air 208
    nitrogen 209
    noble gases 209
    oxygen 207, 209
    reducing air pollution 211
    separating gases from air 208
alanine 270
alcohols 252–253
    alcoholic drinks 252
algae 225
alkali metals 162, 178
    chemical properties 164–165
    physical properties 164
    reactivity 165
alkaline earth metals 162, 178
alkalis 144, 148
    alkalis produce hydroxide ions 147
    ammonia solution 147
    indicators 145
    pH scale 145
    strong and weak alkalis 147
alkanes 246, 248
    isomers 249
    key points 248–249
alkenes 245, 250
    isomers in the alkene family 251
    key points 250
    polymerisation 251
    test for unsaturation 251
allotropes 57, 226, 230
alloys 171, 193, 197, 199, 200
alumina 196
aluminium 178, 203
    alloys 199

corrosion 187
extraction 196–197
ore 191, 193, 196
properties 197
reactivity 187
uses 196, 198
aluminium oxide 191, 196
amide linkage 262, 270
amino acids 259, 268
ammonia 221
    ammonia production 122–123
    ammonia solution 147
    covalent compound 52
    fountain experiment 221
    how amino acids join up to make proteins 270
    making ammonia in the lab 221
    Haber process 222–223
ammonium 224
amylase 269
anions 282
    tests for anions 282–283
anodes 100, 101, 102, 104, 106, 107
aqueous solution 13
argon 44, 168, 169
atmosphere 206
    atmospheric pollution 210–211
atoms 3, 26, 28, 30
    atomic number 28
    atomic weight 174
    atoms of the first 20 elements 29
    covalent bond 50–51
    development of model of the atom 34–35
    inside the atom 36–37
    relative atomic mass 66
    why do atoms form bonds? 44–45
Avogadro constant 72
Avogadro's Law 76
aquifers 214

bacteria 31
barium sulfate 156–157
bases 148, 150, 151, 178
    proton acceptors 151
bath salts 48
batteries 98, 118
    button batteries 119
    car batteries 118–119
    lithium-ion batteries 119
    negative terminals 98, 118
    positive terminals 98
    torch batteries 118
bauxite 191, 196
biofuels 252
biological detergents 137, 139
bio-polymers 267
blast furnaces 194
    basic oxygen converters 201
    charge 194

products 195
    reactions in blast furnace 195
boiling points 4, 5, 7
    covalent compounds 55
    gases in air 208
    ionic compounds 55
    purity 15
bonding
    atomic bonds 44–45
    bond energies 112
    covalent bonds 50–51
    double bonds 260
    ionic bonds 46–47
    making and breaking bonds 112
    metals 58–59
bonding diagrams 46
branched chain structures 240, 249
brass 199
bread 138
brine 104–105
bromine water 251
bronze 202
Brownian motion 2, 35
butane 248

calcium 89
cancer treatment by radioisotopes 31
carbohydrates 268
    complex carbohydrates 268
carbon 182, 230
    allotropes 57
    carbon atom 30
    carbon compounds 230, 232–233
    carbon dating 31
    metal extraction 192
    reaction between carbon and oxygen 64
carbon cycle 230
    adding carbon dioxide to atmosphere 231
    fossil fuels 231
    removing carbon dioxide from atmosphere 231
carbon-12 66, 72
carbonates 148, 232
    carbonate ions 283
carbon dioxide 153, 232
    covalent compound 53
    carbon cycle 231
    greenhouse gas 234
    properties 232
carbonic acid 144
carbon monoxide 55, 193, 210, 232
carboxyl functional group 254
carboxylic acids 254–255
catalysts 123, 136, 140, 171
    catalysts in the chemical industry 137
    decomposition of hydrogen peroxide 136

enzymes 136, 137
    how do catalysts work? 137
catalytic converters 211
cathode rays 36
cathodes 100, 101, 102, 104, 106, 107
cations 280, 282
    chemical tests 280–281
    flame tests 280
cellulases 269
cellulose 259, 268, 269
cement 237
changes of state 5, 6–7
    heat 7
    reversing changes 7
    sublimation 5
charcoal 219, 230
chemical change 42
    signs of chemical change 43
chemical properties 38
chemical reactions 43
    elements always react in same ratio
        68
    equations for chemical reactions
        64–65
    total mass does not change during
        chemical reaction 68, 74
chlorine 44, 249
    covalent bond 51
    how chlorine atoms gain stable outer
        shell 45
    how sodium and chlorine atoms
        bond 46
    uses 105
chlorophyll 140
chocolates 139
chromatography 19–22
    chromatograms 19
    chromatography and crime detection
        23
    uses 23
chromium 107
citric acid 144
civilisation 202–203
climate change 235
coagulants 219
coal 114, 210, 240
coke 194
collision theory 134
combustion 89, 230, 231, 232
    incomplete combustion 248
compounds 42, 190
    finding formulae from structure of
        compounds 62
    mixtures 42
    names of compounds 62
    reactive compounds 245
    saturated compounds 248, 250
    unsaturated compounds 250
concentration 78, 130
    finding amount of solute in solution
        79
    finding concentration in moles 78
    finding by titration 158–159
    sample calculations 79
concrete, sulfur 226

condensation polymerisation 259, 262,
270
    making nylon 262
    making Terylene 263
condensation reactions 255
conductivity 59, 170
    testing for conductivity 99
conductors 98–99
Contact process 228
copper 58, 69, 170, 184
    copper ions 171
    refining copper 106, 107
    uses 198
copper(II) oxide 152
copper(II) sulfate 120
corrosion 144, 187, 212–213
covalent bond 50–51
covalent compounds 52–53
    ionic compounds 54
    properties 55
covalent giant structures 56–57
cracking hydrocarbons 222, 244–245
crude oil 18, 240–241
crystallisation 16
crystals 54, 58
cysteine 270

decay, radioactive 30
decomposition 100, 103
    hydrogen peroxide 136
denim, stone-washed 139
density 164, 170, 178
detergents, biological 137, 139
diamond 56, 57, 58
diffusion 2
    gases 9
digestion 271
digital cameras 157
digital technology 202
disaccharides 269
displacement 167, 181, 183, 220
dissolving 12–13, 231
distillation 18
    fractional distillation 18
    fractional distillation in industry 18
DNA testing 139
ductility 59, 170, 178
dynamic equilibrium 121

Earth's atmosphere 206
Earth's crust 190
electric charge 28
electricity 98, 116
    electric circuits 98
    electricity from redox reaction 116
    hydrogen fuel cells 117
    International Space Station 217
    simple cells 116, 186
electrodes 100, 106, 186
    half-equations for electrode reactions
        103
    lead bromide 102
    negative electrode 118
    sodium chloride 102–103
electrolysis 99, 203

aluminium extraction 196–197
aqueous solutions 101
brine 104–105
electroplating 106, 107
extraction of metals 193
lead bromide 100
molten compounds 100
refining copper 106
electrolytes 99, 101
electrons 28, 36, 50, 98, 100
    delocalised electrons 58
    electron shells 32, 37
    electron shells for first 20 elements
        32
    electron transfer 90
    Group VIII 33, 44
    patterns in Periodic Table 33
    valency electrons 33, 162, 166
electroplating 106, 107, 213
elements 26–27, 190
    artificial elements 163
    atoms of the first 20 elements 29
    electron shells for first 20 elements
        32
    elements after calcium 33
    elements always react in same ratio
        68
    ions of the first 20 elements 48
    relative atomic mass 66
    symbols for elements 26
    transition elements 49, 137, 163, 178
empirical formula 80
    experiment to find empirical formula
        81
    finding empirical formula 80–81
endothermic reactions 111, 112, 113,
120
energy changes in reactions 110–111
    making and breaking bonds 112
    bond energies 112
    calculating energy changes in
        reactions 113
    reversible reactions 120
    starting a reaction 113
enzymes 136, 137
    denatured 137
    how enzymes work 139
    hydrolysis 271
    making enzymes 138
    modern uses 139
    traditional uses 138
equations
    calculating gas volumes from
        equations 77
    calculating masses from equations
        75
    half-equations for electrode reactions
        103
    mole 73, 74–75
    state symbols 64
    symbol equations 64
    working out equations, from masses
        75
    writing equation for reaction 65
    writing half-equations to show

electron transfer 90
ionic equations 91, 150
equilibrium 121
shifting equilibrium 122–123
esters 254, 255
ester linkage 263
ethane 245, 248
ethanoate ion 255
ethanoic acid 254
ethanoate ion 255
typical acid reactions 255
ethanol 18, 99, 252
fuel 115, 253
two ways to make ethanol 252–253
ethene 250, 251
covalent compound 53
hydration of ethene 253
evaporation 4, 17
exothermic reactions 110, 112, 113, 120, 182
Contact process 228
experiments 274
example: comparing kitchen cleaners 276–277
skills required 275
variables 274
explosions 133
extraction of metals 192
carbon 192
extraction and the reactivity series 192
ore extraction 193
extremophiles 139

feedstock
naphtha fraction 243, 245
fermentation 138, 252, 253
acid fermentation 254
fertilisers 224
synthetic fertilisers 224, 225
what plants need 224
fibre 269
filtering 16
filtrate 16
flame tests 280
flammability 242
flotation tanks 215
flue gas desulfurisation 211, 237
food, natural polymers in 268–269, 270–271
formulae 42
empirical formulae 80–81
finding formulae from structure of compounds 62
ionic compounds 82
molecular compounds 82–83
structural formulae 240
writing formulae by balancing charges 63
writing formulae using valencies 63
fossil fuels 114, 210, 231, 240
fountain experiment 221
fractional distillation 18, 242
liquid air 208
refining petroleum 242–243, 244

fuels 114, 248
different amounts of heat 115
ethanol 115
fossil fuels 114
hydrogen 115
nuclear fuels 115
what makes a good fuel? 114
functional groups 247
COOH 254

galena 226
galvanising 187, 213
gamma rays 30, 36
gas chromatography 22, 23
gas syringes 128
gases 4, 6–7
boiling points 208
calculating gas volumes 77
collecting gases 278
compression 8
gas pressure 8
heat 8
preparing gases in the lab 278
rate of diffusion 9
reaction that produces a gas 128–129
reactions between gases 134
reactions involving gases 76–77
separating gases from air 208
tests for gases 279
Geiger counters 31
giant covalent structures 56–57
global warming 234, 235
glucose 259, 268, 269
glycine 270
glycogen 269
gold 179, 191, 203
grains 58
graphite 57
pencils 57
grease 213
greenhouse gases 234
groundwater 214
gypsum 237

Haber process 222
conditions 223
improving yield of ammonia 223
raw materials 223
half-equations 90
half-life 30
halides 101, 167
halide ions 282
halogens 162, 166
chemical properties 166
physical properties 166
halides 167
reactivity 166–167
heat 7
heating curve 5
helium 168, 169
hematite 194
homologous series 247
hydrocarbons 82, 240
cracking 222, 244–245
hydrochloric acid 181

hydrogen 101
covalent bond 50
displacement 181, 220
fuel 115, 117
Haber process 222
hydrogen ions 146
making hydrogen in the lab 220
Periodic Table 163
reaction between hydrogen and oxygen 64, 89
uses 105
hydrogen chloride 52
hydrogen peroxide 136
hydrogen sulfide 63, 226
hydrolysis 271
identifying products 271
hydroxide ions 147
hydroxides 148
hypotheses 274

impurities 14, 15
indicators 145
indigestion 151
inorganic compounds 233
insulators 98
International Space Station 216–217
electricity 217
oxygen 216
water 217
invertase 139
ionic bonds 46–47
ionic compounds 46, 47
covalent compounds 54
formulae 48, 82
names 48
properties 55
ions 3, 45, 99, 100
complex ions 171
compound ions 49
finding mass 67
ions of the first 20 elements 48
metals in competition 183
spectator ions 156
transition elements 49
iron 137, 170, 179, 202
cast iron 195
extraction 194–195
ore 194
pig iron 195
rust 187, 193, 212–213
steels 200
iron(III) oxide 152
isomers 249, 251
isotopes 30–31
mass 66
reactions of isotopes 33

keratin 259
kinetic particle theory 7
krypton 168, 169

laboratory work 274, 275
anions 282–283
cations 280–281
working with gases 278–279

landfill sites 266
lattices 6, 46, 47, 54
lead 210
lead bromide 99, 100, 102
lead sulfide 203
Le Chatelier's principle 122
light 140–141
lime 149, 236
limestone 149, 185, 194, 236
limewater 12, 237
liquids 4, 6–7
    separating solid from liquid 16–17
    volatile liquids 13, 242
lithium 165
    lithium-ion batteries 119
litmus 144, 145

macromolecules 56
magnesium 181
magnesium chloride 47
magnesium oxide 47, 55, 152
magnetism 179
malleability 59, 170
maltose 268
masses 66
    calculating masses from equations
        75
    empirical formulae 80
    isotopes 66
    relative atomic mass 66–67
    finding mass of mole 73
    finding masses of molecules and ions
        67
    working out equations, from masses
        75
mass number 29
measuring cylinders 278
melting points 4, 5
    covalent compounds 55
    giant covalent structures 56
    ionic compounds 55
    metals 58, 59
    purity 15
metallic bonds 58
metalloids 172
    change from metal to non-metal
        172–173
metal oxides 148, 152
metals 38, 39, 47, 101, 148, 198,
202–203
    chemical properties 178–179
    compared with non-metals 38–39
    comparing the drive to form ions
        183
    competing to form ions in solution
        183
    competing with carbon 182
    competing with other metals, for
        oxygen 182
    Earth's crust 190
    extraction from ores 192–193
    key properties of metals 59
    ores 191
    Periodic Table 178

physical properties 178
    precious metals 190
    reaction with hydrochloric acid 181
    reaction with water 180
    reactivity series 184–185
    stability of compounds 185
    structure of metals 58
    transition elements 170–171
    uses 171, 198–199
methane 233, 246, 249
    covalent compound 52
    fuel 240
    greenhouse gas 234
methanoic acid 144
methanol 53
microbes 138, 214
    extremophiles 139
microscopes 35
    field-ion microscope 35
    scanning tunneling microscope 35
mining 191
mixtures 12
    compounds 42
molar volume 76
molar solution 78
moles 72
    Avogadro constant 72
    calculating gas volumes from moles
        77
    empirical formulae 80
    equations 73, 74–75
    finding mass 73
molecules 3, 50, 99
    calculating molecular formula 83
    finding mass 67
    finding molecular formula 82
    molecular covalent compounds 54,
        55
    shapes of molecules 53
monomers 251, 258, 260, 261
monosaccharides 269

naphtha fraction 243, 245
natural gas 114, 222, 233, 240, 246
negative pole 116, 118
neon 44, 168, 169
neutralisation reactions 88, 149
    neutralisation of acid by alkali 150
    neutralisation of acid by insoluble
        base 151
    proton donors and acceptors 151
neutral substances 145
neutrons 28, 30, 37
nickel 170
ninhydrin 21
nitrates 224
    nitrate ions 283
nitrogen 220–221
    covalent bond 51
    fertilisers 224, 225
    Haber process 222
    uses 209
nitrogen oxides 210
noble gases 44, 162, 168

uses 169
non-conductors 98
non-metals 38, 39, 47, 101
    compared with metals 38
    in Periodic Table 31
non-renewable resources 241
nuclear fuels 115
nucleon number 29, 162
nucleus of atoms 28, 37
nylon 262, 265

Octaves, Law of 174
oil (petroleum) 114, 210, 240–241
    refining of 242–243
    sulfur extraction 226
OILRIG 90, 94, 102
ores 191, 192–193
organic chemistry 233
organic compounds 233, 240
    alkanes 246
    comparing families 247
    functional groups 247
    homologous series 247
    names 246
outer-shell electrons 33
oxidation 88–89, 90, 91
    oxidation states 92–93, 171
oxides 152, 178
    acidic oxides 153, 172
    amphoteric oxides 153, 172
    basic oxides 152, 172
    neutral oxides 153
oxidising agents 94, 254
    oxidising and reducing agents in the
        lab 94–95
    oxidising agents outside the lab 95
oxy-acetylene torches 209
oxygen 207
    covalent bond 51
    International Space Station 216
    measuring percentage of oxygen in
        air 207
    reaction between calcium and oxygen
        89
    reaction between carbon and oxygen
        64
    reaction between hydrogen and
        oxygen 64, 89
    uses 209
oxygen masks 209
oxygen tents 209

paint 157, 213
paper chromatography 19
    identifying substances in colourless
        mixture 20–21
    key ideas in chromatography 22
    techniques 22
particles 2–3, 6–7, 37
    alpha particles 30, 36
    beta particles 30, 36
    evidence for 2, 3, 35
    kinetic particle theory 7
    sub-atomic particles 28

percentages 68
    percentage composition of
        compounds 69
    percentage purity 69, 84–85
    percentage yield 84
Periodic Table 27, 162, 174–175
    groups 27, 162
    hydrogen 163
    metals and non-metals 163
    patterns and trends 33, 163
    Period 3 172
    periodicity 162
    periods 162, 163
    transition elements 163
petroleum (crude oil) 114, 210, 240–241
    refining of 242–243
    sulfur extraction 226
pH scale 145
penolphthalein 155
phosphorus 224
phosphorus(V) oxide 153
photochemical reactions 140, 249
    film photography 141
    photosynthesis 140–141
photography 141, 157
photosynthesis 140, 230, 231, 268
    changing rate of photosynthesis
        reaction 140–141
physical change 43
physical properties 38
phytoplankton 231
plastics 264
    changing the properties 264
    degradeable plastics 267
    non-biodegradeable 266
    pollution 266
    properties 264
    recycling plastics 267
    uses 265
pollution
    atmospheric pollution 210–211
    plastic 266–267
polyamide 262
polyester 263
polymerase 139
polymerisation 251
    addition polymerisation
    condensation polymerisation 259,
        262, 263, 270
polymers 251, 258
    natural polymers 259, 268–269
    reactions that produce polymers 259
    synthetic polymers 258, 264–265
poly(propene) 261
polysaccharides 269
polystyrene 265
polythene (poly(ethene)) 251, 258, 260
    chain lengths 260
    changing properties 264
    pollution 266
potassium 165
    fertilisers 224
potassium iodide 95
potassium manganate(VII) 94

power stations 114, 115, 210, 214, 237
precipitation reactions 88
    making insoluble salts 156
    precipitates 282
    steps in making barium sulfate
        156–157
    uses 157
pressure 123, 223
    Contact process 228
propanone 13
proteases 271
proteins 220, 259, 268, 270, 271
    how amino acids join up to make
        proteins 270
    when you eat proteins 270
proton donors and acceptors 151
protons 28, 37
    proton number 28, 29, 162, 175
Proust, Louis 35
purity 14
    melting and boiling points 15
    separation 15
    percentage 69, 84, 85
PVC 261

quantities, calculating 68
quartz 56
quicklime 236

radiation 30, 31
radioactivity 30–31, 36
radioisotopes 30, 31, 115
radiotherapy 31
rates of reaction 126
    catalysts 136–137
    changing concentration 130,
        134–135
    changing surface area 132–133
    changing temperature 131, 135
    collision theory 134
    measuring 128–129
    rate of chemical reaction 127, 135
reactants 64
    Haber process 222
reactions, equations for 64–65
    energy changes 110–111
    reversible reactions 120
reactivity 173
    alkali metals 165
    halogens 166–167
    metals 180–181
    noble gases 168
    reactivity series 184–185, 186–187,
        192
    unreactivity 162, 168
reagents 64
redox reactions 89
    acids and redox reactions 149
    displacement of hydrogen 181
    electricity from redox reaction 116
    electron transfer 90
    oxidation states 93
    redox without oxygen 91
reducing agents 94, 182

oxidising and reducing agents in the
    lab 94–95
reduction 88–89, 90, 91, 182, 227
    metal ores 192
relative atomic mass 66, 174
relative formula mass 67
relative molecular mass 9, 67
residues 16
respiration 230, 231, 269
reversible reactions 120
    back reaction 120, 123
    Contact process 228
    copper(II) sulfate crystals 120
    equilibrium 121
    forward reaction 120, 123
    shifting equilibrium 122–123
rock salt 191
rotary kilns 236
rusting 187, 193, 212
    experiment to investigate rusting 212
    how to prevent rusting 213
    rusting and salt 212

saccharides 269
sacrificial protection 187, 193, 213
salt see sodium chloride
salts 148, 178, 255
    insoluble salts 156–157
    precipitation 156
    solubility 156
    starting with alkali 155
    starting with insoluble base 154
    starting with metal 154
    steps in making barium sulfate 156
    steps in making sodium chloride 155
    testing a salt 280
    transition elements 171
sand 56, 190
saturation 248, 250
scientific method 274
semi-conductors 173
separation 15
    separation methods 16–19
silica 190
silicon (IV) oxide 56, 57, 62
silver 170
    electroplating 107
silver bromide 141, 157
simple cells 116, 186
slag 201
slaked lime 149, 237
sodium 44, 165, 179
    charge on sodium atom 28
    how sodium and chlorine atoms
        bond 46
    how sodium atoms gain stable outer
        shell 45
sodium chloride 44–45, 46, 54, 58, 62,
191
    electrolysis of solutions 102–103,
        104–105
    rusting 212
sodium hydroxide 144
    uses 105

solids 4, 6–7
    comparing structures of solids 54
    separating solid from liquid 16–17
    separating mixture of two solids 17
solutions 12–13
    concentration of 78–79
    electrolysis 101, 102–103, 104–105
    saturated solutions 16
    standard solutions 158
solvents 12, 13
sonorousness 178
Space see International Space Station
stability 184
    stability of metal compounds 185
starch 259, 268, 269, 271
state symbols 64
steel 171, 195
    electroplating 107
    galvanising 187, 193
    manufacture 200–201
    mild steel 200
    stainless steel 171, 200, 212
stomata 140
substitution reactions 249
sublimation 5
sugar 268, 269
sulfates 229, 283
    sulfate ions 283
sulfites 283
sulfur 26, 226, 228
    extraction 226
    properties 226
    uses 226
sulfur concrete 226
sulfur dioxide 69, 153, 227, 228
    pollutant 210, 211, 227
    uses 227
sulfur trioxide 228
sulfuric acid 228
    concentrated sulfuric acid 229
    Contact process 228
    dilute sulfuric acid 229
    uses 229
sulfurous acid 227

Teflon 265
temperature 9, 122, 131, 223
Terylene 263
thermal decomposition 185, 236, 244
thermite process 186
tin 99, 213
titanium 203
titrations 155, 158–159
total mass 68, 74
tracers 31
transition elements 49, 137, 163, 170, 178
    catalysts 137, 171
    chemical properties 170–171
    physical properties 170
    uses 171
troposphere 206

universal indicator paper 145

unsaturation 250, 251

valency 62, 173
    valency electrons 33, 162, 166
    writing formulae using valencies 63
vanadium(IV) oxide 137
vinegar 254
volatility 13, 242
voltage 186
vulcanisation 226

water 4, 62, 228
    covalent compound 52
    distilled water 18
    heating curve for 5
    reaction with metals 180
    solvent 12, 13
    water of crystallisation 120
    water vapour 4
water supply 214
    fertilisers 225
    International Space Station 217
    is it clean? 214
    modern treatment plant 215
    providing water supply on tap 214
    where does water come from? 214
white spirit 245

xenon 168, 169

yeast 252
yield 84
    ammonia production 122–123
yoghurt 138

zinc 193, 213
    uses 198
zinc blende 193

First published in January 2010

British Library Cataloguing in Publication Data
A catalogue record for this book is available from the British Library

ISBN 978 1 84425 520 7

Library of Congress control no. 2009936760

Published by Haynes Publishing,
Sparkford, Yeovil, Somerset BA22 7JJ, UK
Tel: 01963 442030 Fax: 01963 440001
Int. tel: +44 1963 442030 Int. fax: +44 1963 440001
E-mail: sales@haynes.co.uk
Website: www.haynes.co.uk

Haynes North America Inc.
861 Lawrence Drive, Newbury Park,
California 91320, USA

Designed by Lee Parsons

Printed and bound in the UK

TO CONVERT FROM IMPERIAL MEASUREMENTS TO METRIC

Pounds (lb) to kilograms (kg)	multiply by **0.454**
Pounds-force feet (lb/ft) to Newton metres (Nm)	multiply by **1.356**
Inches (in) to millimetres (mm)	multiply by **25.4**
Feet (ft) to metres (m)	multiply by **0.305**

mad Minis

The crazy world of modified Minis

IainAyre

Contents

Introduction — 6

1 Minissan: God Save The Clown — 8

2 The Suicide Chop — 14

3 Chuck's Fire Engine — 20

4 Orange and Juicy — 24

5 The Magic Bus — 30

6 Turbo Triple Miniac — 34

7 The GT40 Mini — 38

8 International Rescue — 42

9 The Miniha — 46

10 Moking — 50

11 The Rotary Club — 54

12 Offroad — 60

13 Street and Strip — 64

14 Carmelie — 68

15 Trikes — 74

16	The Multicultural Mini	**78**	
17	Varnishing Point	**84**	
18	The Roofless Cooper	**88**	
19	The Madmen	**92**	
20	Pooh and the Hunny Pot	**96**	
21	Mad Frank	**102**	
22	Admirable Nelsons	**108**	
23	Z-Cars	**114**	

24	Ice Racing	**120**
25	Mini Limo	**124**
26	Mini Marcos	**128**
27	The Psycho Twini	**134**
28	Mini Kit Cars	**140**
29	Mini Mouse	**146**
30	Posh Minis	**152**
	More Mad Minis	**158**

Introduction

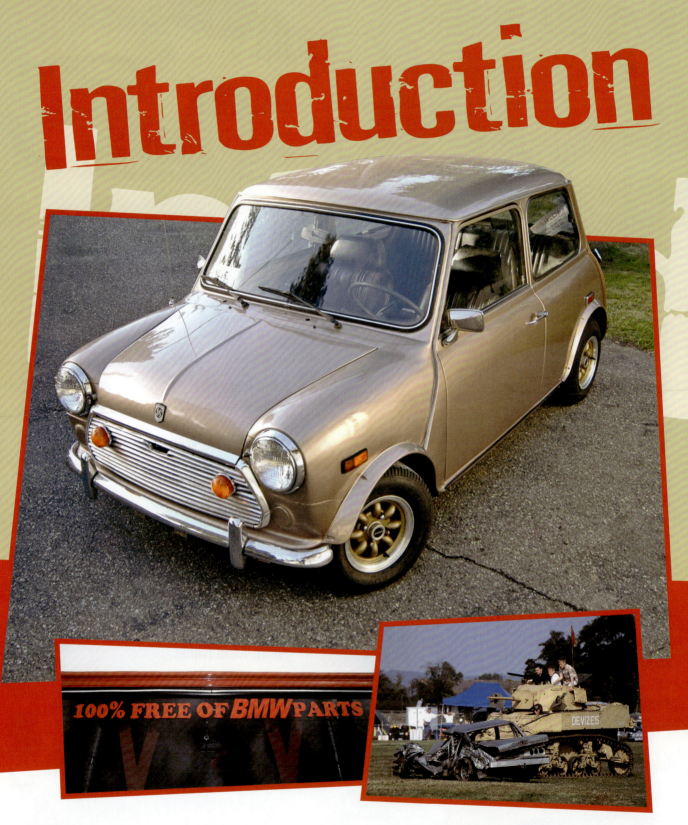

100% FREE OF *BMW* PARTS

↑↑ Bubble is my current Mini. It might finish up as a unique MG Mini. Or a unique MG Turbo Mini. Or a 4x4 Minzuki. Or a stretch limo. Or a Z-Car.

↑ '100% free of BMW parts'… many classic Mini enthusiasts really don't like BMW's Mini-inspired pastiche. In a *MiniWorld* survey 50% liked them, 50% hated them. I'll start writing about them as soon as somebody fits one with an interesting engine.

↑ The Volvo Crush: the Southern Mini Club know just how much Mini owners used to hate Volvos, driven by smug people who feel safe and don't bother to look where they're going. It's harder to spot Volvos now, as they look like BMWs or Hyundais.

I've been *MiniWorld*'s North American correspondent for a few years now since moving to Canada, but before that I was writing about Minis for a long time in the UK. I started with *Mini Magazine*, and later graduated to the more senior *MiniWorld* magazine.

Some of the Minis featured in this book have been seen in *MiniWorld*, though for their appearance in *Mad Minis* their stories have usually been extended from my original notes, and offer significantly more detail than is possible in a magazine. Also, magazines are edited to be more polite than books, so if I think something's bollocks I can say so here.

A few of the stories have been seen in the Australian magazine *The Mini Experience*, which is also a top read.

The reason I'm still writing about Minis after more than a decade is mostly because of their owners. Mini people are sociable, eccentric, opinionated, cheerful, creative, helpful, and endlessly interested and interesting. They also keep coming up with new and fascinating angles on the Mini theme.

As with most Brits, there are several Minis in my own history – my mother bought one new in 1971, I designed and prototyped a three-wheeled Mini-based kit car during the 1990s, and my last Mini in the UK was known as Old Scabby and yielded regular technical repair stories each time it broke down. It's now treasured by a friend's daughter and has been painted Ford 1600E purple and rechristened Lady Scabs.

I now run a Mini in Canada, known as Bubble because it's the colour of champagne and because it has a few bubbles in the paint. Minis in North America aren't just budget fun cars, though – they're rare, foreign, exotic, and valuable. Rather like running a '60s Lancia in the UK. For a sensible daily car I drive an immense but surprisingly economical 1958 Chevrolet with a 5.7-litre V8 from a Z28 Camaro, on which the tailfins are about the same length as a Mini. In fact I sometimes take my Mini with me in the boot. Less sensibly, I'm also manufacturing low-budget lightweight turbocharged Mazda MX5-based AC replicas under the Ayrspeed. com banner. I don't seem to be able to leave ACs behind. And I've just taken on a 1950 Roll-Royce project. Couldn't resist it.

Having recently driven a Mini Marcos halfway across Europe, I find I badly need one of those too. Fortunately they're still available as a new kit, and by the time I've finished this book I might even have made some progress with my Cunning Plan to race one at the Le Mans Classic. Race a Mini at Le Mans? What? The trouble with Mini-based eccentricity is that it's infectious. I've also recently driven a couple of Z-Cars Minis too, so of course I need one of those as well.

The Minis in this book range from barking mad to quite sensible in a psychotic sort of way. The owners all look and seem quite normal, and some of them even hold down conventional jobs. On the other hand, many of them, your humble narrator included, are completely unemployable.

Here's what some Mini enthusiasts did during their holidays.

↑An increasing number of people feel that a ten-foot-long four-seater Mini wastes too much space, so this is how they cure it. Not many cars will do a wheelie backwards…

↑Team Arse was one of the teams of Mini enthusiasts competing in deeply silly games at a typical Mini meet at the Hop Farm in Kent. Seen here undertaking the co-ordinated walking-the-planks. They're still sober, too.

↑The obsession with Minis doesn't just include full-size cars: concours judge Fiona Mannion's extensive collection of model Minis is fairly typical.

Minissan
God Save The Clown

↑ The dynamics of Minissan were unbelievably well sorted, straight out of the box. It only gets rude when you boot it.

→ At first glance the car still looks like a Mini. Retaining the chrome bumper helped with that.

Mike Guido may sway towards the madder side of the Mini continuum, but he also sometimes suffers from a very sane clarity of vision. He thinks that tobacco and alcohol companies marketing lethal drugs to children is a bad thing, morally not far from premeditated murder for financial gain. If you follow his logic, it's hard to argue with him: just reflect on the real purpose of alcopops. He decided to do something about it, and created an anti-drug campaign featuring The World's Fastest Clown. As the Clown, he got a factory ride with Nissan driving 240SX racing cars, but he drove them in full clown make-up and a fireproof clown suit. He says that gave him a real edge on the track: if he nudged somebody from behind and they looked in their mirror to see a clown grinning at them, it was really hard for them to concentrate on the next corner, which is all it would take for the Clown to get past.

The mad racing Clown then went to junior schools and put on an anti-drug magic show. As he's mad in a cool way, he got through to children where advice from parents and teachers never will, and he has probably saved a lot of young people from unpleasant, lingering but profitable deaths.

Nissan pulled out of racing, and the Clown resorted to competing in an innovative American budget race series where you're only allowed to spend a couple of grand on your car. In 2009 you're allowed to spend $2,009 on the car.

Minissan was originally intended for this, but instead it evolved into a show-and-go car that is now intended to attract as much attention as possible to get his campaign back on track, hence www.GodSaveTheClown.com on the windscreen.

Four years and 4,000 hours later, here it is. The shell is from 1962, and is the remains of Mike's first Mini that he couldn't bear to part with. The reason why it wandered so far off the true Mini path was to correct the two main dynamic deficiencies of the original Mini: the engines are too wee, and the drive is at the wrong end.

With enough money, you can persuade an A-series to kick out remarkable power, but 130bhp is only remarkable because the starting point is a little 1930s engine, and 350bhp from an A-series in practical and economical terms is simply not going to happen; whereas a second-hand Nissan V6 engine with cams and a nitrous bottle costs about the same as a week's holiday in Florida, which Mike doesn't need to pay for as he lives in Florida anyway.

The front-wheel-drive problem is even worse. More power and more acceleration means that more weight shifts to the back of the car, so there's even less weight on the front tyres, and you just get smoke and noise without much progress until the vehicle eventually collects enough forward speed for the tyres to stop spinning.

The front end geometry and suspension of Minissan remain related to Minis, but Mike had to learn a lot about steering and suspension theory to get the rack height perfect and avoid thumb-breaking bump-steer. He used a standard right-hand-drive Mini steering rack, which had to be mounted under the sump. Rack height is critical in any car, and getting it wrong results in bump-steer and vicious steering kick. He even got the Ackermann angle right, which is the degree to which the inner front wheel steers more than the outer front wheel when cornering. That angle is not critically important as it mostly means minor

← How many Minis have two spinning cam gears under the bonnet? The V6 fits in there very comfortably.

↑ Mike left some of the original spot welds in the shell just to prove the car is a real Mini.

← Much of the front end of the car could still be a Mini – there's no sign of its real character until you get near the back.

↓ The area of holes matches the area of the radiator fan, which blasts the heat out of the cooling system in seconds.

↑→ The interior fittings gleam. Everything is either black, red or gleaming polished metal, much of it stainless steel.

tyre scrub when parking, but it's revealing that he took the trouble to research it and sort it out. Mike's first crack at steering design beats TVR, who in 1967 were still building in rack-height bump-steer after designing cars for 20 years.

The rear end of Minissan took a solid month of work. When Mike does reluctantly spend time on his day job, he has to build adventure playground towers, platforms, rope bridges and so on. If he can slide out of work and spend

his days in the garage playing with mad Minis instead, he goes for that option every time. Minissan's rear end is the independent diff, shafts and subframe from a Nissan 240SX, narrowed by 12in. The half-shafts were each chopped by 6in: they're so short that the rubber boots are touching in the middle. Rear shock struts are 240SX front strut cartridges modified to work as rear struts, and the front shocks are from a Yamaha R6 bike. The springs are a massive 650lb front and 550lb rear, so it's weird that Minissan feels so drivable. It's also funny watching the car trying to get up a sloping driveway at the oblique angle required to avoid wrecking the front air dam: the car is so stiff that a wheel comes off the ground, and it's best to go up backwards and fast.

When power is dumped in, the back of the car is designed to squat a little to maximise grip. With Mike in the car, the front-to-back weight distribution is a virtually perfect 50% front, 50% back.

With a mid-front mounted V6 and independent rear-wheel drive, the original monocoque structure of the Mini was to some extent rendered irrelevant, and was replaced by a rollcage/spaceframe chassis that provided mountings for everything, which was constructed in and around the Mini shell. The body is not the usual decorative and detachable GRP shell found on most spaceframed Minis, it's the original steel body integrated with the cage, which passes through the body in several places. As much of the original Mini was used as possible – the Mininess of the car was important to Mike, contradictory though that may sound. The remaining inner wings still feature spot-welds from 1962.

I haven't seen another car of this type with this level of detailing: each of the welds that passes through the Mini shell was ground, sanded, wet'n'dried, guide-coated, primed, wet-sanded and finally covered in rich red paint.

There isn't anything else in the world quite this shape, so the fat wheel arches and other body panels had to be made from scratch. That meant mentally creating the shapes, then sketching them out, then fabricating them in 3D in the garage out of whatever material came to hand, then perfecting them, then duplicating the whole lot as mirror images for the other side of the car, then taking moulds and finally making the panels. It's making me tired just thinking about it.

Mike was paranoid about engine cooling, which was very sensible as the boot of a car is the worst possible place to put the radiator. There isn't a great deal of spare real estate under the bonnet, and Mike wanted to move some weight to the rear anyway. His response was overkill. The bootlid has more holes in it than global warming babble, the radiator is vast and the electric cooling fan is

18in in diameter. The fan is so powerful that it blew big fuses until sorted out. It now runs at half speed, which is more than enough. The cooling is amazing – on the car's first outing, a blown fan fuse allowed the engine to brew up. A replacement fuse was fitted to the circuit. When it was turned on, there was a blast of hot air out of the bootlid, which faded to warm and then cool air in about ten seconds, as the temperature gauge needle swivelled back to room temperature literally as you watched it.

The most remarkable thing about this car is the way it all worked straight out of the box. As ever with these projects, it was completed at 4:00am the night before its photo shoot. The car was taken off its jacks, fired up and driven down the street, and apart from the monster

↑ Early stages – sit in the car, build it around you.

↓ The elaborate frame in the back supports the independent rear axle and suspension.

← Mike did so much paint rubbing that he wore his fingers down to the elbow.

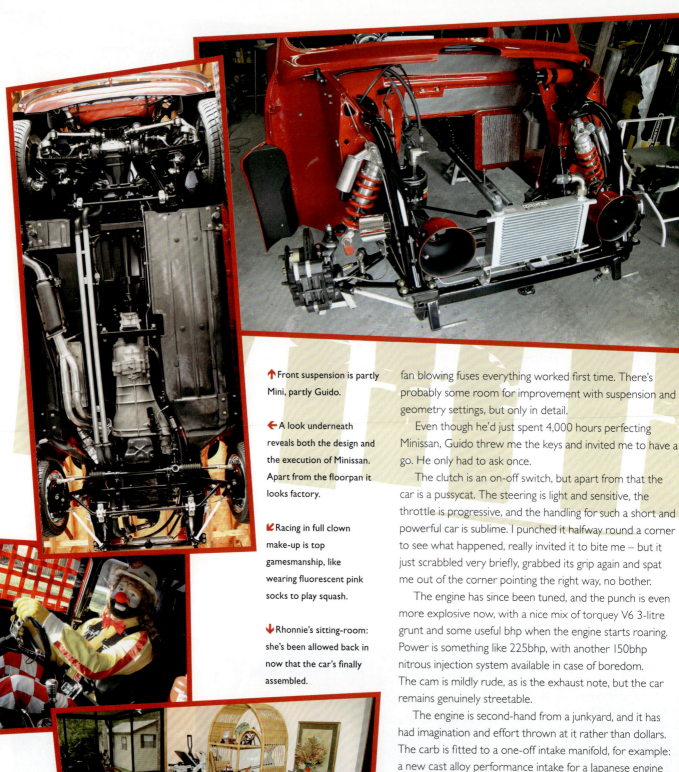

↑ Front suspension is partly Mini, partly Guido.

← A look underneath reveals both the design and the execution of Minissan. Apart from the floorpan it looks factory.

↙ Racing in full clown make-up is top gamesmanship, like wearing fluorescent pink socks to play squash.

↓ Rhonnie's sitting-room: she's been allowed back in now that the car's finally assembled.

fan blowing fuses everything worked first time. There's probably some room for improvement with suspension and geometry settings, but only in detail.

Even though he'd just spent 4,000 hours perfecting Minissan, Guido threw me the keys and invited me to have a go. He only had to ask once.

The clutch is an on-off switch, but apart from that the car is a pussycat. The steering is light and sensitive, the throttle is progressive, and the handling for such a short and powerful car is sublime. I punched it halfway round a corner to see what happened, really invited it to bite me – but it just scrabbled very briefly, grabbed its grip again and spat me out of the corner pointing the right way, no bother.

The engine has since been tuned, and the punch is even more explosive now, with a nice mix of torquey V6 3-litre grunt and some useful bhp when the engine starts roaring. Power is something like 225bhp, with another 150bhp nitrous injection system available in case of boredom. The cam is mildly rude, as is the exhaust note, but the car remains genuinely streetable.

The engine is second-hand from a junkyard, and it has had imagination and effort thrown at it rather than dollars. The carb is fitted to a one-off intake manifold, for example: a new cast alloy performance intake for a Japanese engine is expensive, so Mike chopped up an intake from a late 1980s throttle body injected Nissan pickup and started modifying that instead. He welded a Holley spacer plate to the top of it and then ported and polished it, then had it ceramic coated. It's probably within a few bhp of a seriously expensive performance item, but the price was paid in skill and patience.

Mike knows from experience that the stock Nissan V6 can handle up to 450bhp, so it can definitely take a joke.

Even with the nitrous, the engine is well within its limits, and even if he got a lot more cheeky with this engine and blew it up, a replacement would only cost a few hundred bucks.

Remarkably, the entire budget for Minissan was around four grand in pounds, because of the car being originally conceived for low-budget racing. Most of the money has been spent on paint and chrome plating, so it's 4,000 hours of hard graft you're looking at, not a chequebook street rod.

Mike's wife Rhonnie has been behind him all the way, including making the carpets, hand-polishing several hundred little stainless steel screws, helping lay the fibreglass moulds and final pieces, and putting up with her kitchen and house crammed full of Minissan bits for several years: so she absolutely deserves the shorty Mini she's been promised.

The Clown has been unable to get enough funding or official support for his drug prevention campaign in America, as the less effective but more conventional programmes are firmly in place: the individual towns and schools to which he has taken his show always ask him back, though.

If he got the opportunity to bring his campaign to British children he'd be happy to do so. If you're able to put The World's Fastest Clown in a professional motor racing series and into schools, get in touch with Mike at mike@worldsfastestclown.com. American parents may lack the imagination and commitment for this, but perhaps the Brits can do better? It wouldn't do any harm to British companies' media profile either, come to think of it.

ENGINE – Nissan 300ZX VG30 2998cc V6, steel block, alloy OHC heads. Lightened harmonic balancer, Schneider cams with 0.427in lift, 270 duration, advanced 3°. MSD Digital 6 distributor with magnetic pickup, NGK plugs, Accel leads, Holley 350CFM two-barrel carb with K&N filter on modified throttle body injection inlet manifold from 1980s Nissan truck. Standard ceramic-coated exhaust manifolds, single 12in silencer, 150bhp nitrous oxide system, Mocal 18-row oil cooler, 19in x 27in x 3in radiator with 18in electric fan. Holley 7psi race fuel pump, 100amp Mini alternator, aluminium flywheel, ACT 6-puck race clutch.

TRANSMISSION – 1990s 300ZX five-speed manual gearbox. Narrowed independent Nissan 240SX diff, driveshafts and subframe with viscous LSD, 4:1 final drive ratio.

SUSPENSION – Front, Mini subframe with 1.5in removed from towers. Standard upper arms, 1.5° negative camber lower arms, HD adjustable tie rods. Yamaha R6 monoshocks with 650lb springs. 0.75in adjustable anti-roll bar. Rear, ex-240SX front Koni strut cartridges, modified. SPL rose-jointed upper and toe arms, aluminium uprights, 550lb springs. 0.75in adjustable anti-roll bar.

BRAKES – Mini pedal box, Geo Metro (Chevrolet) master cylinder and servo. Adjustable front/rear bias. Handbrake from 1967 Austin-Healey Sprite, cables from 240SX. Front, Outlaw four-pot aluminium calipers with 9in vented discs, Performance Friction pads. Rear, 240SX non-vented discs.

WHEELS AND TYRES – All Diamond steel. Road wheels: front, 7J x 13 with Sumitomo HTR200 175/50 x 13; rear, 7J x 15 with Sumitomo HTR50Z 205/50 x 15. Race wheels: front, 8J x 13 with Hoosier 225/40 x 13; rear, 8J x 15 with Hoosier 225/45 x 15.

INTERIOR – Aluminium race seat, five-point Safe Quip racing harness, custom cage, Smiths oil pressure, water temp, fuel pressure in custom pod. Auto Meter tacho, water temp warning, MSD shift light. Hurst gear lever. Brake bias handle. Detachable steering wheel. Detachable nitrous bottle and frame alternate with Rhonnie's passenger seat and harness.

EXTERIOR – Fibreglass doors and flip front, home-made flares/air dam/side scoops/bootlid/bonnet scoop.

← With four years of design and construction out of the way, it's time for some fun on dragstrips and racing circuits.

The Suicide Chop

Jason Neumann was headed for a career in graphic design, and to pay his way through college he worked at a Costco wholesale supermarket. Trouble is, Costco liked him and kept giving him more money, so by the time he finished his degree he would have had to take a stonking pay cut to start being a designer.

As he enjoys the good things in life such as food and having a house to live in, Costco won. However, you can't keep artists down, as their ideas will emerge in another form – and after four years of imagination and application, this Mini is the result.

The shell is a 1963, but it was only a shell, so all he knows is its date and that it was an Austin. Could be it was an 850, possibly a Cooper S – but in its current form it's certainly transcended its original purpose during Jason's metamorphosis of it.

The roof had already been chopped by 4in when Jason bought the car as a rusty shell, but hadn't been chopped very well. It needed considerable work, so he decided to get radical and fit a big glass sunroof out of a BMW Mini while he was at it. In the end he cut a large chunk of the roof off a BMW Mini and welded that into the Mini shell, which in itself must have taken a while. This gave the roof a flatter look than the usual rather domed Mini roof panel, and made the car look even lower. It also allows Jason to drive it in the 100° Florida summer without air conditioning, which would otherwise be pretty brutal.

There was much grunt work in this project before anything creative could take place: the shell was de-seamed, which if done properly means many, many hours of patient work. The strength originally provided by the exterior seams needs to be replaced with more welded steelwork along every inch of the seams, or the structure ends up with the strength of a wet paper bag. The front wings also needed to be extended, and Jason chose the best but most difficult approach, which is to cut them in half at the arch tops and add sheet metal in the middle.

Body stiffness loss from de-seaming was not a risk here anyway, as Jason made a ten-point rollcage from steel tubing that connects the front and rear subframes and the body. Rigidity in this car is not going to be a problem.

The doors and bootlid are shaved, which with this level of paintwork is in itself a big job, as it means replacing holes and door-handle dips with new metal so that the change is invisible. Thick blobs of filler just won't cut it.

More than that, the doors and bootlid all open the wrong way. Changing ordinary MkI doors to rear-hinged suicide doors isn't too bad a job because the geometry of the opening angle is relatively simple, but a side-opening bootlid has to swing directly outwards before it starts turning, and the geometry of that can get weird. I ran into that problem designing bootlid hinges for replica Jaguar XK120s, and it was a bitch. Jason had to scratch-build hinges with sketches, cardboard templates and finally steel, but the result rewards his efforts – the bootlid fits a treat. Of course, it helped that he filled all the door and panel gaps, block-sanded the whole thing and recut the gaps to

← My first front-cover shot for *MiniWorld*. Much location hunting to find the right alleyway, then a two-hour wait for the sun to arrive at the right place.

→ The bonnet is the only panel on Jason's car that opens the way you would expect it to.

finer tolerances. Which, as you can imagine, took a while.

Some nice touches include the rear lights, which you can't see. Not until they're lit, anyway: they're just thin strips cut in the bodywork and filled with translucent red lighting plastic the same colour as the paint. They were heated into shape, pushed through the holes to stick up proud by quarter of an inch, then rubbed down flat with the body and finally lacquered over after painting to make them more or less disappear. Very cool, that. The rear window is also plastic – it's unbreakable Lexan. It's unbreakable but not unscratchable, and has to be polished rather than cleaned. The glass front screens tend to crack when cut down, and it was Jason's penultimate screen that survived: he now only has one spare.

The front end is also quite dramatic, and involved interesting styling choices as well as subtle but extensive fabrication. The front wings are extended slightly by 2in, and the front bulkhead would no longer be familiar to a Mini owner. The bonnet scoop is from a Toyota 4-Runner pickup truck, and the front bumper is an amalgam of various shapes, blended together and finished as a moulding buck, then moulded and reproduced.

The door inners were constructed using the same technique, so Jason can make more of any of the scoops, front panel, dashboard or door inners if you like the look of them (jaynu@aol.com).

The dashboard was also made in the same way, with wood, chicken-wire, plastic foam gloop, whatever, then a mould was taken, then out came a silky smooth final panel of even thickness, ready to paint. Jason estimates that the dash and door panels took 500 hours…

The mad but brilliant gauges on stalks are a combination of bits of plumbing pipe and top graphic imagination. Having the steering on the right isn't anglophilia gone mad: right-hand drive just made it much easier to get everything in because the bulk of the Honda engine is on the left side of the engine bay, looking at the car from the back. There is a massive sound system, but it's just for the show stage of the car's career, and the speakers will probably be replaced by a back seat before too long.

That candy pearl paint job is quite a big deal. It's not really paint as such – it's more like lightly tinted clearcoat or varnish. The base coat is a gold metallic, and then the coats of House of Kolor Brandywine Kandy start being laid on. It takes many, many coats of it to get a colour this deep, and each of those coats has to be flawless. If that weren't tricky enough, there's pearl in there as well, and pearl paint is much worse than ordinary metallic paints for not matching.

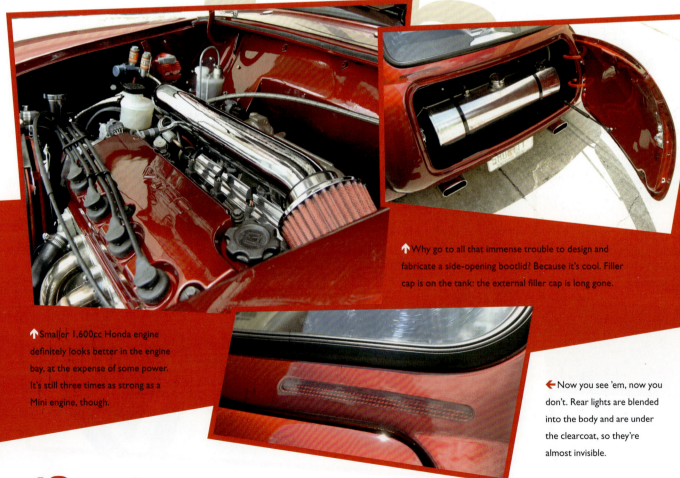

↑ Smaller 1,600cc Honda engine definitely looks better in the engine bay, at the expense of some power. It's still three times as strong as a Mini engine, though.

↑ Why go to all that immense trouble to design and fabricate a side-opening bootlid? Because it's cool. Filler cap is on the tank: the external filler cap is long gone.

← Now you see 'em, now you don't. Rear lights are blended into the body and are under the clearcoat, so they're almost invisible.

You really have to paint the whole car and all its panels at the same time, which means all holes masked off and the doors and panels mounted on stands, centrally supported – if you hook them up in the usual way, the hooks will disrupt the paint spray pattern and the anomaly will be visible. The surface needs to be prepped and completely dust-free between each coat too. As you can see from the pics, Jason managed to achieve the state of stubborn, anal-retentive obsession necessary to get the paint right.

The choice of a smaller Honda engine, the 1.6-litre single-cam rather than the usual 1.8-litre twin-cam, was a matter of both physical size and prettiness. That smaller 1,600cc Honda Vtec engine will go into a standard Mini front end, and it doesn't have to be jammed in, although it has to be said that the alternator was repositioned and an inch was cut off the tubular stainless exhaust manifold. Even so, the manifold position made Jason change from left-hand to right-hand drive just to make sense of the engine bay. It had to look good as well as being functional. The intake pipe was custom-made to pull cooler air into the engine, clear of the radiator. A detail, but worth a few extra horsepower.

The 1,600 looks a lot more visually comfortable in there than the bigger 1,800cc versions of the engine, and

there are plenty of surface pieces on the engine available to be painted in the same colour as the body. If you're used to an A-series, going from 1,000cc and 50bhp to a Honda with 1,600cc and 150bhp is going to be a fairly dramatic change even if you leave the engine standard, but of course Jason didn't leave it standard.

Tuning Honda engines is a different world from tuning the Mini's A-series, but LCBs and Swiftune are just changed for VAFCs and Apexi. The principles remain the same, though – get more air in and out, match the fuel supply to the air supply and away we go. The engine bay was designed to leave room for a turbo, which will make it quite nippy. Even with this level of planning there were issues: with the shell painted and the engine in, Jason's intended big radiator wouldn't fit and had to be substituted for a stock Honda rad. Fortunately it can handle the job, no prob.

Jason's engine is a Vtec, which has variable cam profiling to give a shopping cam for supermarket action at 1,200rpm, and a race cam for track day action at 7,000rpm. With a heavy Honda shell the cam profile change isn't too brutal, but in a Mini it chucks you half off the road, so Jason used an Apexi Vtec controller. This allows you to bring in the Vtec effect lower down the rev

↖These might be mirrors, but they're also electric door handles.

←Most cars this radical have some feature that doesn't sit right, but this one has no faults. It all works together seamlessly.

↑Magnificent interior, created from imagination, bits of plumbing, wood and chicken wire and then replicated in gleaming GRP.

↑ Roof chop gives the car an uncompromising attitude. Removing the gutters usually makes Minis look bald, but in this case it works. Flat BMW sunroof helps.

↓ New front subframe was custom-built around the engine to get it all in without looking too crowded.

← The bumper buck was constructed from bits of different air dams blended into one and then reproduced.

← Door mouldings were scratch-built, entirely to Jason's design.

range where its kick is less nasty, and allows more control over fuelling, making it easier to match the fuel supply to the extra air coming in through the uprated air filter and intake. There are also bigger injectors from an 1,800cc engine. All of this is good sense and sound tuning practice, and the result according to Jason should be about 150bhp at the crank: that sounds reasonable. There's much more still to go into the engine later – plans include a Gude cam and cam gear – but the engine is very amusing as it is.

One side effect of the fairly brutal cam profile change is that the engine moved too far on its original mountings for the clearances available. It's okay if it flaps about in a Honda shell as there's plenty of room, but in the small Mini engine bay it had to be kept tied down tighter, with stiff custom engine mounts. That's even with the mods Jason made to the bulkhead to give himself clearance to work on the engine. The stiff engine mountings have added to vibration and cabin noise, but it can't be helped.

The Honda motor is bigger than the A-series but is all aluminium, so it doesn't upset the car's weight balance too much. The NOS system is really just there to frighten people: it's not really part of Jason's approach.

It's an interesting thought that the old A-series Mini engine tops out at about 130bhp, which is about where the smaller Honda engine bhp numbers begin.

The gearbox is a stock Honda five-speed, and as a Mini fanatic from way back, Jason enjoys upgrading to one-finger Japanese gearchanges and the additional cog. The uprated clutch master cylinder and ultra-lightweight JUN flywheel sharpen up the gear changes to match the engine's general tight feel, and sticking with the original Mini rubber cones, although tamed to some extent by Spax adjustable shocks, means the hard ride also matches the hard engine.

Freeway speeds in the USA reflect the lack of speed-tax cameras, and 80mph is usual. A Mini engine makes a fair old racket cruising at 80mph, and Jason does enjoy the luxury of changing up into fifth gear without handing over several grand for a custom A-series five-speed. He also enjoys the 40mpg+ cruising economy provided by the combination of the five-speed, the low aerodynamic profile and the car's weight of 1,000lb less than the donor Honda.

He doesn't enjoy the ferocious steering kick so much – the driveshafts are part-Honda and part-Mini, and the rest of the suspension and the brakes are either Mini or Metro – after all, if it ain't broke don't fix it, and Mini suspension design definitely ain't broke. It all works rather well considering that it was designed for 34bhp rather than 150bhp. The nitrous oxide system is for real, but has never been used: it would certainly provide some bicep-building torque steer if it was turned on.

One of the cool things that he and everybody else enjoys is opening the doors. With the handles shaved,

ENGINE – 1.6 Honda Vtec SOHC, with 1.8 Integra injectors, MSD distributor cap, MSD SS coil, APEXI valve-timing controller, high-flow air filter and intake, FAL Performance cooling fan. Nitrous oxide injection system. Shortened and re-welded stainless exhaust manifold, custom twin-tailpipe exhaust. Custom air intake pulls cool air. Gude cam and cam gear due soon, currently about 150bhp at the crank with nitrous switched off. Custom front subframe and engine cradle.

TRANSMISSION – JUN ultralight flywheel, Stage Three clutch and master cylinder. Honda 1.6 Vtec gearbox, standard with shortened gear linkages. Honda/Mini custom driveshafts.

BRAKES – System, Geo Metro (US car) master cylinder and servo. Stainless braided lines. Front, Metro Turbo vented four-pot. Rear, Cooper S drums.

SUSPENSION – Rubber cone, with adjustable Hi-Los. Spax adjustable shocks all round.

WHEELS AND TYRES – Superlite deep dish 7J x 13, with Yokohama 175/50 R13 tyres.

INTERIOR – Cobra seats, retrimmed with Mini logos. Ten-point handmade rollcage connecting both subframes, Auto Meter gauges and shift light, custom GRP dashboard and door trims, electric door opening via mirrors, Sony flip CD, Earthquake 3,000W amp, two Earthquake 12in BBX-12DR bass speakers, two 6.5in and two 5.75in MB Quart high end speakers.

EXTERIOR – 4in roof chop, 2005 BMW Mini electric glass sunroof, suicide doors, de-seamed shell, shaved rear lights, door handles, filler neck and bootlid, custom side-hinged bootlid, recessed licence plate mount, flush rear strip lighting, extended front wings and bonnet, Toyota bonnet scoop, one-off GRP front air dam/bumper, House of Kolor Brandywine Kandy paint.

➜ Jason's car is cotton-budded perfection at the moment, but is also designed for practicality and performance when he gets bored with winning show prizes.

door and bootlid operation is electric, and the doors are operated by pushing the door mirrors down, at which point the doors click and pop open.

Jason's car was built mostly as a sculpture, and mostly to win awards, which it has done with tremendous success. However, it was also built to provide a bundle of more active fun once the initial impact had quietened down, and he will gradually begin to use the Mini rather than show it: he just has to accept that the paint job that

took hundreds of hours of rubbing will collect more road rash and scratches the more fun he has with the car. Some sort of better rear lighting is also going to have to be devised, as the cool rear light strips look wonderful but don't actually work very well. Unlike some Honda installations, the oil filter location was designed to be easily accessible: Jason expects to be changing the oil quite frequently as the years go by, particularly when the matching Mini trailer has been constructed.

Chuck's Fire Engine

BEXLEY
HEALTH AUTHORITY

madMinis

At the 2003 All British Field Meet in Van Dusen Gardens in Vancouver, Canada, an elderly expat Brit looked at a Mini, went white and stepped back in amazement. As well he might: he'd more or less seen a ghost.

In around 1968, in Bexley Hospital, England, he'd been the driver of this little red Mini-based fire engine.

Fast-forward to 2003, through an emigration and 35 years, and he was wandering around a classic British car show in Canada admiring some nice pre-war Bentleys and Austins, when he came across the very same fire engine. It had been imported into Seattle, USA, a while back, and had been taken up to Vancouver for the weekend to be displayed at the show by members of the Vancouver Mini Club.

It seems extraordinary that it had survived unrestored for this length of time, but it does make sense when you think about it. The reason there are so many ancient fire engines still around is that they have a pretty easy life. Unless something's on fire, they just sit in a shed. They're serviced, given a little gentle exercise now and again to keep them running, given an annual MOT test and then put back in the shed. This particular vehicle probably didn't see much firefighting action – the main reason it existed was that Bexley Hospital was of a size that meant it was required to have its own fire tender because it was some distance from the nearest fire station. I reckon they would just have called 999 if they smelled smoke.

For 31 years, the Mini Firefly sat in its shed, accumulating mileage at the recorded rate of 51 miles per annum. Its present mileage reading is 1,592.8. The speedo is currently broken, so there may have been a few more miles – who knows?

The Firefly's origins go back 1,600 miles and 40 years to the Angus Fire Armour company, whose London Fire Engine and Sprinkler Department announced a brand new 'compact and capable fire-fighting unit', the Mini-based Firefly. With an optional ladder (sadly lost many years back), pump, water tank, hose, ladder gantries and rear canvas roof or tilt, the converted pickup retailed for the grand total of £829. The Mini-based fire engine was a new concept, but the Angus Firefly was an established brand:

← The doors weren't originally signwritten, but somebody has had some magnetic Bexley Health Authority decals made. It should logically be Bexley Illness Authority, shouldn't it?

↓ The pump, hose and tank fit neatly. Hauling a steel tank containing 50 gallons of water would be hard going for an 850, you'd have thought.

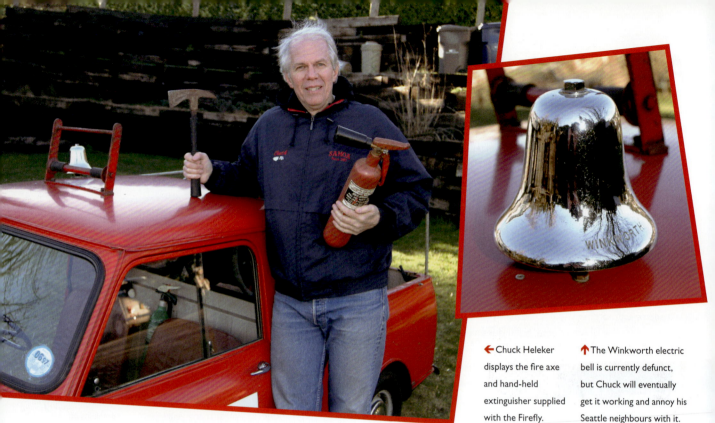

← Chuck Heleker displays the fire axe and hand-held extinguisher supplied with the Firefly.

↑ The Winkworth electric bell is currently defunct, but Chuck will eventually get it working and annoy his Seattle neighbours with it.

the 1960 Angus Firefly was the same sort of design, but based on a Series II 109in Land Rover.

The big nobs at Bexley Hospital, Dartford Heath, Bexley, Kent, saw the demonstrator and what was possibly also the prototype Mini Firefly at the British Fire Services Association Conference at Hastings, and they liked the look of it. They asked for a brochure on 16 January 1964, ordered their Firefly on 20 February, and took delivery in June. For the next 31 years the Firefly pottered around the hospital, occasionally extinguishing a dustbin fire but mostly parked in its shed.

In 1995 it was sold, and any dreams of the open road that it might have had were dashed. Its new British owner registered it and then stored it. Three years later, Seattle Mini enthusiast Chuck Heleker bought it and shipped it from the UK to the Pacific Northwest, where it arrived in February of 1998. The engine still ran, but neither the hydraulics nor the tyres were safe to drive at anything more than walking speed. Chuck describes the brakes as a 'dual leaking shoe design'.

In the twelve years since then, has the old fire truck tasted freedom and the highway and put some miles under its belt? No … it's been stored in the back of Chuck's basement garage with his other 13 assorted Minis.

'What are you going to do with it?' I asked.

'You know, that's a real tough call…' said Chuck.

He's not yet at all sure what to do with the little fire engine, although it's high on his list of treasured possessions. It only has two seats and the back is full of hose, water

tank and pump, so it's not really of much practical use. It's also quite heavy and only has an 850cc engine, so driving it could be fairly hairy on Seattle's 80mph freeways.

Chuck doesn't want to restore the Firefly because that would spoil it. The likely outcome is that when he has some spare time and money, he'll get the engine going properly, overhaul the hydraulics and fill them with silicone fluid, and then just take it out and play with it sometimes. If the hydraulic system is all new, there are unlikely to be problems with swollen seals, and with no hygroscopic brake fluid sucking water into the system there will be no further rust in the brake and clutch cylinders. Maybe Chuck will even finish off running the engine in. Of course something may catch fire some day, in which case it could be very useful to have a fire engine in the middle of the workshop.

One thing that he definitely wants to do is take it to a few fire engine shows. He's not a particular fire engine enthusiast himself, but lots of people are, and they'd definitely love to see it.

The pump works fine: again, it hasn't seen a great deal of action. It's fixed to the pickup bed, but has handles at either end. It was originally designed to be manhandled to a stream or a source of water and then it was supposed to pump water straight from there to the fire. There's a 50-gallon water tank in the pickup bed which would only just be enough to put out a dustbin fire. The hose is probably not the 1964 original, as it still looks serviceable, if elderly.

There may or may not have been a pair of lockers inside

the pickup bed. The brochure shows them but the Firefly doesn't have them.

The chrome-plated brass Winkworth bell was not specified in the original invoice: the Firefly was supposed to have had a 'Du Flash Syren' fitted, but presumably that packed up and was replaced by the electric bell. Which has now also packed up. Chuck expresses no surprise that elderly British wiring on a Mini fire engine bell doesn't work any more, although to be fair, the rest of the Lucas electricals do still seem to work unexpectedly well.

There isn't much known about these little fire engines, other than that there were at least two built – the prototype and this one. Chuck would be fascinated to hear if anybody has any more information, and he would be over the moon if anybody could replace his missing ladder.

Chuck is busier than ever, either fixing Minis or driving them mad distances around America. So the fire engine is still on the lower rungs of his ladder of priorities. Another few years in storage for the Firefly, then.

Chuck would like to talk to the chap he met in Vancouver, whose name may have been Dave Harmer… he has a note reminding him to call, but no number. If anybody knows Dave, Chuck can be contacted via www.SeattleMiniOwners.com.

← … and a bargain at 829 pounds, no shillings and no pence.

↓ KR is a Kent registration. From January 1963 most local councils issued number plates with an A suffix for 1963 and a B suffix for 1964. After 1964, year suffix letters were mandatory.

VERSATILE, ROBUST, POWERFUL, – AND SO EASY TO MANOEUVRE

Tech

ENGINE – Standard 848cc, SU HS2 carb.
TRANSMISSION – Standard four-speed.
IGNITION – Standard points/condenser.
BRAKES – Standard drums all round.
WHEELS AND TYRES – 3.5in steels, with replacement 145 x 10 tyres. Original wheels have possibly the original tyres, now disintegrating.
INTERIOR – Stock original, with additional fire axe and hand fire extinguisher.
EXTERIOR – Fire Brigade Red over original Westminster Beige. Modifications: bell and wiring; ladder gantries; Villiers water pump; 50-gallon water tank; fire hose and reel.

BEXLEY HEALTH AUTHORITY

6077KR

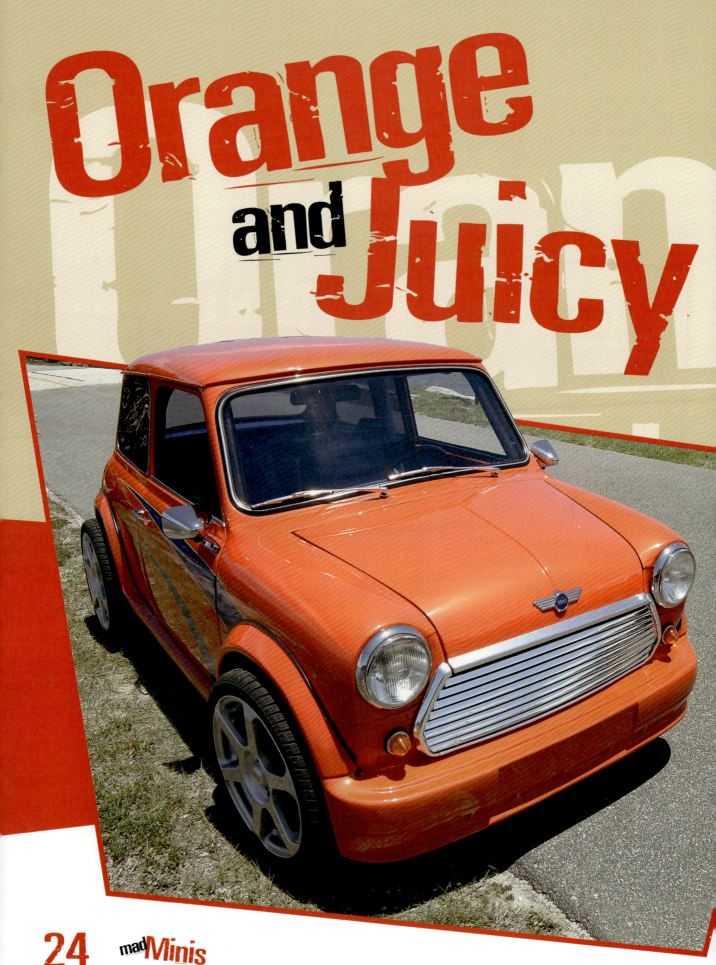

Orange and Juicy

Mark Heller is a retired US Navy bosun, which is the rank between those giving orders and those carrying them out. He's the guy who gets things done. As you can imagine, retiring to potter about in the garden and complain about the youth of today did not appeal to such a man, so he built a mad Mini instead.

He has a history of messing about with British cars, as his father was an engineer and a Brit-car enthusiast, although he didn't usually own a car. The Heller family lived in central Brooklyn, New York, and even 60 years ago good public transport and limited parking space meant there was no need for one. Mark learned to drive in an ancient Frogeye Sprite, and has owned a good few British cars, including a Humber Super Snipe. He liked its walnut and leather, and its general ambience. He also had a Hillman Imp, which he thought was an excellent design but a horrible car. A top buy was a Sunbeam Tiger, basically an Alpine with a 5-litre Ford V8, and one of the undervalued gems among British classics. That particular car has been replaced by his current Austin-Healey replica, which is a Corvette in a Healey-shaped frock supplied by Sebring – a company whose descendants still make Healey replicas today in Norfolk. Mark also likes grotesquely huge cars as well as tiny ones, and owns a stretch Cadillac limo and a Cadillac hearse that has been converted to a mobile bordello with a crushed velvet interior and clinking crystal glasses.

One of Mark's major regrets, and the underlying reason for the existence of his mad Mini, is that he sold the 1967 Cooper S he bought new and later traded in. He regretted that decision for years, but he says he didn't inherit any of his father's talent for engineering, and as a non-mechanic he doesn't have the confidence to drive a 40-year-old genuine Cooper. Being a gentleman he put this politely, but he remembers that reliability was not the most notable attribute of British cars even when they were 40 years newer, so the idea of creating a Mini with a big and reliable Japanese engine appealed. 34bhp Minis are lots of fun, but it has to be said that 195bhp Minis are even more fun, and 195bhp Minis with Japanese engines that start up every day and don't throw a wobbly when it rains are maximum fun.

Mark took his concept to a local rod shop run by Dennis la Monte, who thought it would be a bit of a challenge and would make a nice change from building big-block V8 street and strip Mustangs, and who made an excellent job of realising Mark's plans. Finding the right Mini to work with was tricky, as Mark really didn't want to cut up a good one. After much scratching around,

← ← At first glance this just looks like a Mini with fat wheels.

↓ Looking at the back, the sheer size of the wheels becomes clear – it's not possible to fit anything bigger than 13in wheels with low profile tyres on a standard Mini.

↑ The depth of the front spoiler conceals the ride height of the car: if it just had a thin chrome bumper it would look like a monster truck.

a suitable car showed up in Deland, Florida, only half an hour away from his house in Daytona Beach. It was sunk into a garden, as rotten as an election promise, and the engine was shot as well. He licked his lips and made a suitable offer for it as scrap.

In some ways he would have liked to put the engine in the front, and it is possible to squeeze the Honda engine into the front end of a Mini, but unless you're really obsessed with retaining a standard front end the usual routine is to extend the front wings and bonnet, which didn't appeal to him. There's very definitely no room for air conditioning as well, though, and Mark lives in Florida where the temperature and humidity get pretty extreme. Okay then, if he wanted to keep the standard-length Mini front end, the engine would have to go in the back.

The Mini was stripped to the outer shell with the crumbled floor completely cut out, and a frame was constructed that would fit under the body. It made sense to leave the body unstressed and make the chassis more or less drivable on its own, because of the small space available inside the Mini shell. Being able to fit and remove the body would save time and trouble during the build.

The body kit around the edges of the Mini is British – if you check out the front shot closely you'll see the space left for the large British-sized plate. Florida law doesn't require a front licence plate, which makes a lot of sense. Not many criminals reverse away from the scene of a crime: even crims have realised that cars go better when facing forwards. The only use for front number plates

is speed-tax cameras, which in my view are verging on criminal themselves. Automobile bumf is pretty user-friendly in Florida: this car is titled as a 1974 two-door Austin Mini, with no requirement for inspection or emissions. Mark's Corvette-based kit car is a 1961 Healey, no worries. Much of Florida is pretty relaxed: the restaurant health inspector gets a welcome, a beer, a pizza and some NASCAR chat, and hey, this place looks clean enough.

Even with the engine occupying the whole back end of the car, getting everything jammed inside the bodywork was a challenge. 'You know how it is with Minis,' says Mark, 'every inch is accounted for.'

The chassis was constructed to use the front suspension and brakes from a 1999 Honda Prelude, and the rear end is the complete package from a '99 Integra. Because the rear end is now a front end, it comes with steering – but the way to deal with that is to throw away the steering rack and make solid brackets on the chassis, replicating the trackrod ends' geometry. The ex-trackrods are then attached to the brackets and can be used to adjust the rear tracking rather than steering the wheels. One big advantage of using factory geometry, or of misusing it depending on how you look at it, is that details like anti-dive geometry and the rack height can remain correct. If the rack height is wrong, the steering will be horrible, and even a small error can generate vicious bump-steer and kickback that can break your thumbs. Shortening the wheelbase does mean a change

in the Ackerman angle, but as we've seen before, that can be calculated and corrected, and even if it isn't, all that happens is increased tyre scrub at low speeds.

One area which puts Mark's concept way ahead of most Minis dynamically is the rear engine and rear-wheel drive. Front-wheel drive is excellent for packaging and cheap manufacturing, but as far as performance goes it's crap.

The problem is weight transfer. As soon as a front-drive car accelerates hard, the weight moves to the back and the front wheels lose grip. Much tyre smoke: not much progress. Mark's car, on the other hand, squats, loads up the back tyres and shoots off like a rocket. No fuss, no tyre

↑ Looking at the back, the track is somewhat wider than a standard Mini.

→ Tasty custom paint job has pinstripes, marbling and a base orange colour bright enough even to be seen by centenarians in Cadillacs.

↓ The profile reveals that the Mini's bodywork has been kept to the original length.

radiator is front-mounted, and amazingly even with no engine in there the front boot or ex-engine bay is already full with the rad, the fuel cell and the brake and clutch hydraulics. Fuel cells for racing are a good idea for custom Mini builders – they come in assorted shapes and sizes, they have integral senders and fuel caps, and they do well in a crash.

The gear linkages were a real pain, as the whole mechanism had to be fabricated to work 180° the wrong way, and had to run along for 4ft under the engine in order for the gear lever to come up by the driver. The five forward gears work fine, but reverse has proved to be a problem: the car just sits there while assorted horrible crunching noises emanate from the gearbox. Like JZR Morgan replica trike drivers with only forward gears, Mark learned to avoid parking anywhere that involved reversing.

The brakes didn't prove to be a problem, but a new system had to be designed from scratch. The major issue here is that the car has two sets of front brakes, which is potentially a very dangerous mismatch – you need about 20% of braking going to the back wheels and 80% to the front. In the end a Chrysler master cylinder was used with some custom work on the Mini pedal to get the right leverage ratio, and it was easy to plumb in a brake-proportioning valve and set the pressure to go mostly to the front.

The Honda engine uses an electrical control unit and fuel injection, which led to complications – it's not as simple as recycling the wiring and throwing the Honda away. In the end, the ECU and the engine loom were persuaded to coexist with the rest of the electricals, and they run the engine just fine. There are eight separate fused circuits for accessories, and the wiring is otherwise standard Honda or rewired to a Mini diagram.

The Honda Vtec engine gets 195bhp partly because of variable valve geometry, so you effectively get different camshafts at different engine speeds. The Vtec cam has a shopping-car cam profile at idle, and extra locking lobes turn it into a monster cam at full chat. It's operated by engine oil pressure, but it's controlled by the ECU, as are the rest of the engine's functions – so without the computer, the engine is an ornament. Mark would have preferred carbs and points, but that engine wouldn't get anywhere near 195bhp without electronics.

Mark had his fun with this car, then passed it on to a rabid Florida Gators American Football fan who has added his own details. The colour scheme just happens to be exactly the team colours of the Gators. This is the kind of

squealing, it's just suddenly elsewhere. The 1,800cc Honda Integra package also comes with a limited-slip differential, which means the traction is as good as it gets. The Honda LSD can be quite brutal when fitted to the front of a Mini because of the lack of weight: but the weight at the back of Mark's Mini as it accelerates is quite a lot and loads the diff up nicely, so it just takes off in a straight line. The general handling's pretty good too, says Mark, although once the car gets out of line it tends to snap around and spin because of the rearwards weight balance and the short chassis. The suspension is still Honda, but the spring rates and shocks are designed to make the suspension much more hardcore Mini in feel than soggy Honda. There are also custom anti-roll bars front and back, so I'd suspect that Mark has to push this car pretty damn hard to get it to step out of line.

The wheels are enormous: they're 17in in diameter rather than the Mini's original 10in, which is why the car looks so much like a huge Hot Wheels toy. That's fine with Mark – a big toy is exactly what it is.

The Honda engine and gearbox mounts were recycled as well as the suspension geometry, keeping everything simple and reusing the standard mountings and bushes, so fitting the engine and box was very straightforward. The front and back ends of the chassis were connected by beams, and then an eight-point roll cage was built. The Mini shell was then lowered on to the chassis.

Where the rear seats used to be is now a carpeted box covering the engine and the air conditioning system, and the tunnel on the floor covers the radiator hosing, fuel and brake lines as they run to the front of the car. The

← Engine, transmission, suspension, driveshafts, hubs all bolt to the car's massively strong custom frame.

ENGINE – Honda Integra 1.8 DOHC Vtec B18 C5, standard, factory spec 195bhp at crank. Standard ECU spliced to custom/Mini wiring loom. Fabricated front-mounted cooling system with aluminium radiator. Fabricated exhaust manifold and exhaust, Supertrapp silencer on tailpipe. Remote fuel pump and filter, K&N air filter.

TRANSMISSION – Standard LSD Honda five-speed, with fabricated gear linkages and B&M short-throw gear lever. Fabricated chrome moly driveshafts.

BRAKES – Mini pedal box with altered pedal lever ratios and Chrysler master cylinder. Fabricated handbrake mechanism using additional calipers. Fabricated steel lines, front-to-rear adjustable balancing valve. Front, Honda Prelude discs and calipers with semi-metallic pads. Rear, Honda Integra discs and calipers.

SUSPENSION – Skunk Z Works adjustable coilover shocks all round. Fabricated front and rear anti-roll bars.

WHEELS AND TYRES – 8J x 17in TireWorld six-spokes with Pirelli P7000 225/35Z17 all round.

INTERIOR – Custom, with Honda Prelude seats narrowed by 2in and trimmed in blue vinyl, fabricated engine/air con system cover, centre tunnel, dashboard. Mountney steering wheel, Auto Meter Phantom gauge set, with 10,000rpm tacho and 160mph speedo. Autocraft harnesses. Custom trimming and dash.

EXTERIOR – Standard Mini shell, fabricated chassis and eight-point rollcage. Z-Max body kit reshaped to fit enlarged wheel arches and 17in wheels. Hella 130W headlamps. Paint, House of Kolor pearl orange, cobalt blue stripes with custom finish and pinstriping.

football where they run around for ten seconds wearing armour, then stop for commercials and for the cheerleaders to bounce about. Rather than British football where they run around for ten minutes then foul each other and fall over squealing and pretending they've got a broken leg.

The Magic Bus

After being involved in Mini trike endeavours as described later in this tome (Chapter 15), Alexander Fraser – Sandy to his chums – formed Lion Omnibuses and used the Mini mechanical package to power ⅔rd-scale drivable models of vintage lorries and buses. Among the more memorable ones were wartime AEC Matadors in Royal Air Force livery that were used to tow trailers full of visitors around Duxford aerodrome. Those were fitted with Mini engines, but longitudinally rather than transversely: the driveshafts went to live axles and diffs at either end and thus gave the Matador models four-wheel drive.

Some 40 of these assorted working models were made, and the bus featured is from the 1980s. Its purpose is to trundle slowly round summer fetes, loaded with screaming, sugar-buzzed children. Incidentally, I spotted a cool sign in a Vancouver bookshop recently that said 'Unaccompanied Children will be given a Double Espresso and a Puppy'.

The design of the bus is very practical for its purpose – the suspension is fairly stiff so there's little danger of the bus falling over, and the radiator is many times the size of a Mini one, constructed from rows of tubing like a condenser, and helped out by an electric fan: so it will amble around for hours at walking pace without brewing up.

The engine bay echoes with the emptiness of a cathedral, and the Mini engine lurks right down at the bottom. This particular engine turns out to be a standard 998, with a standard gearbox as well. The relatively huge and very convincing cast wheels result in some peculiar gearing: first gear feels like about third, and the bus is unlikely ever to go fast enough to use fourth.

←← The proportions and detailing look spot-on. Who would ever expect Mini power?

← Elf'n'Safety would have something to say about the open deck and staircase if this were a commercial bus. Rear fog light looks daft but is legally required.

← Standard Mini 1000 engine is dwarfed by the radiator and by the horn. It soldiers bravely on, though.

↓ The driving position is less bus-like than a standard Mini. Weird pedals hinge upwards rather than backwards.

The chassis and body frame are steel, carefully judged to be thick enough to do the job but thin enough to keep the weight manageable. The styling isn't accurate to any particular bus, but a spot of Googling around ancient bus sites reveals that there are hints of the 1929 K-Series and also of the 1930 Leyland Titan. It's rather nice that the Leyland-inspired bus is powered by a British Leyland engine.

The more you examine it, the more the detailing pleases: Hammersmith to Liverpool St is genuinely the route of the No.11.

The bus is currently in the care of Chris Hollier of Stoke Ferry in Norfolk, the engineer who made the chassis and body of my own Mini-based three-wheeler: small world. It's kept MOT'd and street legal, and has a Q plate because it's a 'vehicle of indeterminate age'. It certainly fulfils that description.

Chris fires up the bus so that I can try it out, and there is a satisfactory amount of blue oil smoke: the engine is a little tired, which seems fair given the job it's been doing for 20 years. The gearbox is also tired, and the clutch is very weird. The pedals are now mounted horizontally, so you push them down rather than forwards. With crunching synchros and some very authentic whining from the gearbox, the bus ambles out on to Great Man's Way, the dead-end road Chris lives on, which is gradually sinking back into the fens, some parts faster than others. Slip the clutch to get going and

we lurch along, top deck leaning alarmingly, bell ringing furiously as the bus sways back and forth. Out on to a proper road and the ride is actually not bad. Floor it in first and wait for a while, have a chat, cup of tea, then eventually it's time for second gear. Time passes, Chris's children grow up, go off to work and get married, and by then the engine is finally revolving fast enough for third gear, and we're rolling now. Something of a queue forming behind, but hey ho. First and second gears are all you use most of the time, with third as an occasional treat. I try fourth just for fun, but it chugs unhappily and doesn't like it, way too high-geared even at the heady speed of 30mph.

Is this the most-obscure-ever vehicle powered by a Mini? It's certainly one of the most charming.

← The starting handle, although apposite, wouldn't have much effect on the transverse engine.

→ Cast ID plate says Lion Omnibus 87 CH 009, suggesting it's the ninth, built in 1987.

→ Chris is a big bloke, so he gives you some idea of what size the Mini-bus is.

Turbo Triple Miniac

6

Many old Minis run on three cylinders now and again, but only this one does it on purpose. Canadian Lyle Jansen didn't specifically set out to construct a Miniac from a Pontiac Firefly and a dead Mini: he was looking at various Japanese cars as potential donors for a Mini project that would be utterly different from any other Mini. This three-cylinder turbocharged rocket shopper just happened to turn up and seemed like an interesting option. Lyle's Minis are usually immaculate and standard, so this was definitely an aberration, but what a top aberration.

If you want to build a triple Mini as well, you don't need to hunt down a rare Canadian Pontiac Firefly Turbo Sprint. You can instead look for a Suzuki Swift, Maruti Swift, Geo Metro, Chevrolet Sprint, Maruti Esteem, Suzuki Cultus, Subaru Justy, Suzuki Forsa, Daewoo Kalos or just possibly a Holden Barima. BMC was in the forefront of badge engineering in the mid-'60s, with Austin or Morris Minis also available in Riley or Wolseley versions, and the larger and rustier 1100/1300 Pininfarina-styled saloon range available as Austin, Morris, Wolseley, Riley, Vanden Plas and even MG in the brief period between their construction and their disintegration. Suzuki have elevated badge bodgery to an art form.

Lyle is an automotive painter, restorer and fabricator by trade and by choice, and is one of those people so happy in his work he only stops for food and wine. When it's time to stop work he just says, yeah, whatever, eats some

chips or something and then carries on until midnight.

He does have a 'proper' Mini as well, a standard round-fronted Mini Estate from 1970 with its correct A-series engine. It's now in gleaming red paint and slowly being assembled. Lyle has also worked on a couple of Vancouver Mini Club cars, as well as fellow Vancouver madman Rick Higgs's yellow Mini-and-trailer combo, also featured in this fine volume (Chapter 20).

Lyle's history includes British bikes and Harleys, Plymouth Roadrunners, big-block Chevelles, Corvettes and all sorts of serious and charismatic muscle, so your guess is as good as mine as to why he finished up with a turbocharged three-cylinder Mini.

Building custom cars for other people is fine, but Lyle wanted his own, and he was considering his options when he was offered a 1991 three-cylinder Pontiac Firefly Turbo. In standard form this is a slightly boring modern pocket rocket, but Lyle decided to combine it with a Mini shell to create something amusing, unique and also quite practical. It must be the world's first three-cylinder turbocharged intercooled Mini: anybody who knows different, let me know.

The beginning of the project was fairly simple – the Firefly engine was bored out by 60 thou, then rebuilt to standard spec for reliability. Taking the engine out is easy, as it's all-alloy and only has three cylinders anyway: Lyle just posted it off in a Jiffy bag for a rebuild.

↑ When you look at the car from a different angle it just becomes more confusing. Is it a van? A fastback? Why would a van have a boot? What's going on?

← ← From the front, Lyle's monster looks more or less like a Mini. The odd trim around the screen is because the latter was bonded in to add strength, rather than using traditional screen rubbers. Front end is a steel flip.

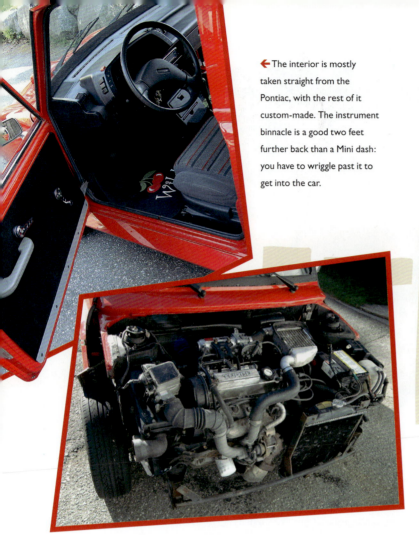

← The interior is mostly taken straight from the Pontiac, with the rest of it custom-made. The instrument binnacle is a good two feet further back than a Mini dash: you have to wriggle past it to get into the car.

↑ The entire mechanicals and floorpan are from the Pontiac. The lack of underbonnet shininess reflects daily use. EFI is multi-point, with separate injectors. The tiny turbocharger produces only limited horsepower, but the torque figure is quite good at 107lb/ft.

He then took an air chisel to the Firefly, cut it off at the ankles and threw most of it away above sill level. Next up for the *Brrraaaap Brraaaaap* treatment with the air chisel came the Mini, which lost its entire floor structure and both subframes. Rust had already achieved most of that anyway, so it was largely a matter of cleaning what was left back to solid metal. When the skin of the Mini was lowered on to the Firefly floorpan, the Mini shell was 18in too short. No problem: he just cut it in half and chopped the roof off another scrap Mini to make up the difference. Lyle now admits he was a little hasty with the air chisel, as the original roof was cut too far back. Mini roofs actually have quite a pronounced crown, and the roof panels should ideally have been cut across the middle at the B-post to retain the original proportions.

As it happens, the final downwards-sloping slightly fastbacky look to the bodywork is part of its character – it's weird, but weird interesting rather than weird horrible… but doubtless everyone will have their own opinion. The amount and the quality of the steelwork is impressive. The rear windows have gone, plated over in sheet steel. To do that without visible wrinkles is not as easy as it sounds.

The flip front is also a labour of love, very neatly executed. The easy way out is a GRP flip front available by post for a few hundred bucks, but Lyle welded this one together from Mini panels, and designed and fabricated the necessary framework, hinges and detachable mounts. There's a lot going on under the bonnet, so fixed steel Mini wings weren't really a practical option for accessibility reasons. In any case, a smooth, well-made steel flip front really looks the business.

The back end of the Mini is just that, a standard rear end pressing and bootlid welded back on, but the rear panelling and the quarter-panels were fabricated from sheet steel, quite impressive work and a brave decision, as there's no way of disguising mistakes. Quite a lot of the Mini panels are new – it's much easier to build cars from new panels than to repair old ones.

The dash from the Pontiac was left in place and the Mini was then reconstructed around it. The dash is enormous and crammed full of stuff, and reaches way back into the Mini's cabin, which means you sit in what would have been the back seat. It's also rather fun that the driver is invisible – when the car's rolling it seriously looks as though there's nobody driving it, and the number of double-takes and dropped jaws is entertaining.

'It's not a Morris any more, but it's still a Mini,' says Lyle. 'People like it.'

The Wal-Mart plakky hubcaps are slightly embarrassing, but the Firefly's stud pattern is weird, and the available choices for it in North America are limited, nasty, and covered in cheap chrome, and mostly look worse than the hubcaps. Think loser pimp, think budget bling. Lyle will replace the wheels when something genuinely suitable turns up.

On the road, the independent suspension and the extra 18in on the wheelbase mean the ride is smooth and civilised, although the turbo whistle and blowoff-valve noises replacing the A-series gearbox whine sound pretty strange in a Mini.

There were no complications with changing the suspension to MacPherson struts at the front and multi-link at the rear, as the Pontiac's floorpan was retained as far up as the suspension mounts – Lyle just welded the Mini shell to the Pontiac pan and some of the substructure so that the stock suspension mounts were properly braced in the new structure, and left the geometry and mechanical components exactly as they came out of the Pontiac. The handling may not be as sharp as a proper Mini, but the trade-off in ride and comfort suits Lyle. The car is a daily driver and even tows a small trailer, so it has to work for a living and is not trailered to shows cradled in a nest of cotton buds and rose petals – it's a trailer truck rather than a trailer queen. The car's regular 40mpg

↑ The roofline on a Mini drops quite dramatically after the midpoint, which is why the rear end droops the way it does.

plus – even on smaller American gallons – is very useful in protecting Lyle's wine fund, Californian Fisheye Cabernet Sauvignon being his tipple of choice. He says the Miniac just seems to run on fumes rather than using any petrol to speak of.

'The power is about the same as a good 1275 Cooper S,' he says, 'and that's plenty.' Max power wasn't the point, or he'd have put a V8 in it. It's interesting to note the torque figure, which at 107lb/ft is good going for a small turbo engine and explains why the car drives so well: usually a small turbocharged engine is pretty feeble at low revs and only comes alive in the upper reaches of the tacho when the compressor is really screaming. Top speed is as yet untested, but even at licence-losing speeds there still seems to be plenty of power left.

The driving position is well weird, because the Pontiac's enormous dashboard containing the electronics was retained: the steering wheel has moved back 2ft from its original position, and so has the driving seat. Most of the time this works fine, as you just lean forward if you need to look out of a side window.

I'm keeping an eye on Lyle – I have no idea what he's going to do next, but it will probably top this.

Tech

ENGINE – 993cc three-cylinder all-alloy OHC multipoint fuel-injected Suzuki engine, RHB32 turbocharger with intercooler, bored out +60, 70bhp, 107lb/ft torque.

TRANSMISSION – Pontiac Firefly five-speed manual FWD transmission.

SUSPENSION – Complete floorpan from 1992 Pontiac Firefly Sprint Turbo. Front, independent with MacPherson struts, coils over, driveshafts through hubs. Rear, multi-link independent with MacPherson struts, separate coils, wide wishbones, additional stabilising arms.

BRAKES – Pontiac powered discs front, drums rear.

WHEELS AND TYRES – Standard Pontiac 4.5J x 13 steels, 165/65R13 Westlake H600 tyres. Wal-Mart plastic hubcaps: unusual PCD stud pattern limits wheel choice.

INTERIOR – Pontiac seats, entire dashboard, steering column, pedals. Rear deck constructed and trimmed, carpet-material headlining.

EXTERIOR – Mini sedan shell stretched 18in to fit Pontiac floor and wheelbase, rear window converted into van sides. Flip/detachable steel front constructed from new Mini panels. All panel work is steel. Paint, Ford Vermilion Red.

The GT40 Mini

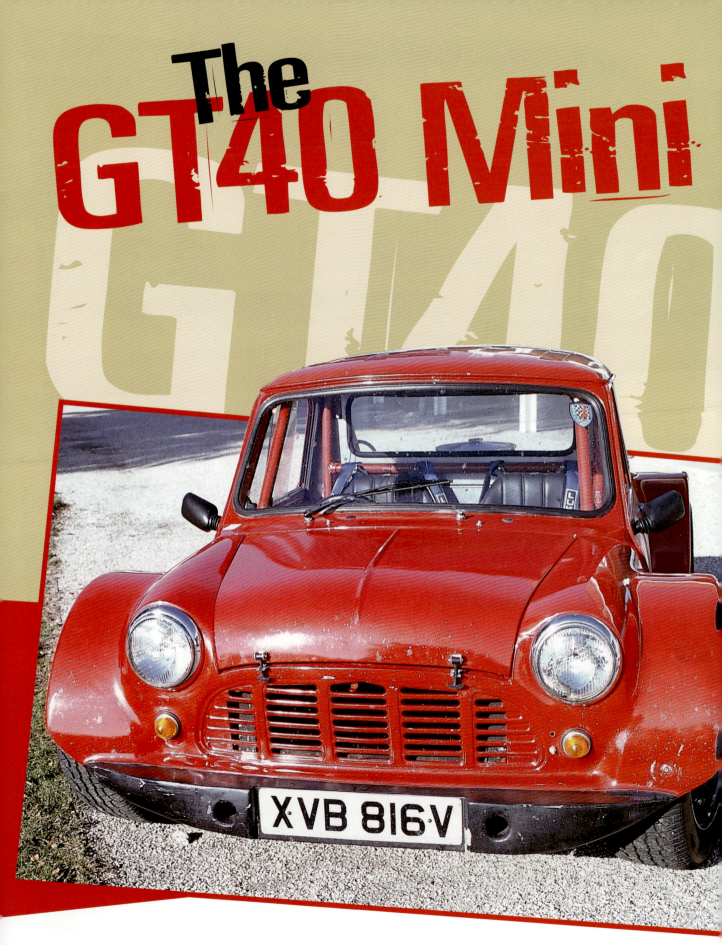

XVB 816V

One of the maddest Minis ever is Ray Christopher's pickup, which has a Ford GT40 racing engine in the back. Ray used to run a kit car company called GTD, which made very good replicas of the Ford GT40. These were so successful that his life became deeply dull, and he found himself spending way too much time stuck in an office shuffling bits of paper and sorting out production trivia.

His response was to build this Mini. It has caused as much trouble as it has hilarity, and Ray has been banned from every race series he's ever managed to get it entered in. The reason for the multiple bans is simply that the thing is genuinely dangerous. It has a short chassis, the aerodynamics are crap, and as Ray says, you're always steering into a skid: if it gets more than 20° out of line it's going to spin off into the kitty litter. It's like steering a block of flats, he says, because it leans as well as drifting. The power is the same as a GT40, but while the GT40's gorgeous curves serve to keep it stable and smooth at high speeds, the Mini's half-brick aerodynamics have exactly the opposite effect. Despite that, a 1 minute 38 second lap time around Goodwood is respectable, and repeatable as long as the car can be kept pointing mostly forwards.

'Some racing people were saying, you've got to have innovation, got to let him do it: and the others were saying get the loony off the track before he kills himself.' A picture in *Autosport* magazine of the Mini kicking up a 20ft high rooster tail of dirt as it spun off a circuit at full speed did its reputation no good at all.

Banned from the track, Ray took to hillclimbing until the rule books were changed to stop him. '*Que sera*, innit.'

The starting point for this car was a V-reg Mini Traveller with a replica GT40 frame under it, but the rear end of the roof soon came off to be replaced with a pickup bulkhead. That was because driving the car meant you were actually inside the engine bay along with a full-race V8 Ford engine and eight red-hot snake-orgy exhaust pipes: it got a little warm in there. The original Traveller front was also lost in a tyre-wall wipeout early on, but the current van front end has survived for a while. The GT40 Mini has a deep and irresistible attraction for tyre walls, and has snogged them big-time on three occasions so far.

The arches are GRP, but most of the body is still steel and still Mini, although not many of the original car's panels survive. The rear end had to be significantly reconstructed after a substantial crash left a rear wheel sitting on top of the engine.

The chassis is a cut-down replica GT40 frame, with 7in taken out of the sill area. Take it from me, it could use the lost seven inches.

The engine is a period Ford Motorsport 302 (five-litre) with something like 480–500bhp available, more or less what was used in the real GT40s. It's strangled by

← The front end of the Mini contains nothing much, which is one reason why it gets light at 150mph+.

↓ The back end of a pickup offers plenty of room for an engine, and is about the right size for a 302 V8.

↑ The frame of the pickup is an altered GT40 replica chassis.

↑ The only reason the car didn't get its usual huge audience is that the only people present at the track were testing racing cars.

↓ The pickup is MOTable, but not a relaxing motorway cruiser – so it lives in the toy hauler.

the huge silencer needed to drive on obediently shushed British racing circuits – you can make more noise in a library than you can on a British motor racing circuit these days – but even so it packs a fair old punch. The rev limit is now 7,500rpm. It used to be 8,500 but the engine's getting a little tired by now and is worth preserving, rather than risking blowing it up.

Preserving the engine is becoming a moot point these days, though, as Ray has found a new place to race – the Three Hills Challenge on the Isle of Man. Three days of full-on racing, starting at 10:00am and finishing at 4:30pm, after which you go off to the pub with the marshals and the Clerk of the Course. If you want to push your luck and crash into a dry stone dyke you're free to do so, and if you want to make a lot of noise that's OK too.

As Ray says, the car gets lots of attention and lots of smiles from the public and spectators, but not quite so many smiles from drivers and track management if it's spinning down a racetrack among a bunch of expensive racing cars.

As mad Minis go, it's up there.

A lap of Goodwood

A cruise around the Goodwood circuit with Ray in the driving seat starts off with a blast of aural thunder and a brutal kick from behind as the clutch is dumped, then up to 75mph for the first corner. By Frith we're up to 130, and approaching No-Name it's 150–160 on the way in and 105 through the corner. St Mary is slow and difficult, with Ray working the wheel and pushing as hard as he dares – the car is getting very twitchy.

Lavant has new tarmac and sod-all grip, so all four tyres are screaming all the way round. The straight is taken at 170mph, and Ray bawls over the hammering, howling engine, 'THE FRONT GETS A BIT SQUIRLY OVER 160.' I already know that, thanks, I can feel it.

The speedo's not connected because the needle just flaps about at speeds over 150mph, but the car's in top gear and the tacho needle's on the red line so it's not hanging about. More stomping on the brakes, the nose thumps down and you bang through the next lot of corners at around 100mph. You can smell the brakes frying as the speed is clawed down for the last corner and the chicane, and a violent wriggle gets us past that and past the pits in a blur.

I didn't get many laps in, as somebody else trying out the car got more than the critical 20° out of line and flat-spotted all the tyres in a huge cloud of burnt blue rubber smoke before he went off the circuit into the cornfield. He's probably still picking bits of stalk out of his ears.

Tech

ENGINE – Ford Motorsport 302CI 5-litre V8. Balanced, twin-bolted mains, big valve Motorsport heads, roller rockers. Holley 750 double pumper on Edelbrock manifold, MSD ignition, custom headers and exhaust. Oil cooler, Facet fuel pump.

TRANSMISSION – Four-blade paddle clutch, Renault 30 gearbox and transaxle.

BRAKES – Ford Scorpio standard brake system with twin servos ('almost adequate').

SUSPENSION – Experimental alloy GTD GT40 hubs, Spax adjustable race coilover shocks, 320lb front, 450lb rear. Double wishbones front, four trailing radius rods rear with reverse bottom link and single top arm.

WHEELS AND TYRES – GTD GT40 alloys, 15 x 6J front with P215/60 BF Goodrich Comp T/A, 15 x 10J rear with P265/50 BFG.

BODY and chassis – Modified GTD40 tubular steel chassis, mostly steel body from Mini Traveller. Fuel tanks in sills, piping passes through central floor tunnel. Door hinges from a Peterbilt truck.

INTERIOR – Unidentified but comfy race seats sourced from Beaulieu Autojumble. Luke harnesses. GT40 pedals. Clocks: tacho, speedo, fuel, volts, oil pressure, coolant temp.

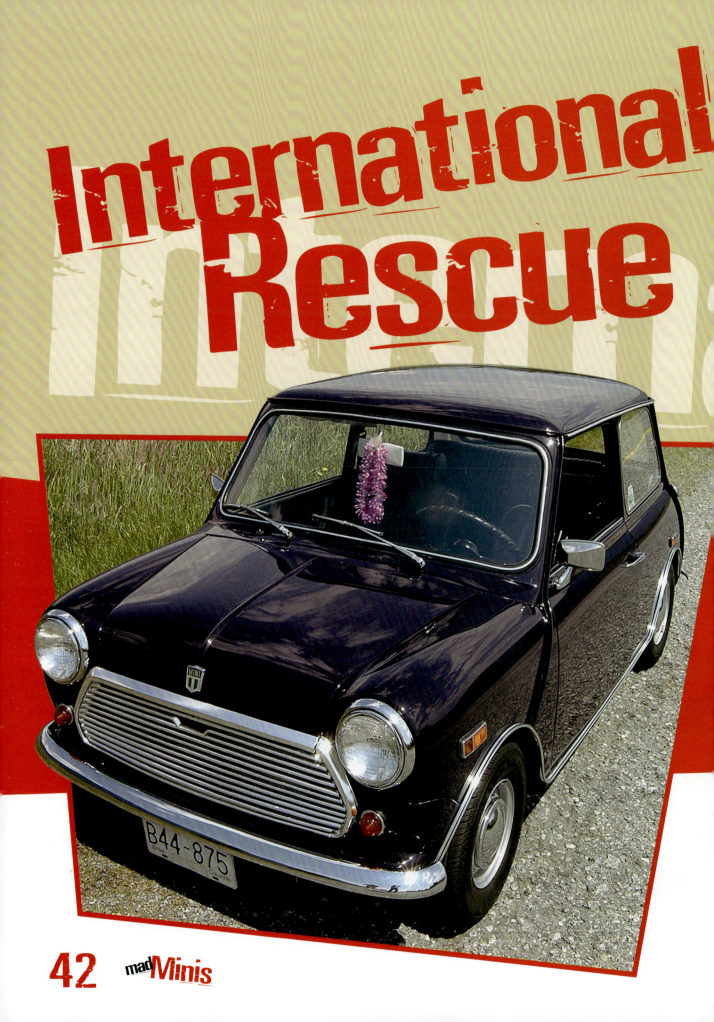

International Rescue

John Goolevitch isn't so much mad as just extremely stubborn. Like most members of the Vancouver Mini Club, John paid around $6,000 for this Mini: that's what it costs for a drivable example in North America. He bought it from the Bolivian Consulate in Vancouver, so it had diplomatic plates. He didn't get to keep them, which was a shame – they'd have been useful for sidestepping parking and speeding fines. 'Scusi, Senor Policia – I are deeplomat, no spik English very bueno.'

The car had been a Valentine's Day present to John's wife Jacquie, but she never got near it. She also lost her husband to the garage for a year and a half, but from her point of view she knew where he was and what he was doing, so he says she didn't mind too much.

The Mini had a few rust bubbles here and there but didn't look too bad. He realised that there were some issues, as the Californians say, during a club outing to Oregon in heavy rain, when his feet began to get wet as he drove along. When the convoy stopped, John's carpets were floating in two inches of water, and little waterfalls were pouring out of the sills. His club colleagues, with rather British-style sympathy, bought him a yellow rubber ducky to use as a water depth gauge inside the car. At this point he realised he might be in trouble.

Sure enough, each rust bubble that was poked with a screwdriver turned into a gaping hole, and the sills literally came off in his hands when he started pulling at them. The front end of the car came away more or less in one piece, and not because it was a flip front. The A-panels and

front wings disintegrated, revealing that the ends of the dashboard were no longer solidly connected to anything and had been held in place largely by the windscreen and what was left of the scuttle.

The floors mostly disintegrated when kicked, and so did the back seat platform. The transverse seat support panels weren't too bad, but the rusty seat runner securing nuts snapped off rather than coming undone, so that whole area was cut out as well. When all the rust was finally cut back to clean metal, John was left with a solid roof, good roof pillars, some solid areas of floor and most of the rear wings.

In the UK, the car would have been scrapped when the front end fell off. In Canada, the situation is different and the options less palatable. John had lost the significant amount of dollars he had put into buying the car and would have to find the same amount again to buy another usable Mini just to get back to square one, so he decided to build a mostly new Mini shell from repair panels. It would cost plenty, but at least that way he would definitely finish up with a really good rust-free Mini.

He managed to borrow a huge 10,000amp spot welder, which would enable him to restore the Mini not just to new condition, but to concours condition with the right sort and number of spot welds. Mind you, the number of spot welds to be found in a Mini depends on the day of the week on which it was built: Friday cars generally have fewer and more randomly placed welds.

One piece of advice that John would offer to anybody

← John Goolevitch's Mini finished up much better than new and has won prizes.

↑ Sills? What sills?

→ The doors are hinged further out than this gaping hole, or they would literally have fallen off.

← The body is inverted, the boot lid, panels and floor are gone, so we should be seeing some solid metal…

↑ …but the floorpans are also deeply nasty. That alone would have been terminal for most UK Minis.

tackling a major Mini restoration is to consider replacing major panels rather than repair-patching and reassembling them. He also says that you don't need an expensive welder and can get by just as well with a cheap one, but in the absence of talent and experience I tend to disagree: I have a new, small, cheap MIG welder which is barely able to cope, and I recently got a bigger, older and more serious welder, and with the bigger one welding is much easier. There's no need to go mad, but buying something in the middle of the amateur welding market is worthwhile, to my mind. John's first go at welding was plating over the speaker hole on the back shelf in this Mini, a good place to start because it's invisible. Success with that encouraged him to carry on.

Having recommended buying whole replacement panels, in John's case financial acuity dictated that anything that could be correctly made from folded and welded sheet metal would be created from scratch, to keep the expensively-imported parts bill down to just massive rather than truly gargantuan. Buying cleverly can be productive, too: the whole rear end of a Mini, from the valance to the window, only cost $80 more than the repair panels needed to restore the original one – and all he had to do was spot-weld the whole assembly on to the main shell. Alignment, he says, wasn't a problem, partly because Minis are small. It's uncharitable but true to point out that they weren't always put together with much precision anyway, particularly if another British Leyland strike was brewing. Or if it was a Friday. Or a Monday.

Sandblasting has pros and cons. John's sandblasting cabinet, he says, is the best tool he ever bought. Conversely, having large thin flat panels sandblasted is a bad idea – they warp unless approached from a shallow angle, gently and patiently and with a less aggressive medium than grit. Classic American cars tend to be made of much thicker steel than British cars, so American shotblasters are used to blasting away with cheery brutality and suffering few disastrous consequences.

Tech

ENGINE – 1,293cc, .020 oversize bores/pistons, Pertronix ignition module, LCB manifold, RC40 exhaust, otherwise standard. Original engine was 998cc.

TRANSMISSION – Heavy-duty clutch, otherwise standard four-speed all-synchro.

SUSPENSION – Standard dry rubber cones.

BRAKES – Non-assisted, drums all round.

WHEELS AND TYRES – Standard restored 3.5J x 10 steel rims, Kumho 145 x 10in tyres.

INTERIOR – Just Stitching Around piped seats, Newton Commercial carpets with extra binding.

EXTERIOR – 1972 Mini 1000 to Canadian spec, extensively restored and repainted in Black Tulip. British spec sidelight/indicators and bumpers fitted.

The doors on John's car looked OK, but he stuffed a small camera inside them and took some flash pictures – then having seen the resulting images he pulled the old skins off the doors and replaced them, fortunately quite an easy and cheap task and a much better idea than repairing them. I've tried both, and reskinning takes about 10% of the effort and time of repairing rusty door panels.

The whole underside of the car has been coated in Gravel-Guard, a thick and permanent coating. The topside got ten coats of hand-rubbed primer before the shiny paint went on, but you can probably see that by looking at the pictures.

The rest of the components of the car looked quite nice because they were generally still in good condition – it had just been the rust problem that was severe. The wiring was dismantled, cleaned, wire-brushed and gently media-blasted where appropriate, then rewrapped and clipped back in. A new headlining was fitted – a pain-in-the-bum job every time. The brake and clutch cylinders were cleaned, blasted, re-rubbered and clear-coated, then refitted. The 998cc engine was replaced with a bored-out (to 1293cc) 1275. Another tip from John, on reassembling a restored and painted car – it's safer to drop the car on to the engine rather than to jack the engine up into the car – there's less risk of scratching brand new paint.

The Black Tulip paint in question, a deep purple, very '70s and quite cool, had originally been matched with orange nylon velour seats – also very '70s but hideously uncool. Part of John wanted to refit the concours-correct seats, but when he put the rear seat back in the car he had to take it straight out again – the colour was just too upsetting. The seats were stripped, re-foamed and re-trimmed in black with purple piping.

The overall cost was pretty brutal. Remember that in Canada, Minis are expensive foreign exotica, and so are the spares – a new front wing that can be bought for 20 quid cash at a UK car show, or a tenner if you wait until they're loading their van at the end of the show, costs $90 (£50) by the time it gets to John's door, and a front apron is $120 (£66).

The total cost? 'I didn't want to keep track, because it would be too depressing,' he says, 'so I don't really know.' But we all know roughly what a full restoration costs, so why do it in the first place, rather than just throw the scrap Mini car away and buy a good one?

'Because it was *my* Mini.'

Fair enough.

↑↑ After a year and a half of hard work with spot welder and rubbing block, John is a happy bunny.

↑ Purple piping on black seats looks nice – the original orange nylon was in the end just too indigestible.

The Miniha

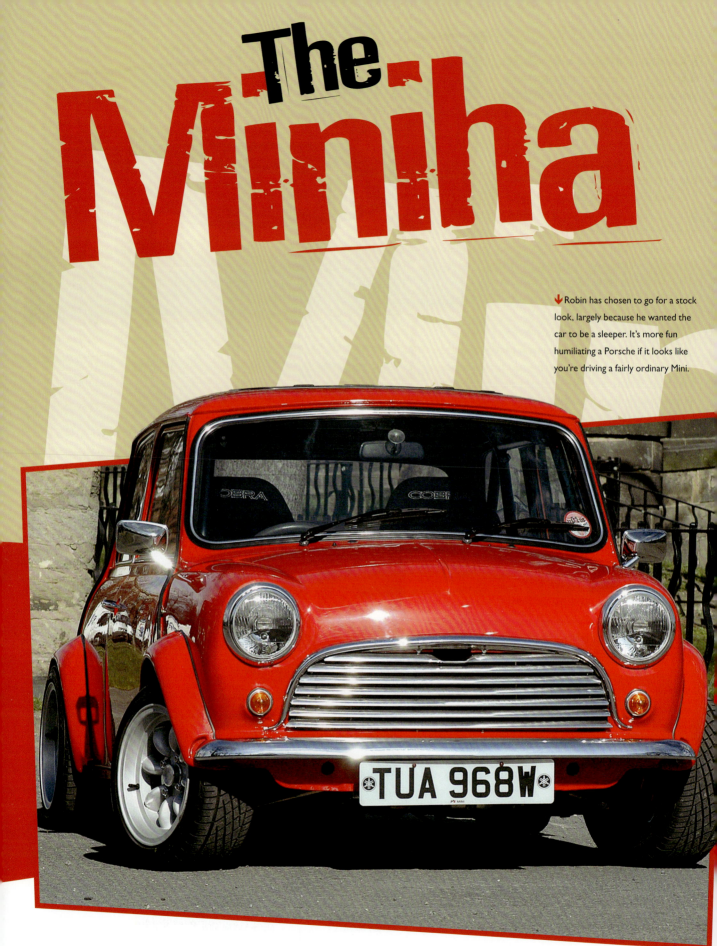

Robin has chosen to go for a stock look, largely because he wanted the car to be a sleeper. It's more fun humiliating a Porsche if it looks like you're driving a fairly ordinary Mini.

TUA 968W

Robin Knapton is well pleased with his Yamaha R1-powered Mini. When he began the project the R1 was the fastest bike in the world, so that's a pretty good starting point for a powerplant for a bike-engined Mini. The R1 motor has a reputation for being indestructible, and it's also less expensive than a Hayabusa engine. It may not have the torque of the bigger Hayabusa engine, but 285bhp/ton and 4.5 seconds to 60mph is quite frisky.

With the engine and the drive at the back, all the power goes into forward motion rather than rubber smoke: the car is extremely fast. It's still a real steel Mini, though – or to be more accurate, what remains of the Mini's shell is still steel. It's a Z-Cars conversion, but it's not one of their full-on terror machines, which are a bike engine in a roll cage with a plastic Mini-shaped hat on it.

As a Mini enthusiast and a superbike enthusiast, Robin successfully combined both in one vehicle. He likes the way the engine will rev to double the rpm of a Mini engine, and he likes the way it spins its wheels on take-off but still has plenty left when you plant it, whatever speed you happen to be going. He also enjoys humiliating Porsche 911 Carrera Turbos: he can take them, no bother, up to 120mph. After which he can legitimately back off anyway, as the job is done.

One of the other things he really likes about this car is the visceral excitement it generates, even at legal speeds. He's owned Subarus and big bikes aplenty, but any serious throttle action lasting more than five seconds on either of those could cost him his licence. The Miniha, on the other hand, feels about 100mph faster than it actually is. The engine doesn't just start like boring old production cars: it explodes into life, and then it sits there growling urgently, twitching and nervy, gagging to go, like a greyhound in its trap. Robin just sits there waiting for it to warm up, as the engine trembles and twitches in its kennel behind you.

When he lets it out of the trap, the noise is appalling. At 9,000rpm you're thinking for God's sake please change gear before it throws a rod, but that's when the cam comes in and it really gets going: the rev limiter cuts in at 14,200rpm. By the end of second gear, the scenery is a blur, the entire universe has gone screaming mad, your head is having trouble staying on, every muscle is rigid, and the speedo reads… 60mph.

Oo-er. Felt like 150mph. *Click click click* up the gearbox, and the engine is now producing a loud tenor rumble as we cruise to an empty road to give it some proper exercise. There's no rubber cush drive as there is in the bike's rear hub, so there's lots of clacking and banging through the transmission, amplified by the biscuit-tin acoustics of the bare Mini shell.

The suspension settings are pretty hard-core, and ride comfort is not helped by the cast-iron A-series anchor in the front being entirely missing. The tyres are pumped to 18lb, as the car only weighs around 600 kilos, and the

9

⬇ The whole front end is very nicely custom-made in steel from Mini panels.

↑ Digidash avoids the need for any other instruments, and also avoids Yamaha/BL sensor-to-instrument clashes. It runs through its functions and settles on road-speed display as its usual default function.

↑ Even with no engine it always seems to be crowded in a Mini's engine bay. Fuel tank, brake servo and radiator are at the front.

↑ The engine bay looks almost empty: the tiny engine is tucked away at the front. The rear suspension offers adjustable coilovers, an anti-roll bar and disc brakes.

↑ The chain drive connects to a custom differential provided by Z-Cars.

→ → This is the view of the car that most people see as it screams past them. There are clues as to why – the exhaust, the side-window air intakes, the yammering scream from the exhaust.

The stance with the low back end isn't a styling choice: Z-Cars have found that lowering the rear end to this height achieves the best handling.

already stiff Mini ride is stiffened yet further by the weight reduction. If you imagine being towed along a rough road in an engineless Mini by a jet fighter, that's more or less what this thing feels like.

With a stretch of clear road ahead, Robin punches it properly, and when the tyres get a grip, it goes scream SCREAM clack, scream SCREAM clack, much like before, but after the sixth click it's time to back off as all the needles have swivelled a very long way past sensible. The second big scream is caused by the cams coming on: the power smacks in with such a bang that I can't imagine you would be able to stay on top of a Yamaha R1 bike at full throttle in the lowest gears. It must feel like being rear-ended by a dump truck when the cams hit on an R1 bike. In a steel Mini, though, you can actually use the full power in all gears, and of course being a Mini it can corner much, much harder than a bike. I reckon this Miniha would cream an R1 on a racetrack, even if the rider had *cojones* the size of grapefruit.

After a short while this level of noise gets pretty tiresome, and Robin keeps a pair of earplugs in the ashtray for going home after a blast. There's no soundproofing at all, and the engine bulkhead, an inch behind your shoulder, is a single sheet of aluminium. If you wanted to civilise a Miniha with sound deadening and carpets, you could – but I don't know that you would necessarily want to. At full welly the exhaust screams like a dozen electrocuted sopranos, and the crackling, spitting and popping of half-burnt fuel on the overrun sounds like a gaggle of Type 35 Bugattis cornering at the Le Mans Classic.

Getting the car to this point was… easy is the wrong word, as it's a big project, but it was straightforward and fun. The Mini was stripped to the shell, the rust was repaired with new panels, the rear end was chopped out and it was painted ready for the Z-Cars kit. Robin rates Z-Cars highly. He bought the kit and also the engine from them, guaranteed to be a good runner for £1,600, and fitted the kit to the car with no problems at all.

The kit starts with a roll hoop in the usual position behind your head, from which the suspension and engine are suspended. There are new trailing arms with disc brakes, and driveshafts lead inwards from the hubs to a chain sprocket and a Z-Cars differential. You can add an electric reverse gear, or get out and push, whichever. Oddly enough, while a reverse gear is not currently required for road cars it's compulsory for racing. If you finish up stranded on the apex of a corner after a misjudgement and a spin, there may not be room to get out of the way of approaching competitors by going forwards. The Z-cars electric reverse system is basically a converted starter motor, but it's £400 and Robin decided to skip it.

There's an anti-roll bar and a set of light alloy coilover shocks, and most of the joints are rose-jointed and adjustable. The engine sits as far forward as it will go in order to achieve the best possible mid-rear-engine weight distribution – you could almost bang it with your elbow – and the six gears are sequential. The gear lever simply clicks forward six times to get into top. You don't really use the clutch going up the gearbox, just lift off the throttle a little,

although it's kinder to the engine and box to blip and declutch on the way down.

At the front, there's a crash bar that's attached to the rollcage, to the Mini subframe, and to the body, and there are brackets for the adjustable front lightweight coilover shocks, the radiator, and the racing fuel cell. The suspension is adjustable, and the brakes are Mini four-pot discs.

Many people choose a plakky flip front, but Robin welded together a top-class steel one. Somehow steel flip fronts always look just a little sharper, a little better: they're worth the effort, considerable though it is.

Six months of happily busy weekends saw the rest of the car's assembly and the plumbing and electrics going in, and quite quickly the big day came when the engine was fired up for the first time. On its first drive there was a nasty vibration, for which Robin couldn't find a reason or cure. Z-Cars took the car in and sorted it out over the next couple of weeks, free of charge, and gave it back driving perfectly.

Since then Robin has sorted out better cold air flow to the carbs, and has changed the glass rear side-windows for drilled Perspex ones with a custom air scoop that helps to feed cold air to the carbs and to keep the engine cool. He's also devoted a lot of dedicated effort to learning to get the best out of an evilly fast rear-wheel-drive Mini while retaining his driving licence.

If you ever see this car for sale, buy it: there won't be anything wrong with it, but Robin has the Need for Speed, and he's seen Z-Cars' rollcage-framed GRP Monte Carlo which is even faster than this one.

Tech

ENGINE – Yamaha R1, 2002: 998cc, four-cylinder transverse liquid-cooled EFI DOHC 20-valve 170bhp aluminium bike engine. Pipercross air filter, NGK Laser iridium spark plugs, Z-Cars wiring loom. Bosch universal fuel pump. Custom-tuned exhaust. 1970s Fiat Uno radiator, Kenlowe 12in fan.

TRANSMISSION – Multi-plate clutch, six-speed sequential integral gearbox, Z-Cars custom chain drive and diff. Reverse gear operated by manual owner-limb contra-progression, ie get out and push.

BRAKES – Servo-assisted Mini pedal and master cylinder, braided lines throughout. Front, MiniSport four-pot calipers, solid Mini discs. Rear, Ford discs and calipers. Custom handbrake.

SUSPENSION – Front, adjustable tie bars and control arms, coilovers. Rear, independent axle, fixed chain drive and diff, halfshafts, trailing arms, coilovers, rose-jointed adjustable anti-roll bar. Pro Tech alloy coilover shock absorbers all round, adjustable for compression and rebound.

WHEELS AND TYRES – 7J x 13 Minilite alloys, with 175/50 x 13 Yokohama A539 tyres, set at 18psi.

EXTERIOR – 1981 Mini shell, Sport Pack arches, MkI grille surround, custom seamless steel flip front, Perspex rear windows with air scoops to cool engine, Ford Rosso Red paint.

INTERIOR – Monaco Sport bucket seats, three-point Sabelt harnesses, Z-Cars full rollcage, EVO Digidash.

PERFORMANCE – 0–60: 4.5 seconds. BHP per ton: 275. Top speed: 120mph so far. Fuel economy: 25–30mpg.

Moking

Many North American Mini Mokes are used for autocrossing, which is local-level, budget motorsport that anybody can enjoy. Most American Mini meetings feature an autocross – all you need is a car park and some cones.

Minis, of course, are brilliant at autocrossing. They're short, with a wheel at each corner, they have firm suspension and they handle well. They also allow even a novice to make excitingly rapid progress because front-wheel drive is quite amateur-friendly: the front wheels tend to drag you in the right direction even if they're skidding a bit. Push your luck as a normal driver in a rear-drive car and you'll usually spin out very quickly.

Minis are top autocross tools, then. So imagine a Mini with a longer wheelbase for extra stability, with the heavy roof, glasshouse and doors removed, with a strong engine and a stiff boxed perimeter chassis – it would be the perfect autocrosser.

That's exactly what a Mini Moke is.

For its original purpose, as a military jeep, it was comically useless. The cunning plan was that Mokes would be stacked in a Hercules transport aircraft with parachutes attached, and would then be thrown out to land on the battlefield, where troops would jump into them and drive off to kill people. The stacking idea worked brilliantly – with the windscreens down, the flat mudguards of the bottom Moke are ideally placed for the wheels of the Moke on top.

The fatal flaw was the total lack of ground clearance. If the parachuted Moke happened to land on a road, no worries, jump in and drive off. If it landed on a flat dry field with a convenient gate leading on to a road, you might get away with that as well if it wasn't too bumpy. If it was raining, forget it: if one wheel spun in the mud, the Moke would go nowhere. Even military purchasing authorities spotted this flaw, and they formed an orderly queue not to buy Mokes. The only people who did buy them were hippies, and latterly Caribbean hotels.

That lack of ground clearance made the Moke one of the most useless military vehicles ever, but the same

← The author spent a cheerful few days driving this restored standard 850cc Moke around Tennessee in 100° heat, and enjoyed every moment: you don't need air-conditioning in a Moke.

↓ The interior of a Moke is a fairly Spartan environment. This one has a 118bhp engine and many more clocks than the standard lonely speedo.

→ Most Mokers stick with the traditional A-series Mini engine, but one can't help wondering about a Honda Vtec. You wouldn't even need to alter the bodywork…

10

↑Vancouver Mini Club's Rob Fram shows how it's done. Strong engine, sticky tyres, throw the back end out just enough and use the power to haul the Moke around the cones. A fair amount of blue smoke from tailpipe and tyres, but a tidy run with no squished cones.

low ground clearance, combined with its lack of a roof and windows, gave it a centre of gravity lower than the ethical standards of a cabinet minister. For something to throw around a small car park, at max speed, slaloming between cones, you really can't get anything better. Even if you designed a pure autocross racer it would finish up looking like a Mini Moke.

The other bonus of the Moke is that you can provide excellent entertainment for yourself, a few friends and the general public simply by driving one around the streets, and you get to join the Mini world, which is full of amusing, eccentric, friendly and helpful people who energetically organise themselves to have all sorts of harmless fun.

Mokes have even been credited with curing shyness: a Moker friend used to suffer from painful shyness until he bought a Moke, after which people wouldn't leave him alone, so he had no choice but to become gregarious and friendly.

Mokes everywhere get a huge amount of attention, but the questions in America are always the same. You can have fun making up new answers to the same old questions, though.

Is it legal? (No, I made the licence plates myself.)

Is it electric? (Have a listen to the engine noise and take a wild guess.)

Is it a golf cart? (No, I'm way too young to play golf.)

What does it do to the gallon? (About four times as much as yours.)

Can I see your licence and registration, please? (Um, no problem, officer. I do apologise if I was mistakenly going a bit fast back there, this is a 1966 Moke and the speedo's not very accurate.)

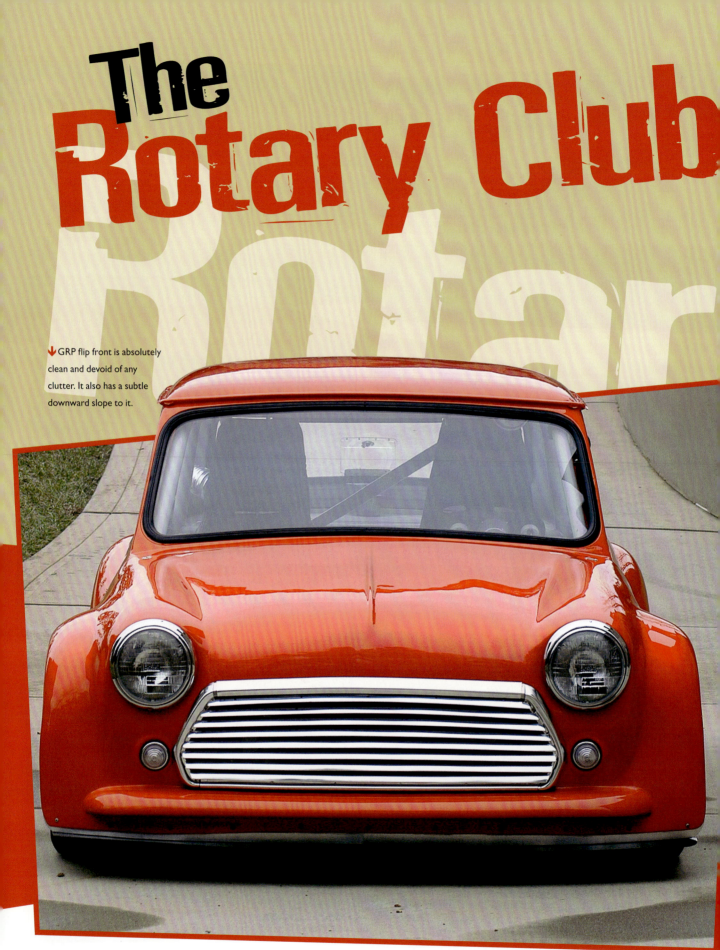

The Rotary Club

↓ GRP flip front is absolutely clean and devoid of any clutter. It also has a subtle downward slope to it.

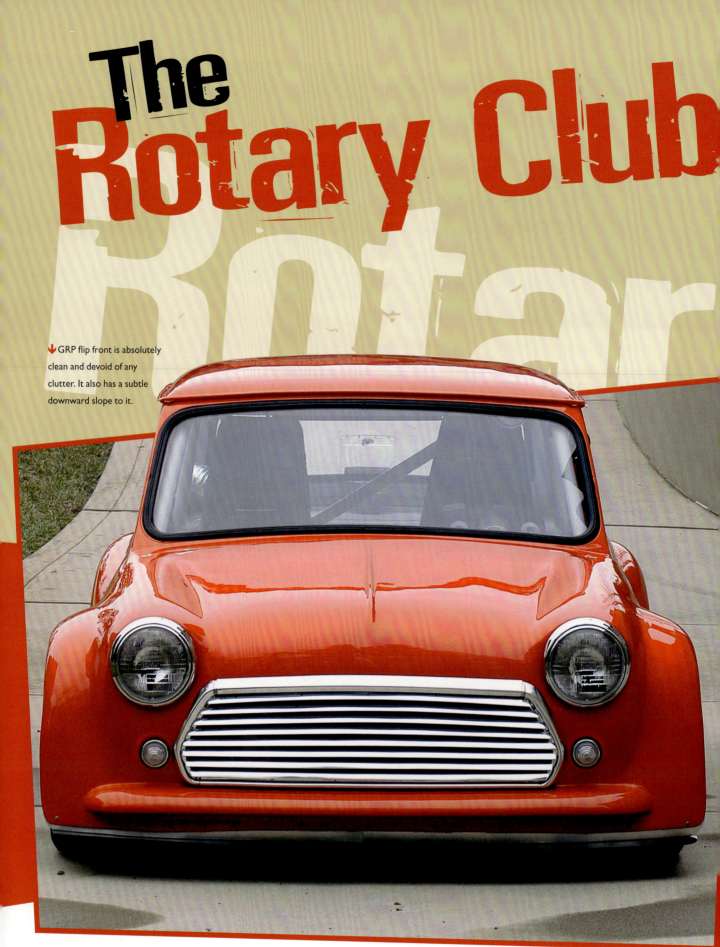

The Rotary Club

↓ GRP flip front is absolutely clean and devoid of any clutter. It also has a subtle downward slope to it.

Jeff Griffin has used a race-derived spaceframe chassis, a Mazda rotary engine, rear-wheel drive and a custom plastic front in his Mini, but for all that it's still a Mini.

It was a guy called Kool-Aid who started the project a good few years back. Kool-Aid is a fizzy American powdered sugar drink, but it's also Kool-Aid's nickname, and a top nickname it is too.

The chassis was based on the design concept of a successful racing GT5 Mini built by veteran racing car driver and designer Joe Huffaker, who was to American Minis what John Cooper was to British Minis. Joe won three SCCA Championships in his Mini from 2000 to 2002. He started with a racing Healey in 1954, and went on to build racing versions of MGBs, Jaguars, Healeys, Austin-Healeys and Jensen-Healeys, and then moved on to Minis later as the competition possibilities offered by the Mini format became clear. So the Huffaker-inspired chassis made by Kool-Aid was authentic Mini history, although it wasn't British Mini history.

Joe Huffaker is still making old British cars go faster, and still sells stonking dry-sumped race engines for Spridgets, which of course almost fit Minis… check out www.huffakerengineering.com.

This particular Mini was a massively ambitious project, and Kool-Aid gradually came to a stop as he realised he was never going to finish it. Having made his very major contribution, he passed the project on to his friend Jeff Griffin, who managed to make quite reasonable progress over the next four years. However, an increasingly successful kitchen cabinet business began to interfere more and more with building the car, and Jeff's progress also ground to a halt. However, during his four years of patient but stop-start development, Jeff had figured out exactly what needed to be done to finish the car, and how. He couldn't do it himself any more, but he could now afford to have somebody else do it on his behalf. Fortunately, Jeff's home town of Charlotte, North Carolina, is an excellent place to be inventing new cars, as there are numerous racing shops around to serve the local NASCAR and SCCA racing scene. Racing engineer Todd Hayworth was entrusted with the task of carrying on where Jeff had left off.

The car was actually quite well on the way by that point: the rollcage and internal body bracing had been constructed, and the roof had also been chopped. The roof chop is interesting – the height of the roof has been cut by an inch and a half, but the windscreen remains the standard size, although made of Lexan rather than glass. Jeff cut up the entire A-pillar and scuttle assembly, moved the bottom of the screen forwards and pulled the top of the screen backwards, and refitted the original screen surround so that a standard screen could still be used – an excellent trick for anybody considering a relatively subtle roof job on a Mini. It's not a big chop compared to Jason Neumann's four-inch roof amputation in Chapter 2, but Jeff's roofline still looks just that bit cooler, reduces

 Rear-wheel-drive Minis offer very useful rearward weight transfer, so rather than spinning the front wheels they just grab and go.

↑ The entire car displays clear thinking, ergonomic design and neat execution. Big rad ensures no cooling probs.

↗ The engine is small and mounted right back, so the engine bay is tidy, neat and uncrowded. Carb, alternator and triple Tilton hydraulic cylinders all look very much at home in there.

→ Enormous double-pumper Holley has four huge chokes to feed three sort-of cylinders. It doesn't flood the engine, just makes it go faster.

aerodynamic drag just that extra bit, and rakes the windscreen back just a few degrees – all of which makes a worthwhile contribution to gaining a few extra seconds on a racetrack lap. The car may finish up just a track toy rather than in actual competition, but that's not the point – if it's a race car let's get it right.

In this case there's no point in getting aerated about the chopping-up of an original MkI Mini, because all that existed of the Mini in question was a percentage of a shell, and it was in nasty condition to boot. There wasn't even enough of the car left to establish what year it was, although it had external hinge holes and inner door frame structures confirming that it had been a MkI. The worthwhile bits of the old Mini were dutifully recycled into the new monster, although to be honest not much was left in usable condition: the roof and quite a lot of the rear exterior panelling is still original, but that's about all. The wheel arches are by Fortech, altered and blended in.

The doors look like Mini doors, but they're actually handmade from sheet steel with tubular frames, constructed from scratch. The one-piece and one-off detachable flip front carries Mini lights and grille, but it's been reshaped to droop downwards for better aerodynamics: this was possible because the Mazda engine

is a small barrel-shaped device mounted approximately beneath the windscreen and scuttle, and it doesn't require the high bonnet line that's dictated for most Minis by the height of the tall, long-stroke A-series engine that also has a gearbox beneath it. The less frontal area your car has, the faster you go – so the raked-back screen and droopy front all help.

The lack of windscreen wipers, an impractical idea in Britain, is usually not a problem in Jeff's locale: Georgia and North and South Carolina in 2008 were experiencing a long and serious drought, which naturally broke for the day to allow enough rain to make a Brit feel very much at home, shooting a car with drips going down his neck and not enough light. The drought then came back the next day and carried on afterwards. Normally most days are dry, so a fun car with no wipers makes sense out there.

Getting a longitudinally-mounted engine, gearbox, axle and Jeff all jammed inside a lowered Mini shell took some doing, and involved much sitting, shifting things, sitting again, shifting things, and in the end a custom gear linkage was still required to put the gearknob within comfortable reach: the gear lever had to end up somewhere fairly natural, because awkward gearchanges make for slow gearchanges. The gearbox didn't have to be a cool piece of sculpture hand-carved from turned aluminium billet, though – making it that way was just more fun. All the work ethic, perfectionism and money that went into getting the gear lever right did not continue when it came to the steering wheel: it came from a go-kart, cost all of ten bucks and plopped straight on. Tell you what, it looks and works just fine too.

Something that's not usually used by longitudinal Mini artists is the Mini-based idea of using the radiator sideways on, beside the engine if room is tight – the automatic tendency is to put it in front where it looks right. A big enough side-mounted rad, properly ducted and fanned, can work as well as it does in a Mini. Jeff's front-mounted radiator arrangement works just fine and looks nice too, so no worries. The fan rarely turns itself on if the car is moving, which is the most reliable indication that the system is working well.

The Mazda rotary is a hot-running little engine, even if the cooling system and the internal engine temperatures are correct. One of Jeff's future tasks is to get the cabin heat down a little, as it gets a bit tropical in there when the engine gets up to its full working temperature.

Would he use another rotary engine in a Mini? Having tried this one out, he probably wouldn't. It's hot, it's frenetic, it guzzles fuel, but the worst thing about it is the noise. Rotaries rev high, and they make a nasty and ear-splitting noise. There isn't room for any more silencers under the car, and the existing pair of silencers, although

they're proper silencers rather than straight-through Cherry Bombs or whatever, simply can't shut much of the noise down. Jeff was quite relieved when the photo shoot was over, as he had just moved into a new house and didn't want to annoy his brand-new neighbours too much by wasping noisily up and down the road.

On the track the rotary would be great, screaming its nuts off and generating even more adrenaline, but on the

↑↑ The Bilstein shocks aren't adjustable, but were custom-built to Jeff's spec for the car.

↑ The hefty anti-roll bar runs through the main front chassis beam: very elegant design and engineering.

↓ In order to get the turned billet gear lever in the right place, the driver has to sit in the back seat.

track you're focused on pushing it as hard as you can, hopefully just short of pushing it 1% too hard, and you're also wearing a crash helmet which would cut some noise. On the road the raspy buzz would soon become a pain.

Rotary engines are also smelly and have emissions issues – they have to burn some oil, like a two-stroke, for internal lubrication of the rotor tips. The principle of their operation is that a spinning triangular rotor rotates inside the oval tubular engine casing (it's epitrochoidal, to be exact, if you want to be flash) and creates three spaces that act as combustion chambers. There are no camshafts and no pistons or rods, so there's a limit to what you can do to a rotary engine to make it more amusing. Spiky cams and balanced rods and pistons are not options, as there are none of any of them. All you can really do to a rotary is to get the porting perfect and then try to jam as much air and fuel mixture into the engine as possible.

There's a fan-base for Mazda's rotary sports cars who enjoy playing with these engines: they frequently achieve more power by turbocharging, and sometimes by twin-turbocharging and by nitrous injection, but in Jeff's case the engine is topped by a cavernous normally-aspirated Holley double-pumper four-barrel carb that lets in a massive 650cu ft a minute. It's intended for something like a five-to-seven-litre V8 and is about four times the size of a Mini carb. The Mazda rotary engine displaces 1.3 litres, although that number doesn't mean much unless it's applied to a piston engine. The Mazda engine in the Mini is still standard apart from the carb and some porting, and from experience Jeff is expecting something over 225bhp when the car is dynoed. For a 1,450lb car, 225bhp is going to be quite nippy.

I suspect that a bike engine would do the same job better, although it's not cheap to get 225bhp out of a bike engine. I wouldn't be surprised to find either a Hayabusa engine or a violently modified MX5 or Honda 2000 engine in the car next time I pass through North Carolina.

The rest of the car's drivetrain is thoroughly sorted. Rear-wheel drive for a fun performance car is obviously way better than front-wheel, and the Mazda in-line gearbox is strong and quite light with slick, quick gearchanges. It's an RX7 casing with MX5 gears, which have closer ratios than those of the larger RX7.

The rear axle is also Mazda, and has a very useful 'posi' limited-slip diff.

The suspension and steering are really race-derived, largely because of the availability of good racing gear within easy reach. Bilstein have a location 20 miles away, so Jeff's shocks were made by them to his specification. The steering rack is from a Legends racer, and was designed into the front suspension way back. The front suspension is double wishbone with a monster aluminium anti-roll bar that attaches to the very front of the chassis.

The rear end is controlled by a four-bar set-up, which inevitably has short bars because of the shortness of the car. The two lower trailing arms are set at 45° rather like a Lotus Cortina, and also, anachronistically, like my 1958 Chevy Delray. These allow vertical movement but restrict lateral movement. The upper trailing arms just go straight backwards, keeping the axle vertical. As a system it's not ideal, but given the space available there weren't many options. Using something close to the original Mini suspension combined with rear-wheel drive would be ideal, but it would be very complicated and would require castings and special hubs. In any case, this is a fast street and track car with very limited suspension travel, so any changes in axle geometry due to short suspension arms wouldn't be a big deal anyway. Anything Mini-shaped with stiff suspension is going to have pretty impressive handling in any case, certainly when you put it up against anything else of its type that it's likely to meet on an American racetrack.

So – another top mad Mini. The engine hasn't in the end turned out to be ideal, but every aspect of the rest of the car is spot on, and Jeff has plenty of time to play with it and consider potential powerplant improvements as time goes on.

⬇ The stance and ground clearance of this Mini is spot on. It looks mean and moody but doesn't scrape its bodywork as much as you would expect.

ENGINE – Mazda 13b 1,300cc rotary, with fast road porting and a 650CFM Holley double-pumper four-choke carb. Custom aluminium radiator, custom rear-mounted oil cooler, dual silencers, expected bhp +/–225.

TRANSMISSION – In-line Mazda RX7 gearbox casing with close-ratio Mazda MX5 gear cluster, propshaft to narrowed RX7 rear axle with 4.11:1 limited-slip Positraction diff.

SUSPENSION – Front, double wishbone with ¾in alloy anti-roll bar, Bilstein custom-valved shocks with 250lb springs. Rear, 45° lower trailing arms provide lateral and longitudinal location, straight upper trailing arms provide longitudinal location; Bilstein custom-valved shocks with 200lb springs. Steering rack from Legends race car.

BRAKES – Wilwood four-pot aluminium calipers all round, adjustable balance.

WHEELS AND TYRES – Panasport 7J x 13 wheels with 215/50 x 13 Sumitomo tyres.

EXTERIOR – Floorpan and chassis resembles a Joe Huffaker racing Mini design extended by 2in. Cage is Jeff's own: body is partly MkI steel Mini with a 1.5in roof chop and partly GRP replica with Fortech arches and modified nose. 1,450lb all-up weight. House of Kolor pearl orange paint.

INTERIOR – ATL ten-gallon fuel cell. Race seats, Simpson harnesses, $10 go-kart steering wheel, Auto Meter tacho, oil pressure, water temperature, voltmeter, fuel pressure, speedo. Lexan windows.

Offroad

↑ Functional rather than elegant – the box on the front bodywork is where your foot goes.

← When driving in the rough, the road wheels are changed for the offroad ones: the spare wheels and tyres add weight over the driving wheels.

→ The white Mountaineer is a development of the earlier Mosquito in the background, resplendent in attention-grabbing pink. Hollier apparently has no worries about his masculinity.

The Mini Moke wasn't an auspicious start for Minis and offroading, but one or two people have made the Mini concept work quite well in the kak.

As we've established elsewhere in this compendium (Chapter 10), Mini Mokes, although cool and top fun, are crap unless they're on smooth tarmac. However, there have been a few successful offroad Minis. Two in particular are worthy of a mention, one because it was actually rather effective, and the other because although it was as good an offroader as it's possible to construct with a compromised FWD-only Mini drivetrain, it was adopted by a particular collection of loonies who got more fun out of it than anybody is really entitled to.

The very competent Mini-based offroader was designed and built by Chris Hollier, already mentioned earlier in this very book (Chapter 5), and it was called the Mountaineer. It was a development of his Mosquito, and both that and the Mountaineer were basically rollcages with a wheel at each corner and initially a Mini engine and box in the back. The pink Mosquito in the background had a tiny but loyal band of enthusiasts, but it was still really a slightly mad road car rather than an offroader. The Mountaineer was definitely an offroader, sold in small numbers for competition, and was a much more serious bit of kit.

With the engine and everything else at the back, all the weight was on the driven wheels, so traction was excellent, and the rack on the back was designed to carry the spare tyres. The idea was that you would drive on road tyres to your chosen patch of mud or offroading

event, change them over for tractorish offroad tyres with about 20psi of air in them, and then strap the road tyres on top to add weight over the back wheels. The overall weight was low, and as it happens four-wheel drive is not the most important feature of successful offroaders – shortness, high ground clearance and low overhangs are more important, and 4WD is only really crucial for wading though deep goop where you're more or less swimming. I've been amazed at how far a 2WD Cortina-based kit jeep would go on a purpose-built offroad track.

The Mountaineer was a top tool for offroad racing, but sadly not many people are queuing up to buy vehicles specifically built to do that, so it joined Chris's extensive list of interesting past projects.

↑ The Mountaineer is as happy airborne as earthbound, and is built to take hard landings in its stride. The Mini componentry evolved out of it: you need long, soft springs for yumping, because landing on Mini rubber-doughnut suspension takes your fillings out.

→ The Scamp's ground clearance is certainly better than a Moke, but its body design shares the Moke's styling subtlety.

↓ The roof frame is also the rollcage: doors and roof are optional.

The more practical Scamp was a much bigger commercial success, although less agile in the kak. Success in kit car terms means several hundred were made per decade, rather than just… several.

Mini Scamps were built for many years in a collection of rural sheds by the charming Andrew McLean, who lived in a beached boat in a farmyard. He's still making them now, but with Suzuki Jeeps as a base rather than Minis.

The Scamp achieved offroad progress by being indestructible rather than by having top traction and ground clearance. If you only have two-wheel drive, an effective technique is to select second gear, grip the steering wheel firmly (keeping your thumbs clear of the spokes), mash your foot down flat on the throttle, and don't back off until you either get to flat dry ground again or hit something.

Andrew challenged me to give his demo Scamp a proper test on a nearby muddy forest track, suggesting that most visiting journalists were pussies and wouldn't give it any real exercise, so I applied the above technique with enthusiasm. A small tree had fallen down across the path, but it looked small enough to be either pushed out of the way or driven over the top of, so I kept my foot down. Yee-ha. The Scamp seriously impressed me, only losing its number plate and exhaust system along a route that would have given a Range Rover a hard time.

The Scamp Club, at least the members I used to talk to, were barking mad. They genuinely thought it was top fun to roll a Scamp down a mountain, and during their big summer camping get-together their favourite game was field-surfing, which involved towing each other at 40mph on dustbin lids across a wet field behind a Scamp. Most car clubs have amusing little games involving driving slowly and skilfully with fishing rods and rubber ducks, but the Scamp nutters considered that their annual summer camping holiday had been a big success if nobody was seriously injured. Top bunch

of chaps, up there with the Dutton club in the stratosphere of kit car eccentricity.

The Scamp itself was fully Mini-based, but many of them had different rear trailing arm arrangements or spacers to allow bigger wheels and chunky tyres. The chassis/frame was made from light welded steel tubing, well braced and very strong, and skinned in aluminium sheeting. There was no wussy nonsense about curves, and the whole approach was utilitarian in the extreme. The back end was a box that you could fill with seats, and the options list was only limited by the builder's imagination. Some roofs and sides were aluminium, some were canvas, some were missing.

Although the standard Scamp is attractive in a functional sort of way, many home-built Scamps are truly hideous, and reflect their owners' idiosyncrasies: some have elevated idiosyncrasy to an art form, and a total lack of artistic talent never stopped any of them. Scamps are available from a few hundred quid, if you think you can handle this level of fun – check out the club's site at www. scampownersclub.co.uk.

↑ With the doors and the roof attached, the Scamp is actually surprisingly weatherproof and mudproof.

→ Andrew McLean suggested I was too wimpy to drive on his offroad test course. Here he is picking up the exhaust pipe I tore off while doing it.

↓ Wheel arches are legally required to cover the wheels, so why waste time and metal doing any more than that?

Street and Strip

At first glance it's just a very red and shiny Mini pickup. Hang on a minute, though…

Lynn Schoonover's Mini pickup is mad, but like the best mad ideas it has been carefully and seriously executed, and on closer examination it turned out to be astonishingly competent. It really shouldn't be. If the chassis had been built for proper drag racing it would have been a lot bigger and longer, and jamming the sort of mechanicals and chassis you would expect to see under a street/strip '57 Chevy into a tiny Mini pickup should have compromised it into relative uselessness.

The car is at heart a joke, but it's a beautifully crafted joke with a hell of a punchline. Lynn offered to light up the tyres and burn some rubber for the shoot, because a red pickup emerging from huge clouds of blue rubber smoke would have looked way cool, but Dinkee doesn't do a smoke show, it just takes off like a *Daily Mail* reporter on the scent of a paedophile story. Every time Lynn stamped on his throttle the pickup just hooked up, the rear end squatted down and it took off. It just refused to indulge in any cheap showing off, and was only interested in serious business.

It's actually quite drivable and tractable on the road as well, although not very comfortable as the enormous transmission tunnel takes up most of the cabin, leaving only a few inches on either side for feet and pedals. There are no seats as such, because there isn't room: there's a sort of padded bench and that's about it.

The Fuellie cam in the Chevy engine was originally designed to work with early fuel-injection Corvettes, so it doesn't have a particularly fierce profile. The cam isn't spiky enough to make the engine wobble and splutter at idle, it just gives the V8 a bit of an attitude at tickover and allows more action higher up the tacho.

The reasons for the pickup's astonishing competence start at the back, with a pair of huge Mickey Thompson street-legal slick tyres with treads cut in them, mounted on American Racing wheels which are bolted more or less straight on to the brakes and the diff, which is the traditional extremely strong 9in Ford.

Dragstrip traction requires monster tyres, but street rod style requires that the tyres remain inside the original sides of the car's bodywork. Obviously most of the pickup bed is occupied with the tubs required to cover the tyres, but by deleting most of the rear axle, the tyres are not just inside the arches but quite significantly inside them.

The front tyres are as thin as the rear ones are fat, and they're the same size as those fitted to a Citroën 2CV: compared to the massive rubber on the rear they look like moped tyres. However, they're not just stylistically and functionally correct for drag racing, they also allow lots of room between them to get the 5.7-litre Corvette spec engine and its headers inside the Mini shell. Not much of the engine and box are actually inside the engine bay, but they've all been stuffed inside the shell of the Mini, even leaving room for a radiator at the front, although only just enough room: clearance is in millimetres.

The front has been extended and louvred and reshaped to droop a little, partly for a little extra speed advantage and partly for style. As speeds during the last stages of a dragstrip run are well into treble mph figures, the half-brick aerodynamics of a Mini pickup have a real negative impact on speed during the later part of the quarter-mile, so a smooth droopy front is worth the effort. Fitting louvres in the bonnet right above the exhaust manifolds is also a smart move and helps to

↓ The side view reveals something of what's really going on. It's still largely a Mini, but the wheels and the front end don't look right at all.

13

← From behind, the extent of the changes begins to become clear. This isn't really a Mini at all, it's a psycho dragstrip monster masquerading as a Mini.

↙ 327CI Corvette V8 is neatly tucked under the scuttle. Detailing throughout is to a very high standard.

↓ The extent of the front modifications can be seen here: about 5in has been added to the rear edges of the bonnet and wings.

↑ The flat sheeting of the pickup bed has been replaced with a stonking chassis and a 9in Ford diff, with huge wheels and drag slicks bolted to it. Handling will obviously never be as good as a Mini again, but isn't bad at all.

keep engine bay temperatures a little more reasonable. The radiator and cooling system on the Mini proved very effective on the day of the shoot: it was 105° in Riverside, California, and we were asking the Mini to launch under full power, reverse back up the road for a hundred feet and launch again, maybe twenty times, one after the other. The engine was certainly getting hot, but it wasn't getting bothered, and wasn't throwing the hissy fit to which it was fully entitled after half an hour of that sort of treatment.

The engine is a blueprinted 327 Chevy, originally from a '65 Corvette, bored out plus 30 thou, fairly stock but for Edelbrock aluminium heads, a Corvette Fuellie cam and a fat Carter AFB carb on a Weiand inlet manifold. The headers are obviously custom – there are plenty of Chevy header options and plenty of Mini exhaust manifold options, but off-the-shelf Mini V8 headers are rarer than honest politicians.

Lynn says the engine produces 370bhp at the wheels, and there is certainly some major grunt going on. There wasn't room to let the beast fully off its leash in rural Riverside, but Lynn was able to give it some throttle and some space to make its point a few times. It's as you would expect: a blast of action and noise, hit the back of the seats, bang into the next gear, same blast of noise and fury, same push, then bang into third and time to hit the brakes before somebody calls the cops.

Retro drag rods are often finished to a very high standard, and this one is no exception. The extensive steel fabrication on the front end has all been carried out to perfection inside and out, with the front-end panels all welded together first and then the whole assembly altered as one unit. Welding the wings, valance and bonnet together has also given a seamless line to the whole front of the car: it's a Mini, Jim, but not as we know it.

Apart from the front end, the shell remains stock 1961 Mini. It must have been tempting to stretch the pickup bed and sides to get a bit more chassis length, and to take a couple of inches off the roof when the car was being built, but the temptation was resisted and the Mini remains spiritually a Mini, although it's definitely speeding off to the madder end of the spectrum.

ENGINE – 1965 Chevrolet Corvette 327 V8, overbored 30 thou, blueprinted, with Corvette Fuellie (early fuel injection) grind cam, Edelbrock aluminium heads, Carter AFB carb with K&N filter on Weiand inlet manifold, high-output 50K coil, custom headers and cooling system. 370bhp at the wheels.

TRANSMISSION – General Motors 200RH overdrive autobox by Art Carr. Rear end, Ford 9in diff with 4.11:1 final drive ratio.

SUSPENSION – Front, single transverse spring, adjustable shocks, leading links. Rear, four-bar link with coilover shocks. Rack and pinion steering.

BRAKES – Front, 11in discs. Rear, standard Ford drums.

WHEELS AND TYRES – Front, American Racing rims, 135 x 15 Michelins. Rear, American Racing wheels, 26 x 10.5 x 15in Mickey Thompson ET street-legal cut slicks.

INTERIOR – Custom vinyl bench seating, custom steering wheel and pedals, Auto Meter gauges, Genie shifter.

EXTERIOR – Louvred and drooped one-piece flip bonnet and front end extended by 5in, tubbed rear pickup bed.

↑The custom-trimmed cabin is mostly stuffed full of the back end of the engine and a huge automatic gearbox. Drivers with big feet need not apply. Half-cut steering wheel is inconvenient but cool. If you get far enough out of line to need much steering input on the strip, you're in the poo anyway.

↑Traditional California rodding themes and quality standards have been applied. They'd be a cliché on a '57 Chevy but they look amazing on a Mini.

←No poncy drama-queen clouds of blue smoke and black lines on the road: just a brief roar, and the Mini is gone.

Carmelie

This is top-class bodywork. All the panels fit perfectly, the paint is fab and the pinstriping finishes it off just right.

Most Minis that start out being built for wives and girlfriends finish up with 1380 engines, rollcages and one seat, and are taken off the road each winter for a rebuild to go even faster. Donnie Neron's wife Carmelie LaFlamme is a lucky girl – she's actually going to get her hands on this one. Pretty soon. Honestly.

Like many of the best mad Minis it started off as scrap. This approach relieves the builder of any guilt for chopping it up, and confers a real freedom. Rather sad that California rod builders require immaculate classic cars to scrap: Mini enthusiasts are very reluctant to chop up a good Mini.

In this case, the scrap Mini's bottom half was see-through, but the roof was still solid. That's just as well, because the roof has now become the tonneau panel at the back. There's not much of the original shell left – replacement panels include the wings, nose, bonnet, A-panels, sills, floors, quarters, rear panel and valances.

A Mini is a deceptively simple-looking shape, and you find out just how subtly complex it is when you start cutting and shutting any aspect of it. Mini roof chops are inevitably more complex than they look, and although the roof looks about the same size as the back end of the car, it's not. It's a lot smaller. So when Donnie cut the roof off and tried to lay it across the rear end, it fell right through. To make it fit, he had to weld in a strip several inches wide across the middle, and butt-welding sheet metal through a long, smooth curve without warping it is a long and tricky bodywork task. He also had to follow the quite dramatic double curve of the roof panel, and deal with the significant amount of droop towards the back edge.

The old corner fuel tank has been replaced with a 15-gallon racing fuel cell in the safest place on the car, in the middle behind the bulkhead, and it's now reached by a flush-mounted filler cap. When the hole for the cap was originally drilled in the freshly-finished tonneau panel, it was about ¼in off to one side, and there was some argument about whether that mattered. 'Let's go get Carmelie, and see if she can spot it,' said Donnie. Carmelie was duly fetched and invited to inspect the back end.

'Why's the hole not in the middle?' she asked.

They can be so cruel, can't they?

Much of the rest of the custom rear bodywork was made from bits of surplus roof, but a lot of it was scratch-built from flat sheet steel literally beaten into shape with hammers and dollies on a bag of sand.

Donnie recommends a plasma cutter as a top tool, because it allows you to make tidy, accurate and smooth-edged cuts, rather than the usual chewed edge you get from trying to cut curves with shears, or the rather approximate shortened and frayed edge you get from cutting with a grinder.

⬇ Small fat bum and small fat wheels are a combo that works a treat from a styling viewpoint.

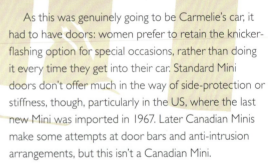

As this was genuinely going to be Carmelie's car, it had to have doors: women prefer to retain the knicker-flashing option for special occasions, rather than doing it every time they get into their car. Standard Mini doors don't offer much in the way of side-protection or stiffness, though, particularly in the US, where the last new Mini was imported in 1967. Later Canadian Minis make some attempts at door bars and anti-intrusion arrangements, but this isn't a Canadian Mini.

With the roof becoming history but doors and therefore door apertures still in the picture, the monocoque was as floppy as a wet paper bag – so proper side bars were inevitable, and they're at a good height for protection. Donnie carried on with the safety/ structural cage throughout the car, connecting the front, back and sides of the frame. The final structure of this convertible is a lot stiffer than a standard Mini monocoque, and the bodywork is attached to the internal frame to stiffen it up even more. Good structural stiffness means better suspension location and therefore better handling as well as a safe structure, as the suspension is mounted to a stable and rigid platform and operates exactly as designed.

Many mad Mini makers go for much bigger wheels, but this is ultimately a girly car, and the retention of fat little 10in wheels accentuates the Mini's fat little bum. I don't know whether Donnie figured this out consciously or subconsciously or just got lucky, but cars with fat bums make girls' bums look slimmer by comparison. I have yet

→ The rebuilding had to start before the car was completely dismantled, or it would just have been a pile of disconnected bits that would have needed a jig to assemble it.

↘ Here's the pile of new panels. That'll just leave the screen frame and the roof from the old body, then.

↓ Rear-end metalwork complete, and the door gaps are spot on – time for the paint job.

to meet a woman who objects to something that makes her bum look slim.

The finish and paint job are seriously nice, and the choice of colours is interesting and also effective. A little delicate pinstriping adds to the effect.

As the inside of any Speedster is very much on display, it had to be special. The main focal point is the dashboard, which has a cast aluminium Dakota Digital hot-rod instrument set. This offers speedo, tacho, oil pressure, water temp, volts and fuel, as well as more esoteric electronic functions – gearshift light, maximum rpm recall, max speed recall, 0–60 time and quarter-mile time. Donny's custom GRP dash in which the Dakota gauges are mounted has two mid-range speakers pointed right at you, which are backed up by a 12in sub behind the seats. So while a couple of hundred watts of Nelly Furtado at her finest might send a shiver up your spine, the same spine will get a nice massage from the bass.

The colour of the Mountney steering wheel was chosen to complement the paint colours, and the wooden door tops and trim are stained to match. Matching the colours and tones of steering wheels and other interior woodwork is something that many people don't bother doing, and if they don't make the effort, any darker wood colours in the dash or trim have the unfortunate effect of making the wheel rim look very much like the cheap plywood from which it's actually made. A little matched staining and we're sitting in expensive furniture rather than a DIY superstore.

Recycling the roof into a new tonneau panel is not as easy as it sounds, particularly with the swage line running round the edge. That must have taken a while.

Three-piece wheels are painted in the same colour as the tonneau and the engine: neat detailing.

The shell had to be painted before assembly, as the inside as well as the outside is show quality. Putting it back together without scratching the paint must have been a bit hard on the nerves...

The steering column in this car is also cool – it's been reduced to a minimalist steel tube, and the stalk functions have been relocated to the gear lever knob, which has microswitches operating the indicators, horn and fan. More electrical fun includes mirror-mounted indicator repeaters, billet number-plate holders with built-in lighting, and a third brake light. The wiring loom is custom, which is a scary idea for a lot of people. It's built around a 14-fuse panel with built-in flashers and relays, so there are no Lucas wiring parts in the car at all – the ghost of Joseph Lucas has been completely exorcised. In truth, it's only corroded and elderly British wiring that should scare anyone, as electricity running along new Lucas-spec wires in a newly built car running between new components nearly always follows the laws of physics – not always, but nearly always. Whereas the kind of grumpy old electricity that inhabits ancient Mini looms regards the laws of physics as a set of slightly impertinent suggestions, and frequently ignores them just on principle.

Donnie has remained faithful to the A-series engine, and Carmelie's powerplant is an overbored 1275 Austin America engine, which we would know in the UK as the Morris/Austin 1300. In this case it's much improved for reliability, and given enough power to provide plenty of entertainment to match the character of the car.

The state of tune is actually an excellent balance, taking the engine as far as it will go before it starts becoming temperamental and fussy. The two carbs, the big bore, the high compression ratio and the 266 cam get the most out of the engine but leave it usable in traffic with a reasonable tickover, and planting your throttle foot gets an enthusiastic response. On the other hand, the electronic ignition, forged pistons, balancing and even the electric fan will mean few worries about reliability even when the car's given the spanking it asks for. I can't think of anything I don't like about this car, apart from it being in Carmelie and Donnie's garage rather than mine.

Tech

ENGINE – Austin America block with centre main strap, 1,275cc plus 40 thou rebore. 11.5:1 compression ratio, forged pistons with 6cc dish, lightened and balanced internals. Kent 266 cam, Cooper S forged rockers, Isky dual valve springs, ported and polished Cooper S 11-stud head (12G 940) with big valves. Pertronix electronic ignition, Accel high-energy coil. Taylor 8mm silicone plug leads, side entry distributor cap, NGK BP6ES plugs. Dual 1.5in SU carbs on ported intakes, K&N air filters, LCB exhaust manifold, custom welded piping, stainless Stebro muffler, Ractive stainless oval tailpipe. Two-core rad, electric fan, polished alloy overflow tank.

TRANSMISSION – Lightened backing plate and flywheel, Rally clutch plate, Grey diaphragm pressure plate, central oil pickup, 3.44:1 final drive.

BRAKES – Single line with servo, braided lines, adjustable proportioning valve. Front, 7.5in grooved discs, Mini Spares four-pot calipers, Green Stuff pads. Rear, Cooper S drums.

SUSPENSION – Alloy box section suspension frame in rear, Spax coilovers all round, quickrack.

WHEELS AND TYRES – CMS 10in three-piece modular alloys, anodised centres, polished rims, stainless nuts, Yokohama A008 165/70 x 10.

INTERIOR – Cobra seats, Schroth three-point detachable harnesses. Nine-point custom rollcage with removable chrome hoops. Mountney wheel. Dakota Digital instrument cluster, smoothed steering column, handmade door inserts, Racerboy gear-lever-mounted indicators, horn, fan switches. Newton Commercial carpets. Alpine sound system: CDA9811, Orion four-channel amp, Earthquake SW5 12in sub, Clarion Pro Audio 5.25in mids, custom amp box, sub box, head unit mount. Remote-controlled LED lighting, microwave alarm system, custom 14-fuse wiring loom, billet aluminium illuminated licence plate frames.

EXTERIOR – Virtually all panels replaced. Chopped top, de-seamed shell. Shaved bumpers, door handles. Custom bootlid, floor, inner quarter panels, flush fuel filler, modified valance, chromed hinges. Paint, Plymouth Prowler Orange and Audi Hibiscus red, with pinstripes.

↓The final result is stiffer than a standard full-roofed Mini monocoque, so hurtling round a roundabout for a photo session is no bother at all.

Trikes

The Mosquito/Triad was interesting as a design, and from some angles looked quite convincing.

I have to admit that shooting it from directly side-on was unkind. The front end is empty, as the Mini subframe ends just in front of the wheels. Still, if you like eccentric fun and attention you couldn't do better.

Q228 FDD

I rely on producing many crap ideas amongst which is the occasional gem. One of the crap ideas was the Brookland Swallow. Actually it was a pretty good idea with one major fault. It was a 2+2 Mini-based three-wheeler, with French-British-Italian design influences: the front was inspired by Facel Vega and by Farina Rileys and Wolseleys, and the back, made from the back end of a crashed Volvo P1800, had Michelotti/Triumph Herald fins.

The Swallow's major fault was legal rather than dynamic (it actually drove very well, and the back end never misbehaved despite enthusiastic test-driving abuse) and its terminal problem became apparent at the marketing stage. The backbone of the marketing plan was to take the prototype to a kit car show and see if anybody fancied it. They did, big time, and that was when the fatal flaw was revealed – it was too heavy to be a lightweight motor tricycle and would have had to be classified as a three-wheeled car. I drive trikes for fun and because the better Morgan-inspired examples are like motorbikes that don't fall over, but many Brits drive trikes and Reliants because they don't have a full car licence. On a bike licence you can legally drive a lightweight trike under 1,000lb, but not a Mini-based affair with a cast iron engine, steel Mini doors, wind-up glass windows and a 7in x 4in perimeter steel chassis with sheet steel bulkheads and floors. Continuing the project would have meant redesigning the whole thing as a two-seater with no doors and a Citroën engine. Then the owners of the remnants of the Brooklands property

sent some greedy patent lawyer creatures after me because of the Brookland name being close to Brooklands, and that was the end of it. I haven't been back to the Brooklands museum since then, as the episode left an unpleasant taste. Cars I've designed and built since then are called Ayrspeeds. The single Swallow made it into history in Chris Rees's book on three-wheelers, but was then scrapped.

Malvern Autocraft's Triad was madder than the Swallow, and had a tiny but enthusiastic audience. It began with a competition one-off that evolved into Robert Moss's Mosquito: six were built and then the project was shelved. A Hereford motorcycle shop, Mead and

15

↙ The cabin of the Triad was a nice place to be, at least in the summer: despite the flat windscreen, the buffeting from sidewinds wasn't bad at all.

← ↑ Realisation of the sketches in 3D with bits of Mini wings and doors and Volvo rear end involved compromises. But with a lighter screen frame, two-tone paint and a chrome bumper it would have looked not bad. Flip front conceals the wipers: I never liked exposed wipers.

↓ Lancia and Facel Vega meets Wolseley, but only on Planet Iain. Initial sketches for the Swallow.

↘ The rear fins were going to be quite subtle, as originally conceived. With trikes, more length means more stability, hence the long tail.

← Later sketches include the Renault 5 screen and frame, Herald-style fins and hooded headlamps.

Tomkins, took over in 1984 and built three more. Then around 1990 Rick Jones and Ian Browse formed Malvern Autocraft, updated the design a little, and relaunched the car as the Triad. The 'Warrior' competition version weighed 826lb, roughly 50% of the weight of a Mini, so driving a lightweight 1275 Triad would be like driving a 2,550cc Mini. I reviewed one for a kit car magazine, and it was a nice little car, although the styling was bizarre. But what the hell – let's have bizarre rather than dull any day.

Sandy Fraser, whom we've already met in Chapter 5, spotted the design opportunities offered by the compact and self-contained Mini engine/box/subframe/axles unit in the late 1960s, and started making three-wheelers with Mini mechanicals and largely wooden bodies in 1969. They had hardwood frames skinned in mahogany-faced marine plywood, and they drove very well.

Not many people bought them, because they were over the weight limit to drive with a motorcycle licence, but the trike, which was called the AF Spider, was slimmed down to scrape under the weight limit and a few were constructed over the next few years. The original flat wooden wings were replaced by cycle wings and the project began to take on the look of a Morgan. About a dozen were made altogether.

I've saved the best for last, though, which is The Stimson Scorcher. Barry Stimson is actually a talented car designer. He has never been in the box in the first place, and didn't need any encouragement to think outside it. He's currently designing very clever mini-motorhomes in which you can open the back window, go outside and use the cooker as a barbie. Previous to that, he was making ingenious ultra-economical little Romahome motorhomes out of Citroën saloon vans and tiny Japanese Daihatsu and Suzuki trucks. He's also now making new buggies and superbike-based trikes.

Barry is bored with hearing about the Scorcher, and wishes everybody would forget about it. Sorry, mate, no chance. The Scorcher is completely, sublimely mad. Most people would have sketched it, sniggered and gone back to work. Barry started manufacturing it instead. A few nutters bought Scorchers, and many later nutters still revere them. They have a Mini subframe and mechanicals at the front, and a single trailing wheel at the back, which is fine, but it all goes downhill from there. The Scorcher is the least practical vehicle ever made, and combines the disadvantages of both a motorcycle and a car. The driver/rider and passengers sit in the open, astride a backbone, getting wet when it rains, and then when they get where they're going the Scorcher is as big as a car so they have to pay for a parking meter.

← Plywood sheets, being somewhat flat, impose both restrictions and opportunities for styling and design.

↓ Mad but cool. If you like the idea of riding a motorbike but would prefer not to be crippled when you T-bone a Volvo, build a Mini trike and get some bugs in your teeth.

↑ A long Mini-based three-wheeler like this may look absurd, but the handling is astonishingly competent, safe and stable.

→ Barry Stimson's Stimson Scorcher, possibly the maddest Mini ever: the only answer to the obvious question as to why, is … why not?
(Courtesy of Barry Stimson)

The Multicultural Mini

New Jersey
QQ52344
HISTORIC

Akshay Crimmins is as exotic a mixture as his car: he's an Irish Indian American, and his car's a mental Japanese British Honda Clubman. His wife Jacqueline is patient and supportive although she doesn't get it, and she's an American American. The Mini was mostly built in Britain, and finished off in the California desert near Edwards Air Force Base. Now it's being written about by a Scotsman who lives in Canada, for a book published in Britain. Got a grip on all that?

OK, the story starts with Jacqueline being posted by the USAF to Norfolk, and naturally husband Akshay went along as well. When in Britain, the natural thing to do is to play with fast Minis, and that's what Akshay did. A Clubman Estate is quite rare and interesting in the USA, but in England it's traditionally been regarded as about as exciting as Wolverhampton. A rusty Clubman Estate is pretty much at the top of the list of stuff that most people don't want to buy, so Akshay didn't have to fight anybody off to get his hands on it. The bodywork was fairly horrible, with much lacework around the bottom few inches of the car, and the floors and A-panels crumbled like damp Hobnobs. The car finished up in the end looking even tastier than Hobnobs, though. The rather fab orange is Lotus Chrome Orange set off with a black roof and a walnut interior. More reminiscent of the colour scheme of a Jaffa Cake than a Hobnob, come to think of it, if we're in a bikky-metaphor mood. The orange paint works nicely with the burr walnut dash and door cappings, and the instrumentation is all Auto Meter and pretty comprehensive. Switchgear looks

remarkably aircrafty, but then Akshay used to work on USAF transport planes for a living, so that makes sense.

The early stages of the build took place in the UK, which was the best place to acquire the car, collect some cheap assorted bargain bits and pieces, and buy the necessary 15 or so Heritage panels – all at reasonable UK prices. Reasonable? Oh, yes. Mint Rover Recaro seats, 50 quid, a Schroth three-point harness for ten quid, and a second-hand Safety Devices multipoint rollcage for a hundred. Being in the UK also meant paying no import duty, shipping and US taxes on the replacement body panels, although VAT in the UK remained pretty brutal.

Also fairly cheaply available in the UK was the Metro Turbo engine, which was the first replacement engine that went in the car. Akshay is a part-time student and also a part-time test driver for Hyundai: test-driving Hyundais doesn't mean pootling along to Costcutter or Lidl the way genuine Hyundai owners will do, it means thrashing the hell out of them until they break. Akshay was actually picked for this job because he's a natural at torturing cars, and also because he's articulate enough to provide useful information as the smoking remains are hauled away. A lady vicar might be able to keep a Metro Turbo going for a while, but Akshay destroyed that engine fairly quickly, and moved on to a Honda on the basis that it would be strong enough to take a joke. The engine has been fine so far, although Akshay has managed to destroy a few springs and poked the back suspension up through the reinforced body mounts.

16

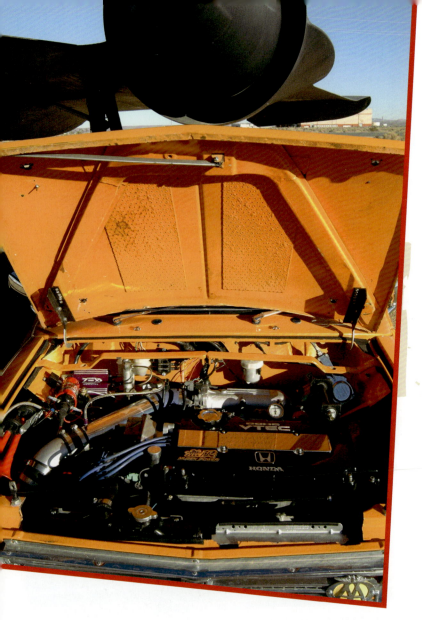

the bigger Japanese motor. In any case, the square front only really looks weird on the round and curvy little Mini saloon shell: it looks fine on a square Estate shell.

The Clubman's inner wings were chopped out and the bulkhead was modified to make plenty of room for the Honda's quite large intake system at the back of the engine. If you're prepared to cut out the bulkhead to make room for the large standard Honda air intake it might be worth doing, as the intake is quite carefully designed and doesn't benefit from having a couple of inches rudely chopped out of it. Shortened B series manifolds still work okay, but not 100%. With the Honda motor we do have a few bhp to spare, though.

The front valance was modified quite a lot to provide better cooling, and there's an extra oil cooler used for water cooling, because by this time Uncle Sam had tapped Jacqueline on the shoulder in the UK and had told her that California was her next home. Edwards Air Force Base is in the Mojave Desert, which gets seriously baked in the summer. It's also seriously empty quite a lot of the time, and there's one small road that just disappears into the desert, which is where we took the car out to play. The road apparently goes in a straight line for a few hundred miles and finishes up in Las Vegas, but nobody was using it.

OK, Akshay, how fast does your Clubman go? First it shreds a bit of tyre tread, then gets a grip and screams up to 5,000rpm, at which point there's a bang, a jerk and the dual cams slap in another big chunk of power up to about 9,000. Several driveshafts have been snapped or corkscrewed in the search for a pair that would survive an Akshay take-off, and the current pair from Watson Rally are surviving well. He slaps it into second and floors it again. More blattering noise, more gear whine, the cams bang in again and torque-steer visibly twitches the steering wheel. Third gear, fourth and fifth, and the entire structure is vibrating and banging as the desert landscape hurtles past in a blur. Can't focus on the speedometer at this speed, because the road's too bumpy: the needle's gone right round and more, though. OK, this is quite fast. Unlike the SR-71 Blackbird, the Clubman is neither stealthy nor radar-invisible, so it's surprising Akshay still has a licence.

In standard form, the B-series 1.8-litre Vtec engine as fitted to Honda saloons such as the Integra puts out between 140 and 215bhp, so in a light Mini shell the engine provides more fun than when Coyote caught the Roadrunner ("Fxxxxxx beep beep now, ya xxxxx"). The variable cam geometry allows for civilised and economical pootling about at low revs, but if you leave your foot on it and let the revs get going, the camshaft profiles change quite violently as the cam lobes change from mild to wild, and maximum power suddenly piles in. The Vtec system works by providing two cam lobes for each valve: at low

↑ The Honda engine looks quite at home. Plenty of room for a turbocharger in there as well. 34,000lb of thrust is quite frisky – oh, hang on, that's the SR-71, not the Honda.

The suspension is by coilovers with Gaz shocks. Tough enough for most people driving a Mini saloon, but with Akshay's driving and the additional weight of the Clubman's Estate back end with its rear doors and all the extra glass, one of the springs couldn't take it and let go. Its replacement has been fine so far.

For a car this quick, decent brakes are a necessity, and fortunately Metro Turbo four-pots with EBC Green Stuff pads and grooved and cross-drilled discs are well up to the task and didn't break the bank. They don't have shiny ally calipers and cool brand badges, but they do slow the car down from jail speeds very quickly, with no bother at all.

There was talk of converting the square front end to round Mini front end panels on styling grounds, but it didn't make much sense to replace the front end with a whole bunch of new panels and then chop it all up to extend it for the Honda engine, when the car was a Clubman already with plenty of room at the front for

speeds, the smaller-profile, rounder cam lobes open the valves a small amount for a short time, providing low power, a smooth tickover and good economy. As the revs rise, a pin driven by rising oil pressure locks in a second cam lobe with a much sharper and higher profile. This holds the valves open wider and for longer, throwing lots more air and fuel through the engine – which is the key to more performance.

The added bang of power provides torque-steer violent enough to rip the steering wheel out of your hands if you're not firmly hanging on to it, and people are now developing ECU software changes to soften the cam change for Mini Vtec conversions.

Naturally this engine at around 170bhp wasn't fast enough or violent enough for Akshay, so he rebuilt his Honda motor with a 20-thou overbore and higher compression Type-R USDM (United States Dealer Market) Honda pistons, changed the cams for a pair of Stage 1 Skunk2 items, ported and polished the head and inlet manifold, gas-flowed the valve seats with three angles and changed the ECU for a Skunk2 that lets the engine get much closer to its real power limits. Even if you don't buy all your performance goodies from the same good-quality tuning company, which is the best option for uprating

↓ It's too risky to let me on an airliner with a nail-clipper, but taking pictures on an active USAF base is no problem. Go figure.

↑ Rear plate bears American, Indian, British and Japanese flags. Akshay is an Irish Indian American, which probably makes him a tribe of one.

↑ The serious rollcage visible in this Mini not only keeps you in one piece if the poo hits the fan, but stiffens the whole shell and sharpens up the handling significantly.

engines, it's a good idea to match components where you can – so getting the cams from Skunk2 and matching them with a Skunk2 ECU was well worthwhile. You already know that they've been developed to work together without any conceptual or engineering mismatches.

With nitrous oxide you need a big fat clean spark, so ignition is now MSD 6-AL with 9mm plug leads and cold platinum plugs. The cat has strayed, as there's no need for a catalytic converter on a 1975-registered car in California, which has ferocious environmental laws on newer cars but so far leaves old ones alone to a great extent. We just have to hope the current scare story switches back from global warming to global cooling and the next New Ice Age before all amusing cars are banned. In ten years our political types will be squealing fashionably for all the carbon dioxide they can get to feed the rainforest and stop the glaciers from covering Alaska.

Akshay's exhaust is an appalling ecological disaster – a simple four-into-two-into-one with a single fat silencer at the end. His final power figure is about 220bhp, as his mods are all designed to be productive rather than cosmetic, and his nitrous set-up is a 50-shot kit with a big bottle adding another 50 usable horsepower, so we're looking at something like 270bhp at full chat. No wonder the Clubman is a bit nippy.

His plan for the next stage with the engine is a turbocharger… so the car's current 4.5 seconds to 60mph time is going to look pretty lazy after the turbo goes in. The exhaust manifold is at the front of the engine, so there are plenty of options for turbo placement. He's serious about this: with a decent-sized turbo it will add another 150bhp to make 400bhp+.

0–60mph times in production cars tend to be achieved by test drivers selecting first gear, bouncing the engine revs off the rev limiter and then sliding their foot sideways off the clutch. They either get a good 0–60 time or a broken car. Akshay has a tendency towards this sort of brutality, but I'm not sure it's as productive with front-wheel drive as with rear-wheel drive. It'll be interesting to see what sort of times he will get with his next project, which will be a mid-engined rear-drive blown Vtec Minivan. That will get him the key 0–60 advantage of rear-wheel drive.

By having the taste and discernment to buy this book, you've made a valuable contribution to the chance of *More Mad Minis* being published, and if that happens Akshay's next build will be high on the list of contents. Watch this space, as they say.

Tech

ENGINE – 20-thou overbored 1,834cc Honda Vtec DOHC B18 block with balanced, micro-polished crank, USDM Type-R high-comp pistons (10.9:1), new stock rods, Stage 1 Skunk2 cams with AEM adjustable timing gears. Ported, polished head with three-angled valve seats, ported inlet manifold, 4-2-1 exhaust manifold, de-catted exhaust with single tail box. AEM fuel rail and pressure regulator. Skunk2 OBD-1 ECU with OBD-II harness by LocashRacing.com, MSD 6-AL ignition with external coil. 9mm plug leads, cold platinum plugs. ZEX 50-shot nitrous oxide kit. Oil cooler. Standard Honda radiator with slim FAL electric fan, additional oil cooler used as booster rad for coolant. External MSD fuel pump.

TRANSMISSION – Lightened aluminium flywheel, Stage 1 race clutch. Honda LS Integra gearbox with high fifth gear. Watson Rally driveshafts.

BRAKES – Metro Turbo four-pot calipers with EBC Green Stuff pads, aluminium Minifins rear with EBC Green Stuff linings. Braided lines, modified brake pedal.

SUSPENSION – Gaz adjustable coilovers with custom springs, custom K-series Metro steering rack.

WHEELS AND TYRES – Custom three-piece 9lb 13in magnesium alloys by Keiser Racing, with 175/50 x 13 Dunlops.

INTERIOR – Recaro seats ex-Rover, Safety Devices rollcage, Schroth full harness, Moto-Lita steering wheel, Auto Meter 5in electronic programmable speedo, 5in monster tacho with shift light, water temp, oil pressure, oil temp, volts, fuel. Modified Honda gear lever. Burr walnut door cappings. Clarion CD, Rockford Fosgate amp, two Sony 6 x 9 speakers.

EXTERIOR – 1975 Clubman shell restored with 15 replacement Heritage panels. Front bulkhead modified, inner wings deleted. Front valance modified but retained.

↓ This straight and empty road goes all the way to Las Vegas, and makes a top test track for mad Minis. Brown racing underpants are recommended.

Varnishing Point

At first glance, Lorraine's car looks like a nicely restored early Woody Wagon.

Michael and Lorraine Mihalik are not mad as such, but they don't think the same way as most of their people-carrier-driving Seattle neighbours.

When I first saw a low-resolution picture of the Mini Woody that Michael had built to present to Lorraine, I wondered why on earth they'd painted the back panels brown. It only sank in later that the panels were actually veneered in African Mahogany.

Surely you can't veneer metal, though, can you? 'Why not?' is the answer. Most people have simply never thought of doing it. Veneering usually means sticking a thin layer of an expensive wood over a piece of cheaper wood, and then varnishing it between ten and twenty times. But if you want to stick veneer to bits of metal instead of bits of wood, there's absolutely nothing stopping you but convention.

Why would anybody want to veneer metal? Well, after you've checked out these pics you might well decide that it actually looks fab, which is a fine reason to do it. Michael's initial reason was rather more pragmatic than that, though. He spent a long time on the bodywork of this Mini, getting it straight and sorted and finally doing all the expensive and time-consuming paint prep himself to save the five grand or so it costs to get a good professional painter to sort the body surface ready for a show quality finish. Then he sent it off to the painter, who sprayed on a lovely job of Spinnaker White. Michael and Lorraine inspected it, admired it and approved it: the finished paint was superb. The next day, as

instructed, the painter went to work again and laid on another beautifully smooth coat of white on the inside of the car. Sadly, it wasn't quite the same shade of Spinnaker White as the previous day's paint on the outside of the car. Maybe it was Mainsail White, or Fore Topgallant Sail White. Pretty though the colour might have been, Spinnaker White it wasn't.

'You're going to have to do it again,' said Michael.

The painter was not entirely over the moon about that idea, as he would have had to duplicate much of Michael's months of patient prep. In the end, he suggested that Michael just take the car away and forget about paying for it. Michael did that, but was still left with a very slightly two-tone car.

That's when the veneering idea presented itself. If the interior were to be completely covered in fine wood veneers and then retrimmed in leather, it would look fabulous, although it would look like no other Mini. Which would in itself be no bad thing.

So that's the way it went. Much of the inside of a Mini

↑ A closer look reveals that what looked like brown paint is actually genuine African red mahogany veneer. It takes many varnish coats and much rubbing down to achieve this sort of depth.

17

← The side view looks strange – why are the side panels painted brown?

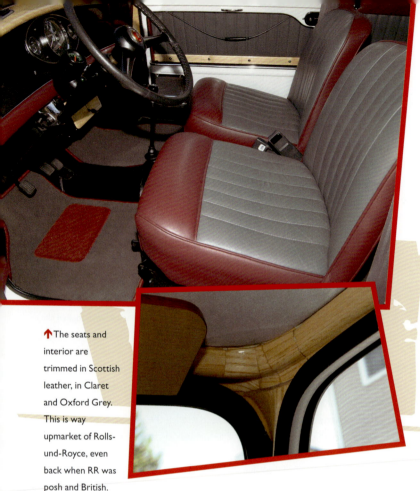

↑ The seats and interior are trimmed in Scottish leather, in Claret and Oxford Grey. This is way upmarket of Rolls-und-Royce, even back when RR was posh and British.

↗ Veneer comes in flat sheets, so covering a complex curve is what you might call challenging.

The skill and experience gained with the MG was applied to the outside woodwork on the Woody. A half-timbered Mini Traveller isn't actually a proper Woody like a Morris Minor Traveller or a Rolls-Royce shooting brake, in which the woodwork is the main structure of the back end. I suppose the Rolls shooting brakes and their descendants must be the origin of the term Estate Car in the first place, as they were actually used on estates. A Mini Woody is actually a steel estate car body, a Minivan variant with some decorative woodwork glued on top. This is good rather than bad news, as the whole structure is less prone to rotting and collapsing. I once had a Morris Minor Traveller with a delightful little glade of mushrooms and mosses growing in the damp and rotting wooden rear window tracks, but I wouldn't have dared open any of the windows, as they were all that was holding the roof up. As the Morris was crumbling audibly with its last MOT on the horizon and was not long for this world anyway, I used to water the little garden if it began to look dry.

Something to remember if you're ever rebuilding a Mini Woody is to allow any new ash parts you've bought to dry out and season for at least a year, and preferably two, because if you varnish the wood with any dampness still inside, it's like sealing fruit into a plastic bag: it will rot from the inside out.

With the exterior woodwork seasoned, sorted out and coated in 13 layers of marine urethane varnish, the next step was to get the interior carpentry and decor up to a matching standard. The result was a retrim in grey and burgundy leather, which looks and smells seriously nice. It has something of a 1960s Wood & Pickett feel to it, but the interior veneering and custom woodwork puts it into a different league. The grey suede custom headlining looks pretty sweet too.

The car was built for local and highway use as well as for show. One reason for the necessary highway capability is that Michael and Lorraine have to drive on Interstate 5 as it passes through Seattle. Speeds on the section of Interstate Highway 5 that stampedes through the Seattle conurbation average a bumper-to-bumper 85mph with seven lanes charging each way – it's illegal by 30mph over the limit, but who's going to stop them? There are some substantial hills, and many dumb and impatient rednecks tailgating you in vast shiny chest-wig pickups. The Mini therefore needed a serious engine, but Michael and Lorraine fancied spending a little time without black fingernails so they bought a built engine from Mini Mania, with a Longman head, an Elgin cam and Powermax pistons, an Aldon Yellow distributor and enough cooling to deal with summer trips to the desert in the southern parts of California. The electric cooling fan is automatic, but has a manual override switch in case of nervousness. Flywheel

is composed of usefully flat or single-curve surfaces, but where the surfaces are concavely curved, the technique is to cut triangular slices out of the veneer until the surface is covered.

For convex curves, you have to add little pizza slices and try to match the grain. The grain won't ever match up exactly when you've done that, but even so it still looks nice.

There are also quite a lot of new solid pieces of wood inside the car, specifically custom-carved for it – the heater panel and the sun-visors are solid ash wood, as are the strips across the roof and in the loading bay. Mind you, nobody is going to be throwing toolboxes into the back, so the ash floor strakes are mostly decorative.

The couple's expertise in woodwork comes from owning and restoring an ancient 1948 MG over a couple of decades. The MG is of the coachbuilt type that was made from a separate steel chassis and a thin metal skin over an ash frame. The couple also run a Land Rover, so they're well versed in the vicissitudes of classic British rust. With the MG, they not only had to deal with the familiar scenario of crumbling steel, but also with woodworm, assorted hungry boring insects enjoying a tasty and nicely-matured wood banquet, dry rot and wet rot – so after dealing successfully with that, a spot of planing and routing action in clean new ash was a walk in the park.

bhp is 113 with 102lb/ft of torque, which is enough to provide relaxed agility and to keep up with even Seattle's mad highway speeds.

The drum brakes also had to be upgraded, but they wanted to stick with the correct steel 10in wheels, so the only option was a set of Cooper discs. Not cheap, but they do the business and still fit inside the little wheels, so job done.

All was sorted, fine and dandy, but then Michael, newly freed from the bounds of convention, had another off-the-wall idea. He wanted to make a roof-rack out of bits of chromed kitchen plumbing tube, which would involve drilling 26 holes in the freshly painted roof to fit it. He approached Lorraine with this scheme.

'I got this idea. I want to add a roof rack, like the old '40s woodies.'

'No way, you're not doing that on my car. You're NOT doing that on my car. I'm having nothing to do with it. Idiot.'

Gentle pressure prevailed, and although drilling 26 holes in the gleaming new paint provided a few nervous moments, you can see how well it worked out. The rack is ingeniously made from copper kitchen piping and fittings. By making split wooden templates and then gradually bolting the two halves together with the piping inside, the copper was gently curved to the right shape. Heat-sink mud was used to stop heat transfer, so that the rest of the roof rack didn't come unsoldered and fall apart as each new joint was soldered. Chrome plating involves a copper coating anyway, so solid copper pipe as it comes is already perfectly prepped to go into the chroming tanks. Solder can also be chromed: the chrome plating process is cold, so the soldered joints were unaffected and the rack came out completely intact, to provide the perfect finishing touch on an impressive Mini project.

Tech

ENGINE – 1,380cc by Mini Mania, 73.5mm Powermax pistons, Elgin cam, Longman GT6 head. Aldon Yellow distributor, HIF6 SU carb with BDK needle, Maniflow LCB, double core radiator, auxiliary electric fan. 113 crank bhp @ 6,500rpm, 102lb/ft @ 4,700rpm.

TRANSMISSION – Standard four-speed, 3.11:1 final drive ratio. Rod change gear lever extension.

BRAKES – Braided lines, plastic Lockheed master cylinder reservoir. Front, Cooper S discs and calipers, Mintex pads. Rear, standard.

SUSPENSION – Standard.

WHEELS AND TYRES – Standard 10in steels, 145 SR 10 radials.

INTERIOR – Most surfaces veneered in English white ash or African red mahogany. Seats reupholstered in Oxford Grey and Claret Scottish leather. Tacho, retractable seat belts added.

EXTERIOR – Rear lower body panels veneered in red mahogany. Custom roof rack in chromed copper tubing with English white ash slats. Thirteen coats wood varnish. Paint, base coat Spinnaker White with clearcoat over.

↑Ashtray is let into the veneered steel dash panel. The Woody is a no-smoking zone, as you might expect.

←Wooden sun visors were made from scratch. The convex mirror is rather unkind: it would make even Helena Bonham-Carter look like Winston Churchill.

↑Veneering the rear oddments box and ashtray panel is one thing, but we all know that the B-post with its awkward bulge is steel – which must have been seriously tricky to do.

The Roofless Cooper

The GRP flip front saves considerable weight and makes engine and suspension access a joy. Double front bumper is detachable and separate from the rest of the front end.

It's Jeremy's party and he can cry if he wants to. Many people would say he was completely mad to chop up an original Cooper S. He would agree with them now, but he says he was young and foolish when this project started.

'At the time I didn't have an ordinary Mini to modify so I chopped the S. Looking back I never would normally have done that, but I was young and had a hacksaw in one of my hands, so off with the top!'

There's also the point that the Cooper S as a model is much more common in the USA as a percentage of Minis than it is in the UK. That may seem odd, but the reason is that nearly all British Minis until the final decade or so were just day-to-day working 998s or whatever, which were bought as cheap transport. In the US, people who bought Minis normally bought them for fun, or for a second weekend car, or just because they were cute – so the Americans were much more likely to buy a Cooper version or a Woody. A Woody is, after all, a half-timbered car, like Shakespeare's house, so their Tudorbethan styling appealed to anglophile Americans, and still does. In the 1960s, American V8s were fast in a straight line, but a Cooper S would run rings round most of them if there were any corners involved. Of the remaining American Minis, many are Coopers, Woodies, Wolseleys, pickups and so on.

However, any functioning Mini is a relatively rare beast in the US. It's still relatively easy in the UK – although increasingly less so – to pop out and buy an MOT-failed 998 scrapper for a few hundred quid, snatch it from the jaws of the crusher and use it as the basis for a Mad Mini, but that option simply doesn't exist in America. Even pretty horrible basket cases are rescued, as described elsewhere in this book.

In any case, before we start sticking pins in Jeremy-shaped voodoo dollies for chopping up a Cooper S, Jeremy has already been punished by the Mini Spirits. When he was loading his freshly finished Speedster on to its trailer to go to Mini Meet West in 2001, one of the extra wide flares caught on the edge of the trailer and got half torn off. The car is normally used on the road and isn't a trailer queen, but it does have a suede pigskin dashboard and no roof, so it usually goes to shows on a trailer.

Jeremy stayed up all night, repaired the wing and charged off to the show as dawn broke, thinking that karma was restored, and that he'd paid the price for chopping the S. Things certainly looked that way at the show – he won People's Choice, Car of the Show, and the autocross.

Evil spirits waited until he was doing quite well in the rally, lulled him into a false sense of security and then went for him big time. A control arm bolt fell off, and he crashed into a ditch. That sounds bad, but it was with some relief that Jeremy checked the car out and found minimal damage, even to the precious bodywork. Ace Mini mechanic Martin Webber (www.westcoastclassics.com) from the Vancouver

18

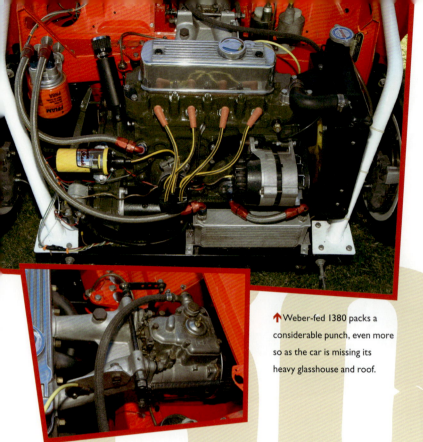

Weber-fed 1380 packs a considerable punch, even more so as the car is missing its heavy glasshouse and roof.

↑ The poor old instruments had nowhere to go, as the Weber and its filter pushed them right back inside the car. Top gurgle-and-suck soundtrack accompanies stonking performance.

→ Second-generation Fortech wheel arches are attached with just-visible rivet heads: this is a deliberate styling choice. Revolution four-spokes in powder-coat white with machined edges are still top lookers as far as wheels go.

Mini Club is the type of bloke who always has a spare control arm bolt in his pocket, so the car was recovered from the ditch and sorted out without too much drama. The front was detached and carefully laid on a grass bank as the car was gingerly winched back up on to the road. Jeremy risked a smile of relief, and the evil spirits spotted him smiling and reactivated his punishment. The wind gusted, flipped the front upside down and dragged it right across the road on its shiny side. He won the Hard Luck award that year as well, but the spirits reckoned that enough karma payback was enough, and he was allowed to proceed with his life. The universe was back in balance again.

If any Mini fanatics are still seething and spitting over the disgraceful scalping of a rare classic, you've probably shredded this book and jumped up and down on its corpse by now, but you do have the option of buying Jeremy's car, importing it and restoring it to its original standard trim, as it's for sale.

Jeremy has owned 43 Minis, which has to make him one of the most enthusiastic American Mini enthusiasts ever. That's probably one reason why his creations get wilder and wilder as the years pass. How many Coopers can you restore without it turning into just work?

Minis offer such a contrast to big American and boring Japanese cars that, once hooked, Americans often become fanatical enthusiasts – somebody once bought a Mini from Jeremy even though a wheel fell off during the test drive. That's a pretty keen buyer.

Removing the roof of a monocoque Mini makes the

whole structure floppy as it does any other chassisless monocoque, and the further removal of the front outer and inner wings to be replaced by a GRP flip front also reduced this car's rigidity to that of a wet teddy bear.

You can put some stiffness back into an open Mini by bracing the floors and fabricating deep sills, but you really need some quite substantial steelwork to get anywhere near the structural rigidity provided by the original steel roof.

Jeremy constructed a massive frame that connects the front and back of the cabin and then extends through the bulkhead to bolt to the front of the subframe, so the car finished up much stronger than a standard shell.

The rollover bar is also part of this structure, and it's genuine rather than decorative – the main hoop is triangulated backwards to the frame, and is also braced sideways so that it can't 'lozenge' or collapse sideways. Single roll hoops behind your head are better than nothing, and useful if you just roll over, but more ambitious and enthusiastic crashing requires a rollover bar braced in at least one additional direction. The car has been reworked for 13in wheels – easy enough at the front but trickier at the back, which has been fully tubbed to suit the bigger wheels.

The car's suspension is as stiff as the rollover bar, with S Racer Red springs and Spax shocks all round. If the original unsympathetic Mini seats were still fitted, Jeremy's inaugural drive in the car would probably have terminated in the car park of his chiropractor's office, but his Corbeaux (The plural of Corbeau would have to be Corbeaux, wouldn't it?) make the ride just stiff rather than brutal. These are rare period low-back Corbeaux, hard to find in the UK never mind in the US, and they are much treasured by Jeremy.

Advice from David Vizard was adhered to when it came to porting the head on the 1380 to suit Rimflow valves, and to reworking the combustion chambers. Vizard remains one of the top Mini tuning gurus, and now lives in the US and pops up at Mini Meets sometimes: he also remains approachable and entertaining.

The big Weber carb and air filter on its huge intake came through the bulkhead and occupied all the space previously inhabited by the speedo, but Jeremy just moved the whole dash plate backwards and covered it in suede, without even boxing it in. It might look wrong in a more conventional Mini, but on this one the idea works a treat.

Braking performance is more ambitious than it is in most Minis. The system remains on an optimistic single circuit, as in the original Cooper, but the front calipers are big four-pot JFZ/SRP numbers. (JFZ were an old and respected American name in performance brakes, like Wilwood, but they were bought out by SRP.) The discs themselves are large vented Brembos. With the original 10in Cooper wheels the choice of brakes is limited and expensive, but Jeremy took advantage of the clearance offered by his 13in

wheels to avail himself of the full range of brake options, and put seriously big brakes on. The car is used for autocross in coned car parks, and with big grunt and big brakes it's almost as fast as a Moke.

The soundtrack of this car is even more hardcore than the springs: it churns over slowly and then barks as it fires up, with lots of sucking and gurgling noises from the enormous Weber. You can actually hear air and fuel being sucked down the seven-inch-long intake trumpets, and the engine shakes and stumbles on idle, partly because of the cam and partly because of the inevitably rich mixture.

The high-ratio quick-rack steering combined with fat sticky Pirellis means that steering is seriously hard work at anything less than 20mph, and the engine is sulky and has to be kept going with careful use of the throttle – but all that's forgotten as the engine clears its throat and lets loose on the highway, providing joyous grunt and howl. In 100° Oregon summer heat the lack of a roof is murderous when stationary, but bliss at speed.

The security system on this car is provided by the detachable steering wheel, which you simply take with you when you go shopping. This does look daft, and there's not much chance of the car being successfully stolen anyway: it takes artistry with the throttle and clutch to get it moving, skills way beyond the driving talents of most modern crims.

Jeremy's penchant for small evil things with fat wheels extends to his Go-Ped fetish, which in a way makes sense. Oregonians can legally drive unlicensed kiddy-style scooters with little petrol engines on them, so Jeremy enthusiastically abuses this privilege by fitting these tiny things with massively powerful motors and tearing around burning rubber and popping wheelies on them. Almost as much fun as a Mini.

Tech

ENGINE – 1,380cc, lightened and balanced rods, Omega pistons and rings, Kent 530 cam. Head by Tony Lewis of Eugene, to David Vizard specs: ported, polished, Rimflow valves. Cooper S no-vacuum distributor, Pertronix ignition, NGK plugs, BumbleBee leads. 45DCOE Weber carb on 7in alloy intake, LCB exhaust manifold and single box RC40 exhaust. S pulley on stock water pump and rad. Large remote oil filter, cooler, aeroquip lines. Maybe 130–140bhp at the crank.

TRANSMISSION – Orange diaphragm with standard clutch, four-synchro close-ratio Cooper S box. 3.44:1 final drive ratio.

BRAKES – Standard single circuit, braided lines. Front, JFZ/SRP four-piston calipers on vented Brembo discs. Rear, Super Minifin drums.

SUSPENSION – S Racer adjustable Hi-Los with Red coil springs, Spax shocks. Quick rack.

WHEELS AND TYRES – 7J x 13 Revolution four-spokes, powder-coated white with polished lip, 175/50 x 13 Pirelli tyres.

INTERIOR – Rare lowback Corbeau GTA seats, 3in lap belts, quick-release Momo suede wheel, Auto Meter Pro-Comp Ultra Lite gauges in pigskin suede extended pod. KAD short shift on Cooper S remote. Newton Commercial trim and carpets.

EXTERIOR – One-piece GRP front end, Fortech Gen II wheel arches, tubbed rear end, roof removed, rollover bar and body brace structure fitted. PPG Pure Toner Orange paint.

The MADMEN

Madman Allen Frost's pickup just looks like a nice Mini…

The Mid-America Diehard Mini Enthusiasts' Network enjoys that very rare thing, an acronym that works perfectly. The club is based in Missouri, inasmuch as it's based anywhere, and Missouri is absolutely mid-America.

Diehard they are too: one of the Madmen's regular trips is to drive, in mostly 40-year-old Minis, from the annual Mini Meet West gathering, usually on the Pacific coast, right across the USA to the other annual event, Mini Meet East, which is always held a few days later. In 2008 this meant driving from Los Angeles, California, to Bethel, Maine, which took the Madmen over 3,000 miles in about four days across some serious mountains in cars with solid rubber suspension. That trip would be hard work in three tons of luxury motorhome, so doing it in a Mini certainly qualifies as diehard. Mini enthusiasts? We can't argue with that description either – these people are Mini enthusiasts in the same way that Attila the Hun was a fighting enthusiast. Finally the word network accurately describes the structure or lack of structure of the organisation. The members are scattered all over the USA and Canada, and for that matter the whole world, so they can't pop down to Bubba's Greasyburgers for a club meeting on the third Tuesday of the month.

The organisation was started by Karl Strauch in 1988, when he and his three Minis moved from New England to Missouri, all piled in one moving truck along with his furniture. Three more Minis soon turned up locally, which was enough to form a club. Karl published newsletters which finished up going all over the place, and the informal club developed into a structure that Karl compares to Alcoholics Anonymous, except that the idea is to encourage new people to develop an addiction to charging around in Minis, rather than curing them of it. Like all Mini clubs, they organise treasure hunts, but there could be a thousand miles between clues.

The big idea for 2009 was a trip to Alaska, where the mosquitoes are bigger than Minis, followed straight afterwards by attendance at the 2009 East Meets West Mini Meet in Minnesota to celebrate the Mini's 50th birthday. Just going a couple of thousand miles to Minnesota and back would have been too easy.

You can't just apply to join the Madmen, though – you need to complete a continuous 5,000-mile cross-continental single trip in a Mini powered by an A-series engine, backed up by pictures and receipts. For people living in the UK, driving east for 2,500 miles would get you deep into Russia or Kazakhstan. How many Brits would casually leap into their Minis and head for Moscow?

Mini guru and author David Vizard is the honorary patron of the club, and well qualified to be a Madman, having driven a Mini on a circuit of America getting 55mpg at 55mph. A top man, he has strong and amusing opinions on pretty well everything you care to mention.

This Madness may actually be infectious: for 2008

← The Madmen's route across country from Mini Meet West in Los Angeles to Bethel in Maine covers slightly over 3,000 miles in just a few days. Pedal to the metal and an oil change at either end.

↑ …until you notice that it tows a custom-built mini-trailer with a proper gooseneck mounting in the pickup bed. It's really for rescuing the occasional dead Mini rather than a serious load carrier.

→ When the bike bounced off the side of the pickup it took out the tyre but missed the rare and expensive rim, which was a bit of luck.

I planned a trip in the Marcos Heritage Mini Marcos which would have taken me up and down the UK for a week or two touring sofas and visiting chums, then to Belgium to shoot Mad Frank's V8 big-block Mini (Chapter 21), then southwards to check out a spot of vintage racing at the Circuit des Remparts in Angoulême, then to Barcelona to check out some top mad Gaudi architecture, then along the French Riviera to Italy, then via Florence and Rome to Bari on the Adriatic coast to catch a ferry to Dubrovnik in Croatia and thence to visit the in-laws in Split, then north through Slovenia and Austria to Switzerland and Germany. Travel in Germany would have been kept brief: their roads are either hurtling tailgating nightmares or stationary car parks as tailgating's inevitable results are cleared and all the squashed slow-learners are airlifted to hospital.

Then back across blissfully empty middle France on a bit of an eating and wine-tasting tour back towards the UK (Macon burgundy tastes particularly good in Macon; I suspect they keep some of the good stuff back for the locals) to return the Marcos demonstrator and catch a plane home. Which some might say makes me as mad as the Madmen. I wouldn't disagree. In the end circumstances conspired to torpedo that plan and all I managed was the sofa tour, the Twini Mini in Birmingham, the Goodwood Revival and Mad Frank in Zeebrugge. Maybe I'll do the full tour next year: the Mini Marcos went very well.

Long-distance Mini action is not without its dramas. During the 2007 Mini Meet East near Knoxville, Tennessee, many people went for a play on a road called the Dragon's Tail. It's very wriggly, rather like your average Welsh or Scottish Highland mountain road but with a better surface and less sheep poo. There are 318 curves in 11 miles. Americans regard corners as a fascinating novelty, and have a go at the Dragon's tail in huge numbers, frequently on big bikes that they can't handle. They fall off this road quite a lot.

Mini drivers are usually fast and competent, and of course they have the huge advantage of driving the best car in the world for whizzing round tight corners. They didn't have any problems at all on the Dragon's tail until an American biker whose confidence exceeded his competence went round a corner too fast and fell off his bike, which then hit Madman Allen Frost's pickup-and-trailer rig. The bike bounced off the back of the Mini and then went under the trailer, while the rider went flying over the trailer. His head was still attached to his body when he stopped sliding down the road, but only because he was very, very lucky. Incredibly, the Americans still allow articulated 18-wheeler trucks on this road: I suppose it's good business for the local private ambulance service and the coroner.

The Targa Newfoundland (www.targanewfoundland.com) is a marathon rallying event that attracts assorted Madmen and usually features several Minis. Even going along as part of a support crew is demanding, as Madman Rick Higgs and his wife Elaine discovered. The event covers 1,394 miles: 1,056 are transits between stages, but 310 of those miles are full-speed competition with normal legal speed limits suspended and the roads cleared of traffic. For Vancouver entrants, that all happens only after you've driven 4,500 miles from Vancouver just to get to the event. When you get there, Newfoundland doesn't even welcome you with nice weather: it's like Scotland's Atlantic coast, very pretty when you can spot a glimpse of it through the usual sheets of driving rain.

Mad? I should say so.

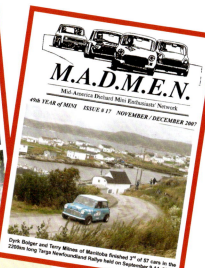

Dyrk Bolger and Terry Milnes of Manitoba finished 3rd of 57 cars in the 2200km long Targa Newfoundland Rallye held on September 9-14, 2007.

↑ The Targa Newfoundland is becoming increasingly popular. Local roads are closed and used for stages: half the fun is that you can legally drive down village main streets at 100mph during the stages that go through inhabited areas. Many competitors use relatively cheap cars, on the basis that they may not survive the rally.

↑ The Madmen's newsletter has been produced for decades: founder member Karl is taking a rest from it, so new blood will be needed.

➔ The body was slightly bashed, so it was decorated with a very apt dragon design at Mini Meet East by future Mini enthusiast Emily Caldwell.

↓ Once you've proved yourself with receipts and pics, you're entitled to display the coveted badge on your Mini.

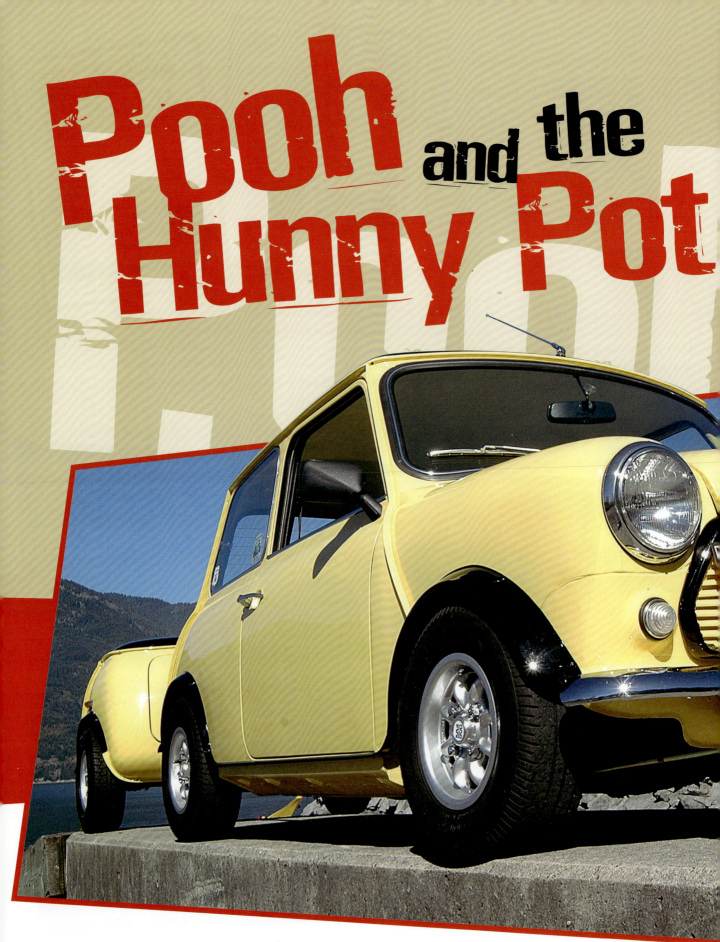

Pooh and the Hunny Pot

Rick Higgs looks and sounds entirely normal and sensible, and so does his Mini. Until you notice the trailer on the back, which is used to pack enough stuff for Rick and his wife Elaine to drive multi-thousand-mile trips around America.

Rick is a top Madman, and likes nothing better than charging around on the more interesting routes in North America. He lives in Vancouver, which is in the far north-western corner of the continent, so he has to drive thousands of miles more than people who live in the middle: it's just as well that he likes driving, then.

North America offers a lot of Mini fun, but the distances between venues are vast. For a Vancouverite to go to the Spring Thing in Florida and back is 5,600 miles: that's about the same distance as driving from London right through Europe and all of Africa to Cape Town. Pooh's first trip, 28 days after the engine was first fired up,

was to the Mini Meets on both sides of the continent, a trip of 8,000 miles plus between Vancouver, Arizona and Maryland.

Pooh came into Rick's family in 1995, having only covered 25,000 miles in its first 15 years, so the poor thing must have suffered a horrible fright when it found out how many thousands of miles of highway it would be required to drive when it got to middle age.

Rick's Mini started off as a 1979 998cc Canadian-spec Mini Sedan with huge Canadian-spec 5mph impact bumpers. Canadians traditionally hated these bumpers: as soon as they bought a Mini they would take them off and throw them away, then replace them with English-style chrome bumpers, so Canadian bumpers are actually quite rare now. Rick kept his, and recently gave them to me: as a Brit I think they're weird and cool, so they're going to be fitted to my own Canadian Mini.

Rick's car was named Pooh by his wife Elaine, because it was originally poo brown, or as BMC would have put it, Russet Brown. The 1970s saw some pretty indigestible colours in the trendy British spectrum: some of the BMC colour combinations, particularly when juxtaposed with stripy orange nylon seats, were quite emetic. Reliant really took the biscuit for kak colours, though – the entire 1970s Scimitar paint colour palette is based on human body fluids, mostly beginning with the letter p. Poo is one of the more attractive options. My own late and lamented Scimitar was Poo Brown with an interior in Babypoo velour, yum yum.

Rick's whole rig was conceived and designed for hard travelling. The mechanical base of his car is an engine powerful enough to deal with some fairly extreme temperatures and gradients. The little thermometer on the dash is a reminder of a long US road trip, with fellow Vancouverite and convoy companion John Goolevitch

↑ The Mini got the name Pooh from Rick's wife Elaine. Naturally, Pooh comes with a Hunny Pot.

20

← 'I've got a brilliant idea! Why don't we balance Rick's Mini up on that bit of concrete and take pictures of it?' This spectacular location is halfway up British Columbia's Sea to Sky Highway – rather like the Scottish Highlands but more so.

↑ Interior and matching trailer interior are black and yellow, just like the bumblebee technicians involved in honey-production hive facilities.

calling regularly from an air-conditioned Jeep 50 feet away to say 'Hey, the temperature just topped 90 degrees,' then ten minutes later to say 'Hey, it's 95 degrees now!' Yeah, thanks for that, John.

The original 998 is long gone, replaced by a 1275 A+, offset-bored to 1,380cc with a Swiftune cam and roller rockers. Ain't no substitoot for cubes, say the Americans: the Canadians concur with this approach, but they tend to recommend increasing engine capacity to achieve more torque. This is particularly important with Pooh because of its high 2.95:1 final drive ratio, chosen for civilised highway cruising. The spec on the engine tells a tale of bulletproof strength rather than frisky power, with a polished crank, balanced Metro rods and Powermax pistons all held together by Caterpillar and ARP bolts and studs. Everything that rotates is balanced, and carburation is a single HIF44 with a BDK needle. Electrical improvements include the uprated fuel pump, the switched cooling fan, and more and brighter lighting, and there is a relay box that reduces the demands placed on elderly Lucas wiring. The toughest electrical problem to solve was the side marker lamps on the trailer, which were positive-earthed LEDs. There's an oil cooler, and Rick is thinking about mounting a heater core from a

modern pickup as an additional front-mounted cooling radiator to support the custom two-core main rad.

Overall, Rick is well pleased with the engine. It achieved the anticipated power band with lots of low-down torque, it's demonstrably efficient as shown by its excellent fuel mileage, and it goes straight through British Columbia's fairly tight Aircare emissions testing every year, no bother. The differential ratio could possibly be improved to deal better with the weight of the trailer, but changing from the current high 2.95:1 ratio to the ideal 3.1:1 can happen next time the diff is disassembled.

Pooh's structure and suspension have received some beefing-up too, during the total stripdown and rebuild that started in the dead days between Christmas and New Year of 2004. The hard-mounted subframes have additional stitch-welding and gussets, the tie bars are heavy duty, the lower suspension arms are rose-jointed and poly bushes rule throughout. There's no rust left in the shell either – the kind of demands Rick makes on the car don't allow for structural weakness, so every panel in the car that showed even the faintest signs of bubbling was chopped out and replaced.

Where there's a Pooh there's inevitably a Hunny Pot, and Rick has followed an honourable Mini tradition of

making trailers out of the back ends of dead Minis.

Hunny Pot the trailer represents some really original thinking. It's relatively easy to cut the back off a scrap Mini, fabricate a bar to hold a hitch, and weld plates over the front and top. However, Rick's trailer was made from the last foot or so of the back ends and roofs of *two* scrap Minis. The two Mini butts were, aptly enough, butt-welded together by Craig McGuire and finished off by John Goolevitch, who restored his rotten Mini properly and is therefore now an expert welder. Hiding the join along 6ft of butt-welding is no mean feat, although it helped that most Minis in Canada old enough to be scrapped are made out of fairly thick steel.

The frame and the leaf-sprung beam axle on the trailer were fabricated as a completely new structure rather than using the usual recycled Mini subframe, and the trailer's loaded weight with camping gear is about 380lb. This makes the whole Pooh-and-Hunny-Pot outfit around 2,500lb and 16ft long, still a couple of feet shorter and a thousand pounds lighter than most North American cars. And a lot cheaper to run as well – on its longest (8,000-mile) trip Pooh and the Hunny Pot achieved a remarkable 41.5mpg.

The bumblebee yellow and black trim job runs through the trailer as well as the car. The cost of trimming the trailer as well went over budget, but trimmer Murray

↑ Rather than a bodged Mini subframe, the trailer has its own unique custom chassis and suspension.

→ A wise man always carries a couple of spare bootlids with him. Hunny Pot is the back ends of two scrap Minis, welded together.

not only put up with being sidelined during the two years of this Mini build, but actively supported him, and now happily charges around the continent with him, camping out of the trailer and only complaining when faced with ice and snow, which don't impress her much.

When the project was finished, Pooh and the Hunny Pot won every show they went to, although the combo was never particularly intended for showing. Rick isn't particularly proud of the finish: as far as he's concerned it was simply done as it should be done. Boyd Coddington used to produce $150,000 hot rods that were built the same way and for the same intrinsic reason. If it's worth doing, it's worth doing well. The rig won First in Class at both Mini Meet East and Mini Meet West, having been driven more than 8,000 miles to get to both events, and also won a First in Class and an Honourable Mention at the Portland All-British Field Meet out of a field of 805 cars.

After several multi-thousand-mile trips it now has a few paint chips, and there was a minor mechanical sulk after a Mini Meet, involving a failed drive flange. When it came to the photo session Rick was embarrassed about the paint chips, but for a well-used fun car it still looks pretty good.

Griessel said he'd do the work free if Rick paid for the vinyl, just because he wanted it all to look just right. It certainly does, doesn't it? Both Pooh and the Hunny Pot have many happy thousands of touring miles behind and ahead of them, and the black over BMW Liquid Yellow colour scheme is designed to be safe as well as cheery – most people will see it coming. Halogen headlights also help with that, and Canadian law requires that headlights remain switched on whenever the engine is running, quite a good safety idea.

Yellow and Minis are themes that have now crossed the generation gap. Rick's son Jeremy drives a Liquid Yellow BMW Mini, and was also involved with his dad in a previous classic Mini restoration in the early '90s. Rick is also very lucky to be married to his wife Elaine, who

Tech

ENGINE – 1,275cc A+, offset-bored to 1,380cc. 1275 crank, ground, polished, balanced, with centre main strap, Caterpillar mains bolts. Resized Metro rods, balanced, ARP big end bolts. 73.5mm Powermax pistons. Swiftune SW5 cam, 1.5:1 roller rockers, Mini Spares big bore 1380 head, ARP studs and bolts, lightened and balanced flywheel. Aldon Yellow distributor with SW5 advance curve, NGK BP6ES plugs, Kingsborne leads. HIF44 with BDK needles, K&N filter with sub-stack, milled aluminium spacer, 5/16in double-filtered fuel line. Ceramic-coated Mini Spares big bore intake manifold and big bore LCB, side-exit RC40. Two-core, ¾in dimpled-tube radiator, 11-row oil cooler, electric fan.

TRANSMISSION – Orange diaphragm, standard clutch plate, A+ gears and ratios, central oil pickup. 2.95:1 final drive ratio, CWP gear set. Quickshift.

BRAKES – Cooper S uprights and discs, standard pads. Rear S drums, standard linings. Braided lines, tandem master cylinder.

SUSPENSION – Hard-mounted subframes, adjustable HD tie bars, rose-jointed lower suspension arms, adjustable camber/castor, poly bushes. Hi-Los, KYB shocks. Trailer uses dead axle, leaf springs.

WHEELS AND TYRES – Minilite 5J x 10, Yokohama A008, 165/70 x 10.

INTERIOR – Mini seats custom-upholstered, with custom door cards, visors, rear arch covers, parcel shelf, dash pad. Tacho. Door handles/winders from MGB and E-Type Jag. Pioneer 200W four-speaker sound system.

EXTERIOR – Reversion to chrome bumpers, roof drains revert to earlier style. Replaced floor, outer sills, heel board, A-panels, wings, front valance. Halogen headlamps and driving lamps, roof antenna. Black over BMW Liquid Yellow 902.

← Engine has a sting in its tail, with 1,380cc, Swiftune cam and Mini Spares big-bore head.

↓ A Mini with a trailer is unusual, but a matching Mini and trailer rig constructed with this amount of care and attention is another thing entirely.

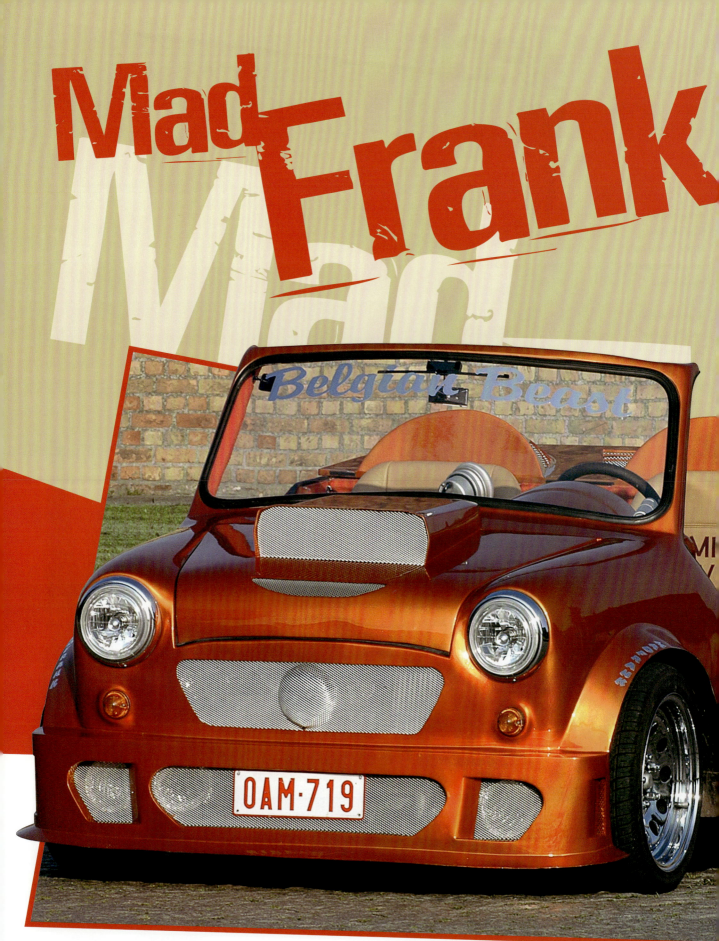

Mad Frank

Belgian Beast

OAM·719

I first met Zeebrugge's Mad Frank a decade ago, when he was showing his previous car, a Minivan, at the Hop Farm Mini meeting in Kent. The van was a work of art in sheet steel, and also had an immensely powerful sound system in the back that used to win sound-off competitions at shows. The system was so powerful it literally blew the roof off the van: during one competition the sound suddenly went weird and flat, and when Frank investigated he discovered that the seam between the side and roof on one side of the van had split wide open. That van also had a Perspex panel in the bonnet, and beneath that a Perspex panel let into the rocker cover through which you could see the rockers working, illuminated by little lights.

After the van, the V8 Mini was Mad Frank's next creation. His madness actually conceals much method, as he's a full-time carrossier and his Belgian Beasts are rolling advertising for him. His day job is massaging the bodywork of boring Audis and Citroëns so that they look indefinably different, but without making any identifiable changes. Belgium is firmly under the thumbs of EU penpushers,

and it's illegal to make any significant changes to TUV-approved standard vehicles. So Frank adjusts the shape of a wheel arch, drops a bonnet line by a centimetre or two, massages the door gaps, takes half a coil off the springs. If the cops stop somebody because their car looks a bit sexier than the other Euroboxes, it's hard for them to identify exactly why the car looks better.

Before the dead hand of TUV stopped the fun, there were no formal specifications for officials to use to regulate Belgian car enthusiasts, so with no written standard specs owners can do what they like with earlier cars. The V8 Mini represents Frank doing just that. His Belgian Beast is legally a 1965 Mini 850. As long as the brakes stop it and the lights light up, that's more or less all the law requires from it.

Originally he acquired a Dodge V8 out of a Dutch fire brigade pickup, laid it on his garage floor, then welded the front and back axles of a Ford Granada together with the engine in the middle. He then chopped the roof off a Mini shell, cut the roofless body in four, dropped the four bits of bodywork on top of the engine and chassis and filled in the gaps with new steel. The back end is 34cm wider than the standard Mini, but the front end, weirdly, is very slightly narrower than a standard Mini. The headrest shapes were formed by wrapping sheets of steel round a big industrial pipe. The air scoops down the side are real, and feed the rear radiator with cool air.

An interior structure was formed that allowed enough foot-room to drive the car, and when the bodywork was finished it was coated in a paint commissioned specially from Du Pont, using more flip-flop colours than had ever been used before. The engine was so big that the water-

A stonking Dodge V8 is just the thing for livening up a dull Mini. Short exhausts and minimal silencing mean it sounds like a thunderstorm having a psychotic episode after somebody said it had a fat bum.

21

← Mad Frank's Belgian Beast is now in its second incarnation with a new chassis, new paint and new interior after another few thousand hours of work.

pump pulley stuck out through the radiator grille, so it was also painted in four-way flip-flop, and it changed colour as it revolved. Very fetching.

The radiator was relocated to the back as there wasn't even room for an ant in the engine bay. Amazing though the car looked, it handled like a farm trailer and after a few years Frank pulled the shell off, scrapped the original chassis and started again. This time he paid more attention to the geometry, and made much of his own suspension from scratch. It's not necessarily particularly elegant, but it works a lot better. Weight is not really an issue compared to the importance of strength, so the suspension components Frank has fabricated are roughly the same size as the I-beams that hold skyscrapers together … they're not going to break. The rear end is still independent but it's now designed rather than just built, and when you give the car some welly it hooks up much better than before. Obviously it will still pull an instant wheelie as soon as the tyres get a grip, but the difference now is that Frank has some idea what's going to happen

when the front end comes back down again: previously it was pretty random as to whether it would lurch to the left or the right, depending on which front wheel came down and hit the ground first.

The new interior is the result of more thought than the earlier version, and is almost entirely scratchbuilt. It was made of various wood, plastic and metal pieces, with a collection of bits of other car dashboard parts such as the vents, and then the basic shape was coated with a thick layer of GRP. The rough surface of the fibreglass was then sanded back, filled, sanded and filled until it finished up ready for paint, and ready to be fitted with an impressive collection of instruments.

The car's old paint job is now some very expensive dust on Frank's garage floor, and the new finish is a deep metallic yellow-gold candy, which involves many layers of translucent paint, each meticulously wet-sanded between coats. The deeper the colour of a candy finish, the more weeks of work have gone into the paintwork prep.

Frank's Mini is now a more tractable and comfortable

The headrest shapes were created by wrapping sheet steel round a big pipe.

beast, and possibly even more fun that it was before. The V8 makes all the right noises, and its enormous torque means that Frank can put it in gear, lock up the front brakes and let the engine revolve the rear wheels slowly round as the car remains still. Cheaper than burnouts, and also more cool.

Cooling has always been a problem, as these huge Mopar V8s normally sit in the vast engine bay of a full-size pickup, where you could hold a dinner party for eight and still have room for a dance afterwards. Jamming one of these monsters into a tiny engine bay means that the

→ The water-pump used to stick out of the front of the car and was painted in flip-flop colours so it changed colour while spinning. It's now discreetly concealed behind a new grille.

OAM-719

← The back streets of Zeebrugge are Frank's personal racetrack. The exhaust thunder bounces off the walls, and all the babies and BMWs wail in panic.

Complete new interior and dash were more or less crafted from scratch.

New central console carries DVD player, screen and sound system. Colour-coded instrument bezels are a nice touch.

whole engine heats up, not just the coolant. There is some air-cooling effect on the whole engine block in most cars as moving air passes through a car's engine bay, but if there's no room for air movement the engine tends to cumulatively heat up despite the efforts of the radiator. My 1990s Chevy V8-powered Cobra was dreadful for this: as soon as I saw traffic on the horizon I had to jump out of the car, open the bonnet an inch or two and rest it on the bonnet pins to provide an exit for the hot engine-bay air, or the thing would brew up big time and spit the rad hoses off. Uncool in every sense.

Frank's solution was to wrap the manifolds to shift some heat out of the engine bay, and then to jam an air-to-air turbo intercooler into a corner of the bay with the engine coolant running through that as well as the radiator. That did the job, and the cooling is now under control even when the V8 heat-sink effect gets going in traffic. The headers themselves are stumpy and short, and they don't have elegant and ideal matched primary pipes snaking towards collector boxes. You can tune engines for more torque or more bhp by adjusting the length of the primary pipes coming out of the cylinder heads, but not if you've squeezed a barrel into a pint pot and only have a couple of inches before the exhaust manifold hits the chassis. The current arrangement probably loses a few bhp off the top end as a result: well, boo-hoo. Mad Frank's Belgian Beast has enough power to be going on with.

Tech

ENGINE – Dodge V8, bored and stroked to the max. Steel crank and rods, fast road camshaft. Carter AFB four-barrel carb with B&M ram air filter. Custom wrapped headers, radiator relocated to rear, additional front intercooler in water cooling system.

TRANSMISSION – Six-speed high/low range Mopar automatic gearbox.

SUSPENSION – Front, custom-made double wishbones with coilover shocks. Rear, independent, custom setup with modified Ford suspension components.

BRAKES – Ford Granada system and solid discs all round.

WHEELS AND TYRES – American Custom Chrome with Dunlop R/T Qualifiers rear, Pirelli P5000 front.

INTERIOR – Custom GRP console and floors, Auto-Style programmable tacho with shift light, Autostyle water temp, oil pressure, voltmeter, ammeter, oil temperature, vacuum and boost gauges. JVC KD-SH707R head unit, Marquant DVD player.

EXTERIOR – Custom steel body based on original Mini panels. Orange candy paint.

⬆ The original Belgian Beast was a cool all-steel Minivan that literally blew its roof off during a sound system competition: the vibration and air compression from the speakers split the gutter seam all the way along.

⬇ The Beast hooks up much better with its new chassis design, and also steers better when the front wheels are touching the road.

Admirable Nelsons

↓ With the outside temperature at a nasty 11° below zero, shooting the Turbo Woody in a heated hangar seemed a smart move.

← Rachel and Andy with most of their collection, which has since expanded.

Andy and Rachel Nelson are enthusiasts' enthusiasts: their entire lives revolve around playing with Minis. Currently they own a '64 Cooper S, '65 and '67 road sedans, a '65 Cooper racer and a BMW Mini. Not a huge number, but each Mini is gleaming perfection.

Andy is a commercial pilot, and Rachel is a manager for Harbour Air, which flies seaplanes from Vancouver BC to Victoria harbour, a short but spectacular flight carried out at 1,000ft or so over rocky little islands and beaches.

They're currently leaving Vancouver and heading for Toronto to find enough space for their Mini collection. Surely a pilot and a seaplane company executive could afford a house with a huge garage, even in expensive

Vancouver? Yes, they could, but they've always spent all their money on playing with Minis and latterly on racing them, and if they stayed in Vancouver too much of their Mini budget would go on boring mortgage payments. They've already sorted out a huge Toronto workshop for the Minis. How about a Toronto house? Whatever, they'll sort one out when they get there, or just live in the workshop.

Andy was a mechanic for 12 years before becoming more ambitious and training to be a pilot, and the combination of good piloting practice and good mechanical practice has been very effective in building a string of top-class roadgoing Minis and then successful track Minis. Everything is prepared in advance and checked over thoroughly and in the correct order.

The 1965/61 Turbo Woody is a good example of a Nelson car. Andy bought it as scrap when he was 19, during his apprenticeship in a British car specialist garage. The Woody was intended to provide space for the couple's two English Bulldogs. That was in 1989, so the bulldogs were in doggy heaven way before the car was finished.

The car's engine was long gone, and the woodwork was in a pile in the back: the Woody was fast approaching scrap status, but Andy decided to bring it back from the edge with help from another dead 1961 example. The project was sidelined for years on a few occasions, and was only really finished quite recently.

The detailing inside and out is exquisite, and there is

22

← Pilots are usually nit-picky, methodical perfectionists. The live ones are, anyway. The same approach pays off when building Minis.

ENGINE – Metro Turbo A+ 1275, bored to 1,312cc. Balanced Metro crank. MEGA turbo pistons with chrome rings. Mini Spares turbo cam, Vernier adjustable timing gearset. Sodium-cooled 29mm Turbo exhaust valves, 35mm inlet valves, Mini Spares uprated anti-coil-bind springs, 1.5:1 roller rockers. Stage 3 polished head. Avonbar/Aldon turbo distributor, Lucas coil, Pertronix electronic ignition, NGK plugs, copper leads, NGK motorcycle caps. MG Metro Turbo 1.75in HIF SU carb, with Avonbar high boost needle and dashpot spring, K&N air filter. Metro Turbo high-volume fuel pump with special regulator. Metro T3 turbocharger with 10psi wastegate actuator. Metro Turbo inlet and exhaust manifold, Maniflow downpipe and 2in Maniflow twin bore exhaust system. Estimated 120–130bhp at the wheels.

COOLING system – Mini Spares two-core rad with six-blade tropical fan and extra electric fan, additional rad in o/s inner wing. Sixteen-row oil cooler with twin electric fans and braided lines. ERA replica cooling duct at rear of bonnet to cool turbo area. Large-impeller water pump.

TRANSMISSION – Jack Knight five-speed straight-cut close-ratio, with matching drop gears. Turbo clutch, Orange diaphragm, lightweight steel flywheel and backplate. 3.4:1 final drive with four-pin diff. KAD quickshift.

BRAKES – Non-assisted, silicone fluid. Front, Metro Turbo 8.4in vented discs, Metro four-pot calipers, standard pads. Rear, standard cylinders and linings, Super Minifin alloy drums.

SUSPENSION – Hi-Los, standard rubber cones with Spax shocks. Front, 1.5° lower arms, adjustable tie bars, poly bushes. Rear, adjustable camber brackets, new subframe with aluminium tie-down plates. MkIV rack with S-type steering arms.

WHEELS AND TYRES – Minator 5J x 12 alloy wheels, 165/60 R12 Yokohama A510 tyres.

INTERIOR – Cobra Le Mans recliners in Blackrock fabric, with rear seats trimmed to match. Newton Commercial/custom trim and carpets. Driver's side Leaf six-point harness, passenger side lap-belt. Standard main dash dial with MG oil-water gauge, Auto Meter boost gauge, period Smiths tacho.

EXTERIOR – ERA replica bonnet scoop, Wood & Pickett front arch extensions, bulkhead modified to clear turbo. Cibie halogen headlights. Viper red paint. Window tints, Cooper grille.

no sign of any short cuts. Replacement panels include the front and rear floors, sills, rear valance, front wings and grille panel, and numerous hand-fabricated repair sections. Paint prep on the inside rear floorpans is as good as that on the front wings, a standard that requires literally hundreds of hours of fine-grade wet and dry.

Both Rachel and Andy get stuck in – they're always out in the garage, rubbing happily away from after work to midnight, both obsessed and both happy to be fiddling about with Minis rather than watching telly or eating food or doing boring house stuff. It helps that North American TV is mostly 200 channels of crap, so it's not tempting even if you're tired.

I suppose the Turbo Woody is a bit mad. It has the performance of a serious nutter Mini, but it still looks like a gleaming half-timbered estate car, which would normally have a picnic basket, some prize plaques and some polish in the back rather than a can of race fuel. The body is visually standard apart from the scoop and the wheel arches, although there have been some bulkhead mods to make room for the turbo.

When Rachel invited me to try this car out, I was in there like a rat up a drainpipe. Do I like playing with turbo Minis? Do bears poo in the woods? After a few years living in Canada I can confirm that bears do indeed poo in the woods on a regular basis, and cheerfully mix their metaphors as well.

The first thing that grabs you about this car is how unexpectedly fast it is, and the second is how smooth. Turbo Minis don't usually feature subtle power curves – mostly it's nothing much, nothing much, nothing much, *blam*. This one gave nice silky smooth punch up to 7,000, and the

↑ Reverse scoop as fitted to ERA turbo Minis doesn't spoil the car's looks, and helps to keep engine bay temperatures down.

→ Racing cars with gleaming show-condition engine bays actually do go faster: it means the whole car has been built to this standard.

fifth gear allows for civilised cruising, apart from the noise. A whining Mini gearbox is either knackered or it has seriously trick guts in it, and this one turned out to be a straight-cut five-speed Jack Knight number. I've been hearing recently that slant-cut gears work just as well as straight-cut, but if you've paid as much for your gearbox as some people pay for their entire car, it's nice to hear the straight-cut whine to remind you that you're getting value for money. There's also the funky *pllffff* noise from the dump valve on the turbo, so the soundtrack is pretty cool; there's no sound system in the car at all, as the in-car entertainment is all provided by what goes on under the bonnet.

The little replica ERA backwards turbo scoop is the only outward sign that the car has been radically interfered with, so it's top fun on the street if you come across somebody at the traffic lights who previously thought they had a fast car.

The engine started off as a Metro Turbo unit, but was inevitably beefed up. It's balanced throughout, and with its light steel flywheel and turbo it whips up to 7,000 no bother. It all went together very well, with a few hairy moments while fitting the ultra-hard chrome piston rings: they can only be bent open a small amount before they snap. The engine has been completely reliable, and with the boost at a reasonable 10lb and good balancing there's no reason why it shouldn't keep going for a long time.

The list of engine mods is pretty hardcore but not too extreme, so the car is still usable in the city as well. The power band is wide and the original flat spot at zero boost was sorted out by Avonbar turbo carb mods. There are enough gears and enough power to allow quite serious track speeds, so the Yokos and the decent brakes are used to their full capacity.

The comprehensive cooling system is designed to deal with track use in a hot Canadian summer. There are in total four fans, two water radiators and one oil radiator. For autocrossing between cones on hot summer airfields, this is great, because everything stays cooler than Clint Eastwood. On an uncharacteristically freezing Vancouver day with the wind chill at the airport taking the temperature down to -11°, the engine is so well cooled it still needs the choke after it's been idling for 20 minutes.

Brakes are 8.4in vented discs rather than the original feeble drums, and the suspension has also been upgraded with Hi-Los, poly bushes, 1.5° lower arms, adjustable tie bars and whatever other subtle and barely visible improvements were available.

↖↑ All the Nelsons' cars, race or road, look better than new. Everything is squeaky clean and sorted to perfection before it goes in.

⬆ Five minutes before the race, and Andy is relaxed and ready. A 7-litre Cobra and a 5-litre Tiger don't know they're about to be humiliated by a 1.3-litre Mini.

The longer wheelbase may also give the car more of a performance edge over a Mini saloon on corners due to the extra stability of the longer wheelbase, but I have to admit that pushing a Woody to the limit isn't a concept that had crossed my mind before I found myself looning around in this car. Sorry, conducting a sober professional journalistic evaluation of it.

The Turbo Woody was around for a while, but it attracted the attention of somebody who had enough taste to realise how fab the car was, and who also had enough money to make an unrefusable offer for it. The Nelsons were getting seriously stuck into vintage racing at the time, which is more expensive than a Keith Richards drug regime and just as addictive, so they took the offer, and the money went straight into the next racing Mini.

A high point of the Nelsons' vintage racing career so far has been the 2006 Rolex Monterey Historic Automobile Races. A few Mini-Coopers were grudgingly tolerated at this posh and expensive event, but only because Cooper racing cars were the featured marque. That's proper Cooper single-seaters, you understand, not our low-rent Cooper-badged BMC Minis. Minis will probably never be invited again.

Ironically, the Americans failed to understand that Minis were and are among the very few completely classless cars you can buy. Mother's cleaning lady had one, Mother had one and so did many British aristos, who wouldn't have owned anything as vulgar as an American car. Or a Rolex, come to think of it.

The thought of driving a Mini at the Monterey Historics appealed big-time to Andy and Rachel. They needed a Cooper with racing papers for the event, and tracked down a '65 Cooper S that had been raced from new. After 30 years of being either raced or neglected, it was nasty enough to present a worthwhile challenge even to the intrepid Nelsons. There wasn't much rust, but there had been a few rugby-style incidents on the track, and the car had been both rolled over and heavily thumped in the right rear quarter.

They removed the roof, then the rear quarter, then scraped the whole shell down to bare metal and rust. Where possible the original panels were saved. Dents and rust were hammered out and cut out, and the fuel cell was sealed in the boot. US vintage racing requires originality apart from safety mods, and the Nelsons were happy to comply, leaving the S as original as possible.

The car was resprayed at home. Hundreds of hours were spent on perfect prep, then lots and lots of paint was sprayed and laboriously cut back to a gleaming finish.

The Monterey rules require a full interior, so although the driver's seat is a racing shell, it's been trimmed in the same colours and material as the rest of the Newton Commercial replica interior.

The engine build proceeded at the same measured pace as the body, so when one of the cam bearings turned out to be conical rather than tubular, there was time to replace it. The engine is genuinely A-series with as much development and strengthening as the spirit of the rules will allow. The intention was to build the optimum racing Mini engine – fast enough and strong enough. A delicate balance. The original crank had come out in several pieces, illustrating just that point. The new crank was wedged, balanced, cross-drilled and nitrided. The rods remain S, but well worked over – lightened, balanced, polished and held together with ARP bolts.

The head is a Longman GT14, much ported and polished, with big 37mm intake valves and standard 29mm exhausts. It's shaved to create 16.5cc combustion chambers with a compression ratio of 12.2:1, so it's just as well the crank is uprated. The AEG 649 cam's timing was perfected with a Swiftune alloy-centred vernier cam gear adjuster.

There are some nice details – the original S rockers were shaved and lightened to cut unnecessary reciprocating weight, and they run with light brass spacers separating them. Reliable sparks were also very important: if the engine splutters when you floor it on the track, you might as well be running an 850. The Cooper S distributor was retained, but fitted with lighter springs to speed up the ignition advance, and Pertronix were brought in when it came to the actual sparks.

Proper cooling for Monterey was crucial: the circuit is inland from the California coast, and the weather can get pretty hot. A double rad is cooled by a plastic multiblade fan and an uprated water pump, and a damaged piece of original inner wing in front of the fan was trimmed off and left off for more airflow. Just to be sure, a 13-row oil cooler was added as well – hot oil is good, but boiling oil is bad: just ask a medieval castle-besieger's ghost.

The suspension remains hydrolastic, which is unusual for racing. However, Andy has found that with a controller to adjust pressure back to front and with a factory rear anti-roll bar, it all works very well. Corners are taken on three wheels anyway, so keeping the car smooth enough for him to be able to keep his feet on the pedals may be a very smart move.

GBA 688 made Mini history first time out after its rebuild, and performed faultlessly. Andy rather tactlessly beat two Shelby Cobras, a Lotus Seven, a Ginetta, an Alfa Romeo and a Sunbeam Tiger. That'll make sure Minis are never invited back to the Rolex Monterey Historics.

To keep up with their future exploits check out Andy and Rachel's website at www.arperformance.citymax.com.

ENGINE – 1275S, now 1,293cc. Rotating assembly balanced. EN40B wedged, cross-drilled, nitrided crank. S rods, lightened, polished, ARP bolts. AE Mega pistons, chrome top rings. Swiftune alloy-centred Vernier adjuster on an AEG 649 cam. Longman GT14 ported, polished race head shaved to 16.5cc combustion chambers, 12.2:1 compression, 37mm inlet valves, 29mm exhaust valves. Pertronix ignition. S distributor with light advance springs. NGK BP8ES plugs, Bumblebee motorcycle leads. Generic alloy inlet manifold, 2 x 1.5in SU with Keith Calver ITG filter. LCB Maniflow, RC40 centre exit exhaust. Mini Spares two-core radiator with multi-blade fan, oversize water pump pulley, oversize water impeller, 13-row oil cooler. Holley Red 97GPH fuel pump, low pressure regulator.

TRANSMISSION – Ultralight steel flywheel, AP bonded race/rally clutch, orange spot diaphragm; three-synchro straight-cut gears, EVO straight-cut drop gears, roller bearing idle, central oil pickup.

SUSPENSION – Hydrolastic, stock hydro return springs, adjustable front/back bias valve, Koni shocks, factory rear anti-roll bar, upgraded bushes throughout.

BRAKES – 7.5in Cooper discs, Porterfield carbon metallic pads, Cooper S drums rear.

WHEELS AND TYRES – 5J x 10 Minilite alloys, with Hoosier TD165/70 x 10 vintage-racing bias-belted (crossplies).

INTERIOR – Cobra racing seat trimmed to match Newton Commercial replica interior, five-point harness, eight-point custom TIG welded rollcage; period aftermarket gear lever extension, headlining signed by Michael Cooper, grippy pedal pads, oil/water combo gauge, oil temp gauge.

EXTERIOR – New roof, rear quarter and front end panels, RH sills, Old English White with black roof.

← The race Mini's engine performed perfectly. Luck is involved as well, though: Rachel's engine was equally well prepared, and ran beautifully until it threw a rod through the crankcase.

Z-Cars

↓ This steel-bodied Z-Cars Mini looks innocent enough: in fact it could still pass for a stock Mini.

G545 FDA

Chris Allanson, boss of Z-Cars, is one of Britain's top kit car madmen. He's an ex-cop who's old enough to remember the grainy 1960s black-and-white television series *Z-Cars*, which was rather fab. Lots of Ford Zephyrs with fins screeching around, and Brian Blessed's shouting-based acting career was launched. The crims used to say 'It's a fair cop, guv,' when caught bang to rights, and the traffic police didn't have to wear stab vests. Z-Cars is a top name for Chris's company, then.

Chris no longer charges around in cop cars, he charges around in his own cars. He tried to scare the hell out of me but failed. It's not that I'm particularly brave: I'm not. Too much imagination, and I enjoy being alive. However, I've been writing about fast cars for 20 years now and experience tells me when I should be scared. Inexperienced and overconfident young drivers in fast production cars with no rollover bars scare me, drunks in pickups scare me, but Chris doesn't.

He makes a good effort, of course, and takes me out on his regular scary route at mad speeds. This works very well at terrifying and impressing customers: he scared the poo out of my brother, who now wants a Z-Cars Mini.

The company's impressively big and clean premises is situated on farmland on the remote coastline of Yorkshire (which has to mean very low overheads and a good chance of surviving recessions: Chris may be mad but he's not daft).

The tiny local roads are empty. It's wet, and Chris fires up a Honda-powered Z-Mini with a crackle from the exhaust. He potters out of the workshop on to his long,

splashy private driveway and floors it. The engine howls, we hit the seat backs, and after a few engine screams and gearchanges he's heading for a T-junction at about 100mph on a ten-foot-wide wet surface and it looks like we're heading for a very big accident. However, no worries. Chris used to hold the world road-car record for 0–60mph in a twin-engined Tiger kit car; he currently races a very nasty Z-Mini; the car we're in has huge brakes and a substantial rollcage; there's only a shallow ditch and a field beyond the junction; and he knows every inch of this road in all conditions. We slam into the straps as the brakes bite, and I just enjoy the ride. I keep a grip on the harness, just in case he hits some diesel or something and a ditch flips the car, because in a rollover you don't want your arm getting outside the car if the door or window comes off. There's not much margin for error here, as Chris is probably going quite close to too fast. Out on the public road, and he doesn't even have to break the speed limit on his little terror tour: taking a 20mph corner at 60mph makes his point very effectively.

Most people aren't used to cars and drivers like this, so they scream in terror for the first few corners, then begin to scream with excitement from the visceral thrill of it, and by the time the demo drive is over they're gagging to buy a kit.

And so they should, because these things are very fine toys.

Some of them are still basically Minis, with either bike engines or Honda engines in the back. The Mini normally weighs around 1,500lb, and used to win rallies with a 65bhp 1,275cc engine. In fact, most of us are aware that in the

Ride height and stance still maintain the illusion that this is a mildly cheeky Mini, rather than a psycho beast.

23

← Chain drive goes from the bike gearbox to the diff. The visible starter motor is actually the electric backup motor.

→ 160bhp+ and rear-wheel drive reminds you that FWD is only good for packaging and cost savings. The Honda B-series Civic engine is top fun in Minis, but double the fun if you put it in the back.

← Even the BUSA lettering on the back won't mean much to most people.

↓ The rollcage is part of the structure, but you can stick with a more traditional dash if the digital variety doesn't suit you.

↘ The Honda car is naughtier than the posh red ZBusa, with visible air intakes and outlets. Racetrack sound level rules are now tougher than noise regulations for street cars, as bureaucrats achieve tighter control. You would not believe how fast this thing goes, and it's not even blown or turbocharged.

Monte Carlo rally, Mini Coopers wiped the floor with the French so brutally that they had to cheat and disqualify the Minis for incorrectly-switched headlights to get a French car into the top ten.

So take this 1,500lb go-kart and fit it with a 225bhp Honda Vtec Type R engine and it's now three and half times as powerful. If you want it even more exciting, fit a blown Hayabusa engine.

But that's the slow version of the Z-Car Mini. A faster Z-Car, but still a regular roadgoing production vehicle, now beats the old world-record 3.15-second time with an incredible 2.6 seconds, and 5.7 seconds to 100mph. 2.6 seconds is Formula One acceleration, leaving pedestrian McLarens struggling at 3.2 seconds and donkey-cart Ferrari 360s and geriatric Shelby 427 Cobras crawling along way behind at 4.6 seconds.

The faster Z-Car Monte Carlo option is a spaceframe with a fibreglass hat on it. It's 30% lighter than a standard Mini at 1,000lb, retaining the unbeatable handling, and something like seven times as powerful for its weight as the cars that humiliated the French in a world-class rally.

I tried two Z-Car types. The first was a steel Mini, owned by the boss of the doomed Rover group, and was still very much a Mini, but instead of the cute but wheezy 1,275cc A-series it had a 1,300cc Suzuki Hayabusa engine in the back. Only the cubic capacity is the same: rather than a plodding cast-iron shopping engine, the Hayabusa motor is a manic superbike screamer with a redline in five figures and power options from 160bhp to 400bhp. The Hayabusa engine and box doesn't weigh much and is in any case mounted in the perfect mid-rear position for weight-transfer-induced rear wheel loading and

maximum grip. The chain from the bike engine goes to a diff and then to the road via independent suspension and driveshafts. The rear seats, floor and subframe of the Mini are filleted and replaced by the Z-Cars custom frame that mounts the engine and suspension. The outer shell of the Mini remains intact, as does everything from the B-pillars forwards, although the engine bay is obviously just used for the fuel tank and radiators. So one of the cool things about retaining the Mini shell is the sleeper aspect: you can wave cheerily to a Porsche or Ferrari driver at the traffic lights and then humiliate them at will, driving a cute little Mini. How cool is that?

It may strike you that superbikes have no reverse gear: this is taken care of by a starter motor attached to the

Tech

Steel Z-Cars Mini

ENGINES – Bike engines: Yamaha R1, Suzuki GSXR1000 or Suzuki Hayabusa. Car engines: 2-litre Honda Civic Type, or other 1,600cc/1,800cc B-series.

TRANSMISSION – For bike engines, six-speed sequential with electric reverse; for car engines, standard Honda five-speed.

SUSPENSION – Front, adjustable double wishbones, rose-joints, Protech adjustable inboard coilover shocks. Rear, independent fully adjustable multilink, Protech coilover shocks.

BRAKES – Various options, rear discs and calipers supplied.

WHEELS AND TYRES – 13in and larger, as Mini rear trailing arm wheel size restriction is removed.

KIT components – Fully assembled roll cage/subframe, rear disks and calipers, gear lever and cable, fuel tank, radiator and mounts, front subframe. Options include aluminium firewalls, limited slip diff, electric reverse, 200 more bhp, brake upgrades, etc.

WEIGHT – With Hayabusa, 1,268lb. With Honda, 1,433lb.

The Monte Carlo is also available with either Honda or bike power. This one has Yamaha R1 power.

This is really a street-legal racing spaceframe that wears a Mini-shaped hat.

chain wheel. You just press a button to back up.

This rear-bike-engine format when applied to a steel Mini is still a usable daily car, if you use enough soundproofing and carpeting to muffle the shriek of the engine, but the 'busa motor at 1,300cc has enough cubes and thus enough torque to cruise well at lower revs, unlike some smaller bike engines that have to be screaming their plums off at 10,000rpm before you get any power out of them. The engine's efficiency also gives good gas mileage if you can keep your foot out of it. Other advantages of the rear-mid-engine configuration include better general balance, better brake balance and stopping power, and, of course, no Mini-style torque steer, as the front wheels are just used for steering and braking.

The more extreme Z-cars Monte Carlo model almost completely disposes of the Mini: it retains the paperwork, glasshouse, lights and a 2in lowered Mini visual silhouette, but that's about all. There's a rear-mounted bike engine or a Honda Type R engine, a spaceframe rollcage, an aluminium floor and some GRP Mini-lookalike panels. The wheels stick out a foot on either side, so the grip is phenomenal. Power to weight ratio is between 250bhp/ton and 600bhp+/ton with a still-streetable Hayabusa engine. These are the Z-Cars with sub-three second

0–60 times that would keep up with Michael Schumacher in a full Formula One car until air resistance slowed them or Schumacher rammed them.

You can pick'n'mix too, of course – and you can get Z-cars to change things and make things for you, at a rate of £40 an hour. In the big workshop there's a TVR with a Nissan Skyline 4WD drivetrain, there's a Citroën 2CV that's been Zedded with a spaceframe and a bike engine, and there's a BMW 'Mini' with a decent engine in it, a tuned Vtec out of a Honda rather than the original shrunken Dodge Neon device.

Most of Z-Cars' business will continue to be mad Minis, and their level of expertise and professionalism accelerates as fast as the cars – the steel-Mini build manual is on DVD and available from Z-Cars or from the producer www.Minifilms.co.uk, who also have other interesting Mini film material.

The build DVD is excellent and comprehensive, and not only tells you everything you could want to know but includes some top footage of Z-Cars on the track.

These are not cheap kits, but you get what you pay for, and the quality of the engineering is obvious. Given their genuine supercar performance, they cost peanuts. A Bugatti Veyron will match a Z-Cars Monte Carlo's performance, but it costs $1.7 million and needs every last one of its 1,001bhp to keep up.

Check out the Z-Cars website at www.zcars.org.uk or type 'YouTube Z cars' into Google if you want to see them in action. If that last sentence is gibberish, find a ten-year-old and ask them to help. Turn the speaker volume right up to get the full effect.

Tech
Z–Cars Monte Carlo

ENGINES – Bike engines: Yamaha R1, Suzuki GSXR1000 or Suzuki Hayabusa. Car engines: 2-litre Honda Civic Type, or other 1,600cc/1,800cc B-series.

TRANSMISSION – Bike engines, six-speed sequential with electric reverse; car engines, standard Honda five-speed.

SUSPENSION – Front, adjustable double A-arms, Protech coilover shocks, custom steel uprights, optional sway bar. Rear, adjustable multi-link trailing arms, Protech coilover shocks, optional sway bar.

BRAKES – Wilwood bias control, custom pendulum pedals. Front, Wilwood four-pot calipers, 235mm vented discs. Rear, Sierra Cosworth 235mm calipers/discs.

WHEELS AND TYRES – Front, 8 x 13 alloys, 195/50R13 Avon ACB10. Rear, 10 x 13 with 245/45 x 13 Avon ACB10.

KIT components – Full spaceframe, 38mm and 45mm CDS round tube steel. (FIA spec optional.) GRP body with detachable rear end. Otherwise similar to steel car.

WEIGHT – With Vtec, 1,111lb. With Hayabusa, 1,023lb.

← The yellow Monte is powered by Honda's Vtec, which is both reliable and frisky in Type R form.

Ice Racing

Nolan's Mini may look like an abandoned wreck, but it's just resting before next season.

olan Kitchener seems like a nice, sane, sensible bloke. He runs Cedar Valley Alignment in Mission, British Columbia. I've never heard him swear, or say anything unpleasant about anyone. However, if you take a quick look at the state of his ice racer, it doesn't look as though it's used for friendly and polite competition, does it? Ice racing is quite far along the mad scale, although being Canadian it's all organised and licensed madness, with rules.

There are very different climates in different parts of BC. The Vancouver area's climate is relatively balmy, and quite British apart from having about 60% more sunshine. I think of it as Baja Canada, and the reasonable winters and very pleasant summers are two of the good reasons why I live here. There's usually a week or two of snow annually, during which many shiny pickups fall into ditches and Hondas are scattered sideways across the roads. Snow is infrequent enough that nobody knows how to drive on it, just as in the UK.

However, start heading northwards into the interior and the picture rapidly changes. Away from the Pacific coast, quite a lot of BC has a continental climate, which tends to mean almost desert conditions, with 35° heat in the summer; but for the entire winter the temperature can remain below freezing.

Barnes Lake is a top ice-racing venue, and every winter the ice racers just wait until the lake water has frozen thick enough to carry the weight of a few dozen charging cars, and then they get on with the fun.

Although the cars tend to look disreputable, this is proper FIA-sanctioned racing, with safety rules and a code of conduct: you have to obtain a specific ice-racing licence. There's no boozing allowed, and no contamination of the ice by engine oil or antifreeze. Make a mess by tearing a driveshaft out and you have to clean it up.

Ice racing is not a free-for-all either. There's a formal one-kilometre circuit with corner marshals, and although a certain amount of automotive rugby is tolerated, this isn't banger racing.

Canadians tend to play hard when it comes to their sports, to relieve the tensions caused by being so nice all the time. Ice hockey is a good example – if you bash somebody on the head with your hockey stick, that's known as 'roughing' and is fine. If you rip his helmet off

← It's a good idea to have big brake lights to penetrate the blizzard of ice and snow your rear wheels kick up.

24

← Nolan's engine is a turbocharged 1275 – functional rather than concours. The big plastic bottle is an oil catch tank, the smaller one covers the air intake to reduce the amount of ice spray that gets sucked into the engine.

→ Studded wheels have a tendency to shred bodywork if there's any contact.

→ The interior features a full rollcage, a tatty old racing seat and a large second modern heater to supplement the Mini system.

and them bash him on the head with your stick, that's a foul and is not encouraged. You have to go pretty berserk to get into trouble, although a Vancouver Canucks player managed it recently by jumping on a guy from behind at full speed and breaking his neck. That was a going a bit far, everybody thought.

There's a parallel in Britain: go and see some Citroën 2CV racing, I strongly recommend it. 2CVs are driven by vicars and social workers who wear wholemeal pullovers and shoes that look like Cornish pasties, and because they have to be sympathetic, warm and fluffy 99% of the time, they go barking mad when they get out on the track in 2CVs. It's literally death and glory, because even the best rollcage in the world is useless if it's attached to something

flimsier than a biscuit tin. Some drivers don't survive big crashes such as being T-boned at the bottom of a hill by one of the other 40 hurtling biscuit tins in the race.

Anyway, Minis are smaller and much stronger – Nolan's ice racer has already survived eight seasons. Like all ice-racing cars, it's fitted with a race-legal FIA rollcage, a full harness, fire extinguishers and tow hooks. The survival of this car for eight years is remarkable, because the mechanical stresses in ice racing are very high even if you don't get bumped on a corner and go flying up the banking and off the track. You'd think ice racing was like Speedway bikes or drifting, a matter of controlled sliding around the corners. But far from it. Novices run on studded street tyres, which are quite grippy but can be drifted on, and the more ambitious ice-racing classes run on seriously grippy tyres they have to make themselves. They glue an extra tread on to an existing tyre, just like a remould. But before the new tread is heat-glued or Vulcanised on to the tyre carcass, dozens of holes are drilled in the new tread cap and long-shanked pop rivets are inserted through the holes from behind. Then wood screws are driven into the tops of the rivets. 10in tyres that you can fit studs to are hard to find, as are directional 10in winter tyres.

So old Mini tyres with screwed recaps is what they race on, and their grip on the ice is better than a sticky Yokohama on a tarmac track: suspension, driveline and driver take quite a pounding. The reason for the bodywork looking a bit shredded is that occasionally it gets a scrape from one of these lethal studded tyres: you can see the results.

⬇ Nolan's Mini hurtling across a frozen lake in northern BC during a race. The Mini still does well against more modern vehicles.

➡ Nolan makes his own studded tyres with remould treads, rivets and screws.

ENGINE – Turbocharged Metro 1275 A+, bored +040. Austin America rods drilled for oil splash on piston bottoms. ARP rod bolts. Hepolite thick-crown Turbo pistons. Cam, one-off Colt Cams regrind for turbo engine. Head, fully ported, large intake valves, 1.7:1 ratio aluminium roller rockers, Isky high-lift springs, ARP head bolts. Garret turbocharger, SU sealed turbo carb with BBC needle, custom turbo distributor, manifold heating disconnected for colder air-fuel mixture, dual radiators. Power 100bhp+.

TRANSMISSION – Standard, with welded diff, 3.76:1 or 4.13:1 depending on circuit. CVs and driveshafts are regarded as disposable.

BRAKES – Standard drums. No, really. Braking is for wimps.

SUSPENSION – Fully adjustable for height, camber, toe-in and caster. Hi-Los, Spax springs and shocks.

WHEELS AND TYRES – Steel Mini wheels, tyres are recapped 145 x 10 with ¼ hex head screw into a pop rivet in the tread, or a 16mm truck stud in the tyre.

INTERIOR – One elderly racing seat, full rollcage, fire system, five-point race harness, additional heater.

MiniLimo

Jan Harde's life changed dramatically for the worse when he was involved in a crash and got his back broken – so he redressed the karmic balance by buying, dismantling, rebuilding and customising his Mini Limo. This has gone a long way towards making him feel better, because he's had an enormous amount of fun out of it and still does on a daily basis. It can't help but cheer you up when many people grin in delight as you drive past.

Jan also owns a seriously nice 1970s Mercedes convertible, but Los Angeles is full of nice open Mercs, whereas they have only one Mini Limousine. What makes the Mini Limo even more delightful is that Hollywood is crawling with seriously huge stretch limos, some based on American saloons and some now based on chopped-up obese pimp jeeps such as Cadillac Escalades, and some even based on Hummers. I mean the fake chromed Hummer lookey-likeys on Chevy Suburban chassis, not the real army ones. Fake Hummers have to be the gayest vehicle in the world: they remind me of studded black leather posing pouches.

The drivers of the LA big boys' limos, naturally enough, scowl at the Mini Limo because its very existence takes the piss, and these people take themselves seriously. They charge a hundred bucks each to take 300lb hillbillies (actually, it's polite to say 'Appalachian-Americans' these days rather than hillbillies or rednecks) from Cow Ass, Tennessee, on a tour of the Stars' Homes, although most of that's probably bollocks: they can just point at any high wall in Beverley Hills with security cameras on it and say 'That's George Clooney's house.' Who's going to argue?

The Mini was originally built in the UK a while back by Lindsay Hayes, who took the safety and structural aspects of the project very seriously and welded in a set of beams along the sills, which make the Mini far stronger than most railway carriages. There was probably no need for such extensive strengthening, as the final car is only about as long as a Mondeo, but it's nice to know that you're riding round in something that would dent a Chieftain tank. The long wheelbase and the stiff chassis also improve the ride: it's still a Mini, but when you're riding in the front seats it doesn't feel bouncy and lively the way Minis usually do. The LA freeways are rather like the M11 but worse – they're just slabs of cracked concrete loosely laid in a general north-south direction, and in anything but a Cadillac you get a constant *badump badump* as you crash from slab to slab. Minis go *badumpbadumpbadumpbadump*; but the Mini Limo goes *badump badump badump*, which is much better.

A total of three scrap Minis were chopped up to provide all the bodywork for the Limo, and the square rear door windows had to be specially cut: the window frames can be welded together from two sets, but not

← Making a Mini Limo look right is no easy task, as the body both droops and narrows quite dramatically towards the tail: just adding a slice of dead Mini isn't enough.

↓ The office is very nice, with most of the electrics and dash guts from a 2000 Rover. Fat speakers, two of fourteen, make a fat noise.

↑ Rear windows are not just tinted, but electrically operated – a Mini first?

25

blow up female frocks a bit, and if you can't work any magic in a tiny limo with *Casablanca* on the video screens, a bit of cheek and a bottle of Moët, something's far wrong with your mojo. The Mini Limo is also unique among Minis in that making lurve in it is possible without risking a visit to the chiropractor afterwards. However, I digress. As Gorgeous Wife had been left behind in Canada, all I did was ride around in the back waving at the public as they jumped up and down, pointing and squealing with delight.

Jan rather modestly says he's only put a thousand hours in to the Mini Limo, and that has been largely in getting it exactly the way he wants it. Working on the car comes at a cost for him, though – after a few operations he still has to chew morphine sticks when his back gives him a hard time, so there's a limit to how much he can achieve. He can't exactly go clambering about like a monkey and has to think strategically before doing something to the car. Fortunately he didn't get any grief at all from his wife Cathy for spending so much time fiddling with the car – if he's happy, she's happy.

The audio and cinema system is pretty impressive, with a beefy pair of pre-amps and amps, surround sound and a total of 14 speakers, so you get every nuance of *The Italian Job*. 'You're only supposed to blow the bloody doors off!'

If you wanted to do the Hollywood film cliché thing of driving through, er, Hollywood, standing up and poking out of the Limo roof with a lush filmy soundtrack belting out of the 14 speakers as your hair blows in the slipstream, you could do it: but it would be best to be a woman, young, and not bald. The Mini Limo is fully equipped for this sort of sunroof cliché action, with a Webasto sliding roof above the rear cabin, and if you want less of a breeze you can slide the roof shut and open the side windows – by pressing a button, as they're electrically operated. Have to admit the lecky rear side windows are an impressive bit of bling, although they do rather stray from the Issigonis function-and-form concept. For yet more cooling breeze, there's air conditioning front and rear.

The original engine wasn't up to hauling this sort of weight around as well as powering the air conditioning compressor, so it was replaced with the current 1,430cc MED motor, which is well capable of keeping up with the traffic. In Hollywood the traffic cruises, but on the LA interstate highways they do give it some stick. Ten lanes of Appalachian-Americans in pickups wearing wifebeater vests, and estate agents in cheap suits and BMWs, all hurtling along at 80mph and more, when they're not jammed solid in sweaty M25-style gridlock. The cops just come by to clear up the crashes, rather than trying to impose any speed limit.

Jan has just fitted a five-speed gearbox to get the

↑ Most of the crowd outside the Marriott entrance were admiring the Mini rather than going in and out of the hotel.

↓ 1,430cc MED A-series is more than strong enough to haul the Limo around. Fat intake tube and thick Metro Turbo plenum on carb reduce intake noise.

the glass. A Limo is not just a Mini chopped in half with an extra bit bunged in the middle. There's a lot more to it than that.

Riding in the back is a whole different world from joining Jan in the office up front. He could have gone for the whole Hollywood pimp look with pink feather boas and Hummer chrome, but has restrained himself to a colour scheme of grey Connolly leather, walnut fittings and deep Wilton carpeting decorated with white sheepskin rugs. The centrepiece of the rear compartment is the drinks table with clinking crystal champagne flutes and a bottle of Moët et Chandon on ice.

It's just occurred to me that this has to be one of the top seduction machines ever. From experience, Ferrari Testarossas are a waste of time, but small cute cars do noise level down to Limo rather than Mini levels, and also because the engine will last longer at more reasonable revs. It has enough bhp to charge around a bit, but the 1,430cc capacity gives it enough torque to make good use of a fifth gear.

The major donor Mini was a 1969 Mini 1000, which had a beans-on-toast electrical system, compared to the *estouffade haricots blancs frais à l'occitane* wiring demanded by the Mini's current electrical and electronic bean-feast. Most of the electrical system was harvested from a 2000 Rover Mini, and it works pretty well most of the time, although the ghost of Joseph Lucas does occasionally haunt it, as it does most Minis – during our shoot the lighting fuse declined to pop and referred all the current to the light-switch instead, which resigned in a puff of rather worrying smoke. Such is Mini life, however, and the hazard lights still worked so no worries.

Jan wanted to achieve the ultimate luxury Mini, and I reckon he's probably done it. If anybody knows of a posher one, let me know.

Many thanks to the friendly, helpful and patient Woodland Hills Marriott staff, for the use of their forecourt for the shoot.

← Here's the money shot. There's Connolly-leathered and Wilton-carpeted room for two in comfort, with a cooled bottle of Moët in the ice bucket and *Casablanca* on the DVD screen.

Tech

ENGINE – MED 1,430cc based on 1275, polished and ported head, belt-driven MD276 cam, Aldon Yellow distributor, Metro Turbo cast plenum chamber with single SU 1.75in HIF, Range Rover VO air filter with Saab 900 fuel filler pipe used as intake tube, twin lower steady bars, custom upper steady bars, custom street-rod chromed alternator to power air conditioning system. 2000 model year electrical system. Radex aluminium racing radiator. 13-row oil cooler. Electric cooling fans. Estimated 115bhp at the crank.

TRANSMISSION – Mini Sport five-speed, Saab gear lever.

BRAKES – Stainless braided hoses, servo, DOT 5 fluid, four-pot calipers, drilled and vented discs.

SUSPENSION – Hi-Los all round, Gaz adjustable shocks over coil springs.

WHEELS AND TYRES – Minilite 6J x 12 with Yokohama A539, 165/60 x 12.

INTERIOR – Connolly leather seating, Mini Mania wheel, walnut dash and console with six gauges, Saab gear lever, Wilton carpets with sheepskin rugs. Electrically operated rear windows, champagne cooler.

EXTERIOR – Main conversion by Lindsay Hayes in UK. Three Mini bodies were used, with significant side and floor bracing steel beams. New roof paint is deep purple candy with 24-carat gold flakes. Webasto sunroof at rear.

ICE – Extreme Audio 800W system with dual pre-amps and amps, surround sound with 14 speakers, three video screens, Alpine ten-disc CD changer, in-dash Phillips Carin GPS.

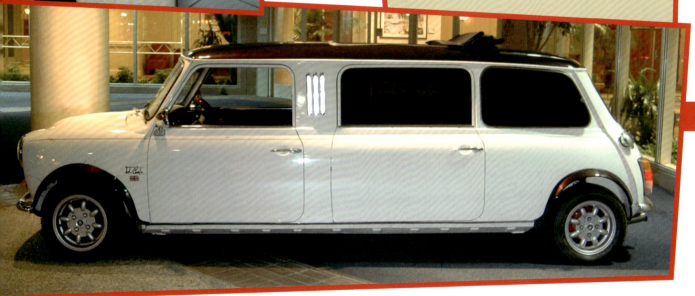

Mini Marcos

↑ The front end of the Mini Marcos is either cute or grotesque depending on your mood and aesthetic sensibilities. Personally, I think it takes ugly into a whole new dimension: I like it a lot. But then I'm possibly a bit mad.

→ The rear end with its stumpy Kamm tail was shaped purely for speed, but offers lots of interior space and interesting lines. A car with a Minivan wheelbase is crowded for four, but with just two it feels like a cathedral.

En route to shoot Mad Frank's Belgian Beast in Zeebrugge, I gave a Mini Marcos a proper European road test, during 2,300 miles and three weeks. Marcos Heritage is still very much in business and can still provide almost any body parts for any Marcos from 1963 to 2002, including bodyshells for all models and a complete Mini Marcos kit.

The Mini Marcos is a racing kit car developed in the early 1960s, based on the entire mechanicals of a Mini. The Mini in standard form wiped the floor with many racing and rally cars in the early 1960s mainly because of its incredible handling, and Minis stormed the Monte Carlo Rally with a 1-2-3 win: the French cheated but we all know who won.

A Mini Marcos did extremely well at Le Mans in 1966, coming 15th overall and being the only British entry to finish. It was technically a French entry in order to avoid being disqualified for being the wrong colour or whatever. It also achieved a flat-out speed of 140mph+ along the Mulsanne straight – impressive, as the engine was an off-the-shelf 1,275cc Mini Cooper S, fresh from the BMC competition department.

The Mini Marcos was and is so successful because it improves significantly on the abilities of the already remarkably competent Mini. It's lighter, its fibreglass monocoque is stronger and stiffer, its glasshouse is smaller and much lower, its centre of gravity is way down, and its longer Traveller/Van wheelbase aids stability.

The shape is either cute or horrible depending on your personal opinion, but it does the job: it's aerodynamically

very slippery and stable in aerodynamic terms, and the high rear roof and chopped-off Kamm tail work well at high speeds.

The Mini Marcos is still as good a proposition in 2009 as it was in 1965, and it's still the same car, apart from a hatchback and more interior space.

The Mini as a donor is far from outdated, and is even making a resurgence. Mini bits are more available now than they were a decade ago and remain cheap, at hobby prices rather than main-dealer prices. More new replica parts and performance goodies appear all the time.

Minis are also rusting as enthusiastically as ever, and from around 1990 the strapped-for-cash Rover company made them out of thinner metal but still failed to provide adequate rustproofing, so rotten later-model donor Minis with excellent mechanicals abound.

The Mini club scene, and for that matter the Mini

↑ The high bonnet line is caused by the Mini engine sitting on top of its integral gearbox. The view from the cockpit looks like that of a much larger and more expensive car.

26

Marcos club scene, are another excellent reason for getting involved with Minis – Mini people are enthusiastic, sociable, intelligent, helpful and just slightly crazy in the right sort of way.

So how did the three-week road test go? On the first day things did not look good. Britain's roads are in poor shape, and the seat in the Marcos had no lumbar support. The suspension is also brutally stiff, partly because the shell is lighter and stiffer than a Mini and partly because of hard low-profile tyres. The Mini suspension components, like the rest of the Mini parts, are standard and are just transferred from the old to the new shell. This car was deliberately built from a second-hand 60,000-mile 1991 Cooper without any improvements or refurbishment at all.

I finished the first day with a sore back, tired, and damp, as it was raining heavily. The car doesn't leak much, but it does steam up during monsoon weather. However, on day two I bought an inflatable neck pillow to use as a back support, and felt much better. Marcos seats can be ordered with lumbar support, so that's definitely worth doing. I also got used to the suspension, and when the sun came out and I found a big roundabout with nobody on it, the fun started.

When it comes to corners, the Mini Marcos is hilarious. None of this boring slowing-down business, you just pile in at full speed, turn the wheel and keep your foot down. I think that on the twisties a Mini Marcos would be very hard to beat with anything but a fairly serious competition car. It just goes round bends completely flat, no leaning, no squealing, just goes wherever you point it, at whatever ridiculous speed you choose. The only risk is that although a Marcos driver can charge into a roundabout at 60mph, everybody else has to crawl round at 25mph, so you have to go around them, and bear in mind that few of them are capable of paying attention or signalling.

Given some empty country roads, as instructed by Marcos Heritage boss Rory, I gave the Mini Marcos some revs and have seldom had so much fun from a cheap car. Being right down on the deck like a go-kart, with the fruity exhaust roaring away and the hedgerows hurtling past my ear in a green blur, it feels even faster than it is. A BMW type, upset at being overtaken, tried to intimidate me by tailgating, but chickened out after two corners: he didn't have a hope of keeping up, even with 200bhp to my 60.

There's something of a Tardis about the Mini Marcos – it looks tiny from the outside, but once you get inside there's an astonishing amount of room. A six-foot passenger with the seat right back couldn't touch the front firewall with his feet, it was so far away. I was

↑ LED rear lights add sparkle and update the decor, but that Kamm tail is timeless anyway: it doesn't look any better or worse than it did in 1966, unlike more fashion-driven designs.

→ Interior space in the Mini Marcos really is remarkable considering the size of the car, and the very low driving position is almost too comfy. Fortunately the stiff suspension will keep you awake.

travelling with a couple of wheely suitcases and there was plenty of room in the back for them, and when I took a friend with me to Belgium there was plenty of room for his kit as well.

The noise levels inside this particular car were quite high, but with plenty of padding under the carpet it wasn't interior noise so much as the roar from the air filter at the front and the sportsy exhaust tailpipe at the back. The car is a demonstrator, built to provide thrills and fun over a short blast on local country roads, not for touring Europe. Even so, I'd happily use it again on the same trip, although I would definitely soften the suspension if I was going to build one. Even if you compromised the razor-sharp handling a little with downrated suspension, a Mini Marcos would still handle better than pretty well everything else on the road.

Only having four gears might have been an issue, but in the event it was fine: the UK admits to having 63 million people jammed into a pretty small island, to which you can add an unknown but sizeable number of illegals, so the roads are crowded and slow even with petrol tax at piracy levels. The later Marcos shell allows the use of 13in wheels rather than the earlier 10in Mini wheels, so the gear ratios were just fine. Cruising at 70mph the engine was working but not howling, and the

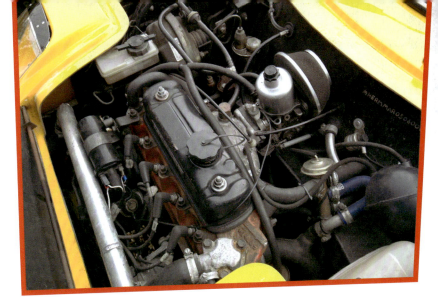

1275 Cooper's 60bhp has enough muscle to carry quite a high final-drive ratio.

There's very little air resistance due to the excellent aerodynamics, so you could cruise at higher speeds, no bother. I'd use higher-profile tyres on 12in rims for a better ride, though.

The same aerodynamic properties that allowed a 140mph+ top speed at Le Mans can also give you astonishing fuel economy. The Cooper engine in the Marcos I was using gave pretty good fuel efficiency

↑ The transverse 60bhp Mini Cooper engine and transaxle has more room in the Mini Marcos engine bay than it does in a Mini, and the radiator goes in front rather than at the side. If your life is not scary enough, a 225bhp Honda engine can be used instead…

← The hatchback reveals plenty of luggage space. It's now mandatory to carry a reflective triangle, fluorescent vest and spare bulb set while driving in France, or face fat Euro fines from French cops, who still don't like the English much.

considering I was either giving it some exercise or stuck in a traffic jam, but if you used the smaller and cheaper 1,000cc A-series engine, with the right carburation and with the crank well balanced to allow you to use the revs when you wanted to charge about a bit, you could still enjoy pretty good performance… but you could be looking at 50mpg. Fuel mileage for fun weekend kit cars may not be an issue now, but, looking a few years ahead, a practical two-seater that sips petrol could be a very useful vehicle indeed as 'green' anti-car taxes go through the roof along with oil company profits.

The cost of a new Mini Marcos kit is currently around

five grand, but all you then need is a complete scrap Mini and you're in business – the kit is comprehensive and there are no further costs. You can also knock off a couple of grand for a paint job, as the kits are supplied in coloured gelcoat. The car pictured is gelcoat-finished, and still gleams after three hard years of life as a demonstrator.

If you have doubts about the performance and reliability limitations, and the attention demands of ancient British engines and Lucas electricals driven hard and daily, that would not be unreasonable. What you need to know in that case is that lots of people are currently removing Mini engines and integral four-speed gearboxes, and are replacing them with 225bhp Honda Type R engines and five-speed boxes. The resultant performance is terrifying, partly because the power increase from the variable-profile Vtec cam comes in with a huge bang, and partly because the Type R diff is a brutal affair that needs the bulk of a heavy Honda shell to tame it. The cures are to use an ECU modification that brings the cam change in earlier, and to use a standard Honda LSD rather than the type R.

Nobody's done it yet, but if you can get a B-series Honda engine into a Mini you can get it in a Mini Marcos too. *Woo-hoo.* Top speed unknown, storming acceleration, and more or less the same astonishing handling. If I didn't already have too many projects…

Check out the Marcos Heritage website at www.rory.uk.com.

↑ Off to Belgium and France to shoot Mad Frank's V8 Mini in Zeebrugge. Looming over P&O's ferry terminal are the Green Cliffs of Dover.

→ The Mini Marcos was the only interesting car in the entire town of Diksmuide, Belgium: the crushing hand of the TUV and EU have banned everything but dullmobiles.

← The spats and the frontal styling on this racing Mini Marcos are there purely for more speed, but they look good too. (Courtesy of Billy Dulles)

Tech

ENGINE – Standard 1991 Mini Cooper S 1275, front-mounted radiator.

TRANSMISSION – Standard Cooper 1275 S four-speed with quickshift.

SUSPENSION – Standard rubber cones, uprated by shell weight reduction.

BRAKES – Standard Cooper S powered discs.

WHEELS AND TYRES – 6J x 13in Minilite alloys, with 175/50 x 13 Yokohamas.

INTERIOR – Marcos trim kit, seats, walnut dash, Moto-Lita steering wheel: donor Mini provides instruments, controls, steering column, pedals, door furniture and window winders.

EXTERIOR – Full GRP Marcos monocoque shell with glazing and some lights. Donor Mini provides mechanical and electrical components, subframes, brakes, suspension, steering, exhaust, door handles, lights etc.

Performance and reliability!

Performance et endurance!

Les vingt-quatre heures du Mans

MARCOS CARS COMPONENTS LIMITED
Greenland Mills,
Bradford-on-Avon,
Wiltshire, England
Telephone: Bradford on Avon 2971-2972

↑ For Marcos in 1966, selling cars paid for going racing, but racing also helped sell cars. (Courtesy of Billy Dulles)

The Psycho Twini

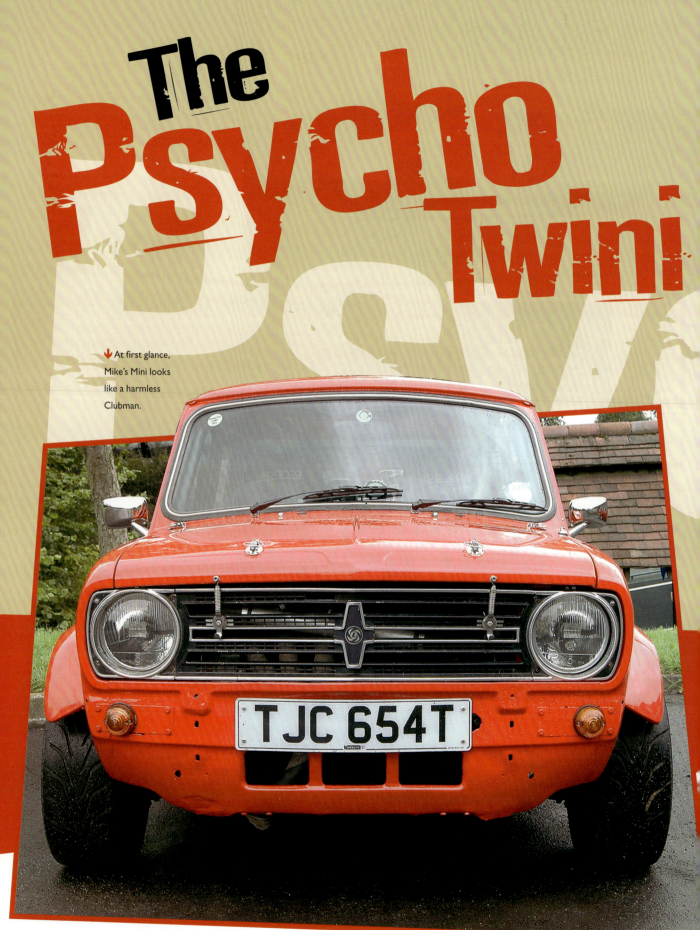

At first glance, Mike's Mini looks like a harmless Clubman.

TJC 654T

'**W**OOOOOOOO! WAAAAAAH! NAAAAAAAH! SIDEWAYS AAAARGH! WOOOOOOO! GO ON, BOOT IT! YEE-HAAAAA!'
That was my Writer's Opinion box for Mike Smith's twin-Vauxhall-engined Clubman estate when I wrote about it in *MiniWorld* magazine: you won't have read it, though, because the Opinion box became a casualty of a redesign.

Mike's Clubman suffered the same fate. It was previously a harmless shopping trolley, minding its own business and crumbling gently into old age when it was rudely Frankensteined back to life as an unruly monster.

Vauxhall-engined Minis are scarier than Icelandic bank accounts. Despite that, just one 180bhp Vauxhall engine in his Clubman wasn't fast enough for Mike, so he added a second one. I've been frightened/exhilarated by Vauxhallated Minis before, even when driving on the track with a cage and good harnesses: they're either skittering towards the kitty litter or taking off like greased squirrels whenever they get a proper grip. You need to be well strapped in.

To achieve grip despite big power, front-wheel drive and no weight isn't easy. Quite an effective approach is to get rid of the suspension. On smooth straights, no worries: on bumpy racetrack corners with a cavalier approach taken to the corner kerbs, this means some time is spent airborne until the tyres grip again. Mike's car actually does have some suspension, although he's already changed 375lb AVO springs for 500lb ones in the search for grip.

The tyre choice is sorted, though – Yoko 888R cut slicks provide excellent grip even in the wet, but at a price. It's not that they're stupid-expensive, it's just that Mike only gets about 4,000 miles out of a set. But that's mostly because he's a maniac.

This Clubman Estate has 360bhp and four-wheel drive, and Mike is still getting the hang of learning to drive it in a straight line, never mind turning corners.

The noise when both engines are started is incredible. The exhausts both come out the back, and oddly enough it's the longer front-engine exhaust pipe that's the noisier. Off we go for a spin on damp Birmingham rural roads, looking for a place to play. Having found somewhere without anything hard to hit if the car decides to take a side excursion into the scenery, Mike checks behind, revs both engines and dumps the clutches. The stereo roar is ear-shredding, and the kick in the back is exhilarating: this is dragstrip acceleration, not road-car power. When the ferocity of the acceleration lifts the front of the car more or less off the deck, Mike's home-made traction control reduces power to the front engine, to stop it grenading itself. As the passing greenery blurs, more of the wheels get a grip, and Mike snatches gearchanges while sawing wildly at the steering wheel to keep the car pointing forwards. The engines yammer dementedly, and the car squirms and wriggles as the diffs brutally and randomly direct power to assorted wheels, according to whim.

↑ The car has a bit of a stance and big wheels, but still looks like a nice little Clubby rather than a scary psycho beast. The rollcage peeping through the rear windows suggests that all is not as it seems.

27

← Even the twin exhausts look like minor showing-off, but there's one for each engine. One Vauxhall engine in a Mini is scary enough, without another one at the other end.

↙ The second engine peeps coyly between the seats.

The Clubman's best quarter-mile time is 12.69 seconds at 117mph, and the best 0–60mph time so far is 3.9 seconds. Top speed is governed to 155mph. Fuel efficiency is excellent with cruising economy of up to 50mpg, but that's with the back engine switched off.

Birmingham is the birthplace of Balti cuisine, which may have provided Mike with the original subconscious inspiration for the car – Balti chili Jalfrezi is hot and tasty stuff, and like this Mini it makes you sweat and helps you to develop strong sphincter muscles.

The Clubman replaced Mike's previous car, a Nissan Skyline. He developed the Skyline to 550bhp, but they're heavy beasts at 1,600kg and need 1,000bhp to be properly lively. His Skyline's 3.5 seconds to 60mph wasn't bad, but getting much more out of it would have become disproportionately expensive. Getting 1,000bhp reliably out of one Skyline engine is very expensive indeed: it pushes many other components past breaking point and ultimately produces less speed than using an 850kg car and building two 355bhp engines.

Mike likes Vauxhall engines: 200 fairly bombproof normally-aspirated bhp is comparatively cheap and easy with forged pistons and steel H-beam rods, and Vauxhall spares are reasonably priced.

The Vauxhall gearboxes and final-drive ratios were designed for 13in wheels rather than 10in, but even with 60% profile tyres there's enough torque to get away pretty sharpish, and of course if you can save one gearchange on the dragstrip with tall gearing, that's worth a big chunk of a precious second.

Conveniently, the Clubman shell was already halfway converted to take a Vauxhall engine, with an AMT engine frame fitted. AMT no longer make engine frames, but www.16Vminiclub.co.uk has alternative sources if the idea of a Vauxhall or dual-Vauxhall Mini blows your frock up.

The front engine arrangements were therefore fairly straightforward, although extra tasks such as shifting the alternator to the other side of the engine added much time to the build. A more demanding matter was designing and fabricating the arrangements for fitting the second engine. It's a similar arrangement to the front engine/gearbox/suspension setup and also uses an AMT frame, except that instead of a steering rack, the redundant rear steering arms are connected to the chassis and used to adjust the rear tracking rather than to steer the car. It was also a challenge figuring out what to attach the frame to, as an estate Mini with no rear subframe is basically an empty tin box. Most of the rear floor went, along with half of each rear suspension turret. The rollcage helped by providing places to weld to, and a subframe was also made and welded to the floorpan.

The main challenges came with the electronics and the gear linkages. Mike used a DTA E48EXP ECU on a custom loom with custom traction control, which he made himself for a budget of 50 quid. If one engine speeds up dramatically compared to the other, its power is automatically reduced. The traction system also lights up LEDs on the dash, which is functional as well as cool. The main instrument is a Digidash data logger/ speedo that can be switched to display and record the activities of either engine. Other instruments are simply duplicated, including the big analogue tachos.

The gear linkages on the front engine were straightforward, and the clutch pedal simply works twin master cylinders like a normal street/competition-biased brake system, with a servo on the rear clutch. The rear engine's gear lever, if left unaltered, would be a foot or so outside the back doors, so both planes of gear lever movement had to be replaced with a cable arrangement going to the single central gear lever in the cabin. That was hard going, but it now works pretty well. More bracketing had to be made to get rid of flex in the cable system, and much of the rear gearchange trouble was finally cured by replacing the front engine mounts with harder ones: it was movement of the engine (and thus the gearbox and gear lever) that was fouling up the linkage to the rear gearbox. There's a pin that disconnects the cable and leaves the rear engine in neutral when it's switched off.

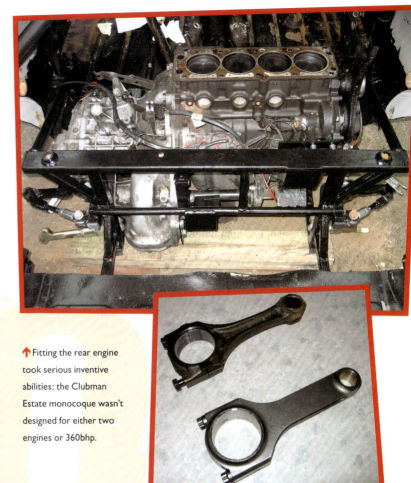

⬆ Fitting the rear engine took serious inventive abilities: the Clubman Estate monocoque wasn't designed for either two engines or 360bhp.

⬇ Two injected ECU-controlled engines meant miles of multi-coloured electrical spaghetti.

⬆ Substantial conrods with ARP bolts reduce the likelihood of a rod doing an Alien through the crankcase, even with nitrous and cheeky throttle action.

Many problems [these are called *Issues* if you're from California] arose during the build, but they were all sorted. The handbrake required thought, because there wasn't one. Using four front hubs means no handbrake or mounts exist, so a mechanical caliper system and cables had to be designed, sourced and fitted. The double clutch was a bear to use until Mike fitted a servo to the rear clutch: now it's fine. The cylinder head on the front engine went porous after a few thousand miles and had to be replaced, which was just boring and annoying. Fortunately Mike keeps a spare engine in the back. Likewise when a throttle trumpet nut came loose and fell into the engine, wrecking a head and piston, it was a

← Starting the engines involves flashy switchgear and drama, and why not? Just using two ignition keys would be dull. Black box is the LED display for traction control.

↖ Sorting out the gear change was tricky: the gear lever operates the front box conventionally and the rear one via a cable system.

↙ Fuse and relay board would frighten an IBM engineer. On the other hand, Joseph Lucas wasn't involved in any of it.

↓ Even without nitrous, the acceleration of a 360bhp, 4WD Mini is breathtaking.

Tech

ENGINES – Two Vauxhall C20XE 16V 1,998cc units, uprated oil pumps, Eagle H-girder pattern steel rods with ARP rod bolts, Mahle pistons with 2mm valve cutouts for future high-lift cams. ECU, DTA E48EXP in custom loom, wideband closed loop Lambda systems, home-designed traction control. Vauxhall beige injectors, QED DTH throttle bodies, ITG foam air filters. Custom 2.25in stainless exhaust systems. Front engine radiator, South African spec Cavalier Turbo. Rear engine radiator, Cavalier GSi. Two Jaguar XKR fuel pumps. About 180bhp at each flywheel, 355bhp each with nitrous.

TRANSMISSION – F20 Vauxhall clutches and gearboxes with standard ratios. Rear gear selector forks reversed, custom cable system replaces rear gearbox shaft-based gearchange. Front diff, Quaife ATB LSD with 3.42:1 final drive, rear diff standard Vauxhall 3.55:1. Driveshafts 4 x AMT custom one-piece, with Vauxhall inner CVs and Austin Allegro outer CVs.

BRAKES – No servo, Goodridge braided hoses. Custom pedal box with front/rear bias adjuster bar. Metro Turbo four-pot calipers all round, vented discs. Custom handbrake calipers and cable.

SUSPENSION – Coilover adjustable AVO shocks all round, with 500lb springs. Rose-jointed adjustable arms top and bottom. Mini steering rack.

WHEELS AND TYRES – White Weller steels 5.5J x 13, with Toyo 888R 185/60 x 13.

INTERIOR – Sparco buckets, harnesses, FIA-spec rollcage, two AMT engine frames. ATB Digidash II connected to both engines. RPM x 2, oil temp x 2, water temp x 2, oil pressure x 2. Custom relay and fuse panel.

EXTERIOR – Gp5 arches, steel flip front. Round nose front signals.

good job he had a spare engine in the front to drive home on.

Mike's now fitting a nitrous setup from the Wizards of Nos, with a 150bhp shot at the front and a 200bhp shot at the rear, adding 350bhp at the crankshafts. The engines are built to handle this, but the gearboxes may not take a joke that funny. If you want to keep up to date with Mike's progress, check in with www.twinenginemini.co.uk now and again and see what happened next. The dashboard is due to be flocked soon, and there is talk of carpets.

Mini Kit Cars

← ↑ Domino Cars saved many a moribund Mini from death by oxidisation: simply dismantle the Mini, move all the bits a few feet over, bolt them into the Pimlico or Premier's immortal GRP tub, fold the roof down and enjoy.

Minis have always provided an excellent package for kit car designers and enthusiasts. The whole powertrain, steering and front suspension assembly comes out in one go, and is even mounted on a convenient subframe, on to which you can bolt your own ideas. Several of the top mad Minis in this book are or were available in kit form.

Some, such as the Domino family, simply repackaged the whole Mini mechanical set and wiring into a new shell that was more amusing than the original Mini, and 100% more rust-resistant. Minis were never properly protected against rust, which is one more reason why they appealed, and still appeal, to kit car builders: many an otherwise perfectly good Mini has crumbled beyond hope of even a pub-bought MOT and has been rescued and recycled into a kit car. The price of a seriously crumbled Mini, up until very recently, has been in the low two figures. Even now there are MOT-failed basket-cases available for the asking.

A Domino Pimlico/Premier is basically an improved Mini. It's convertible, lighter, stronger, stiffer and cheaper than a nice standard Mini, but it's still recognisably and spiritually a Mini. It has deep, strong sills, wheel arches designed for 12in or 13in wheels and no body seams, as they're irrelevant on GRP cars. Dominos have been in production off and on since 1986, and only just finally died in 2008 – the last owners of the company couldn't even make the project break even, due mainly to current levels of UK commercial premises rates, Health and Safety expenses, and liability risks.

Most designers using the Mini abandoned the original Issigonis body shape and went off in their own direction. The GTM uses quite a lot of the Mini donor, including the engine/powertrain package, but it's relocated to the back of the car to create a rear-Mini-engined two-seater sports car with excellent handling and grip. The steering arms are simply connected to brackets on the chassis and become adjustable track-control arms, and quite often cars like the GTM use a second front subframe in the front of the car to provide steering and suspension. The GTM pictured here will also be reviewed in the Fresh Ayre section of *Kit Car* magazine.

The Mini Marcos is another example of the best features of the Mini design being hijacked and re-shelled into something faster and more amusing, with even better handling than the Mini. The GTM might have the edge over a Marcos on the track, being mid-engined – it would be fun to find out. The GTM certainly has the edge over the Marcos as far as looks go, but unlike the Marcos, the Mini-based GTMs are no longer in production, having evolved into a Rover Metro-based descendant.

Getting well away from the Mini shape was the Hustler, from late car designer William Towns. He used Mini donor packages for several city car projects, but the Hustler took off and became a relatively good seller in kit car terms, with 500 sold. The Hustler owners' club used to visit William annually for a picnic until he died. I spent a day with him once for a magazine interview, and found him a thoroughly nice man. He advised me that when designing cars one should never use a straight line. I must have looked rather taken aback at that, given that he had designed the notably curve-deficient Hustler and the equally angular Aston Martin Lagonda, but he went on to explain that you can use lines that *look* straight, but they must never actually *be* straight. I've always found his Lagonda aesthetically indigestible, like most people, but his DBS is a top piece of automotive art.

The Mini-based Hustler originally only had four wheels, but could also be bought with six wheels: you simply added a second rear Mini subframe behind the first one. Although the concept evolved to become bigger and finished up with Maxi power, the theme of a mobile greenhouse with huge sliding glass doors was retained. The Hustler is basically a cross between a conservatory and a Mini Moke. I haven't driven a Hustler, but I imagine it must feel a bit like riding a Vespa scooter, apart from being warm, comfy, relatively safe and not requiring a claustrophobic crash helmet and shoes with drain holes. The Hustler's original GRP and steel bodywork was later replaced by plywood, which if cleverly used as in Hustlers and early Marcos GTs is very light and extremely strong.

↑ The Mini's power package has always been an extremely convenient source of power for kit-car use: cheap, plentiful, neat, small, and comes on its own subframe.

← William Towns's Hustler started off as a straightforward GRP and steel rebodying design exercise for a Mini. The enormous glass area would have cooked occupants, but the vast doors can be slid back to turn it into a mobile gazebo.

→ The GTM started off as a mid-engined Mini, and continued to evolve: it's now become the GTM Libra, and uses the Rover/MG K-series engine. It's still a stiff and strong GRP monocoque.

↓ This example of the Hustler has evolved to six wheels, plywood construction and a Maxi engine to pull the increased weight.

The Beauford still has some traces of Mini DNA, but diluted almost to herbalist percentages. If you looked at a new Beauford kit you'd just see something faintly and familiarly Mini-ish. The original rather clever idea of the Beauford was to create a huge 1930s-style limousine for a small budget. A large chunk of capital-hungry development time was saved by nicking bits of other cars and putting them together.

The chassis under the first Beaufords was a very simple but strong ladderframe. The front end of the bodywork, the grille and long bonnet, are inspired by Packard. The whole back end of the car was a Mini shell plonked on top of the chassis. The rolling gear was provided by a Cortina. Job done, basically.

The result was a vast engine bay that would take any engine you cared to fit. A common choice was the Datsun straight six – reliable, smooth and cheap. Most Beaufords are wedding limos: they do the job for peanuts compared to the usual baggy old 1970s Rolls-Royce Silver Shadow with its potentially breathtaking repair bills.

In kit car design, door-hinges, door shuts, window winders and so on are an expensive nightmare, possibly the most complex area to design, so for Beauford to use the Mini shell saved a bundle and worked really well. As time went on the Mini/Beauford shell was remade in GRP, lost its roof, was extended, and had its curved windscreen replaced by two flat screens in basically the same frame. The doors on their two-door models remain modified steel Mini doors, applying the sound design principle that if it ain't broke, don't fix it. The glass is different and quarter-lights have been added, but the doors are still Mini. I have road-tested a Beauford, and it

was thoroughly nice: smooth, comfy, solid, and silent apart from a baritone rumble from the exhaust.

There is something of a divide in the Mini world between those who revere proper Minis and those who throw the rulebook out of the window, but as far as most Mini people are concerned you can enjoy both. I write for *MiniWorld*, which loves mad Minis as long as they retain the Mini shape. So you'll find stories about Z-Cars with GRP shells, which are kits rather than Minis, but you'll only rarely find a mention of Marcos or GTM.

I also write for *Kit Car* magazine, which is a different world: modern kit cars are regarded as a bit pussy if they can't do at least 150mph, get to 60mph in three seconds and crash survivably into the Armco at 130mph. *Top Gear* avoids mentioning kit cars because they make production sports cars look girly: Clarkson and *Top Gear* stay safely in the unimaginative Ferrari/Porsche-worshipping mainstream. If you've read this far you're already outside that boring box. In addition to the 200mph Ultima kits charging around there are still hundreds of old Mini kit cars bimbling about – once you eliminate rust, Minis can live forever. Looking at the fun-per-quid ratio, it's hard to beat a Mini-based kit car.

Paul Banham is a notable kit car designer who has used Minis and Metros as donors. He's not exactly mad, but he does insist on doing things his own way: if other people think he's mad he doesn't really care. He could

make serious money if he applied his talents to something commercial, but he keeps coming back to inventing new kit cars, even after blowing himself up in a workshop explosion recently. He can't smell any more after years of breathing fibreglass fumes, and a chemical gas leak made itself known via a huge explosion when he switched on his workshop lights. Fortunately he was still only halfway through the doorway and got blown out rather than up. Although quite significantly sautéed at the time, he's basically okay now and still designing and making kit cars.

↓ The installation of the Mini engine/box/ suspension in the GTM is still as it comes out of the Mini – it's just moved to the mid-rear position for rear-wheel drive and better handling and balance.

He doesn't like mechanicals or electricals, so he usually leaves all that intact, cuts a donor production car down to its sills, adds a steel frame to brace it all and then puts something completely unexpected on top of the result. He's made a Porsche Spyder out of a Skoda, he's made something that resembles an Audi TT out of a Metro, and his most recent project is an XK180 replica made out of a chopped-down XJS. This was quite easy,

as the XKR and the XK180 were built by Jaguar the same way: they cut the panels off an XJS monocoque, attached some new ones and presented it as the shiny new XKR. The main difference between the two is that Paul's entire kit costs about the same as a replacement engine for an XKR.

Paul's most unexpected Mini kit was his Sprint, which looks almost exactly like a Frogeye Sprite. He was unable to resist improving it by adding a bootlid, which is how you can tell the difference between the Frogeye and the Banham. The car uses a Mini floorpan, subframes and mechanicals, cut right down and then stiffened with a steel frame, on to which replica Frogeye panels are mounted. The Mini engine is even correct for a Sprite, but is fitted the wrong way round and on top of the gearbox. I tested one of these, and thought it was fab: it felt and sounded quite like the Sprite I used to own way back. The Mini rack and pedals are still used, but the angles of them are greatly improved over the Mini, as you sit well back from the engine package. Banham's kits generally require a lot of welding and bodywork effort, but are also inexpensive, so abandoned projects abound. If you want one, check out www.autos.groups.yahoo.com/groups/sprintists, which is a club for Banham Sprint owners and builders.

If kit Minis in general appeal to you, my *Kit Car Manual* is also published by Haynes, and it gives you a good general picture of the kit-car world and mentions

◀ The Beauford is a rather magnificent 1930s-style limo with a Mini-based bodyshell at the back. Many of them have a Saturday job as a wedding car. *(Courtesy of Steve Hole, totalkitcar.com)*

▶ The Banham Sprint uses Mini mechanicals for a weirdly authentic replica. *(Courtesy of Madabout-Kitcars.com)*

Mini kits quite frequently. If my smartarse blathering after 28 chapters is becoming too annoying to cope with, ignore the last suggestion and amble off to a kit car show (dates can be found on www.kit-cars.com or www.totalkitcar.com) and visit the club areas, where there will be many Mini-based kit cars to examine, and enthusiasts delighted to tell you all about them. My website www.ayrspeed.com also has lots of kit-car action on it.

▶ The Sprint's engine is the correct A-series for the Sprite, but is used the wrong way round. *(Courtesy of Madabout-Kitcars.com)*

▶ My own Swallow trike design used the complete Mini package from a scrap car, with a single Mini rear trailing arm at the back. That solved many design problems in one go, and with its long wheelbase it handled very well.

◀ If you don't believe me about the Beauford's Mini rear bodywork, just check out the inside of that door. Is that from a Mini or what? *(Courtesy of Steve Hole, totalkitcar.com)*

Mini
Mouse

↓ It's a Mini Van, Jim, but not as we know it. Only the grille remains standard on this van – almost everything else has been rodded. Welded-in bumpers and frenched lights are the primary frontal features.

Paul Saulnier's Mini Van is a serious street rod, in the Boyd Coddington or Chip Foose arena. Paul is a member of both Mini clubs and street rod clubs, so it looks as though he decided to save time and blow all their minds with a car that was both at the same time. Street rods are quite often mid-life-crisis cars ordered by rich men who pay others to design and make their rods for them, so the budgets can be in the hundreds of thousands of dollars, but Mini Mouse was conceived and built with imagination and creativity rather than a bottomless chequebook.

The invoices are nonetheless impressive. The shell only cost a grand but the engine and drivetrain cost $15,000, the bodywork and paint $18,000 and the exhaust system $3,000. The total of around 40 grand is low-budget for a rod, but an impressive commitment for a Mini.

Paul is a gearhead from way back. His first car was a 1928 Oldsmobile, with 16,000 miles on the clock, bought illegally without his parents' knowledge and garaged in the next town. He would borrow licence plates from friends and drive it around at nights and weekends. British cars are in his blood as well – he still owns a 1959 Austin-Healey rebuilt from wreckage, so Lucas electrics and enthusiastic body rusting don't worry him.

Drag racing is also in his blood, and he used to run a 1927 Ford at the strip until the third blown engine, after which he got bored with rebuilding the engine and sold the car. That drag-racing experience with its expensive bangs is the reason for the extremely strong bottom end of the Chevy V8 in Mini Mouse. This motor has definitely been built not to blow up. The other hangover from

dragstrip days is the memory of the rush that the Ford gave him when it got to the other end of the strip in one piece in 9.2 seconds.

The Mini Van showed up in 2003 as a bare shell, spotted by his son Phil on eBay. After the usual finger-crossing tension theirs was the top bid, and the van was soon heading for Massachusetts. Phil's 00101111110101010101 skills on the computer also came in very useful during the hunting of parts, both Mini and custom.

Paul is a civil engineer and usually deals with very big things. Minis, particularly in America, are almost comically small things. Putting a big thing in a very small thing appealed to him, and the fact that Chevrolet's original 1950s 265CI (4,300cc) V8 was known as the Mouse Motor allowed him to christen his car Mini Mouse without sounding embarrassingly cute.

A Chevy V8 will visibly fit in the back of a Mini Van, but in order to see exactly how it would fit and how it would have to be mounted, Paul made a full-sized styrofoam replica engine. He could then get inside the van beside the foam engine and move the mock-up around to figure out the drivetrain and suspension frame design. With the frame conceptualised and measurements taken, the bottom of the van was cut out and the bodywork bracing structures and suspension mounts were welded in. The structure ended up resembling the suspension turrets of the Pontiac from which the rear suspension was taken. The finished rollcage/chassis is mounted to the van's shell and supports the engine and gearbox as well as the Pontiac

↑ You still wouldn't think there was a monster V8 in the back rather than just a picnic basket and a couple of show prizes.

29

→ From the rear three-quarters view, the stance of the Mini is easy to appreciate: a lot of thought went into the wheel/arch/ride height equation.

↓ Frisky Chevy 350 with Porsche G50 gearbox and transaxle is a killer combo that fits the Mini Van very neatly indeed. Suspension turrets were designed and built to suit the tried-and-tested Pontiac Fiero rear suspension assembly.

suspension struts. The engine still has to be removable, as the alternative would be to send in Chevrolet-trained chimpanzee mechanics. Properly trained chimp mechanics are quite rare, although some professional mechanical work I've seen suggests that untrained monkey mechanics are everywhere.

The engine is a monster, although the amount of bang per buck is actually relatively modest. 427bhp out of 5.7 litres is not extreme, but that bottom end with Keith Black pistons, forged rods, ARP bolts and a toughened crank would probably handle twice the current power output. Paul would like more power later, and he has the options of supercharging, turbocharging or nitrous oxide to liven things up. The engine is also classic American muscle with some purist subtleties about it: 'pink' rods are genuine GM factory performance rods, still available but originally fitted

to GM performance engines in the 1960s. These rods had a splash of pink paint to identify them in the factory. So 'pink rods' for a Chevy enthusiast mean special authentic factory goodies, in the same way that 'S brakes' mean something to a Mini bunny but not to most people.

Ceramic-coated Sanderson block-hugger headers will again not mean much to Mini people, but they're free-flow exhaust manifolds that hug the block as closely as possible while still achieving reasonably well-matched primary pipe lengths. The coating keeps the heat in, which helps keep the gasflow fast and the engine bay temperatures lower. The Edelbrock 650CFM carb with mechanical secondaries is a four-choke monster, the second two chokes of which are controlled mechanically rather than by vacuum, so you can feel the change on the throttle pedal. By way of comparison, a 1½in SU flows about 140cu ft per minute as against the Edelbrock's 650CFM. The top end of the engine is pretty high-spec with serious cams and gasflow, allowing plenty of options for more power later. Currently, Mini Mouse is expected to comfortably crack the ten-second barrier on the quarter-mile as Paul's old Ford used to do, which means a 0–60mph time in the three-second area.

The gearbox is a G50 Porsche five-speed gearbox/transaxle from a 911, with suspension and driveshafts provided by a Pontiac Fiero, an underrated mid-rear-engined 1980s sports car. The G50 is a good choice, stronger than the previous Porsche transmission. It's used upside down and the linkage comes out of the side rather than the back, as the original linkages would have finished up poking out through the back doors. The box is also reversed, as the van is mid-rear-engined and the Porsche is rear-engined,

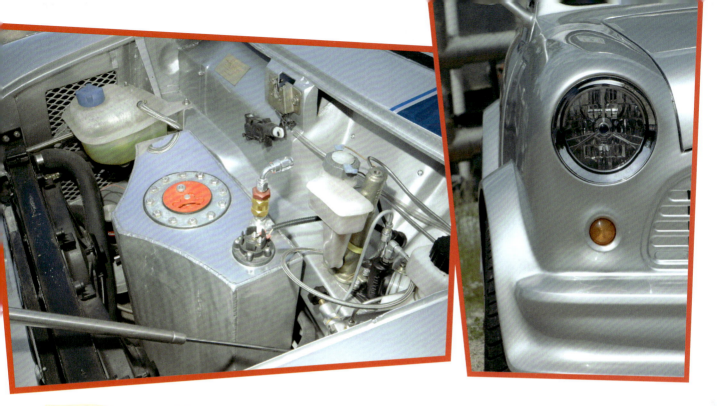

with the weight all behind the axle like a Beetle. Hence its reputation for dodgy handling.

Turning the box upside down and the wrong way round: easy to write, somewhat harder to achieve. The Fiero shafts, struts and hubs have been proven strong enough to handle V8 power, as many people have stuffed big V8s into Fieros with great success. Mini Mouse's half-shafts are hybrid Porsche/Pontiac and quite short. The rear wheels and tyres had to be wide enough to provide grip for 400lb/ft+ of torque, but they also had to fit under the body in the right way stylistically. A tyre-smoke show doesn't do it for Paul, he wants action.

At the front, the cooling system and fuel cell occupy the engine bay. The fuel cell is shaped to allow air to flow smoothly through the radiator and past the tank to the wheel well vents. The Audi radiator is quite small but deep, and its position, its big twin fans and the long coolant lines to the engine make it very efficient.

With the structure sorted, the next task was to get the bodywork up to the same standard as the engine bay. There's major steelwork in this van. Dual chassis frame rails in 2in x 4in steel run along both sides of the car under the bodywork, and connect the rollcage to the front wheel arches. The exterior door hinges were removed and replaced by new rear-mounted hinges, so the remotely controlled doors now open from the front, suicide style. The subtly altered bonnet also pops open remotely and opens the 'wrong' way. The body panel changes are all steel, and the body kit is a one-off. Neat touches include a single screen-top wiper, power windows and the welded-in front bumper. Every seam in the body

was opened up, butt-welded and smoothed over, and the gutters were removed, restyled and replaced. Ron Melanson was the major bodywork dude involved.

The interior had to match the exterior, so it was completely created from scratch. The main feature of the vertical central console is the Art Deco-styled Vintage Air control panel. (Yes, of course it's air-conditioned – what did you expect?) The rest of that central console is minimalist, finished in engine-turned aluminium, and sports just a few lights and the ignition switch: less is more. Dakota Digital provided the futuristic and again minimalist instrumentation – all you see is a dark panel unless something's happening, but the range of information the dash provides is extensive once the engine is started. Very cool indeed. The electrics are custom, starting with a 20-circuit fuse box with ten relays and working outwards from there. Clayton Tomasetti's wiring expertise came in useful here. Steering is by a generic Flaming River rod-style tilt column, a potentially useful idea for UK Mini builders. Apart from anything else, it would be nice to get rid of the bus-like steering wheel angle that comes with a Mini rack and column, and that's instantly achieved by fitting a tilt steering column.

The seats are modified and retrimmed Fiero, and the driver's harness is a five-point drag-legal item so the car will be ready to rock when Paul is. The door action is in tune with the rest of the car. Remote electric suicide doors. Cool. What takes the doors just a little further is the detail. The Mini's original inner door handles are at the top of the door: not good enough. Paul had to rework the whole system to shift them to the bottom so that the

↑ Frenched tripod rod lights and chrome trims bring big bling: they don't light the road up very accurately, but who cares...

↖ The specially shaped fuel cell under the bonnet provides clear cooling fan airflow.

← Suicide doors are cool and unexpected on a Mini. The bonnet opens the wrong way as well.

↓ A blue and silver theme runs through the whole car. The interior is comfy as well as visually very stylish. The Dakota Digital dashboard fits in very well with the dramatic styling.

↑ The central vertical console is a very cool piece of styling, topped by an Art Deco air conditioning control panel.

→ The rear wheels are still reasonably in proportion to the front ones, so it still looks like a Mini rather than a psycho killer beast. The correct tyre contact patch for the weight gets the best grip.

door trim would look better. The door trimming itself is not just matching blue and silver vinyl, though – stitched into the material but without the distraction of a different colour is a trad flame job, a reference to old-school street rod styling. It's sort of compulsory for cars with a rod inclination, but it can be done with subtlety or without.

An interesting revelation is that with such a major project, Paul felt there was no light at the end of the tunnel until one day he actually found himself driving the car. That moment of triumph as Mini Mouse rumbled around the block for the first time was brief, though, as the car ran out of petrol within a couple of hundred yards.

Many, many projects fall foul of the seemingly endless and daunting list of Stuff To Do: kit-car builders suggest a couple of ideas. The first is to ignore the frightening big picture and concentrate on one section at a time, for example the brakes. Second, do something every single day, even if it's just tidying up and cleaning tools. After 365 small somethings, big progress is made. Paul was lucky that his wife Nancy, although oblivious to the dynamics, aesthetics and hoot value of V8 Minis, was supportive in the cheerful Mini-wife way: 'I'm happy if you're happy.'

Paul says the most fun he has nowadays with Mini Mouse is standing back from the car, watching people's reactions and listening to their comments. But maybe the top fun is still to come: he hasn't been out on the dragstrip yet to check out the car's theoretically estimated 9.2-second, 140mph quarter-mile.

Tech

ENGINE – 350CI (5.7-litre) Chevrolet mouse motor (.030 over) with nodular prepared and indexed crank, GM 'pink' (shot-peened and Magnafluxed) forged steel rods with ARP Wavelock bolts, Keith Black hypereutectic 10.3:1 pistons, Childs & Albert ZZ Gapdura Moly file fit rings. Comp Cams Custom series (110 centreline, 488-490 lift corrected 1.6 = 520/522 lift, 226/228 duration). Cloyes true roller rockers, Comp Cams hi-output cam lifter set, Comp Cams XE chrome moly pushrods. Edelbrock Aluminium Performer Heads with revised combustion chambers, severe duty 2.02/1.60 valves. HEI distributor, Spiral Core 8mm plug leads, Edelbrock Performer 650CFM carburettor with mechanical secondary chokes on Edelbrock Performer polished aluminium dual plane intake manifold, Sanderson Block Hugger ceramic-coated exhaust manifolds, custom fabricated 304 stainless steel 2.5in exhaust system with Pacesetter 304 stainless quick-release flex joints and dual Edelbrock SDT 304 stainless steel mufflers, by Dave's Custom Creations, Charlton, Mass. Audi 5000 radiator, Vintage Air condenser, twin 12in electric fans (1400CFM each), vented wheel wells. Porsche side-mount water pump, Edelbrock electric fuel pump and filter. 427bhp, 418lb/ft at the crank. Expected performance 9.2-second quarter-mile at 140mph.

TRANSMISSION – Porsche G50 transaxle converted to run inverted/reversed and converted to side shift: 1st, 3.1538; 2nd, 1.7895; 3rd, 1.2692; 4th, 0.8572; 5th, 0.6774. G50/930 Kevlar high-performance clutch disc. Gearbox parts Renegade Hybrids, Las Vegas: setup. California Motor Sports. Clutch pedal raised 1.5in and lengthened 2.5in to increase throw and leverage. Custom high-strength half-shafts (Porsche/Fiero hybrids) 40% limited slip differential, 3.444 final drive.

BRAKES – Mini dual system, 3/16in stainless lines, some braided hoses. Wilwood proportioning valve, Summit Racing line lock. Front, 8.4in Metro four-pot discs; rear, Pontiac Fiero competition 11in discs.

SUSPENSION – Front, Mini with Hi-Los. Back, Pontiac Fiero Competition coilover struts, bump-steer upgrade kit. Detachable wheelie bars for dragstrip.

WHEELS AND TYRES – Front, 5J x 13 Minilites with diamond-cut lip, P165/65R13 Bridgestone Potenza tyres. Rear, 6J x 15 Minilites with diamond-cut lip, P225/50R15 BF Goodrich G Force TA Drag Radials.

INTERIOR – Pontiac Fiero seats, cut down 6in and reupholstered. Driver's harness RCI five-point drag-racing SFI-rated, passenger harness four-point. Custom rollcage surrounds engine compartment. Scratch-built rear bulkhead with window, insulated for sound, with engine access panel. Flaming River tilt steering column, X-Force 330mm steering wheel, blue leather/silver wood. Renegade custom gear lever and box with X-Force blue and silver aluminium gear knob. Scratch-built all-steel dash and centre console covers. Vintage Air heating and air conditioning unit, carries A/C vents, miscellaneous switches and lights. Dakota Digital VFD3X-1002 instrumentation with Teal lighting: water temp, oil pressure, fuel level, dimmer adjustable, brake failure indicator, high-beams, direction indicators, tacho, speedo. Custom wiring including 'It's a Snap' 20-circuit panel, ten relays, power doors, windows, bonnet, and rear doors.

EXTERIOR – 2in x 4in frame rails from front wheel wells to rear suspension, front bumper welded to body. All-steel ground effect panels with air dam to improve cooling, all body seams cut, butt-welded and smoothed. Roof gutter removed and reshaped over the doors, bonnet corners rounded and hood hinged from the front with remote power opener. Wiper holes filled, one wiper relocated over windscreen. Exterior door hinges removed and doors converted to front opening. Sliding glass windows removed. Power windows and power door latches installed with remote operation. Custom steel sills/ground effect panels and wheel arches, side scoop to bring cool air into engine compartment, rolled rear valance with ventilated centre section. Rear doors modified to internal hinges, Chevy 'Suburban' style windows, licence plate cut-out, power opener. Frenched LED tail lights and third brake light. Frenched Tristar halogen headlights with centre blue dots. Other body lights are LED, including parking, brakes and indicators. DuPont Hot Hues three-stage pearl paint, Psycho Silver and Blue Mood.

Posh Minis

The Wolseley Hornet featured a markedly different front end, with its own bumpers, grillework, front panel and bonnet. Some posh BMC/BL rebadging treatments were grotesque, notably the Vanden Plas Allegro, but this all works rather well.

The Mini was and is one of the few classless cars. Unlike most, it carried no particular status or wealth connotations, and was no more or less of a status symbol than an iPod is today.

The Mini was a direct descendant of the 1920s Austin Seven, which was an even more radical mini-car that also had four seats but was even shorter and narrower than a Mini. Austin Sevens were also radically modified and raced, and that tradition continues in the Mini. In fact the Austin-badged version of the first Mini was called the Seven or Se7en.

As a universally clever, small, cheap, useful and stylish object, and as a design icon in its own right, the Mini has always been able to dance around in the minefield of the British social status system without a care.

However, some people felt that in standard form it did not adequately reflect their superior social or financial status, and some weren't bothered about looking posh but simply wanted something different with more comfort, more style, more chrome, more boot space and more widgets to play with.

BMC themselves provided upmarket versions of the Mini, which were the Riley Elf and the Wolseley Hornet. BMC and then British Leyland had been indulging in 'badge engineering' since the 1952 merger, having acquired many British brands including at various times Austin, Morris, Wolseley, Riley, Triumph, Jaguar, Land-Rover, Austin-Healey, Innocenti, Standard, and the coachbuilder Vanden Plas, as the company mutated through BMC, BMH, BLMC, BL, and Rover before the remnants were bought by Nanjing Automobile. BMW, sadly, owns the Triumph and Riley brand-name rights, so we can expect those respected marques to be exhumed from their graves, spinning or not, and given the Rolls-und-Royce treatment when the BMW badge premium has completely evaporated.

An example of 1950s badge engineering was the mid-sized MG/Wolseley saloon, resembling a gentlemen's club in both pace and décor. The slow 1,500cc B-series-engined version of this hefty car was badged as an MG Magnette, and the painfully slow 1,250cc XPAG-engined version was badged as a Wolseley 4/44… although it was the one with the MG engine. I used to drive one, which is how I learned about the need for percussive maintenance on SU fuel pumps. These cars were upmarket from Austins, but downmarket from Rovers. There was also a similar but larger and faster car known as the Riley Pathfinder, fitted with what could be regarded as the last real Riley engine with twin cams, twin SUs, 2,443cc and 110bhp. My first experience of speeds over 100mph was in an uncle's Pathfinder. The uncle in question later lost the plot and bought an Austin Allegro and a beige cardigan.

The social status of cars is intriguing me at the

⬆ This is one of the rare Heinz 57 Hornets, with its Crayford convertible roof folded down. The extra boot space and foot of bodywork are nice, but rather miss the point of a Mini.

⬉ The Wolseley is posh but not as posh as the Riley, so it only has a small patch of walnut dashboard.

Posh Minis 153

The Wolseley pictured is one of 57 'Heinz' cars presented in 1966 as competition prizes after having been converted to open-tops by body specialists Crayford. The Heinz Wolseleys came with leather seats, insulated food cabinets, kettles and electric points, picnic hampers, radios and make-up trays. These were presumably for mascara repair after a brisk windy run to the British coast with the roof off for a picnic in a seafront car park, where 1960s Brits liked to sit shivering in the car, munching cheese sandwiches, drinking tepid tea from tartan vacuum flasks and watching grey waves thumping onto stony beaches through the drizzle. Mind you, leaving the front and rear windows and frames intact and wound up would have made these pretty civilised as convertible cars go.

Other companies developed even posher Minis to join the Wolseleys and Rileys. Radford and Wood & Pickett were coachbuilders whose market was evaporating as the demand for coachbuilt cars dwindled to nothing. The Silver Wraith was the last coachbuilt Rolls-Royce still supplied as a chassis delivered to a body constructor, and its production ended in 1959.

Radford absorbed another coachbuilder, Freestone & Webb, but the coachbuilding companies were really having to scratch around as the '60s progressed. When people like Ringo Starr with an endless supply of money wanted custom Minis, the surviving coachbuilders cheerfully obliged. Vanden Plas was another previously independent coachbuilding company, bought by BMC and reduced to a posh badge on BMC 1100s and 1300s and

moment as I'm chewing over an idea for a new book called *You Are What You Drive*, which will make some people feel smug, but will seriously upset others. It could be an actively dangerous book to write, but could also be top fun.

Within the rarefied world of BMC's posh Minis, there were subtle gradations. The Wolseley Hornet is just slightly downmarket from the Riley Elf, which you can tell by the smaller acreage of veneer on its dashboard.

Elves and Hornets shared a modified Mini bodyshell, with a little boot added that sported trendy fins, and bonnet and front panel modifications that incorporated a small version of the traditional Wolseley and Riley grille shells. Refinements such as air vents, wind-up windows and concealed door hinges appeared on Rileys and Wolseleys before showing up on ordinary pleb Minis. Around 30,000 of each model were made.

on an upmarket version of the Austin Westminster known as the Vanden Plas Princess 4-litre R, which was fitted with a Rolls-Royce engine and was originally to have been badged as a Bentley.

Radford in 1963 were selling three posh Mini variants – the Radford Mini de Ville, Bel Air and Deluxe models, variously loaded with goodies. Features included Benelite grilles with foglamps, padded and leather-trimmed seats, woolly carpets, Webasto sliding sunroofs, and so on. Instruments were multifarious, chrome-bezelled and set in walnut-veneered leather-bound pods. Headlamps came in for much custom work, as they are the 'eyes' of a car and can make a major styling statement. Lucas tripod P700 headlights in chrome cowls were a favourite, but there were also stacked double headlights and vertical oblong headlights sourced from Facel Vega. Facel was a seriously upmarket French car builder whose cars usually cost as much as several houses.

Radford's electric Mini windows were a novel and dramatic luxury in the UK of the 1960s, and were of the same type used in Rolls-Royces: massive overkill electric motors with steel chains to haul the glass up and down. The doors of a Radford shut with a hefty thud, not with a click or a clonk. Radfords are also very nice to drive, having benefited from all the coachbuilders' noise-reduction tricks. They also weigh a good bit more, and that's part of the formula.

In 1965 Radford's hatchback Mini appeared, the Mini de Ville GT, with most of the rear-end bodywork welded into a single panel that hinged upwards. Peter Sellers ordered one for Britt Ekland.

Wood & Pickett's Margrave offered the same sort of conversions as Radford, using twin sideways headlights with their own small grille, or Mercedes vertical headlights. W&P finished up doing rather better commercially than Radford in the end, despite a tendency to make white leather interiors. Or perhaps because of it.

⬇I have no idea why it required an independent coachbuilder to take the obvious step of changing the Mini into a far more useful hatchback, rather than the car's manufacturer.

↑→ Ogle Design produced a sharp and sexy body on the Mini floorpan, completely obscuring its origin but not its performance and handling. Although proportionally similar to a Mini Marcos, it's as stylistically cohesive as the Marcos isn't.

ERA Minis, converted by English Racing Automobiles in the late 1980s, retained the Mini bodyshell but stuffed in an MG Metro Turbo engine. Modifications included bodyshell changes to the arches to fit 13in wheels, and to the bulkhead to make room for the turbocharger, as well as a reverse scoop at the back of the bonnet to reduce engine bay temperatures, extra water cooling and an oil cooler, and ECU changes. Their interiors featured much leather and instrumentation. Around 480 were built, most of them going to Japan.

The Ogle Mini was a more radical approach to posh Minis, with its own GRP bodywork mounted on the Mini floorpan. It was produced from 1960 through to 1963/4. Sixty-six were built, of which 26 have apparently survived. The Ogle was built from a new Mini Cooper, which was cut down to a floorpan, reinforced, rustproofed and used as a base for a very pretty GRP bodyshell. It was supplied complete rather than as a kit, although it's available in kit form now. Mechanicals remained Mini, although the interior was more like a Mini Marcos with bucket seats, wood dash and steering wheel, and a sports-car rather than shopping-car collection of instruments. Ogle Design was also responsible for the original elegant and stylish Reliant Scimitar, originally intended to be a Daimler. They also styled the Reliant Robin, but let's not rub that in.

New Ogle Minis can now be bought from NostalgiaCars.co.uk. These are entirely approved by Ogle Design and bear continuation chassis numbers: they're unchanged from the early '60s originals apart from better seat belt mountings and rustproofing.

A new idea that has just popped up as a result of writing this chapter is to create a Mini that never was. My own Mini, featured now and again in the Logbook section of *MiniWorld*, is known as Bubble because it is champagne-coloured and also because of some rust bubbles under the paint, which are gradually achieving increased prominence as time goes on. Bubble is a 1978 Canadian Mini 1000, and apart from amusing/grotesque Canada-only impact bumpers, it is not a very exciting Mini as Minis go: the 1978 interior in particular is semi-modernised, plasticky and cheap, so any change would be an improvement.

The MG Mini is the major tooth missing in the Mini's lifetime grin. There were Riley Minis and Wolseley Minis, but there was never an MG Mini. The Morris/Austin 1100/1300 was sold in Riley Kestrel and Wolseley 1100 versions, and there was also an MG 1100/1300, which would offer some guidance as to what the specifications of an MG Mini might have been.

Two options present themselves. The MG Mini might have, or ought to have, happened concurrently with the

↘ *MiniWorld*'s Karen Drury and her partner Jim run a Riley called STelf (stands for Special Tuning Elf) on the track: it gives them an advantage, as less posh Minis instinctively give way to more important people. *(Courtesy of Stefan Gibney)*

Rileys and Wolseleys in the 1960s, so in that case Bubble would get a 1275 with two SUs, disc brakes, possibly hydrolastic suspension, and a bonnet and front panel shape shared with Elves and Hornets but with a slightly different MG grille. I could fabricate that from an adapted MGA grille, cut down to Elf/Hornet size and rechromed. Whether the back end should be extended in Elf/Hornet style or not is debatable. There was a bumless Wolseley variant with a normal Mini rear end sold in South Africa, so the precedent is there.

Seating could be either leather or vinyl, and the dash would definitely need some woodwork and a tacho at the very least. There would be chrome or stainless body trim aplenty, much MG badgery everywhere, and two-tone paint – 'Varitone' paint jobs are an MG tradition.

The other option is an imaginary MG Mini Turbo from the 1980s, which would have been achieved by filleting all the good bits out of an MG Metro Turbo and jamming them into a Mini shell. I don't really understand why the company didn't go for either MG Mini option: but then I also don't understand why they stopped developing MGBs, Midgets and Triumph Spitfires and handed that whole market segment, gift-wrapped, to Mazda. Perhaps all the incompetent managers who were promoted out of the way finally made it to the top.

As well as the missing MG Mini, there is also no Vanden Plas Mini, although the Vanden Plas 1100/1300 is a close cousin. As to whether a Vanden Plas Mini would have been more or less posh than a Radford or a Wood & Pickett, that is a matter for future saloon-bar debate. Although to kick off the debate, it has to be said that tailored cars are a lot posher than even the most upmarket off-the-shelf ones.

More mad Minis

I'm still the North American correspondent for *MiniWorld*, and still coming across excellent Minis all the time. Here are a few more top cars for which we didn't have space earlier in the book to cover them in all their glory.

A few years ago, jamming a big Honda engine into a Mini and making a decent job of it was radical enough to guarantee a feature, but Mini builders have moved onwards and upwards, and have responded to the challenge: the Honda engine is just the starting point now, and more and mad Minis appear all the time. Mini Rules OK, and long may it reign.

↑ Just chopping the roof makes a Mini look fat and squat, but doing the job properly and cutting a slice out of the main bodywork is a massive undertaking.

→ The body is cut quite a long way down – you can see that the lights sit low and the whole bodyshell is lowered compared to the wheels. The car still runs 10-inch wheels and tyres.

Sprint replica

Kenny Benchek's sprint replica is chopped in the same way as the real thing – a chunk is taken out of the roof and also below the waistline. A serious bodywork challenge, many years in the making.

← The engine obviously sits a couple of inches higher in the lowered body, and Kenny had to remake the bonnet frame to allow enough clearance to shut the bonnet. The rocker cover and carb were also shaved.

↑ The lowered roof could make the interior claustrophobic if it weren't for the white Webasto slide-back sunroof.

Posh Riley Elf

At first glance Phil Taggart's Elf just looks nice and shiny. A closer look reveals just how nice, and a chat further reveals that twenty thousand dollars went into making it look this good. It also drives very nicely indeed, and is seriously comfy with its Hydrolastic suspension and fresh ostrich-pattern leather seats.

↑ Ostrich-patterned leather interior could be considered slightly over the top… but if it's good enough for a Bugatti, it's good enough for a Mini.

→ Phil's car looks way better than it did when it was new. It may have been the poshest of the standard Minis, but it never gleamed as it does now.

↓ The chrome-plating bill must have taken quite a chunk of the fat restoration cost: it's up to show standards.

Monster 4x4

This looks quite harmless, but the grey primer finish is because this car is still a prototype. BJ and Norman at Minitec in Georgia can't leave well alone, and after developing a DIY Honda conversion kit for Minis, they dismantled a Honda 4x4 CR-V shopping jeep and got the entire 4x4 drivetrain out of it. Their rear-Honda-engine kit is currently in development too... check out www.superfastminis.com

↓It just looks a bit naughty at a glance, but those wheels are suspiciously big, and isn't the front end longer than it ought to be?

↖ Here's why. Stonking Vtec Honda engine, and six times the original power. With all four wheels gripping and a ridiculous power-to-weight ratio, this has to be a major hoot to drive.

↓The cabin looks a bit sparse, but the shell and floor is still mostly Mini, although gear lever and pedals are custom. Box for feet is interesting: there must be issues with room under the toeboard, or perhaps it was built with Nicolas Sarkozy in mind.